Explo
Nonlin

An Appreciation of How

Exploration of a Nonlinear World

An Appreciation of Howell Tong's Contributions to Statistics

edited by Kung-Sik Chan

University of Iowa, USA

World Scientific

NEW JERSEY • LONDON • SINGAPORE • BEIJING • SHANGHAI • HONG KONG • TAIPEI • CHENNAI

Published by

World Scientific Publishing Co. Pte. Ltd.
5 Toh Tuck Link, Singapore 596224
USA office: 27 Warren Street, Suite 401-402, Hackensack, NJ 07601
UK office: 57 Shelton Street, Covent Garden, London WC2H 9HE

British Library Cataloguing-in-Publication Data
A catalogue record for this book is available from the British Library.

Front cover design by Anna Tong
Back cover image by Nils Chr. Stenseth

EXPLORATION OF A NONLINEAR WORLD
An Appreciation of Howell Tong's Contributions to Statistics

ISBN-13 978-981-283-627-4
ISBN-10 981-283-627-6

Printed by Fulsland Offset Printing (S) Pte Ltd, Singapore

Foreword

This volume is very appropriately titled 'Exploration of a Nonlinear World', because this is just the path which Howell Tong has followed these thirty-odd years: the study of the time series analysis of nonlinear models. I have admired his work and am grateful for this chance to offer my own tribute, even if to do so also means admitting that I have seen the subject vanish over my horizon during those years.

My own work in the nineteen-fifties was firmly rooted in the linear world, which seemed challenging enough at the time. My principal contribution (Whittle, 1951) was to obtain the asymptotic (large n) expression

$$\frac{n}{2\pi}\int_{-\pi}^{\pi}\left[\frac{\hat{f}(\omega)}{f(\omega)}+\log f(\omega)\right]d\omega$$

(modulo constants) for the log-likelihood of a sample of n jointly-normal observations. Here $f(\omega)$ is the spectral density of the process and $\hat{f}(\omega)$ is its raw estimate, the periodogram. A good deal follows from this and its spatial analogue. I obtained the result by the unrigorous approximation of the Laurent covariance matrix occurring in the likelihood by a circulant matrix. The natural statistical path is of course to appeal to an autoregressive representation of the process, and it is just when such a representation exists that the above expression is valid.

I realised very well the importance and ubiquity of nonlinear phenomena, ecology providing one fertile source, but gave up hope of any substantial analysis. Nonlinearity was forced on me observationally, however, when a seiche study (Whittle 1954) revealed the existence of subharmonics. Two principal 'periodic' components appeared (characterised by peaks of a smoothed periodogram) which could be physically identified as distinct seiches. However, components also appeared whose *period* was a sum of integral multiples of these two basic periods. Some thought suggested a physical mechanism, which was in fact a threshold model.

Howell began his career in control theory, but moved within a few years to a mature and individual study of time series and their statistical analysis. By 1978 he had seized on the threshold theme, which he developed progressively to a study of nonlinear models generally, developing a coherent theory and practical methods. These are set forth particularly in his seminal paper with Lim of 1980, and his book of 1983.

Howell's instinct to stretch an envelope so as to bring in subtler features manifests itself repeatedly. In both the linear and the nonlinear case he saw the estimation of order (dimension) as being a crucial problem. A much more radical step was to extend the study of nonlinear models to that of chaotic models (see his paper of 1995 and his books of 1990 and 2001). An essential feature of chaotic models is the discontinuous (but highly structured) dependence of the path upon the initial state value, and it is just this feature which one would expect to be lost in a stochastic version.

The review papers in the present volume bring out Howell's qualities with discernment. Some do so also with charm (Cutler), some with idiosyncrasy (Fomby) and some with substantial technical muscle. As an example of the latter, Cline gives a very thorough account of the way the properties (stability etc.) of a dynamical system transform under stochasticisation of that system. An, Brockwell and Rosenblatt all see nonlinear time series models as stochastic versions of dynamical systems, and concur with Howell on the importance in this context of Richard Tweedie's 1975 paper. Yao and Lawrance attack the difficult chaos/randomness question by asking how one would determine from observations whether the 'deterministic' part of the generator of a random process is or is not a chaotic generator. Leng *et al*, Gao and Tjøstheim all consider the determination of dimension, the latter in particular contrasting the cross-validation method with the methods associated with Akaike, Mallows and Rissanen. Ling addresses the central consistency question.

Howell was always interested in particular applications, and we see Geweke and Stenseth finding continuing interest in the hare/lynx data, and Li and Tsay treating the essential nonlinearities of the financial models which are now so important.

The striking feature of Howell Tong's 150 papers, three books and book contributions and also his personal exposition and engagement is the continuing freshness, boldness and spirit of enquiry which inform them – indeed, proper qualities for an explorer. He stands as the recognised innovator and authority in his subject, while remaining disarmingly direct and enthusiastic. This collection stands as a tribute to his achievements, although there is no expectation that these will cease.

References

1. P. Whittle (1951) *Hypothesis testing in time series analysis.* Almquist and Wicksell,Uppsala.
2. P. Whittle (1954) The statistical analysis of a seiche record. *J. Marine. Res.* **13**, 78-100.

Peter Whittle
Statistical Laboratory
Cambridge University
U.K.
E-mail: P.Whittle@statslab.cam.ac.uk

Preface

This festschrift celebrates the sixty-fifth birthday of Howell Tong. It is a tribute of our admiration to Howell's path-breaking and tireless contributions to nonlinear time series analysis. As one of Howell's students, I have been benefiting from his teaching, friendship and generosity of ideas. In particular, I learn from Howell the significance of nonlinearity and dynamics in statistics and science, the intertwining themes being the linchpin of much of Howell's works.

This volume reprints ten selected papers by Howell and his collaborators. We are grateful to nineteen colleagues for contributing seventeen reviews of Howell's works. Their reviews shed light on Howell's contributions and modern, related developments in statistics and science. We are indebted to Professor Peter Whittle, FRS, for writing the Foreword with an illuminating overview of this festschrift.

Many of us admire Howell's mastery of the English language. In fact, he is also well versed in classic Chinese. The Tang-style Chinese poem (calligraphy by Mr. Yee-Kwong Kwan) at the end of the book is written by Howell. We are thankful to Mr. Kwan for the beautiful calligraphy and his elegant English translation of Howell's poem.

We thank Howell for providing many valuable photographs which enliven this volume, and provide a glimpse of the friendship and joy in the community of scholars. We are thankful to Carol Chan, whose creative editorial design of the photographs enhances the presentation of these memorable images. We thank Anna Tong for her artistic book design, a design that was inspired by Howell's poem. We are grateful to Nils Chr. Stenseth for the lynx photo as the back-cover of the book.

Kung-Sik Chan
Department of Statistics and Actuarial Science
University of Iowa
Iowa City, Iowa 52242
U.S.A.
E-mail kung-sik-chan@uiowa.edu

March 1, 2009

Acknowledgements

Grateful acknowledgements are made to the following:

The *Royal Statistical Society* for permission to reprint the papers "Threshold Autoregression, Limit Cycles and Cyclical Data (with Discussion)", "On Consistent Nonparametric Order Determination and Chaos", "On the Analysis of Bivariate Non-Stationary Processes", "On Likelihood Ratio Tests for Threshold Autoregression" and "An Adaptive Estimation of Dimension Reduction Space (with Discussion)".

The *Applied Probability Trust* for permission to reprint the paper "On the Use of the Deterministic Lyapunov Function for the Ergodicity of Stochastic Difference Equations".

The *Scandinavian Journal of Statistics* for permission to reprint the paper "A Personal Overview of Non-linear Time Series Analysis from a Chaos Perspective (with Discussion)".

Statistica Sinica for permission to reprint the papers "Birth of Threshold Time Series Model" and "Strong Consistency of the Least Squares Estimator for a Non-ergodic Threshold Autoregressive Model".

The *AAAS* for permission to reprint the paper "Common Dynamic Structure of Canada Lynx Populations within Three Climatic Regions" and to reproduce Figure 4 on p. 370.
The *National Academy of Sciences of the United States of America* for permissions to reproduce Figures 1 to 3 on p. 366, 368 and 369.

Contents

Publications of Howell Tong
(In Reversed Chronological Order)

Books

2008: Asset Pricing: A Structural Theory and its Applications. World Scientific 89pp. (with Bing Cheng).

2001: Chaos: A Statistical Perspective. Springer Verlag 300pp. (with K. S. Chan).

1990: Non-linear Time Series: A Dynamical System Approach. Oxford University Press, 564pp.

1983: Threshold Models in Non-linear Time Series Analysis. Lecture Notes in Statistics, No. 21, New York: Springer-Verlag, 323pp.

Papers

(Papers in refereed journals are unmarked, edited volumes are marked e and proceedings are marked p.)

2008

150. Estimation and tests for power-transformed and threshold GARCH models. (with J. Pan and H. Wang). *J. Econometrics.* Vol. 142, 352–378.

2007

149. Estimation of the covariance matrix of random effects in longitudinal studies. (with Y. Sun and W. Zhang). *Ann. Statist.* Vol. 35, 2795–2814.

148p. Exploring volatility from a dynamical system perspective. Invited paper session 64-Stochastic Volatility Modelling: Reflections, recent development and the future. *Proceedings of 56th Session of International Statistical Institute*, Lisbon, Portugal, August 22–29, 2007.

147. Ergodicity and Invertibility of Threshold MA Models. (with S. Ling). *Bernoulli.* Vol. 13, 161–168.

146. Threshold variable selection using nonparametric methods. (with Y. Xia and W. K. Li). *Statistica Sinica*, Vol. 17, 265–288.

145. Semiparametric penalty function method in partially linear model selection. (with C. Dong and J. Gao). *Statistica Sinica*, Vol. 17, 99–114.

144. Birth of the threshold time series model. *Statistica Sinica*, Vol. 17, 8–14.

2006

143. On Bayesian value at risk: from linear to non-linear portfolios (with T.K Siu and H. Yang). *Asian Pacific Financial Markets*, Vol. 11, no. 2, 161–184.

142. Cumulative effects of air pollution on public health. (with Y. Xia). *Statistics in Medicine*, Vol. 25, 3548–3559.

141. On efficiency of estimation for a single-index model. (with Y. Xia). *Frontiers in Statistics*, eds. J. Fan and H. Koul, 63–85.

140. On a simple graphical approach to modelling economic fluctuations with an application to UK price inflation 1265-2005. (with W. S. Chan and M. W. Ng). *Annals of Actuarial Sc.*, Vol. 1, 103–128.

139. Selecting models with different spectral density matrix structure by the cross-validated log likelihood criterion. (with Y. Matsuda and Y. Yajima). *Bernoulli*, Vol. 12, 221–249.

138. Option pricing under threshold autoregressive models by threshold Esscher transform. (with T. K. Siu and H. Yang). *J. Industrial & Management Optimization*, Vol. 2, 177–197.

137. A note on time-reversibility of multivariate linear processes (with K. S. Chan and L.-H. Ho), *Biometrika*, Vol. 93, 221–227.

2005

136. Testing for a linear MA model against threshold MA models. (with S. Ling). *Annals of Statistics*, Vol. 33, 2529–2552.

135. On time-reversibility of multivariate linear processes. (with Z. Zhang). *Statistica Sinica*, Vol. 15, 495–504.

2004

134. Some nonlinear threshold autoregressive time series models for actuarial use. (with W. S. Chan and A. C. S. Wong). *North American Actuarial Journal*, Vol. 8, 37–61.

133. On pricing derivatives under GARCH models: a dynamic Gerber-Shiu approach. (with T. K. Siu and H. Yang). *North American Actuarial Journal*, Vol. 8, 17–31.

132. A note on stochastic difference equations and its application to GARCH models. (with Z. Zhang). *Chinese Journal of Applied Probability and Statistics*, Vol. 20, 259–269.

131. A note on testing for multi-modality with dependent data. (with K. S. Chan). *Biometrika*, Vol. 91, 113–123.

130. Efficient estimation for semivarying-coefficient models. (with Y. Xia and W. Zhang). *Biometrika*, Vol. 91, 661–681.

129e. Statistical tests for Lyapunov exponents of deterministic systems. (with R. C. L. Wolff and Q. Yao). *Studies in Nonlinear Dynamics and Econometrics (Special Issue).* Vol. 8, Issue 2. [Also in *Linear and Non Linear Dynamics in Time Series*, Proceedings of the Cofin 2000 Final Workshop, Bressanone-June 6–7, 2003, pp. 283–301.]

128. Semiparametric nonlinear time series model selection. (with J. Gao). *J. Roy. Statist. Soc. B*, Vol. 66, 321–336.

127. Testing for common structures in a panel of threshold models. (with K. S. Chan and N. Chr. Stenseth). *Biometrics*, Vol. 60, 225–232.

126. A goodness-of-fit test for single-index models. (with Y. Xia, W. K. Li and D. Zhang). *Statistica Sinica*, Vol. 14, 1–28; 34–39.

2003

125. Smoothing for spatio-temporal models and application in modelling muskrat-mink interaction. (with W. Zhang, Q. Yao and N. C. Stenseth). *Biometrics*, Vol. 59, 813–821.

2002

124. Model specification tests in nonparametric stochastic regression models. (with J. Gao and R. C. L. Wolff). *J. Multivariate Analysis*, Vol. 83, 324–359.

123. Single-index volatility and estimation. (with Y. Xia and W. K. Li). *Statistica Sinica*, Vol. 12, 785–799.

122. An adaptive estimation of dimension reduction space-with discussion. (with Y. Xia, W.K. Li and L. Zhu). *J. Roy. Stat. Soc.*, B, Vol. 64, 363–410.

121. A note on the equivalence of two approaches for specifying a Markov process. (with K. S. Chan). *Bernoulli*, Vol. 8, 117–122.

120. Adaptive orthogonal series estimation in additive stochastic regression models. (with J. Gao and R. C. L. Wolff). *Statistica Sinica*, Vol. 12, 409–428.

119. Nonlinear time series analysis since 1990: some personal reflections. *Acta Mathematicae Appllicatae Sinica*, (English Series), Vol. 18, 177–184.

118e. Dynamic model. (with K. S. Chan). *Encyclopaedia of Environmetrics*, Vol. 1, 574-8. John Wiley.

2001

117. On some distributional properties of a first order non-negative bilinear time series model. (with Z. Zhang). *J. Appl. Prob.*, Vol. 38, 659–671.

116. Bayesian risk measures for derivatives for random Esscher transform. (with T. K. Siu and H. Yang). *North Amer. Actuarial J.*, Vol. 5, 78–91.

115. A personal journey through time series in Biometrika. *Biometrika*, Vol. 88, 195–218.

114. Bootstrap estimation of actual significance levels for tests based on estimated nuisance parameters. (with Q. Yao and W. Zhang). *Statistics and Computing*, Vol. 11, 367–371.

113. A conditional density approach to the order determination of time series. (with B. Finkenstadt, and Q. Yao). *Statistics & Computing*, Vol. 11, 229–240.

112e. Advanced methods. (with W. K. Li). *International Encyclopaedia of the Social & Behavioral Sciences*, Vol. 23, 15699–15704. New York: Elsevier.

2000

111. Common structure in panels of short ecological time-series. (with Q. Yao, B. Finkenstadt and N. C. Stenseth). *Proc. Roy. Soc. Lond. B*, Vol. 267, 1–9.

110e. Interval prediction of financial time series. (with B. Cheng). *Statistics and Finance: An interface*, eds. W. S. Chan, W. K. Li and H. Tong, Imperial College Press, 245–260.

109e. A note on kernel estimation in integrated time series. (with Y. Xia and W. K. Li) *Statistics and Finance: An interface*, eds. W. S. Chan, W. K. Li and H. Tong, Imperial College Press, 86–96.

108. Nonparametric estimation of ratios of noise to signal in stochastic regression. (with Q. Yao). *Statistica Sinica*, Vol. 10, 751–770.

107. On the estimation of an instantaneous transformation for time series. (with Y. Xia, W.K. Li and L. Zhu). *J. Roy. Statist. Soc.*, B, Vol. 62, 383–397.

1999

106. On extended partially linear single-index models. (with Y. Xia and W. K. Li). *Biometrika*, Vol. 86, 831–842.

105p. Some recent nonparametric tools for time series data analysis. *Bull. ISI, 52nd Session*, Invited Paper Book 1, 387–390.

104. Common dynamic structure of Canadian lynx populations within three geo-climatic regions. (with N. C. Stenseth, K. S. Chan, R. Boonstra, S. Boutin, C. J. Krebs, E. Post,

M. O'Donoghue, N. G. Yoccoz, M. C. Forchhammer, and J. W. Hurrell). *Science*, Vol. 285, pp. 1071–1077.

103. A test for symmetries of multivariate probability distributions. (with C. Diks). *Biometrika*, Vol. 86, 605–614.

1998

102. Phase- and density-dependent population dynamics in Norwegian lemmings: Interaction between deterministic and stochastic processes. (with N. C. Stenseth, K. S. Chan and E. Framstad). *Proc. Roy. Soc. Ser. B*, Vol. 265, 1957–1968.

101. From patterns to processes: Phase- and density-dependence in the Canadian lynx cycle. (with N. C. Stenseth, K. S. Chan, W. Falck, O. N. Bjornstad, M. O'Donoghue, R. Boonstra, S. Boutin, C. J. Krebs and N. G. Yoccoz). *Proc. National Acad. Sc.*, Vol. 95, 15430–15435.

100. On the statistical inference of a machine generated autoregressive AR(1) model. (with J.-P. Stockis). *J. Roy. Stat. Soc. B*, Vol. 60, 781–796.

99. K-stationarity and wavelets. (with B. Cheng). *J. Stat. Planning and Inf.*, Vol. 68, 129–144.

98. Cross-validatory bandwidth selection for regression estimation based on dependent data. (with Q. Yao), *J. Stat. Planning and Inf.*, Vol. 68, 387–415.

97e. Threshold models. *Encyclopaedia of Statistical Sciences (U)*, Vol. 2, eds. S. Kotz, N. L. Johnson and C. B. Read. New York: Wiley, pp. 664–666.

96. A bootstrap detection for operational determinism. (with Q. Yao), *Physica D*, Vol. 115, 49–55.

95e. Nonlinear time series analysis. Encyclopaedia of biostatistics, eds. P. Armitage and T. Colton. New York: Wiley, pp. 3020–3024.

1997

94e. Some comments on nonlinear time series analysis. *Field's Inst. Comm.*, Vol. 11, 17–27.

1996

93e. A theory of wavelet representation and decomposition for a general stochastic process. (with B. Cheng), In *Athens Conference on Applied Probability and Time Series*, Vol. II: *Time Series Analysis in Memory of E. J. Hannan*, eds. P. M. Robinson and M. Rosenblatt, Lecture Notes in Statistics, Number 115, Heidelberg: Springer-Verlag, 115–129.

92. Estimating conditional densities and sensitivity measures in nonlinear time series. (with J. Fan and Q. Yao). *Biometrika*, Vol. 83, 189–206.

91e. On Delay Co-ordinates in Stochastic Dynamical Systems. (with B. Cheng), In *Stochastic and spatial structures of dynamical systems*, eds. S. J. van Strien and S. M. Verduyn Lunel, Royal Netherlands Academy of Arts and Science, Amsterdam: North-Holland, 29–37.

90. Asymmetric least squares regression estimation: A nonparametric approach. (with Q. Yao). *Nonparametric Statist.*, Vol. 6, 273–292.

1995

89. A personal overview of nonlinear time series from a chaos perspective (with discussions). *Scan. J. Statist.*, Vol. 22, 399–445.

88p. On initial-condition sensitivity and prediction in nonlinear stochastic systems. (with Q. Yao). *Bull. Int. Statist. Inst.*, 50th Session, Beijing, China, Vol. IP 10.3, 395–412.

87e. An overview on chaos. In *Complex Stochastic Systems and Engineering*, IMA Conference Series, New Series, Number 54, ed. D. M. Titterington, Oxford University Press, 3–11.

1994

86e. Akaike's approach can yield consistent order determination. *Frontiers of Statistical Modeling: An Information Approach*, ed. H. Bozdogan, Kleuwer Academic Publication, 93–103.

85. A note on noisy chaos. (with K. S. Chan), *J. Roy. Statist. Soc. B*, Vol. 56, 301–311.

84e. Comments on prediction by nonlinear least squares methods. Chapter 17 in *Probability, Statistics and Optimization: A Tribute to Peter Whittle*, ed. F. Kelly, London: J. Wiley.

83. Quantifying the influence of initial values in nonlinear prediction. (with Q. Yao). *J. Roy. Statist. Soc. B*, Vol. 56, 701–25.

82. On subset selection of stochastic regression model. (with Q. Yao). *Statistica Sinica*, Vol. 4, 51–70.

81. On prediction and chaos in stochastic systems. (with Q. Yao). *Philos. Trans. Roy. Soc. (London) A*, Vol. 348, 357–369.

80. Orthogonal projection, embedding dimension and sample size in chaotic time series from a statistical perspective. (with B. Cheng). *Philos. Trans. Roy. Soc. (London) A*, Vol. 348, 325–41.

1993

79. On residual sums of squares in non-parametric autoregression. (with B. Cheng). *Stochastic Processes and Their Applications*, Vol. 48, 157–174.

78e. Nonparametric function estimation in noisy chaos. (with B. Cheng), *Developments in Time Series Analysis*, ed. T. Subba Rao, London: Chapman and Hall, 183–206.

77. A note on tests for threshold-type nonlinearity in open loop systems (with A. E. Sorour). *Applied Statistics*, Vol. 42, 95–104.

76. Between chance and chaos, *Twenty-first Century*, The Research Institute of Chinese Culture, The Chinese University of Hong Kong, Vol. 20, 90–98.

1992

75e. Contrasting aspects of nonlinear time series analysis. *New Directions in Time Series Analysis*, Part I, *IMA Volumes in Maths & Its Appl*, Vol. 45, eds. D. Brillinger *et al.*, Berlin: Springer-Verlag, pp. 357–370.

74. Some comments on a bridge between nonlinear dynamicists and statisticians. *Physica D*, Vol. 58, 299–303.

73. Likelihood plots, influential data and reparametrization in nonlinear time series modelling. (with K.S. Chan and R. Moeanaddin). *Proceedings of 1990 Taipei Symposium in Statistics*, Taipei, Taiwan, eds. M. T. Chao and P. E. Cheng, Institute of Statistical Science, Taiwan, pp. 37–62.

72. Consistent nonparametric order determination and chaos — with Discussion. (with B. Cheng). *J. Roy. Statist. Soc.*, B, Vol. 54, 427–449 and 451–474.

71. A note on one-dimensional chaotic maps under time reversal. (with B. Cheng). *Adv. Appl. Prob.*, Vol. 24, 219–220.

1991

70. Threshold autoregressive modelling in continuous time. (with I. Yeung). *Statistica Sinica*, Vol. 1, 411–430.

69. Strong consistency of least-squares estimator for a non-ergodic threshold autoregressive model. (with D. T. Pham and K. S. Chan). *Statistica Sinica*, Vol. 1, 361–369.

68. On tests for self-exciting threshold autoregressive-type nonlinearity in partially observed time series. (with I. Yeung) *Applied Statistics*, Vol. 40, 43–62.

1990

67. Is bilinear model an illusion? (with R. Moenaddin). *Statistique et Analyse des Donnees*, Vol. 15, 57–60.

66. On likelihood ratio tests for threshold autoregression. (with K. S. Chan). *J. Roy. Statist. Soc.*, B, Vol. 52, 469–476.

65. Clusters of time series models: An example. (with P. Dabas). *J. Applied Statistics*, Vol. 17, 187–198.

64. Numerical evaluation of distributions in non-linear autoregression (with R. Moeanaddin). *J. Time Series Analysis*, Vol. 11, 33–48.

63. On tests for threshold-type non-linearity in irregularly space time series. (with I. Yeung). *J. Statist. Comp. and Simulations*, Vol. 34, 177–194.

1989

62. A practical method for outlier detection in autoregressive time series modelling. (with M. C. Hau), *Stochastic Hydrology and Hydraulics*, Vol. 3, 241–260.

61p. Strong consistency of the least squares estimator for a non-stationary threshold autoregressive model. (with D. T. Pham and K. S. Chan), *Bull. Int. Stat. Inst.*, 47th Session, C.P. Book 2, pp. 202–203 — full version appeared in Statistica Sinica, see paper 69.

60e. Threshold, stability, non-linear forecasting and irregularly sampled data. *Statistical Analysis & Forecasting Economic Structural Change*, ed. P. Hackl, IIASA, Berlin: Springer-Verlag, 279–296.

59. Non-linear time series models of regularly sampled data: A review (an expanded version of paper 52). *Progress in Mathematics* (China), Vol. 18, 22–43.

58e. A survey of the statistical analysis of univariate threshold autoregressive models. (with K. S. Chan). *Advances of Statistical Analysis and Statistical Computing*, Vol. 2, JAI Press Inc., U.S.A., 1–42.

1988

57. A note on local parameter orthogonality and Levinson-Durbin algorithm. *Biometrika*, Vol. 75, 788–789.

56. A comparison of likelihood ratio test and CUSUM test for threshold autoregression. (with R. Moenaddin). *The Statistician*, Vol. 37, 213–225 (Addendum & Corrigendum in 37, 493-494).

55. On multi-step non-linear least-squares prediction. (with R. Moenaddin). *The Statistician*, Vol. 37, 101–110.

54e. Non-linear time series modelling in population biology: A preliminary case study. *Nonlinear Time Series and Variable Structure in Signal Processing*, ed. R. Mohler. Lecture Notes in Control & Information Science. 106, Heidelberg: Springer-Verlag, 75–87.

1987

53. A note on embedding a discrete parameter ARMA model in a continuous parameter ARMA model. (with K. S. Chan). *J. Time Series Analysis*, Vol. 8, 277–281.

52p. Non-linear time series models of regularly sampled data: a review. *Proc. of First World Congress of the Bernoulli Society*, eds. Y. V. Prohorov and V. V. Sazonov, 2, 355–367. Holland:VNU Science Press. (Note: The expanded version of this paper appeared as paper 59.)

1986

51. On tests for non-linearity in time series analysis. (with W. S. Chan). *J. Forecasting*, Vol. 5, 217–228.

50. A note on certain integral equations associated with non-linear time series analysis. (with K. S. Chan). *Probability Theory and Its Related Fields*, Vol. 73, 153–158.

49. On estimating thresholds in autoregressive models. (with K. S. Chan). *J. Time Series Analysis*, Vol. 7, 179–190.

1985

48. Threshold time series modelling of two Icelandic riverflow systems. (with B. Thanoon and G. Gudmundsson). *Water Resources Bulletin*, Vol. 21, 651–661.

47. On the use of the deterministic Lyapunov function for the ergodicity of stochastic difference equations. (with K. S. Chan). *Adv. Appl. Prob.*, Vol. 17, 666–678.

46. A multiple threshold AR(1) model. (with K. S. Chan, J. D. Petruccelli and S. W. Woolford), *J. Appl. Prob.*, Vol. 22, 267–279.

1984

45. A note on sub-system stability and system stability. (with K. S. Chan). *J. Eng. Mathematics* (China), Vol. 1, Pt.2, 43–51.

1983

44p. Threshold time series models of some riverflow data. *Proc. 44th Session of ISI*, Vol. C.P.46 8.

43e. Threshold autoregression and some frequency-domain characteristics (with J. Pemberton). *Handbook of Statistics*, Vol. 3, eds. D. R. Brillinger and P. R. Krishnaiah, North-Holland, 249–273.

42. On the distribution of a simple stationary bilinear process. (with S. R. Wang and H. Z. An). *J. Time Series Analysis*, Vol. 4, 209–216.

41. A statistical approach to difference-delay equation modelling in ecology–two case studies. (with K. S. Lim), *J. Time Series Analysis*, Vol. 4, 239–267. (Revised version of paper 39).

40. A note on delayed autoregressive process in continuous time. *Biometrika*, Vol. 70, 710–712.

39e. A statistical approach to difference-delay equation modelling in ecology– two case studies (with K. S. Lim). *Rhythms in Biology and Other Fields of Application*, eds. M. Cosnard *et al.*, Lecture Notes in Biomathematics, 49, Springer-Verlag, 319–344. (See paper 41).

1982

38. Some personal experiences in popularising mathematical methods in the People's Republic of China, as a collaborator with the late Professor L. K. Hua. *Int. J. Math. Education in Sc. & Tech.*, Vol. 13, 371–386.

37. A note on using threshold autoregressive models for multi-step-ahead prediction of cyclical data. *J. Time Series Analysis*, Vol. 3, 137–140.

36. Discontinuous decision processes and threshold autoregressive time series modelling. *Biometrika*, Vol. 69, 274–276.

35e. Multi-step-ahead forecasting of cyclical data by threshold autoregression. (with Z. M. Wu). *Time Series Analysis: 'Theory and Practice 1*, ed. O. D. Anderson, North-Holland, 733–753.

1981

34. A note on a Markov bilinear stochastic process in discrete time. *J. Time Series Analysis*, Vol. 2, 279–284.

33. Data transformation and self-exciting threshold autoregression. (with D. K. Ghaddar). *J. Roy. Statist. Soc. C*, Vol. 30, 238–248.

32. A note on the distribution of non-linear autoregressive stochastic processes. (with J. Pemberton). *J. Time Series Analysis*, Vol. 2, 49–52.

1980

31p. Catastrophe in time series analysis? Paper read to Journees de Statistique, Universite Paul Sabatier, Toulouse, France, May 1980. Abstract in *Journees de Statistique*, Resume de Communications, 106.

30. On stability and limit cycles of non-linear autoregression in discrete time. (with J. Pemberton). *Cahiers du CERO*, Vol. 22, 2, 137–148. Bruxelles.

29. Threshold autoregression, limit cycles and cyclical data-with discussion. (with K. S. Lim). *J. Roy. Statist. Soc.*, B, Vol. 42, 245–292.

28e. A view on non-linear time series model building. *Time Series*, ed. O. D. Anderson, 41–56, Amsterdam: North-Holland.

1979

27. Final prediction error and final interpolation error: A paradox? *I.E.E.E. Trans. Inf. Th.*, Vol. IT-25, 758–759.

26. A note on a local equivalence of two recent approaches to autoregressive order determination. *Int. J. Control*, Vol. 29, 441–446.

1978

25e. On a threshold model. *Pattern Recognition and Signal Processing*, NATO ASI Series E: Applied Sc. No. 29, ed. C. H. Chen. The Netherlands: Sijthoff & Noordhoff, 575–586.

24. On the asymptotic joint distribution of the estimated autoregressive coefficients. *Int. J. Control*, Vol. 27, 801–807.

1977

23. Some comments on the Canadian Lynx data-with discussion. *J. Roy. Statist. Soc. A*, Vol. 140, 432–436 and 448–468.

22. On the estimation of $\Pr\{Y < X\}$ for exponential families. *I.E.E.E. Trans. Reliability*, Vol. R-26, 54–56.

21. More on AR model fitting with noisy data by AIC. *I.E.E.E. Trans. Inf. Th.*, Vol. IT-23, 409–410.

1976

20. On Markov chain modelling to some weather data. (with P. Gates). *J. Appl. Meteorology*, Vol. 15, 1145–1151.

19. Fitting a smooth moving average to noisy data. *I.E.E.E. Trans. Inf. Th.*, Vol. IT-22, 493–496.

18. On a statistic useful for dimensionality reduction of linear stochastic systems. (with T. Sugiyama). *Communications in Statistics*, Vol. A5(8), 711–721.

1975

17. Letter to the Editor. *Technometrics*, Vol. 17, 393.

16. A simulation study of the estimation of evolutionary spectral functions. (with W.-Y. T. Chan). *J. Roy. Statist. Soc.* C, Vol. 24, 334–341.

15. Autoregressive model fitting with noisy data by Akaike's information criterion. *I.E.E.E. Trans. Inf. Th.*, Vol. IT-21, 476–480.

14p. On the fitting of non-stationary autoregressive models in time series analysis. (with T. Ozaki). *Proc. 8th Hawaii Int. Conf. on System Sc.*, Western Periodicals, North Hollywood, California, 225–226.

13. Determination of the order of a Markov chain by Akaike's information criterion. *J. Appl. Prob.*, Vol. 12, 488–497.

1974

12. Linear time-dependent systems. (with T. Subba Rao). *I.E.E.E. Trans. Auto. Control*, Vol. AC-19, 736–737.

11. Applications of principal component analysis and factor analysis in stochastic control systems. (with M. B. Priestley and T. Subba Rao). *I.E.E.E. Trans. Auto. Control*, Vol. AC-19, 730–734.

10. Note on the estimation of $\Pr\{Y < X\}$ in the negative exponential case. *Technometrics*, Vol. 16, 625.

9. Frequency-domain approach to regulation of linear stochastic systems. *IFAC J. Automatica*, Vol. 10, 533–538.

8. On time-dependent linear transformations of non-stationary stochastic processes. *J. Appl. Prob.*, Vol. 11, 53–62.

7. Identification of the covariance structure of state space models. (with T. Subba Rao). *Bull. Inst. Math. & Appl.*, Vol. 11, No. 5/6, May/June, 201–203.

1973

6. On some tests for time-dependence of a transfer function. (with T. Subba Rao). *Biometrika*, Vol. 60, 589–597.

5. On the analysis of bivariate non-stationary processes-with discussion. (with M. B. Priestley). *J. Roy. Statist. Soc. B*, Vol. 35, 153–166 and 179–188.

4. Some comments on spectral representations of non-stationary stochastic processes. *J. Appl. Prob.*, Vol. 10, 881–885.

3e. On time-dependent linear stochastic control systems. (with T. Subba Rao). *Recent Mathematical Development in Control*, ed. D. J. Bell, Academic Press.

2e. Identification of the structure of multivariate stochastic systems. (with M. B. Priestley and T. Subba Rao). *Multivariate Analysis III*, ed. P. Krishnaiah, Academic Press.

1972

1. A test for time-dependence of linear open loop systems. (with T. Subba Rao). *J. Roy. Statist. Soc.*, B, Vol. 34, 235–250.

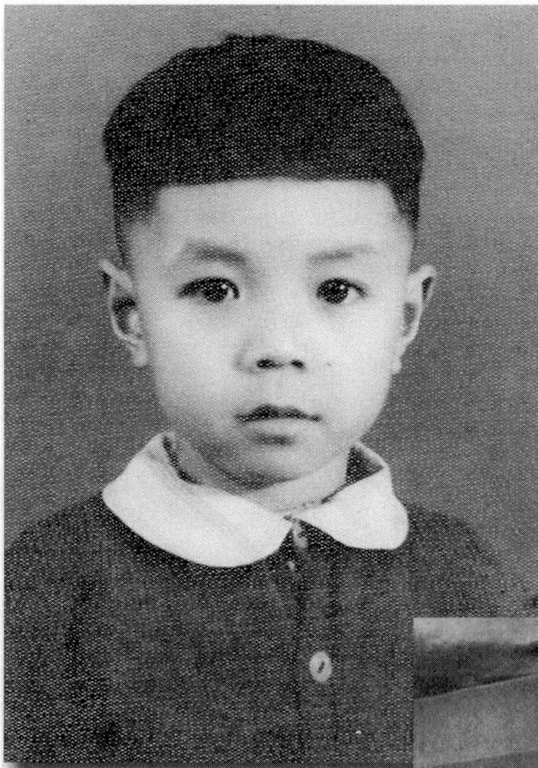

Five Years old in Hong Kong

Sent to UK to Study in the 1960's by my dad

Entertaining my childhood hero, Loo-Keng Hua, in my Manchester home in 1979

With Maurice Priestley at the Valencia Time Series Meeting in 1981

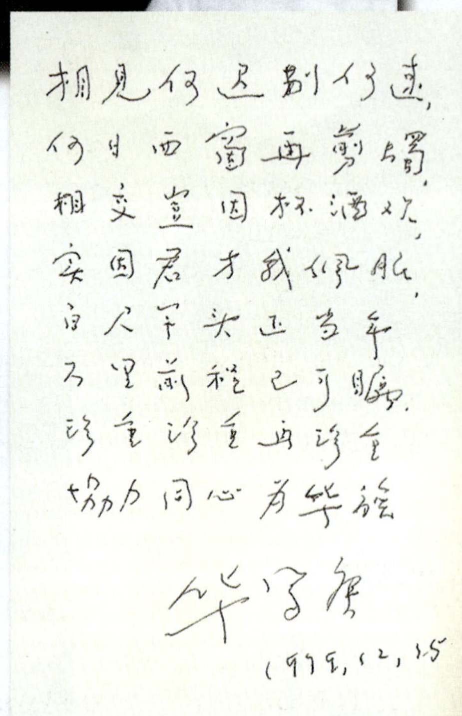

相见何迟别何速，
何日西窗再剪烛，
相交岂因杯酒欢，
实因君才我佩服，
百尺竿头正当年
万里前程已可瞻，
珍重珍重再珍重
协力同心为华族

华罗庚
1979, 12, 15

The Chinese Poem by LK Hua to me

Statistica Sinica **17**(2007), 8-14

Inside Views

Birth of the Threshold Time Series Model

Prologue

In this short note prepared for the theme volume on Threshold Models and New Developments in Time Series[1], I shall start with an account of how the threshold time series model was born and finish with some thoughts on the future directions of nonlinear time series analysis, with some random comments interspersed in between. The style is autobiographical and non-technical.

From the beginning of time series analysis, modeling was dominated by the assumption of linearity. This situation lasted until almost the end of the 1970s. In fact, before 1980, hardly any standard time series textbooks covered nonlinear time series models.

The Year 1977

In the annals of nonlinear time series modeling, I think the first year to remember is 1977. At an Ordinary Meeting of the Royal Statistical Society meeting in that year, Professor (now Sir) David Cox remarked, "all the models for the lynx data considered by Dr. Tong and by Mr. Campbell and Professor Walker are time reversible, …there is a fairly clear evidence from the data of irreversibility....a more likely explanation is the presence of nonlinearity" (See Tong (1977a, p. 453)). At the same meeting, Dr. Granville Tunnicliffe Wilson asked, "would we not prefer a model which....would exhibit stable periodic deterministic behaviour -- a limit cycle? Such limit cycles cannot arise from linear models" (p. 455). As no systematic study of nonlinear time series modeling existed at that time, he concluded pessimistically, "even if we are able to propose a wide class of nonlinear models to be used in fitting cyclical series,

[1] I am grateful to the University of Hong Kong for continuous support leading to the present note.

the problems of identifying, in the sense of Box and Jenkins, a suitable model are enormous"(p. 456).

The Challenge

As events unfolded, this 1977 Ordinary Meeting sparked some extraordinary developments (just as public schools in Britain are not so public, ordinary meetings of the RSS are not so ordinary). The above remarks openly challenged time series analysts to propose a wider class of practically useful nonlinear time series models, to gain a deep understanding of their probabilistic structure, to develop statistical identification/estimation of these models, and to address the general issue of nonlinear forecasting.

To develop useful nonlinear time series models was a daunting task indeed. Where should we start? For, any model which is not linear is nonlinear. To make a good choice we often have to rely on our value judgment, which is often influenced by the philosophy we subscribe to, the culture we have inherited and the taste we have developed. Of course, luck can sometimes come into the picture too.

Philosophy

To take up the challenge, I decided around 1977-78 that I would focus on cyclical animal population and river flow data. I saw at least two main advantages in doing so. First, it is important that the developed nonlinear time series models should be capable of offering insight into the underlying dynamics of the data. In this respect, the deterministic theory of dynamical systems should provide inspiration. Indeed, the reference to "limit cycles" by Granville mentioned above made a deep impression on me. Second, it was sound to have specific data sets in mind for quickly and constantly checking if the methodology under development was headed in the right direction. There is no doubt that I subscribe to the philosophy of the inseparability of theory and practice.

Non-linear Oscillations

Like many statisticians of my generation, I was ill-equipped mathematically because what I had received was predominantly an education in linear mathematics --I was badly taught! This meant I had to teach myself a new subject from scratch, and I started to read (rather slowly) the books by Minorsky (1962) and Andronov and Khaikin (1949). The original text of the latter was in Russian, which I could not read (and still cannot). Luckily, quite by chance, I got hold of a Chinese translation. The copy I acquired was a castaway that arrived in the UK from Shanghai during the turmoil known as the Cultural Revolution. Ironically, I have benefited culturally from the revolution! I should also mention my sense of admiration for Professor Peter Whittle when I saw his reference to exactly the same book in his celebrated paper on the analysis of the seiche record (Whittle (1954)). He noted an arithmetic relationship among the peaks in the power spectrum, explained that this must be the consequence of nonlinearity, and suggested a piecewise linear differential equation model. Of course, I only discovered this gem when writing my Springer Lecture Notes in 1983. Peter seemed (perhaps pleasantly) surprised when he saw my reference to this work because he said, "you know, Howell, you must be the only person who has cited this model of mine."

The Penny Drops

During late 1977 and early 1978, I played around a bit with bilinear time series models after listening to a talk by the Swedish control engineer, Professor Karl J. Aström. I obtained some early results but decided that the approach was not to my taste and abandoned it. Essentially, I could not reconcile the role of the unobservable innovation, used artificially in the univariate bilinear time series models, with the control variable, cited widely in the original control engineering literature. Then one day in 1977, as I was mowing my lawn, the penny dropped: piecewise-linearity was the way! This approach could represent the different phases, increasing and decreasing, in an animal population and the impact of the melting of ice/snow on river flow. Phase transition is, of course, a fundamentally nonlinear phenomenon. Perhaps I was subconsciously reverting to the strategy of "*divide and rule*", which has been so deeply ingrained in both Chinese and English cultures. The curious thing was that I got this idea before reaching the piecewise differential equation bits in Andronov and Khaikin. Would I have had the same idea had I read them first? In fact, while intoxicated by piecewise linearization, I thought I had also invented piecewise linear differential equations. Luckily, that only lasted for a very short time because on turning over the pages, I could see the full glory of these differential equations expounded by Andronov and Khaikin. Clearly I was born at least 40 years too late!

Pride and Prejudice

The threshold idea was thus conceived in 1977 and I recorded it in my contributions to the discussion of a paper by Tony Lawrance in 1977 (Tong (1977b)) and a NATO ASI series in 1978. However, to put the idea into practice meant a huge amount of computer experimentation. I say "huge" because we were in the late 1970s when computers were much slower than they are now. Luckily, a couple of my research students, P. K. Wong and K. S. Lim, were keen to help. I can still remember the joy of seeing the first limit cycle produced by what is now called a SETAR (self-exciting threshold autoregressive) model. Actually, this came in a round-about way. I asked Lim to do some multi-step forecast with a SETAR model via simulation. She misunderstood me and showed the result obtained by recursion of the SETAR model after deleting the innovation, that is, the skeleton in the terminology I introduced later. So, my first glimpse of a SETAR-generated limit cycle was due to my research student carrying out the wrong task. Now, I call that luck!

By the later part of 1978, I had a paper on threshold autoregression written up and submitted to a prestigious journal in the US. As usual with that journal, the review seemed to take ages. When it finally came back, it was basically positive but revision was needed. Alas, by the time I re-submitted the revised paper, there seemed to have been some changes in the editorial board. I cannot remember exactly what happened but the letter of rejection was signed by a different editor and the tone was discouraging. Dejected? Perhaps, but not for long, because I thought I could always try a better platform, namely, a discussion paper read to the Royal Statistical Society. This I did, and the paper was accepted for reading. I read the paper, *Threshold autoregression, limit cycles and cyclical data*, to the RSS on 19th March 1980.

The paper did not attain instant acclaim, although I think there was a "let-us-wait-and-see" welcome. Looking back at my work, I could have polished the paper more. I think the main reason for the hesitant reaction was that the idea was rather new, although its form was deceptively simple. There were still so many rough edges to smooth out (e.g. How to choose the threshold variables? Can the regime switching be continuous rather than discontinuous?), so many unresolved theoretical issues (e.g. What are the sampling properties of the parameter estimates? How to test for linearity within the context of SETAR? How to obtain theoretical multi-step forecast formula?), and so many more data-analytic techniques to develop. In any case, I was spurred on to smooth out the rough edges and to forge an even stronger link between (statistical) nonlinear time series and (deterministic) nonlinear dynamical systems, including chaos. I often collaborated with my students and others. Tong (1990; 1995) and Chan and Tong (2001) give a good summary of our results. Since its publication, the 1980 paper has attracted a great deal of attention and is my most frequently cited paper. What is most pleasing is the fact that many brilliant and mostly younger colleagues have been attracted to the threshold models; their input has gone a long way towards resolving many of the above mentioned issues and beyond.

What Next?

The threshold model as introduced in Tong and Lim (1980) is more general than the SETAR model. This theme issue further shows that the threshold model is still full of vitality and, like its linear predecessor (i.e. Udny Yule's linear AR model), chances are that it might stay around indefinitely. Still, where shall we go next in the wider context of nonlinear time series analysis? As I have said in my book (Tong (1990, p. 345)), he who forecasts does not know. So with this disclaimer, here I go. First, nonlinear time series modeling to-date has focused on the steady state, hence ergodicity/stationarity. The transient state has often been ignored. Nonlinear dynamics tells us that a nonlinear dynamical system can reside in the neighborhood of an equilibrium state for a certain period of time, which can be quite short or quite long, before jumping to another. (Perhaps MCMC enthusiasts can take note!) This prompts me into suggesting that there can be interaction between nonstationarity and nonlinearity, especially if all that we have are the observed data. Can we always tell them apart? Should we unscramble the omelette? If so, how? Next, multiple nonlinear time series analysis is an important area.

It is heartening to see some developments in this volume and elsewhere, but I think much more is waiting out there for us to explore. I do not need to reiterate the importance of multiple time series, linear or nonlinear, in practical applications. Of course, the multi-dimensional world is much richer than the unidimensional one. It is clear that some dimensional reduction is absolutely essential in order to ameliorate the curse of dimensionality. How to best visualize a high dimensional object is not unrelated to the choice of appropriate generalized coordinates in dynamics. It seems to me that the semi-parametric framework is a good candidate, and that there have been some encouraging developments, including at least one paper in this theme volume, but much more needs to be done. There might also be points of contact with the machine learning community.

Last but not least, spatial-temporal data abound. They require spatial-temporal models. There have been some worthwhile developments, including some reported in this volume. One ultimate goal could be some nonlinear/nonstationary spatial-temporal models. Essentially what we want is a discrete time analogue of a stochastic partial differential equation.

Epilogue

Nowadays, seeing that the threshold autoregressive models and the threshold idea have been so successfully applied to many practical problems in diverse fields such as ecology, econometrics, economics, finance, actuarial science, hydrology and many others I think that the efforts have been all worthwhile. Those models are also firmly established in the literature, including textbooks. When I see people using terms or acronyms such as STAR, DTARCH, threshold-ARCH, threshold unit-root test, threshold co-integration, Markov regime-switching (under a different name, for example, in Tong and Lim (1980, p.285)), and the amazing number of citations produced by a scholar.google.com search of these names and their cousins, I cannot help but smile and say to myself, "I bet not many of them know that they are using a US reject!"

References

Andronov, A. A. and Khaikin, S. E. (1949). *Theory of Oscillations.* (first published in Russian in 1937, and translated and adapted by S. Lefschetz), Princeton Univ. Press, Princeton, NJ.

Chan, K. S. and Tong, H. (2001). *Chaos: A Statistical Perspective.* Springer-Verlag, New York.

Minorsky, N. (1962). *Non-Linear Oscillations.* Van Nostrand, Princeton, NJ.

Tong, H. (1977a). Some comments on the Canadian lynx data (with discussion). *J. Roy. Statist. Soc. Ser. A* **140**, 432-436 and 448-468.

Tong, H. (1977b). Contribution to the discussion of the paper entitled "Stochastic modelling of riverflow time series" by A. J. Lawrance and N. T. Kottegoda. *J. Roy. Statist. Soc. Ser. A*, 34-35.

Tong, H. (1978). *On a threshold model.* In *Pattern Recognition and Signal Processing* (Edited by C. H. Chen), NATO ASI Series E: Applied Sec. No. 29, Sijthoff & Noordhoff, Amsterdam, 575-586.

Tong, H. (1983). *Threshold Models in Non-Linear Time Series Analysis.* Lecture Notes in Statistics No. 21, Springer-Verlag, New York.

Tong, H. (1990). *Non-Linear Time Series: A Dynamical System Approach.* Oxford Univ. Press, Oxford.

Tong, H. (1995). A personal overview of nonlinear time series analysis from a chaos perspective (with discussions). *Scand. J. Statist.* **22**, 399-421.

Tong, H. and Lim, K. S. (1980). Threshold autoregression, limit cycles and cyclical data (with discussion). *J. Roy. Statist. Soc. Ser. B* **42**, 245-292.

Whittle, P. (1954). The statistical analysis of a seiche record. *J. Marine Res.* (Sears Foundation) **13**, 76-100.

— Howell Tong

In the enlightened year of 1970, Howell Tong was appointed to a lectureship at the University of Manchester Institute of Science and Technology shortly after he started his Ph.D. program. He received his Ph.D. in 1972 under the supervision of Maurice Priestley, thus making him a student of a student of Maurice Bartlett. He stayed at UMIST until 1982, when he took up the Founding Chair of Statistics at the Chinese University of Hong Kong. In 1986, he returned to the UK, as the first Chinese to hold a Chair of Statistics in the history of the UK, by accepting the Chair at the University of Kent at Canterbury. He stayed there until 1997, when he went to the University of Hong Kong, first as Distinguished Visiting Professor, and then as a Chair Professor of Statistics (and sometimes as a Pro-Vice-Chancellor and the Founding Dean of the Graduate School). He was appointed to his Chair at the London School of Economics in 1999. He has written three books (one with K. S. Chan) and (with collaborators) over 145 papers in Statistics, Ecology, Actuarial Science, Control Engineering, Reliability, Meteorology, Water Engineering, Engineering Mathematics and Mathematical Education. He is a Foreign Member of the Norwegian Academy of Science and Letters, a member of the ISI, a Fellow of IMS and an Honorary Fellow of the Institute of Actuaries (UK). He won a Chinese National Natural Science Prize (Class II) in 2000. He enjoys working with colleagues or students younger and brighter than himself. Having been involved right from the beginning of nonlinear time series modeling in the late 1970s, he is delighted to see that the threshold time series models he created have become an important standard approach and percolated into Econometrics, Ecology and other fields. He enjoys traveling, good food and walking with his wife, admiring theatre sets created by his talented daughter, learning through photographs about the many far-flung places visited by his son and daughter and some solitary reading of things non-statistical.

J. R. Statist. Soc. B (1980),
42, *No.* 3, *pp.* 245–292

Threshold Autoregression, Limit Cycles and Cyclical Data

By H. TONG and K. S. LIM

Department of Mathematics, University of Manchester Institute of Science and Technology

[Read before the ROYAL STATISTICAL SOCIETY at a meeting organized by the RESEARCH SECTION on Wednesday, March 19th, 1980, Professor P. WHITTLE in the Chair]

SUMMARY

The notion of a limit cycle, which can only exist in a non-linear system, plays the key role in the modelling of cyclical data. We have shown that the class of threshold autoregressive models is general enough to capture this notion, a definition of which in discrete time is proposed. The threshold value has an interesting interpretation. Simulation results are presented which demonstrate that this new class of models exhibits some well-known features of non-linear vibrations. Detailed analyses of several real data sets are discussed.

Keywords: THRESHOLD AUTOREGRESSION; LIMIT CYCLE; CYCLICAL DATA; NON-LINEAR AUTOREGRESSION; TIME IRREVERSIBILITY; THRESHOLD AUTOREGRESSIVE/MOVING AVERAGE MODELS; NON-LINEAR VIBRATIONS; JUMP RESONANCE; AMPLITUDE-FREQUENCY DEPENDENCY; SUB-HARMONICS; HIGHER HARMONICS; CANADIAN LYNX; MINK AND MUSKRAT; PREDATOR–PREY; WOLF'S SUNSPOT NUMBERS; RAINFALL–RIVERFLOW; EVENTUAL FORECASTING FUNCTION; STABILITY; AKAIKE'S INFORMATION CRITERION; HOUSEHOLDER TRANSFORMATIONS

1. INTRODUCTION

IT may be said that the era of linear time series modelling began with such linear models as Yule's autoregressive (AR) models (1927), first introduced in the study of sunspot numbers. In the past five decades or so, we have seen remarkable successes in the application of linear time series models in diverse fields, e.g. Box and Jenkins (1970), and the recent Nottingham International Time Series Conference in March 1979. These successes are perhaps rather natural in view of the significant contributions of linear differential equations in all branches of science. In particular, as far as a one-step-ahead prediction is concerned, a linear time series model is often quite adequate.

However, just as a linear differential equation is totally inadequate as a tool to analyse more intricate phenomena such as limit cycles, time irreversibility, amplitude-frequency dependency and jump resonance, a linear time series model should give place to a much wider class of models if we are to gain deeper understanding into the structure of the mechanism generating the observed data. For example, no linear Gaussian model can explain properly the saw-tooth cycles apparent in the Canadian lynx data (see, for example, discussion of papers by Campbell and Walker, 1977, and Tong, 1977a), and many riverflow data (see, for example, Lawrance and Kottegoda, 1977).

The new era of *practical* non-linear time series modelling is, without doubt, long overdue. In this paper, we describe the theory and practice of a new class of non-linear time series models which are based on the idea of piece-wise linearization. Sections 6 and 9 of this paper are due to both authors while the other sections are due to the first author.

We propose the following requirements for our non-linear time series models, in order of preference:

(i) statistical identification of an appropriate model should not entail excessive computation;

(ii) they should be general enough to capture some of the non-linear phenomena mentioned previously;

(iii) one-step-ahead predictions should be easily obtained from the fitted model and, if the adopted model is non-linear, its overall prediction performance should be an improvement upon the linear model;

(iv) the fitted model should preferably reflect to some extent the structure of the mechanism generating the data based on theories outside statistics;

(v) they should preferably possess some degree of generality and be capable of generalization to the multivariate case, not just in theory but also in practice.

Before describing a newly introduced class of non-linear time series models, it may serve us well in recalling some elementary, yet important, properties in the theory of non-linear differential equations or non-linear systems. Here, no stochastic element is involved and only those properties relevant to later exposition are included.

2. NON-LINEAR DIFFERENTIAL EQUATIONS

(i) By definition, the *principle of superposition* does not hold in the non-linear case. In addition, the notion of a "complementary function" and a "particular integral" ceases to be meaningful here.

(ii) Unlike a stable linear system, in which the output (i.e. the solution of the differential equation) dies away when the input is "switched off", the output of a stable non-linear system may contain sustained oscillations which persist in the absence of input.

To illustrate this, let x_1 and x_2 denote the numbers of two species. Kolmogorov (see, for example, Minorsky, 1962, p. 69) has considered the general system of non-linear differential equations,

$$\frac{dx_1}{dt} = \alpha_1(x_1, x_2)\,x_1, \quad \frac{dx_2}{dt} = \alpha_2(x_1, x_2)\,x_2, \tag{2.1}$$

where α_1 and α_2 are continuous functions of x_1 and x_2 with continuous first derivatives. Under very general conditions, he has shown that sustained oscillations (of relatively small amplitude) prevail. It is instructive to quote the following words of Minorsky (1962) in his discussion of the above phenomenon, in which a "common sense" picture of a state of equilibrium is supplemented by relatively small fluctuations :

> "Topologically this ... is precisely a stable limit cycle in the (x_1, x_2) plane onto which wind the spiral trajectories from the outside as well as from the inside. The outside spiral trajectories are those which characterise the establishment of the biological phenomenon and the limit cycle is its representation in a stationary state.... As far as is known, no experimental verification of these results has been made so far. If this is done eventually and the Kolmogorov theory is confirmed, this will give valuable information regarding the actual biological probabilities involved in the co-existence of the two species."

As has been touched on by Tunnicliffe-Wilson (1977), limit cycles will play a central role in the modelling of cyclical data. We may write equation (2.1) in the following form,

$$\dot{\mathbf{x}} = \mathbf{A}\mathbf{x}, \tag{2.2}$$

where the over-dot denotes the time derivative, $\mathbf{x} = (x_1, x_2)^{\mathrm{T}}$, is called the *state vector*,† and

$$\mathbf{A} = \begin{bmatrix} \alpha_1(x_1, x_2) & 0 \\ 0 & \alpha_2(x_1, x_2) \end{bmatrix},$$

where, for greater generality, we may sometimes allow α_1 and α_2 to be discontinuous. The (x_1, x_2)-plane is sometimes referred to as the *phase plane* (or the *state space* in higher dimensional

† ^{T} denotes transpose.

cases). As an example of the phase plane, Fig. 1 represents that of the following non-linear differential equations from the output of an analogue simulation†

$$\frac{dP}{dt} = \begin{cases} -2(P(t)-5) & \text{if } H(t) < 15, \\ 0.5(P(t)-5) & \text{if } H(t) \geqslant 15, \end{cases} \tag{2.3a}$$

$$\frac{dH}{dt} = \begin{cases} (H(t)-8) & \text{if } P(t) < 10, \\ -2(H(t)-8{\cdot}4) & \text{if } P(t) \geqslant 10. \end{cases} \tag{2.3b}$$

Note that the spiral trajectories do not wind, from the outside, into a (singular) point, but they eventually go round and round closed loops, leaving an interior region untraversed, demonstrating the existence of a limit cycle. Note also that as functions of t, $P(t)$ and $H(t)$ are both *periodic* after the transients have died out. In (2.3), the limit cycle is self-excited, while in

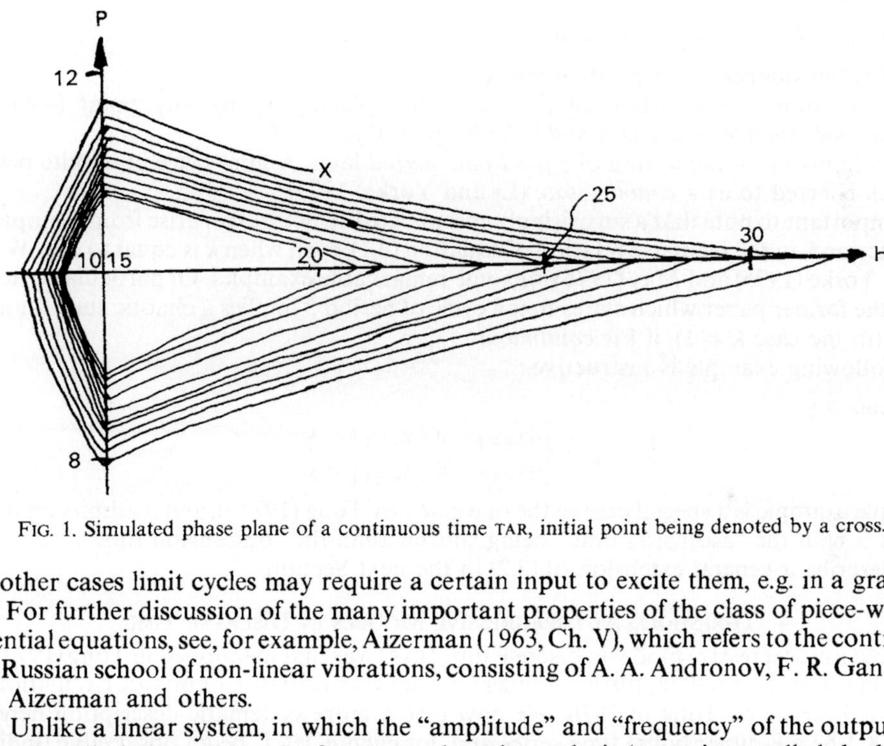

FIG. 1. Simulated phase plane of a continuous time TAR, initial point being denoted by a cross.

some other cases limit cycles may require a certain input to excite them, e.g. in a grandfather clock. For further discussion of the many important properties of the class of piece-wise linear differential equations, see, for example, Aizerman (1963, Ch. V), which refers to the contributions of the Russian school of non-linear vibrations, consisting of A. A. Andronov, F. R. Gantmakher, M. A. Aizerman and others.

(iii) Unlike a linear system, in which the "amplitude" and "frequency" of the output (signal) are functionally independent, the frequency domain analysis (sometimes called the harmonic analysis) of a non-linear system is much more complex. Non-linear vibration engineers have introduced notions such as "amplitude-frequency dependency", "jump resonance" and others.

3. A LIMIT CYCLE IN DISCRETE TIME

The discussion in continuous time of the last section is only relevant in so far as it gives us a reference frame for developing non-linear time series models in discrete time. This situation is not unlike that in which Yule (1927) first developed his celebrated AR models.

In this paper, we focus on the notion of a limit cycle, leaving the mathematical formulations of the other notions for a non-linear system for future developments. We will, however, indicate how the latter notions manifest themselves in the data through some numerical examples in Section 6.

† The unpublished M.Sc. dissertation by Mr P. K. Wong of UMIST (1978) may be consulted for more similar examples.

For each integer n, let $\mathbf{x}_n$ denote a k-dimensional (state) vector, satisfying the equation

$$\mathbf{x}_n = \mathbf{f}(\mathbf{x}_{n-1}). \tag{3.1}$$

Definition 3.1. A k-dimensional vector $\mathbf{x}^*$ is called a *limit point* if there exists an $\mathbf{x}_0$, not equal to $\mathbf{x}^*$, such that starting with n equal to zero, $\mathbf{x}_n$ tends to $\mathbf{x}^*$ component-wise, as n tends to infinity.

Let $\mathscr{C}$ denote the set of k-dimensional vectors $\mathbf{c}_i$ (of finite Euclidean norm), $i = 1, \ldots, T$, T being a positive integer $\leqslant \infty$.

Definition 3.2. $\mathscr{C}$ is called a *limit cycle of period* T if

(i) $\exists\, \mathbf{x}_0 \notin \mathscr{C}$ such that starting with n equal to zero, $\mathbf{x}_n$ will ultimately fall into $\mathscr{C}$ as n increases;

(ii) $\mathbf{c}_i = \mathbf{f}(\mathbf{c}_{i-1}), \quad i = 2, 3, \ldots, T,$

$\mathbf{c}_{T+i} = \mathbf{c}_i, \quad i = 1, 2, \ldots,$ and

(iii) T is the smallest such positive integer.

If, in addition, the assertion of (i) holds on replacing $\mathbf{x}_0$ by any point ($\notin \mathscr{C}$) in its neighbourhood, then $\mathscr{C}$ is called a *stable limit cycle of period T*.

We shall introduce the notion of a *fractional period* later. A limit cycle of infinite period is sometimes referred to as a *chaotic state* (Li and Yorke, 1975).

It is important to note that a surprisingly complicated structure can arise from a simple non-linear function $\mathbf{f}$, in the recursive relation of equation (3.1), even when k is equal to one. We refer to Li and Yorke (1975) and May (1976) for some remarkable examples. Of particular note is the result in the former paper which states that a cycle of period 3 implies a chaotic state for almost every $\mathbf{x}_0$ (in the case $k = 1$), if $\mathbf{f}$ is *continuous*.

The following example is instructive :

Example 3.1.

$$x_n = \begin{cases} 4x_{n-1} & \text{if } |x_{n-1}| \leqslant \frac{1}{2}, \\ \frac{1}{2}x_{n-1} & \text{if } |x_{n-1}| > \frac{1}{2}. \end{cases} \tag{3.2}$$

This simple example is a special case of the one given by Tong (1977b), and it admits limit cycles of period 3 with the "ascension time" being shorter than the "descension time".

We describe a general extension of (3.2) in the next Section.

4. THRESHOLD AUTOREGRESSIVE MODELS IN DISCRETE TIME

A threshold autoregressive model in discrete time (TAR) was first mentioned in Tong (1977b) and reported briefly in Tong (1978, 1980a). A fuller account was available for private circulation in an unpublished report by Tong in 1978. We now give a more systematic description here.

Let $\{\mathbf{X}_n\}$ be a k-dimensional time series and, for each n, let J_n be an observable (indicator) random variable, taking integer values $\{1, 2, \ldots, l\}$.

Definition 4.1. $\{\mathbf{X}_n;\ J_n\}$ is said to be a *general* TAR if

$$\mathbf{X}_n = \mathbf{B}^{(J_n)}\mathbf{X}_n + \mathbf{A}^{(J_n)}\mathbf{X}_{n-1} + \boldsymbol{\varepsilon}_n^{(J_n)} + \mathbf{C}^{(J_n)}, \tag{4.1}$$

where, for $J_n = j$, $\mathbf{A}^{(j)}$ and $\mathbf{B}^{(j)}$ are $k \times k$ (non-random) matrix coefficients, $\mathbf{C}^{(j)}$ is a $k \times 1$ vector of constants, and $\{\boldsymbol{\varepsilon}_n^{(j)}\}$ is a k-dimensional strict white noise sequence of independence random vectors with a diagonal covariance matrix. It is also assumed that $\{\boldsymbol{\varepsilon}_n^{(j)}\}$ and $\{\boldsymbol{\varepsilon}_n^{(j')}\}$ are independent for $j \neq j'$.

We now single out a few interesting special cases of the general TAR for further development.

First, let $\{r_0, r_1, \ldots, r_l\}$ denote a linearly ordered subset of the real numbers, such that $r_0 < r_1 < \ldots r_l$, where r_0 and r_l are taken to be $-\infty$ and $+\infty$ respectively. They define a partition of the real line $\mathbf{R}$, i.e.

$$\mathbf{R} = R_1 \cup R_2 \cup \ldots \cup R_l, \quad \text{say}$$

where

$$R_i = (r_{i-1}, r_i].$$

(A) Writing $\mathbf{X}_n = (X_n, X_{n-1}, \ldots, X_{n-k+1})^{\mathrm{T}}$,

$$\mathbf{A}^{(j)} = \left[\begin{array}{cccc|c} a_1^{(j)} & a_2^{(j)} & \ldots & a_{k-1}^{(j)} & a_k^{(j)} \\ \hline & \mathbf{I}_{k-1} & & & \mathbf{0} \end{array}\right] \quad \text{(a companion matrix)},$$

$$\mathbf{B}^{(j)} = \mathbf{0}, \quad \boldsymbol{\varepsilon}_n^{(j)} = (\varepsilon_n^{(j)}, 0, \ldots, 0), \quad \mathbf{C}^{(j)} = (a_0^{(j)}, 0, \ldots, 0),$$

and $R_j^{(k)} = \mathbf{R} \times \mathbf{R} \times \ldots \times \mathbf{R} \times R_j \times \mathbf{R} \times \ldots \times \mathbf{R}$ is the cylinder set in the cartesian product of k real lines, on the interval R_j with dth coordinate space (d some fixed integer belonging to $\{1, 2, \ldots, k\}$), and setting $J_n = j$ if $\mathbf{X}_{n-1} \in R_j^{(k)}$, we have

$$X_n = a_0^{(j)} + \sum_{i=1}^{k} a_i^{(j)} X_{n-i} + \varepsilon_n^{(j)}, \tag{4.2}$$

conditional on $X_{n-d} \in R_j$; $j = 1, 2, \ldots, l$. Since $\{J_n\}$ is now a function of $\{\mathbf{X}_n\}$ *itself*, we call the univariate time series $\{X_n\}$ given by equation (4.2) a *self-exciting threshold autogressive model of order* $(l; k, \ldots, k)$ *or* SETAR$(l; k, \ldots, k)$ where k is repeated l times. If, for $j = 1, 2, \ldots, l$,

$$a_i^{(j)} = 0 \quad \text{for } i = k_j + 1, \quad k_j + 2, \ldots, k,$$

then we call $\{X_n\}$ a SETAR$(l; k_1, k_2, \ldots, k_l)$. We call $r_1, \ldots, r_{l-1}$ the *thresholds*. Note that an SETAR $(1; k)$ is just a *linear* AR model of order k.

(B) (X_n, Y_n) is called an *open loop threshold autoregressive system* with $\{X_n\}$ as the observable output and $\{Y_n\}$ as the observable input, if

$$X_n = a_0^{(j)} + \sum_{i=1}^{m_j} a_i^{(j)} X_{n-i} + \sum_{i=0}^{m_j'} b_i^{(j)} Y_{n-i} + \varepsilon_n^{(j)}, \tag{4.3}$$

conditional on $Y_{n-d} \in R_j$; $(j = 1, \ldots, l)$, where $\{\varepsilon_n^{(j)}\}$; $j = 1, \ldots, l$, are strict white noise sequences, with zero mean and finite variances and each being independent of $\{Y_n\}$. The l white noise sequences are assumed to be independent of one another. We denote this system by TARSO$(l, (m_1, m_1'), \ldots, (m_l, m_l'))$.

(C) $\{X_n, Y_n\}$ is called a *closed-loop threshold autoregressive system*, or TARSC, if (X_n, Y_n) and (Y_n, X_n) are both TARSO. We assume that all the stationary white noise sequences involved are independent of one another.

5. SOME PERSPECTIVES

The essential idea underlying the class of threshold autoregressive models is the piece-wise linearization of non-linear models over the state space by the introduction of the thresholds $\{r_0, r_1, \ldots, r_l\}$; these models are *locally linear*. Similar ideas were used by Priestley (1965), Priestley and Tong (1973) and Ozaki and Tong (1975), in the analyses of non-stationary time series and time dependent systems, in which local stationarity was the counterpart of our present local linearity.

Of course, local linearity abounds in many practical situations. Indeed, if this were not the case, linear time series analysis could not have survived this long. For example, it may be argued that many real systems are non-linear only in so far as they exhibit the phenomenon of *saturation*. Another example is an electrical relay which is a particularly simple piece-wise linear system. In fact, by some co-ordinate transformations of the relay systems, a general class of piece-wise linear differential equations has been established and studied in depth. (See, for example, Aizerman, 1963.) Tong (1980a) has argued, on physical and biological grounds, for the adoption of piece-wise linear models in the analysis of the Canadian lynx data and Wolf's sunspot numbers. Sugawara (1952, 1961), Todini and Wallis (1977) and Chander in an unpublished Ph.D. thesis (1965) have studied the rainfall–riverflow system from the standpoint of piece-wise linearization. Other related works include Fujishige and Sawaragi (1974), Robinson and Sworder (1974), Rishel (1975), Waltman and Butz (1977), Jacobs and Lewis (1978), Haggan and Ozaki (1980) and Ozaki (1980).

Although at a deeper level, our general TAR must be considered at present as an *ad hoc* class of non-linear time series models, in a certain sense this class is not without some generality. Consider a general first order non-linear autoregressive process, NLAR(1), of the form

$$X_n = f(X_{n-1}) + \varepsilon_n. \tag{5.1}$$

It seems intuitively clear that subject to general conditions on f, an NLAR(1) may be approximated arbitrarily closely by a TAR. We may argue heuristically as follows. Suppose that $f(x)$ is continuous in a closed interval $[x', x'']$. It is well known that $f(x)$ is uniformly continuous in $[x', x'']$ and that the Weierstrass theorem ensures that $f(x)$ may be approximated arbitrarily closely by $\hat{f}(x)$, where

$$\hat{f}(x) = f(x_{(i_{k-1})}) + \alpha_k x, \tag{5.2}$$

for

$$x \in [x_{(i_{k-1})}, x_{(i_k)}), \quad k = 1, \ldots, l,$$

where

$$x_{(i_0)} = x', \quad x_{(i_l)} = x'',$$

and the partition

$$[x', x''] = [x', x_{(i_1)}) \cup [x_{(i_1)}, x_{(i_2)}) \cup \ldots \cup [x_{(i_{l-1})}, x'']$$

is defined depending on the degree of accuracy of the approximation required. Therefore we have obtained an SETAR(l; 1, 1, ..., 1) to the NLAR, with thresholds $\{x_{(i_1)}, x_{(i_2)}, \ldots, x_{(i_{l-1})}\}$. For an NLAR($k$),

$$X_n = f(X_{n-1}, \ldots, X_{n-k}) + \varepsilon_n, \tag{5.3}$$

we may re-write it in vector notation

$$\mathbf{X}_n = \tilde{\mathbf{f}}(\mathbf{X}_{n-1}) + \boldsymbol{\varepsilon}_n, \tag{5.4}$$

where

$$\mathbf{X}_n = (X_n, X_{n-1}, \ldots, X_{n-k+1})^{\mathrm{T}},$$
$$\boldsymbol{\varepsilon}_n = (\varepsilon_n, \underbrace{0, \ldots, 0}_{k-1})$$

and

$$\tilde{\mathbf{f}}(\mathbf{X}_{n-1}) = (f(X_{n-1}, \ldots, X_{n-k}), X_{n-1}, \ldots, X_{n-k+1})^{\mathrm{T}}.$$

A vector version of the Weierstrass theorem will then establish a general TAR approximation of an NLAR(k) under general conditions on f.

A more challenging problem concerns the following non-linear Markovian system (NMS):

$$\begin{aligned} \mathbf{X}_n &= \mathbf{f}(\mathbf{X}_{n-1}) + \boldsymbol{\varepsilon}_n, \\ \mathbf{Y}_n &= \mathbf{g}(\mathbf{X}_n), \end{aligned} \tag{5.5}$$

where $\mathbf{X}_n$ and $\mathbf{Y}_n$ are a k-dimensional unobservable vector and a q-dimensional ($q \leqslant k$) observable vector respectively, and $\boldsymbol{\varepsilon}_n$ defines a zero mean stationary k-dimensional strict white noise sequence and is independent of $\mathbf{X}_{n-1}$.

Suppose that $\mathbf{g}$ is a partition preserving mapping from $\mathbf{R}^k$ to $\mathbf{R}^q$, $k \geqslant q$, in the sense that $\{\mathbf{g}(R_i^{(k)})\}$ defines a partition of $\mathbf{R}^q$ for every partition $\{R_i^{(k)}\}$ of $\mathbf{R}^k$, where for any set A,

$$\mathbf{g}(A) = \{\mathbf{y}; \mathbf{g}(\mathbf{x}) = \mathbf{y}, \mathbf{x} \in A\}.$$

It seems plausible that an NMS with a partition preserving mapping **g** may be arbitrarily closely approximated by a TAR under general conditions on **f**. The problem arises as to the characterization of the class of such mappings. In the case of $k = q$, we know that it contains at least one element, the identity mapping.

Next, consider the recursive relation,

$$x_n = f(x_{n-1}, \dots, x_{n-k}) \quad (n \geqslant 0, k < \infty), \tag{5.6}$$

where f is such that $|x_j| < \infty$ for all $j < \infty$. It is well known that unless some "stability" condition is placed on f, the recursion will diverge. One example is a polynomial in $x_{n-1}, x_{n-2}, \dots, x_{n-k}$. However, in practice, we can usually circumvent this problem by introducing some inbuilt restrictions on the range space of f.

Let d be a pre-fixed integer chosen from $\{1, 2, \dots, k\}$, $k < \infty$. We agree to set

$$x_i = 0, \forall \quad i < 0.$$

Definition 5.1. Let f be a point transformation from $\mathbf{R}^k$ to $\mathbf{R}$ given by (5.6). Let S be a finite interval of $\mathbf{R}$. f_S is said to be a *stabilizer of f induced by S* if it has the following properties:

(i) $f_S(x_{n-1}, \dots, x_{n-k}) = x_n$,
(ii) $x_{n-d} \in S \Rightarrow f_S(x_{n-1}, \dots, x_{n-k}) = f(x_{n-1}, \dots, x_{n-k})$, and
(iii) $x_{n-d} \notin S \Rightarrow f_S(x_{n-1}, \dots, x_{n-k}) = c$, $c \notin S$ and $|c| < \infty$.

Theorem 5.1. f_S defines a stable recursion in the sense that

$$|f_S(x_{n-1}, \dots, x_{n-k})| < \infty, \quad \forall (x_{n-1}, \dots, x_{n-k}) \in \mathbf{R}^k \quad \text{and all } n.$$

Proof. Denote the row vector $(x_n, \dots, x_{n-k+1})$ by $\mathbf{x}_n$. We agree to call x_i an outlier if $x_i \notin S$. Suppose that $\mathbf{x}_{n_0}$ is the first vector with its first component x_{n_0} being an outlier. Under the recursion f_S,

$$\mathbf{x}_{n_0+d} = (c, x_{n_0+d-1}, \dots, x_{n_0+d-k+1}).$$

Obviously, $\forall n \geqslant n_0 + d$, $\mathbf{x}_n$ has at least one component equal to c. It now remains to be shown that the number of components of $\mathbf{x}_n$ equal to c is monotonically nondecreasing as n increases to infinity. There are two possibilities subsequent to $\mathbf{x}_{n_0+d}$. One possibility is that no more outliers will occupy the first component, except for the recurring c, in which case f_S defines a stable recursion. The other possibility is that a new outlier will occupy the first component in addition to the recurring c. Because each outlier will subsequently produce one further component equal to c, we have proved by induction that the number of components equal to c is monotonically non-decreasing. Hence, there exists an $M < \infty$, such that for all $n \geqslant M$,

$$|f_S(x_{n-1}, \dots, x_{n-k})| < \infty, \quad \forall (x_{n-1}, \dots, x_{n-k}) \in \mathbf{R}^k.$$

Therefore, by the finiteness of M,

$$|f_S(x_{n-1}, \dots, x_{n-k})| < \infty, \quad \forall (x_{n-k}, \dots, x_{n-k}) \in \mathbf{R}^k \quad \text{and all } n.$$

This completes the proof of the theorem.

It is interesting to note that f_S corresponds to a *threshold* model.

Finally, the question of stationarity for an NLAR(1), as well as its marginal and conditional distributions, has been studied by Jones (1978), who has applied Tweedie's (1975) general results concerning ergodicity of a Markov chain over a *general* state space to NLAR(1), and has also indicated possible extension to NLAR(k), $k \geqslant 1$. It is possible to show that a sufficient condition for the SETAR $\{\mathbf{X}_n\}$ described in case (A) of Section 4 to be ergodic in the sense of Tweedie (1975) is that the maximum eigenvalue of $\mathbf{A}^{(j)\mathrm{T}} \mathbf{A}^{(j)}$ is strictly less than unity, $j = 1, \dots, l$, and the $\varepsilon_n^{(j)}$'s have absolutely continuous distributions. This is obtained by applying Corollary 5.2 of Tweedie (1975) with $\|\mathbf{x}\|^2 = \mathbf{x}^\mathrm{T}\mathbf{x}$ for a $k \times 1$ vector $\mathbf{x}$, and $\|\mathbf{A}\|^2 =$ maximum eigenvalue of $\mathbf{A}^\mathrm{T}\mathbf{A}$ for a $k \times k$ matrix $\mathbf{A}$. However, it is interesting to investigate if the eigenvalue requirement could be weakened.

6. TAR MODELS AND NON-LINEAR VIBRATIONS

We now describe some simulation results which demonstrate that TAR models exhibit interesting features well known in non-linear vibrations.

(1) *Jump resonance.* It is well known that, unlike a linear system, the output amplitude may have a "resonance jump" at different frequencies depending on whether the input frequency (of constant amplitude) is monotonically increasing or monotonically decreasing. (See Figs 2a, 2b and 2c.)

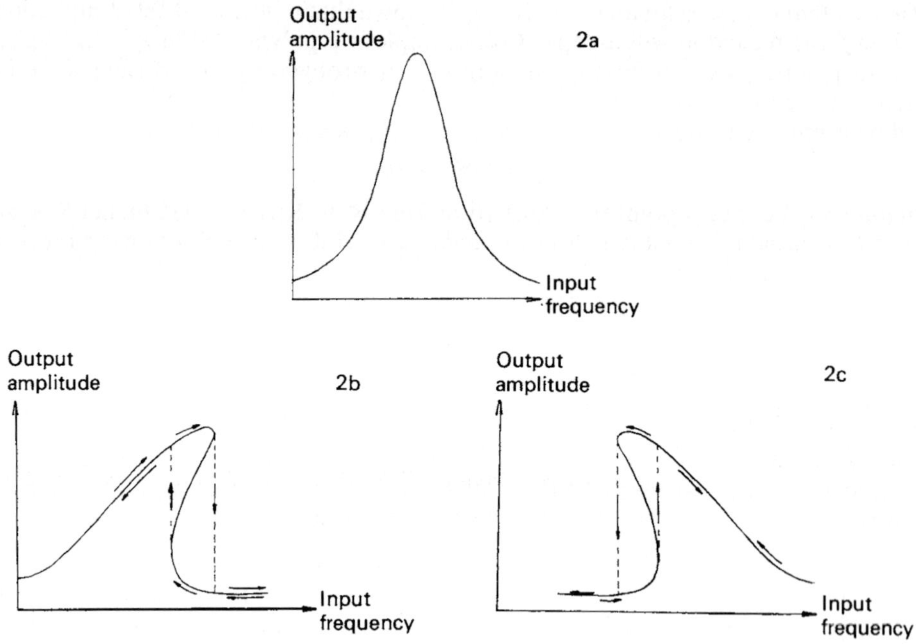

FIGS 2a, 2b, 2c. Jump resonance.

The time plots of Figs 3a and 3b clearly show that our SETAR can capture this engineering notion. The engineering terminology of a "hard spring" and a "soft spring" is an indication of the mode of the "restoring force" of the system. Figs 3a and 3b correspond respectively to the SETAR (2; 9, 3), $d = 5$ and SETAR (2; 3, 8), $d = 6$ given below. (White noise inputs are replaced by sinusoids in this exercise.)

$$X_n = \begin{cases} 0{\cdot}4655 + 1{\cdot}1448X_{n-1} - 0{\cdot}4801X_{n-2} + 0{\cdot}1273X_{n-3} \\ \qquad - 0{\cdot}3580X_{n-4} + 0{\cdot}2565X_{n-5} - 0{\cdot}0781X_{n-6} - 0{\cdot}0493X_{n-7} \\ \qquad + 0{\cdot}2186X_{n-8} + 0{\cdot}0526X_{n-9} + \mathit{input} \quad \text{if } X_{n-5} \leqslant 3{\cdot}05, \\ 1{\cdot}1940 + 1{\cdot}1181X_{n-1} - 0{\cdot}5017X_{n-2} - 0{\cdot}0594X_{n-3} + \mathit{input} \\ \qquad\qquad \text{if } X_{n-5} > 3{\cdot}05, \end{cases} \tag{6.1a}$$

$$X_n = \begin{cases} 1{\cdot}3003 + 1{\cdot}3243X_{n-1} - 0{\cdot}7023X_{n-2} - 0{\cdot}0750X_{n-3} + \mathit{input} \\ \qquad\qquad \text{if } X_{n-6} \leqslant 3{\cdot}31, \\ 0{\cdot}2004 + 1{\cdot}2112X_{n-1} - 0.6971X_{n-2} + 0{\cdot}6191X_{n-3} \\ \qquad - 1{\cdot}0178X_{n-4} + 0{\cdot}9967X_{n-5} - 0{\cdot}7688X_{n-6} + 0{\cdot}6119X_{n-7} \\ \qquad - 0{\cdot}0551X_{n-8} + \mathit{input} \qquad \text{if } X_{n-6} > 3{\cdot}31. \end{cases} \tag{6.1b}$$

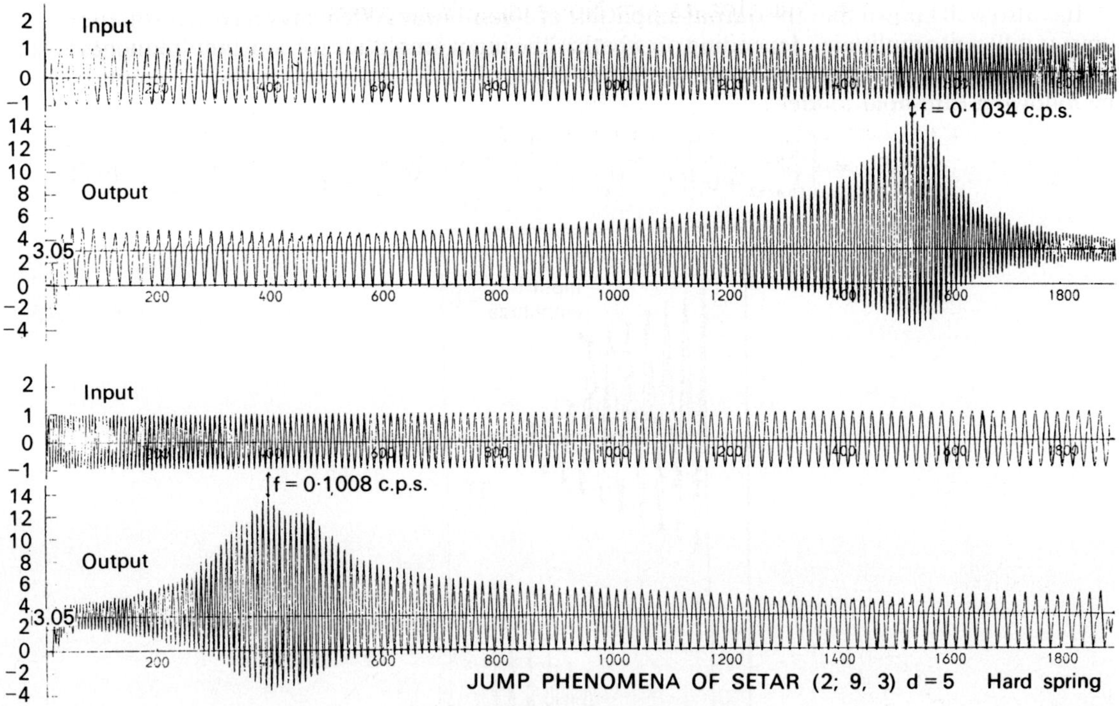

FIG. 3a. Jump resonance, hard spring.

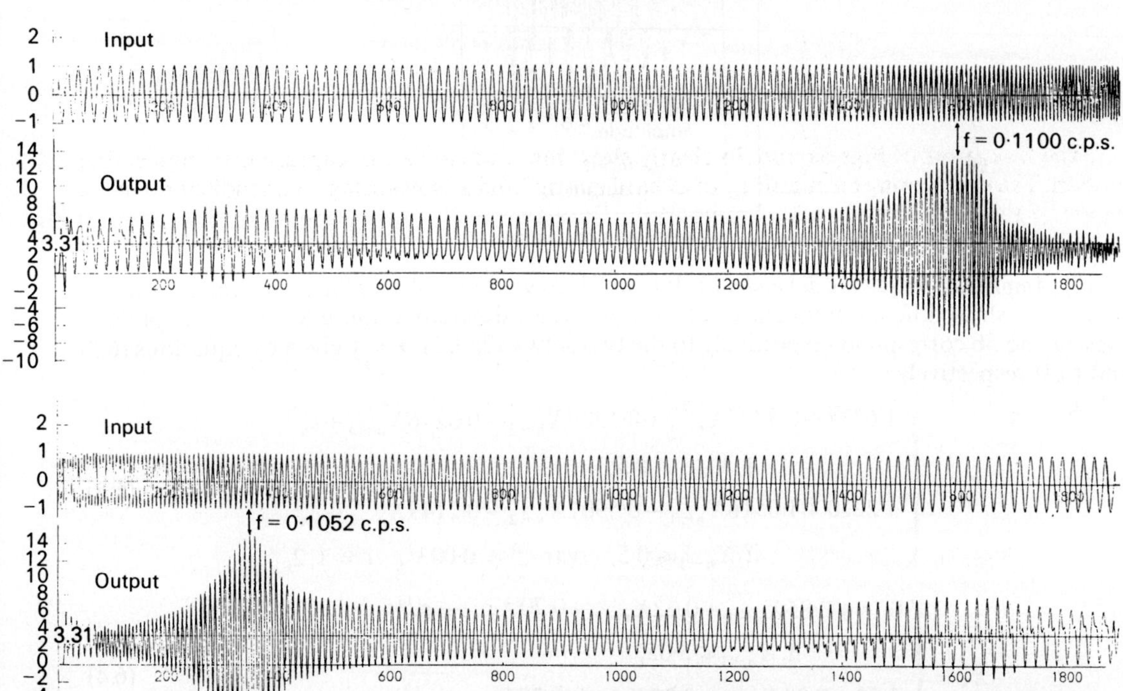

FIG. 3b. Jump resonance, soft spring.

It is also well known that the output amplitude of a non-linear system may have a resonance jump at different amplitudes depending on whether the input amplitude (of constant frequency) is monotonically increasing or monotonically decreasing. Fig. 4 corresponds to the time plots of the following threshold model:

$$X_n = \begin{cases} X_{n-1}+2(Y_n - Y_{n-1}) & \text{if } |Y_{n-1}-Y_{n-2}|>10, \\ X_{n-1}+0{\cdot}1(Y_n - Y_{n-1}) & \text{if } |Y_{n-1}-Y_{n-2}|\leqslant 10. \end{cases} \tag{6.2}$$

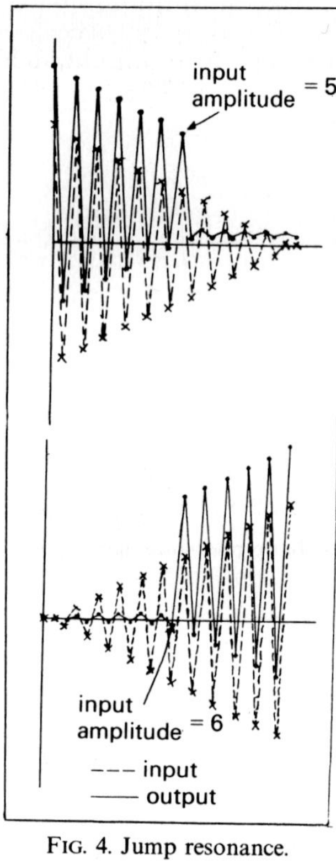

FIG. 4. Jump resonance.

(2) *Amplitude-frequency dependency.* It is well known that, unlike a linear system, the output signal may show different frequencies of oscillations for different amplitudes. The time plots of Figs 5a and 5b correspond respectively to the two SETAR (2; 3, 3), $d = 1$ given by equations (6.3) and (6.4) respectively:

$$X_n = \begin{cases} 1{\cdot}6734-0{\cdot}8295X_{n-1}+0{\cdot}1309X_{n-2}-0{\cdot}0276X_{n-3}+\varepsilon_n^{(1)} \\ \qquad \text{if } X_{n-1}>0{\cdot}5, \\ 1{\cdot}2270+1{\cdot}0516X_{n-1}-0{\cdot}5901X_{n-2}-0{\cdot}2149X_{n-3}+\varepsilon_n^{(2)} \\ \qquad \text{if } X_{n-1}\leqslant 0{\cdot}5, \quad \operatorname{var}\varepsilon_n^{(i)} = 0{\cdot}003^2, \quad i=1,2, \end{cases} \tag{6.3}$$

$$X_n = \begin{cases} 0{\cdot}15+0{\cdot}85X_{n-1}+0.22X_{n-2}-0{\cdot}70X_{n-3}+\varepsilon_n^{(1)} \\ \qquad \text{if } X_{n-1}\leqslant 3{\cdot}05, \\ 0{\cdot}30-0{\cdot}80X_{n-1}+0{\cdot}20X_{n-2}+0{\cdot}70X_{n-3}+\varepsilon_n^{(2)} \\ \qquad \text{if } X_{n-1}> 3{\cdot}05, \quad \operatorname{var}\varepsilon_n^{(i)} = 0{\cdot}003^2, \quad i=1,2. \end{cases} \tag{6.4}$$

Note that Fig. 5a shows the tendency of high frequency of oscillations when the amplitudes are high. Fig. 5b shows the reverse tendency.

(3) *Limit cycles.* Quite a few figures showing limit cycles for SETAR will be given in Section 9.

(4) *Subharmonics.* By a subharmonic it is usually meant an output oscillation at a fraction of

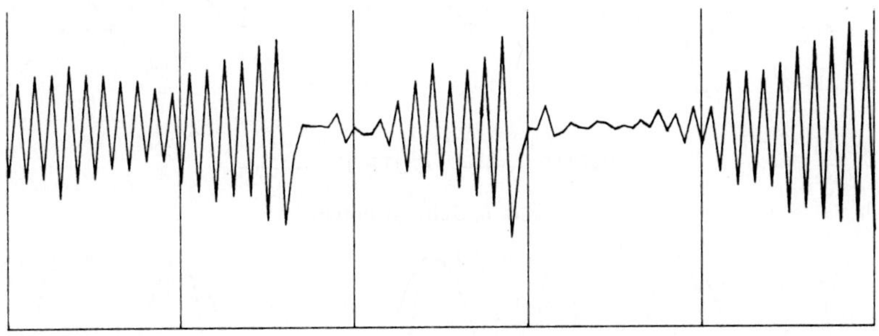

FIG. 5a. Time plots of (6.3) : amplitude-frequency dependency, high (low) amplitudes having high (low) frequencies.

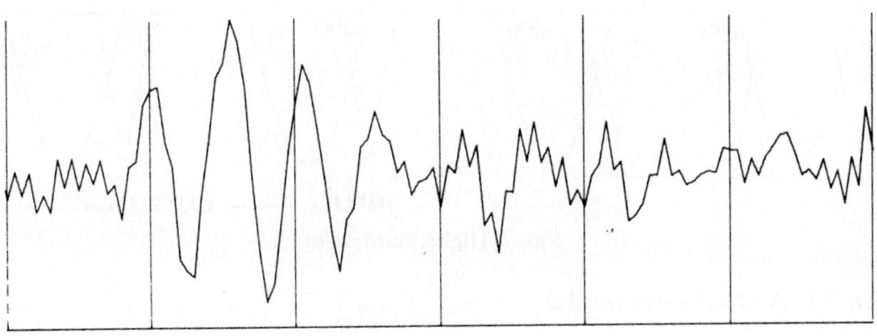

FIG. 5b Time plots of (6.4) : amplitude-frequency dependency, high (low) amplitudes having low (high) frequencies.

the input oscillation frequency. The time plots of Fig. 6 correspond to the following simple SETAR(3; 0, 1, 0) with a periodic input $\{Y_n\}$:

$$X_n = \begin{cases} 2X_{n-1}+Y_n & \text{if } |X_{n-1}| \leqslant 2, \\ Y_n & \text{if } |X_{n-1}| > 2, \end{cases} \quad \text{where } Y_n = \begin{cases} -1 & \text{if } n \text{ is odd,} \\ 1 & \text{if } n \text{ is even.} \end{cases} \tag{6.5}$$

(5) *Higher harmonics.* By a higher harmonic it is usually meant an output oscillation at a multiple of the input oscillation frequency. The time plots of Fig. 7 correspond to the following simple TARSO model model with a periodic input $\{Y_n\}$:

$$X_n = \begin{cases} -(2+\sqrt{2})\,Y_n-(1+\sqrt{2}) & \text{if } -1 < Y_n \leqslant -1/\sqrt{2} \\ -\sqrt{2}Y_n - 1 & \text{if } -1/\sqrt{2} < Y_n \leqslant 0 \\ \sqrt{2}Y_n - 1 & \text{if } 0 < Y_n \leqslant 1/\sqrt{2} \\ (2+\sqrt{2})\,Y_n-(1+\sqrt{2}) & \text{if } 1/\sqrt{2} < Y_n \leqslant 1 \end{cases} \tag{6.6}$$

7. LIMIT CYCLES, LIMIT POINTS AND SETAR

In Section 2(ii) we described a limit cycle as one possible mode of oscillations of a system when the input is "switched off". This motivates the following definition of a *limit cycle* of a stochastic system in which the input may consist of some random and some deterministic components.

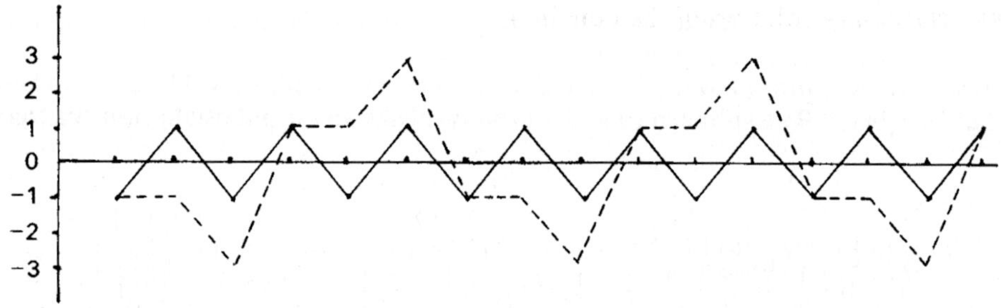

FIG. 6. Sub-harmonics.

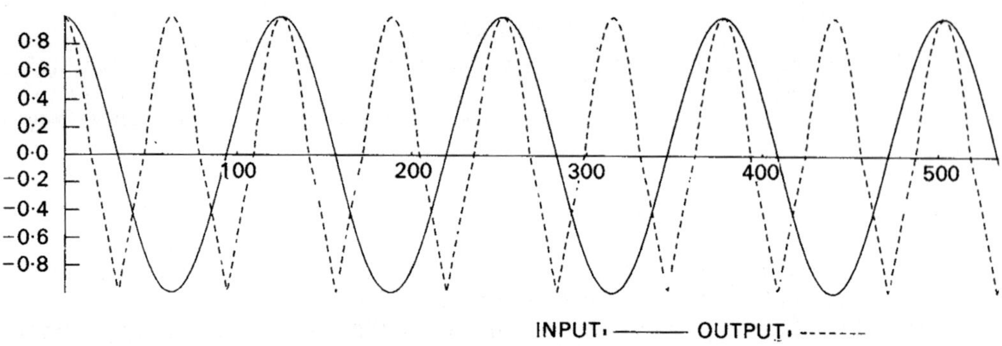

FIG. 7. Higher harmonics.

Definition 7.1. A stochastic model

$$\mathbf{X}_n = \mathbf{f}(\mathbf{X}_{n-1}, \varepsilon_n^{(i)}, \varepsilon_{n-1}^{(i)}, \ldots, \varepsilon_{n-p_i}^{(i)}, u_n^{(i)}, \ldots, u_{n-p_i'}^{(i)}, i = 1, \ldots, q), \tag{7.1}$$

where $\{\mathbf{X}_n\}$ is an observable k-dimensional time series and for $i = 1, \ldots, q$, $\{\varepsilon_n^{(i)}\}$ is an unobservable one-dimensional time series and $\{u_n^{(i)}\}$ is a one-dimensional deterministic sequence, is said to admit a limit cycle if, by denoting $\hat{\varepsilon}_n^{(i)}$ as the conditional expectations of $\varepsilon_n^{(i)}$ given $\mathbf{X}_{n-1}, \mathbf{X}_{n-2}, \ldots$, (assumed to exist),

$$\tilde{\mathbf{X}}_n = \mathbf{f}(\tilde{\mathbf{X}}_{n-1}, \hat{\varepsilon}_n^{(i)}, \hat{\varepsilon}_{n-1}^{(i)}, \ldots, \hat{\varepsilon}_{n-p_i}^{(i)}, \mathbf{0}, i = 1, \ldots, q) \tag{7.2}$$

(where the vector of zeros is of dimension p_i+1) induces a recursive relation in $\tilde{\mathbf{X}}_n$, say,

$$\tilde{\mathbf{X}}_n = \mathbf{g}(\tilde{\mathbf{X}}_{n-1}), \tag{7.3}$$

which has a limit cycle in the sense of Definition 3.2.

Note that if $\varepsilon_n^{(i)}$ is a zero mean random variable and independent of $\mathbf{X}_{n-1}, \mathbf{X}_{n-2}, \ldots$ for each i, then

$$\tilde{\mathbf{X}}_n = \mathbf{f}(\tilde{\mathbf{X}}_{n-1}, \mathbf{0}), \tag{7.4}$$

where the zero vector of dimension $K = \Sigma_{i=1}^{q} (p_i + p_i' + 2)$. In what follows we assume this independence.

Now, for simplicity, consider a (stable) SETAR(3; k, k, k) with $a_0^{(i)} = 0$; $i = 1, 2, 3$ and $-\infty < r_1 < 0 < r_2 < \infty$. In this case, equation (7.3) takes the specific form

$$\tilde{\mathbf{X}}_n = \mathbf{A}^{(j)} \tilde{\mathbf{X}}_{n-1}, \quad \text{if } \tilde{\mathbf{X}}_{n-1} \in R_j^{(k)}, \quad j = 1, 2, 3. \tag{7.5}$$

Let $\rho(\mathbf{A})$ denote the modulus of the maximum eigenvalue of the matrix $\mathbf{A}$. Suppose that

$$\rho(\mathbf{A}^{(1)}) < 1, \quad \rho(\mathbf{A}^{(2)}) > 1, \quad \rho(\mathbf{A}^{(3)}) < 1. \tag{7.6}$$

The only stationary solution of the equation

$$\mathbf{x} = \mathbf{A}^{(j)}\mathbf{x} \quad \text{if } \mathbf{x} \in R_j^{(k)}, \quad j = 1, 2, 3, \tag{7.7}$$

is the zero vector, which belongs to $R_2^{(k)}$. However, that $\rho(\mathbf{A}^{(2)})$ is strictly greater than unity implies that this solution cannot be stable, i.e. there is no stable limit point. On the other hand, the system is stable. Therefore the only stable solutions are periodic, i.e. limit cycles. The extension to an SETAR $(l; k_1, \ldots, k_l)$ with $a_0^{(i)} = 0; i = 1, \ldots, l$, and $0 \in R_i$ for some i not equal to 1 or l, is straightforward. However, the problem of the theoretical classification of solutions, into the number of admissible limit points and limit cycles, for a general SETAR, in terms of the coefficients $a_i^{(j)}$'s, is not completely solved. In practice, this is not necessarily a serious drawback because once an SETAR model has been fitted, we can always check numerically whether it admits a limit cycle with the current observation being the intial point $\mathbf{x}_0$. We develop this point in Section 8.

8. STATISTICAL IDENTIFICATION

Given a finite record, a linear autoregressive (AR) model can be very easily fitted by efficient computational algorithms such as Levinson–Durbin's or the Householder transformation. (For discussion of the former, see, for example, Box and Jenkins, 1970, and of the latter see, for example, Golub, 1965.)

For the fitting of a general non-linear autoregressive model, the above techniques would no longer be suitable, and a much more time-consuming search algorithm would be necessary. However, in view of its piece-wise linearity, a threshold model can still be fitted by the efficient method of Householder transformations. The Levinson–Durbin method cannot be applied here in view of the lack of "Toeplitzian property" of the TAR.

We give only a description of a statistical method of identification. Sampling properties of the estimates of parameters are not included but an application of the recent results of Klimko and Nelson (1978) may prove fruitful. A Gaussian assumption is made on all the white noise sequences. This enables us to write down the likelihood function and derive the maximum likelihood estimates of the unknown parameters, much in the same way as in the linear AR case. It is easy to check that the Jacobian of the transformation from the white noise terms to the observations is unity. The initial part of our identification procedure is based on Akaike's Information Criterion (Akaike, 1973), denoted by AIC, which, for each specified threshold model, takes the generic form,

$$\text{AIC}(k) = N \ln(\text{RSS}/N) + 2k, \tag{8.1}$$

where RSS is the residual sum of squares of the fitted model, based on maximum likelihood estimates of the defining parameters, N is the "effective number of observations" (to be explained later) and k is the number of independent parameters of the model. Equation (8.1) is, of course, strictly speaking, valid only when the "end effects" of the likelihood function are negligible, as are usually assumed in this kind of analysis. (See, for example, Bartlett, 1966, p. 271.) We sometimes normalize the AIC by dividing it by N.

We describe, in some detail, one *computational* procedure implementing the proposed AIC identification for the class of SETAR $(2; k_1, k_2)$. Other classes may be considered in a similar way. First, let d and L be prefixed, where L is the maximum order to be entertained for each of the l piece-wise linear AR models. The choice of L is subjective and usually depends on the sample size. It may be allowed to be different for different regions R_i, but, for the convenience of description, we have set them to be all the same here. (In our program the more flexible alternative is adopted. The programs are obtainable from the authors upon request.) Let n_0 be the maximum of d, L. Let $\{x_1, x_2, \ldots, x_n\}$ denote the observed data and t_q the sample 100qth percentile. Suppose that we agree to use $\{t_{0\cdot30}, t_{0\cdot40}, t_{0\cdot50}, t_{0\cdot60}, t_{0\cdot70}\}$ as a set of potential candidates for the estimation of r_1, the threshold value. Note that this choice is, of course, arbitrary but convenient, and may be changed if necessary. For each choice of t_q, we re-arrange the data set into two sub-sets and set up two sub-systems of linear equations, one for R_1 and the other for R_2.

The following is a typical example. Suppose $x_{n_0-d+1}, x_{n_0-d+4}, x_{n_0-d+5}, \ldots$ are less than or equal to t_q, and the others are greater than t_q. Then

$$\begin{bmatrix} x_{n_0+1} \\ x_{n_0+4} \\ x_{n_0+5} \\ \cdots \end{bmatrix} = \begin{bmatrix} 1 & x_{n_0} & x_{n_0-1} & \cdots \\ 1 & x_{n_0+3} & x_{n_0+2} & \cdots \\ 1 & x_{n_0+4} & x_{n_0+3} & \cdots \\ \cdots & \cdots & \cdots & \cdots \end{bmatrix} \begin{bmatrix} a_0^{(1)} \\ a_1^{(1)} \\ a_2^{(1)} \\ \cdots \end{bmatrix}, \quad \text{i.e. } \mathbf{x}_1 = \mathbf{A}_1 \boldsymbol{\theta}_1, \text{ say,} \tag{8.2a}$$

$$\begin{bmatrix} x_{n_0+2} \\ x_{n_0+3} \\ x_{n_0+6} \\ \cdots \end{bmatrix} = \begin{bmatrix} 1 & x_{n_0+1} & x_{n_0} & \cdots \\ 1 & x_{n_0+2} & x_{n_0+1} & \cdots \\ 1 & x_{n_0+5} & x_{n_0+4} & \cdots \\ \cdots & \cdots & \cdots & \cdots \end{bmatrix} \begin{bmatrix} a_0^{(2)} \\ a_1^{(2)} \\ a_2^{(2)} \\ \cdots \end{bmatrix}, \quad \text{i.e. } \mathbf{x}_2 = \mathbf{A}_2 \boldsymbol{\theta}_2, \text{ say.} \tag{8.2b}$$

We may obtain estimates of $\boldsymbol{\theta}_1$ and $\boldsymbol{\theta}_2$ by Householder transformations of the matrices $\mathbf{A}_1$ and $\mathbf{A}_2$ respectively. For each fixed t_a and d, we use AIC to determine the orders of the two piece-wise linear AR's, k_1 and k_2. Specifically, $\hat{k}_1$ is the minimum AIC estimate of k_1, i.e.

$$\text{AIC}(\hat{k}_1) = \min_{0 \leqslant k_1 \leqslant L} \{N_1 \ln(\text{RSS}_1(k_1)/N_1) + 2(k_1+1)\}, \tag{8.3}$$

where N_1 is the number of elements in $\mathbf{x}_1$ and $\text{RSS}_1(k_1)$ is the residual sum of squares $\|\mathbf{x}_1 - \mathbf{A}_1 \hat{\boldsymbol{\theta}}_1\|^2$. Here $\hat{\boldsymbol{\theta}}_1$ is the least squares estimate of $\boldsymbol{\theta}_1$, assuming a k_1th order AR model, and $\|\cdot\|$ denotes the Euclidean norm of a vector. $\hat{k}_2$ is obtained in a similar way.

Recalling that the computation is fixed at t_q, we may write

$$\text{AIC}(t_q) = \text{AIC}(\hat{k}_1) + \text{AIC}(\hat{k}_2), \tag{8.4}$$

because $\varepsilon_m^{(1)}$'s and $\varepsilon_n^{(2)}$'s are independent of each other. Next, we allow t_q to vary over a pre-selected set of t_q's and minimize the AIC (t_q) over this set. That value of t_q, $\hat{r}_1$ say which is such that

$$\text{AIC}(\hat{r}_1) = \min_{\{t_q\}} \{\text{AIC}(t_q)\}, \tag{8.5}$$

is adopted as our current estimate of r_1, the threshold value, and the $\hat{k}_1, \hat{k}_2$ corresponding to this $\hat{r}_1$ our estimates of k_1 and k_2. Therefore, the minimum AIC model adopted for the fixed value of d is SETAR$(2; \hat{k}_1, \hat{k}_2)$ with threshold $\hat{r}_1$. In all the above searching stages, the *total effective number* of observations remains the same, namely $n - n_0$, while the effective number in each region is smaller. (Care should be taken to ensure that they are sufficiently large.)

Finally, we have to search over d for a set of pre-selected positive integers. The different choices of d may alter n_0 and hence $n - n_0$. In order to get some cross-comparison between the AIC$(\hat{r}_1)$'s for the different choices of d, we normalize the former, Thus, for each d, we write

$$\text{AIC}(d) = \text{AIC}(\hat{r}_1)/(n - \max\{d, L\}), \tag{8.6}$$

where AIC$(\hat{r}_1)$ is defined in (8.5) for this choice of d. After this last search stage over d, we have completed the minimum AIC identification, which will give us estimates of $d, r_1, k_1, a_i^{(1)}; i = 0, \ldots, k_1$, $a_i^{(2)}; i = 0, \ldots, k_2$, $\text{var}\,\varepsilon_j^{(1)}$ and $\text{var}\,\varepsilon_j^{(2)}$.

To complement the final stage of the identification, namely that of d, we also compute the so-called *eventual forecasting function*, eff(d), for each d. Specifically, for each fixed d, we go through all the afore-mentioned search stages, ending with a minimum AIC estimated SETAR$(2; \hat{k}_1, \hat{k}_2)$ with threshold value $\hat{r}_1$. Using the observed data and the fitted model, we may easily obtain the one-step-ahead prediction of X_{n+1}, because the observed value of X_{n-d} determines in which R_i region it falls. Denote this predicted value of X_{n+1} by x_{n+1}. Now, pretending that this x_{n+1} was the observed value of X_{n+1}, we may repeat the same calculation and obtain x_{n+2}, etc. The plot of $x_{n+m}, m = 1, 2, \ldots$, against m is, in fact, just a convenient way of visualizing the "systematic part" of the fitted SETAR$(2; \hat{k}_1, \hat{k}_2)$ model, given the observed data. It should not, however, be confused with the more-than-one-step-ahead prediction function. The eff(d) should therefore either tend to a constant or a periodic function, unless an unstable model has been fitted. The former indicates a limit point and the latter a limit cycle. By comparing the eff(d) for the different choices of d, we may be able to form some subjective judgement as to the preferred choice.

Yet another complementary technique we sometimes find useful in the final stage is that based on a kind of pseudo-cross-validation. We delete the last 10 per cent of observations, say, in the identification procedure, and then compare the one-step-ahead prediction errors on using the fitted model to forecast these deleted observations. Suppose that with d equal to d_0 the total of the prediction errors is a minimum. We then repeat the identification procedure with the complete data set, with d fixed at d_0. If the fitted model using the complete data set does not differ much from that based on the incomplete data set, then adopt the former as our final model with d equal to d_0.

A final check is obtained by studying the fitted residuals and the one-step-ahead prediction errors. The plotting of these is routine in our computer package.

9. TAR MODELS FOR REAL DATA

(A) *The Canadian lynx data (1821–1934).* This set of data has been analysed extensively by many statisticians. (See, in particular, Campbell and Walker, 1977; Tong, 1977a, and Bhansali, 1979.) We now list what we regard as significant features of these data as follows:

(i) obvious cycles of approximately 10 years with varying amplitudes;

(ii) the rise period, from a local minimum to the next local maximum, exceeding the descent period, from a local maximum to the next local minimum, thereby showing time irreversibility.

The proposed identification procedure has enabled us to select the following SETAR(2; 8,3) model as our model for the data which has been logarithmically transformed (to the base 10):

$$X_n = \begin{cases} 0{\cdot}5239 + 1{\cdot}0359X_{n-1} - 0{\cdot}1756X_{n-2} + 0{\cdot}1753X_{n-3} \\ \qquad - 0{\cdot}4339X_{n-4} + 0{\cdot}3457X_{n-5} - 0{\cdot}3032X_{n-6} + 0{\cdot}2165X_{n-7} \\ \qquad + 0{\cdot}0043X_{n-8} + \varepsilon_n^{(1)} \quad \text{if } X_{n-2} \leqslant 3{\cdot}1163, \\ 2{\cdot}6559 + 1{\cdot}4246X_{n-1} - 1{\cdot}1618X_{n-2} - 0{\cdot}1094X_{n-3} + \varepsilon_n^{(2)} \\ \qquad\qquad \text{if } X_{n-2} > 3{\cdot}1163, \end{cases} \tag{9.1}$$

where $\operatorname{var}\varepsilon_n^{(1)} = 0{\cdot}0255$, $\operatorname{var}\varepsilon_n^{(2)} = 0{\cdot}0516$. (The pooled mean sum of squares of residuals $= 0{\cdot}0360$.)

Fig. 8 shows that the eff is an asymmetric periodic function of period ten years (counting minimum year to minimum year *inclusively*), i.e. model (9.1) has a limit cycle of period 9 years as determined by Definition 7.1. The rise and descent periods are six and three, respectively. The limit cycle may be generated from (2·6226, 2·8945, 3·2523, 3·4601, 3·4257, 3·2281, 2·9793, 2·7884, 2·6639). It is interesting to note that a similar limit cycle can be picked up even by fitting a SETAR to just 80 of the 114 observations. The fact that the threshold value depends on X_{n-2} is particularly interesting in view of its implications of a lead-lag relation of approximately 2 years between the lynx population and its prey (cf. Bulmer, 1975). We will consider this point again in Example C.

Tong's AR(11) model (Tong, 1977b) and Campbell–Walker's harmonic-component-plus-AR(2) models (Campbell and Walker, 1977) have been recognized to be inconclusive owing to their linearity. (See Tong, 1980a, and the discussion of the above papers.) Threshold models certainly seem to offer exciting possibilities here. (See also Haken, 1978, p. 9.) The estimated threshold at about 3.1 gives us a rough idea of the critical lynx population in its co-existence with their prey. Figs 9a and 9b show the gain spectra of the fitted model, corresponding to

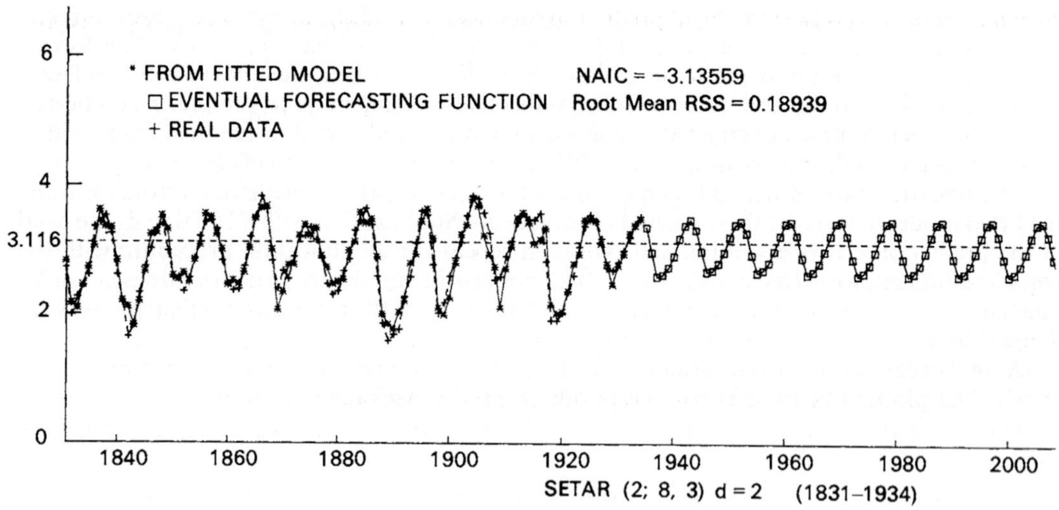

FIG. 8. SETAR for lynx data.

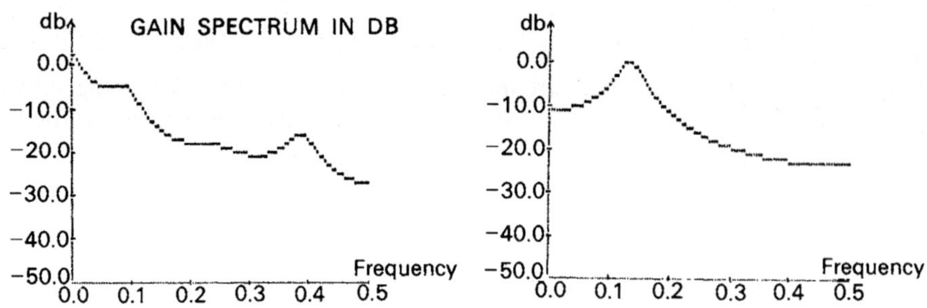

FIGS 9a, 9b. Gain spectra for lynx data.

$X_{n-2} > 3{\cdot}1163$ and $X_{n-2} \leqslant 3{\cdot}1163$ respectively. They appear to peak at different frequencies which might be interpreted as indicating some "amplitude-frequency dependency".

Tong (1980a) has compared the one-step-ahead predictions based on the linear models mentioned in the last paragraph with those based on a SETAR. In particular, the SETAR(2; 6, 3), $d = 2$, fitted to the years 1821–1920 (*op. cit.*) reduces the root-mean-square-error of one-step-ahead predictions (RSME) by 10 per cent when compared with the AR(12) reported in Tong (1977b, p. 466).

At this point we may anticipate a predator–prey system behind the whole scene, for the modelling of which our TARSC may offer interesting possibilities. Unfortunately, we have been unable to obtain reasonably "clean" snowshoe rabbit data in the Hudson Bay area of the same period of time. Some other "dirty" rabbit data of (probably) not exactly the same region were extracted from MacLulich (1937) and discussed in an unpublished report by Tong, which did not give any definite conclusion.

(B) *Sunspot data.* In his discussion of Morris' analysis of the sunspot data, Priestley (1977) has noted that a threshold AR model may be appropriate.

The following SETAR(2; 4, 12) is fitted to Wolf's sunspot numbers $\{X_t;\ t = 1700, \ldots, 1920\}$.

$$X_t = \begin{cases} 10{\cdot}5440 + 1{\cdot}6920X_{t-1} - 1{\cdot}1592X_{t-2} + 0{\cdot}2367X_{t-3} + 0{\cdot}1503X_{t-4} + \varepsilon_t^{(1)} \\ \qquad\qquad \text{if } X_{t-3} \leqslant 36{\cdot}6, \\ 7{\cdot}8041 + 0{\cdot}7432X_{t-1} - 0{\cdot}0409X_{t-2} - 0{\cdot}2020X_{t-3} + 0{\cdot}1730X_{t-4} \\ \quad - 0{\cdot}2266X_{t-5} + 0{\cdot}0189X_{t-6} + 0{\cdot}1612X_{t-7} - 0{\cdot}2564X_{t-8} \\ \quad + 0{\cdot}3195X_{t-9} - 0{\cdot}3891X_{t-10} + 0{\cdot}4306X_{t-11} - 0{\cdot}0397X_{t-12} + \varepsilon_t^{(2)} \\ \qquad\qquad \text{if } X_{t-3} > 36{\cdot}6, \end{cases} \tag{9.2}$$

where $\operatorname{var}\varepsilon_t^{(1)} = 254{\cdot}64$, $\operatorname{var}\varepsilon_t^{(2)} = 66{\cdot}80$. (The pooled mean sum of squares of residuals = 153·71.)

Fig. 10 shows the fitted residuals, the one-step-ahead predictions and the eff. Note that the eff is a periodic function of a 31-year period, consisting of 3 local maxima and 3 local minima, i.e. 3 "local cycles". The local cycles are *asymmetric* with rise (descent) periods being 4 (6), 4 (6), 4 (7). We may regard 31/3 as a "*fractional period*" of the sunspot cycle. We note that the asymmetry of these cycles runs in a reversed direction to that of the lynx. Fig. 11 shows the "high" and "low" gain spectra, which tend to be related to the empirical observation that the skewness of the sunspot cycles depends on their amplitudes. Logarithmic and square-root transformations of the data have been tried but we have not observed any obvious advantage in this case.

Using a method due to Ozaki and Tong (1975), Akaike (1978) has shown that the sunspot data are better modelled as non-stationary over a long period, although they may be regarded as stationary over a shorter period. Some of the non-stationarity must be due to the introduction

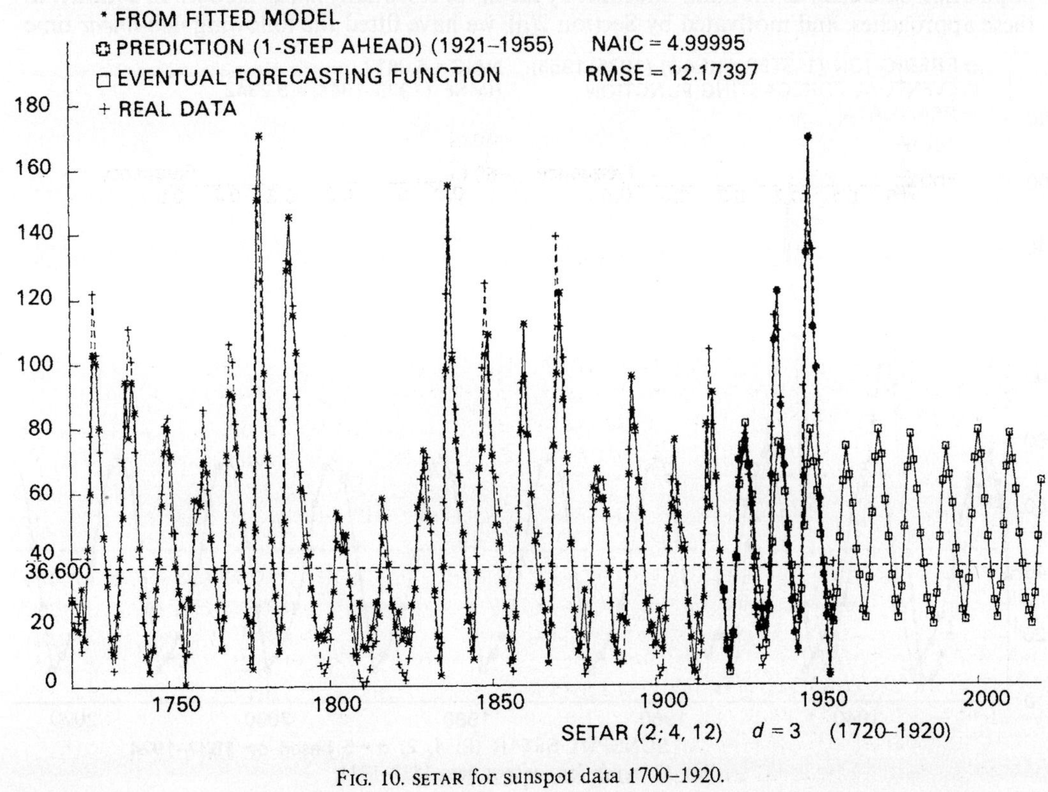

FIG. 10. SETAR for sunspot data 1700–1920.

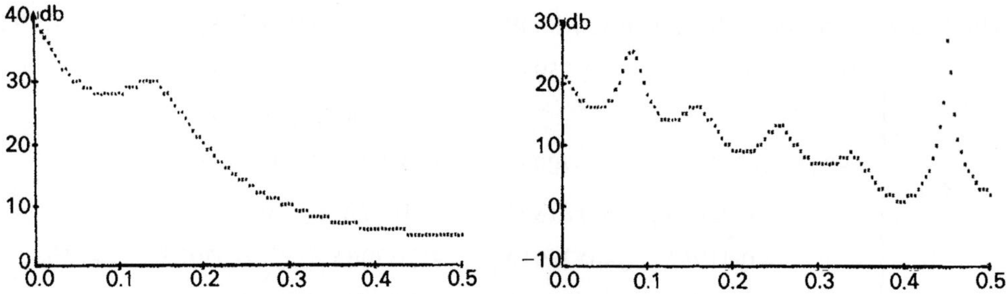

FIGS 11a, 11b. Gain spectra for sunspot data 1700–1920.

of the photographic recording technique towards the later part of the record. We therefore look at the data from 1837 to 1924 more closely and the following SETAR(2; 4, 2) model is fitted :

$$X_t = \begin{cases} 25{\cdot}2120 + 0{\cdot}9820X_{t-1} - 0{\cdot}0377X_{t-2} - 0{\cdot}6378X_{t-3} + 0{\cdot}2454X_{t-4} + \varepsilon_t^{(1)} & \\ & \text{if } X_{t-5} \leqslant 47{\cdot}4, \\ 0{\cdot}3585 + 0{\cdot}7569X_{t-1} - 0{\cdot}0531X_{t-2} + \varepsilon_t^{(2)} & \text{if } X_{t-5} > 47{\cdot}4, \end{cases} \tag{9.3}$$

where $\operatorname{var}\varepsilon_t^{(1)} = 231{\cdot}030$, $\operatorname{var}\varepsilon_t^{(2)} = 63{\cdot}075$. (The pooled mean sum of squares of residuals = 157·819.)

We note that the one-step-ahead predictions for the period up to 1944 are reasonable but deteriorate rapidly from then on thereby suggesting some non-stationarity of the sunspot data.

(C) *Mink–muskrat data* (1767–1849), from Jones (1914). Bulmer (1974, 1975), Jenkins (1975) and Chan and Wallis (1978) have attempted to explain the predator–pray relation of animal population data such as the mink–muskrat by means of essentially *linear* models. In contrast to these approaches, and motivated by Section 2(ii), we have fitted the following *non-linear* time

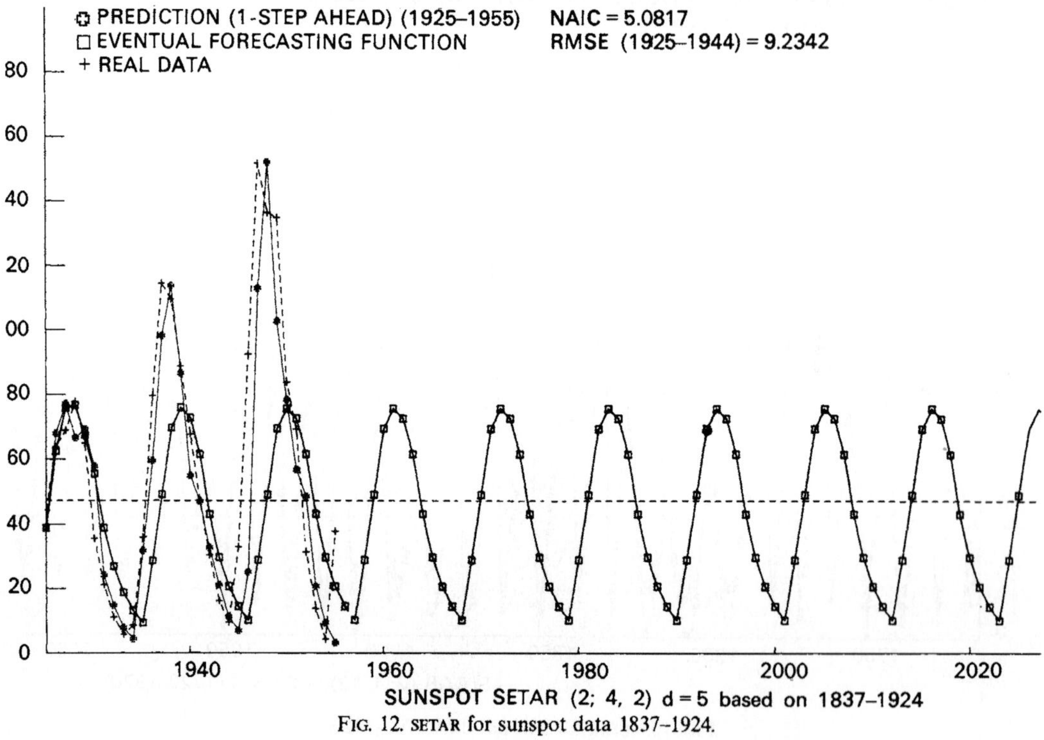

FIG. 12. SETAR for sunspot data 1837–1924.

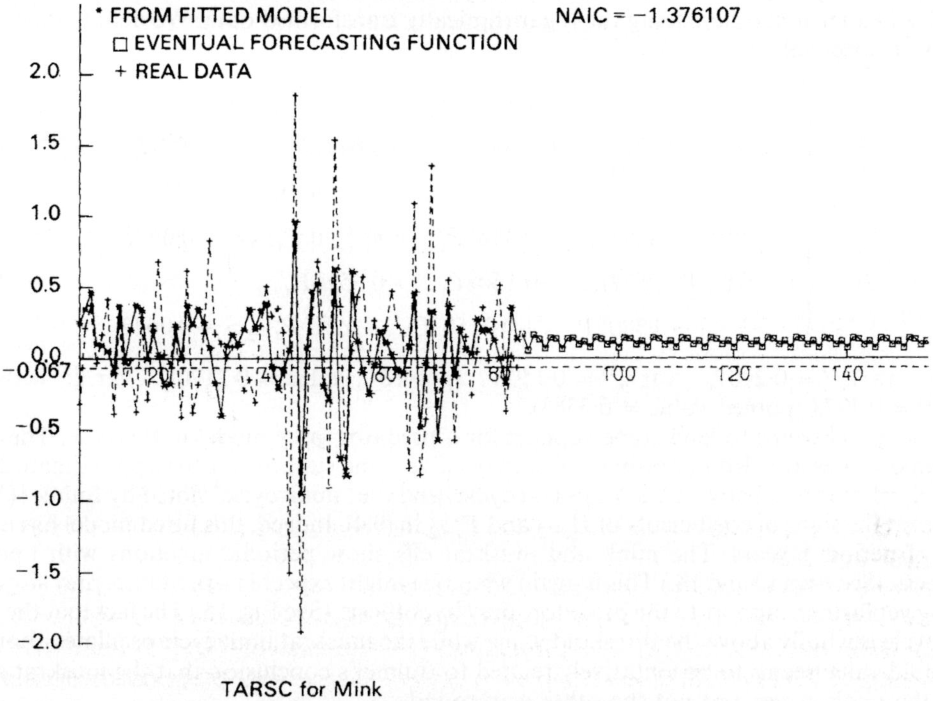

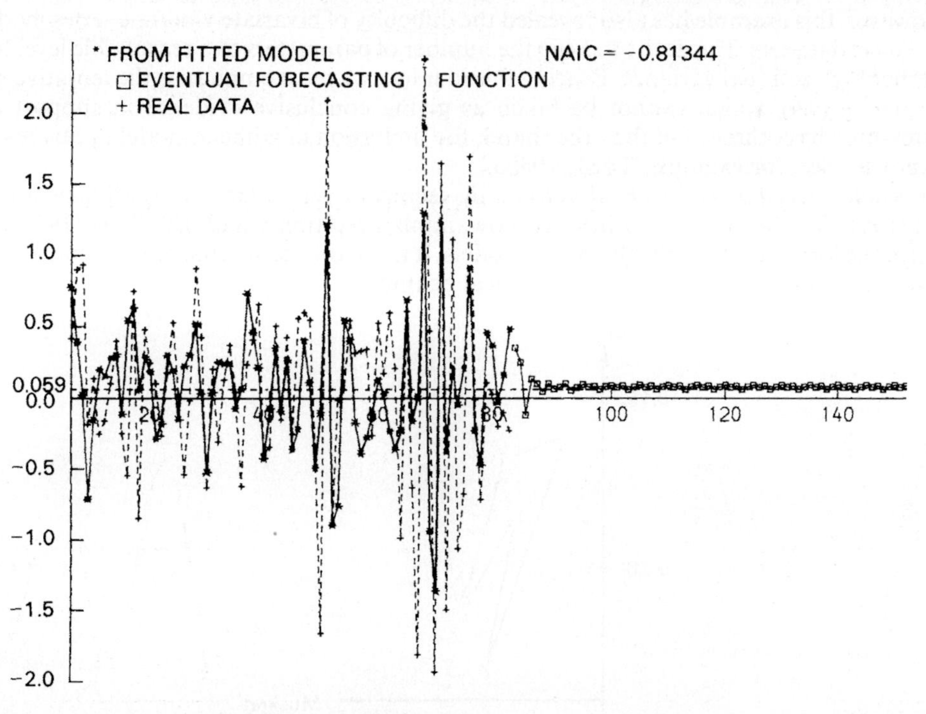

FIGS 13, 14. TARSC for mink and muskrat data.

series model, specifically a TARSC model, to the mink and muskrat data (to the base e) of 1767–1849 after first differencing the logarithmically transformed data. We denote them by P_t and H_t, respectively:

$$P_t = \begin{cases} 0{\cdot}1345 - 0{\cdot}5988P_{t-1} + 0{\cdot}0391H_{t-1} + \varepsilon_t^{(1)} & \text{if } H_{t-2} \leqslant 0{\cdot}0592, \\ 0{\cdot}2326 - 0{\cdot}5272P_{t-1} + 0{\cdot}1047H_{t-1} - 0{\cdot}6445P_{t-2} + 0{\cdot}1002H_{t-2} + \varepsilon_t^{(2)} & \\ & \text{if } H_{t-2} > 0{\cdot}0592, \end{cases} \tag{9.4a}$$

$$H_t = \begin{cases} 0{\cdot}4405 - 0{\cdot}3867H_{t-1} - 0{\cdot}3465P_{t-1} + \eta_t^{(1)} & \text{if } P_{t-2} \leqslant -0{\cdot}0672, \\ 0{\cdot}1976 - 0{\cdot}4967H_{t-1} - 0{\cdot}1608P_{t-1} - 0{\cdot}3516H_{t-2} & \\ \quad - 0{\cdot}3802P_{t-2} + 0{\cdot}0150H_{t-3} + \eta_t^{(2)} & \text{if } P_{t-2} > -0{\cdot}0672, \end{cases} \tag{9.4b}$$

where $\operatorname{var}\varepsilon_t^{(1)} = 0{\cdot}2907$, $\operatorname{var}\varepsilon_t^{(2)} = 0{\cdot}1506$ (pooled value $= 0{\cdot}2170$), $\operatorname{var}\eta_t^{(1)} = 0{\cdot}4616$, $\operatorname{var}\eta_t^{(2)} = 0{\cdot}3073$ (pooled value $= 0{\cdot}3588$).

This model seems to lend some support for a predator–prey model in this case. The fitted threshold values are also interesing and seem to give some support for the approximate 2 year lead-lag relationship between the "muskrat cycle" and the "mink cycle" noted by Bulmer (1975). Note also the signs of coefficients of H_{t-2} and P_{t-2} in (9.4). Indeed, this fitted model has a limit cycle of period 5 years. The mink and muskrat effs show periodic functions with opposite skewness. (See Figs 13 and 14.) This is again what one might expect in a predator–prey situation, adding yet further support to the predator–prey hypothesis. (See Fig. 15.) The fact that the mink limit cycle is wholly above the threshold value while the muskrat limit cycle oscillates about the threshold value seems to be tentatively related to Bulmer's conclusion that the muskrat cycles drive the mink cycles and not the other way round.

However, this example has also revealed the difficulty of bivariate TAR time series modelling to *very short* data sets. The desire to keep the number of parameters to a reasonable level has led to a rather high residual variance. Bearing this in mind, we must emphasise the tentative nature of the model (9.4), which cannot be taken as giving conclusive evidence in support of the predator-prey hypothesis. On the other hand, the limitation of a linear model in this respect is well known. (See, for example, Tong, 1980a.)

(D) *Kanna riverflow and rainfall data* (daily record of year 1956). It was Sugawara's tank model (1961) for the analysis of the riverflow–rainfall relation which led Tong (1977b, 1978, 1980a) to the formulation of the threshold models. It therefore seems appropriate that we should conclude our case studies with a hydrological example.

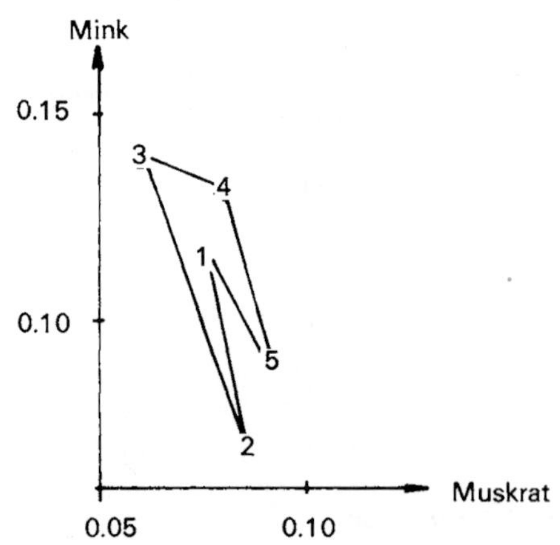

FIG. 15. Phase diagram for mink–muskrat.

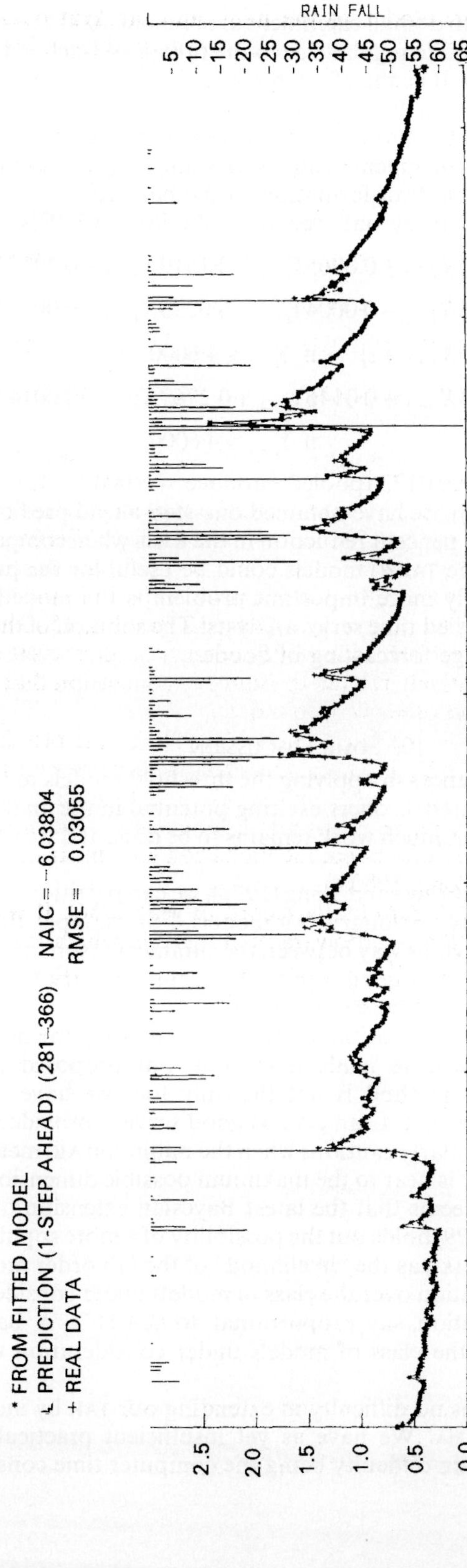

FIG. 16. TARSO(2; (5, 4), (2, 2)) for Kanna riverflow (in mm/day) in the year 1956. The vertical lines give daily record of rainfall (in mm/day) with the vertical scale denoted on the right-hand margin.

The Kanna River is a river with a small catchment area (under 1000 km^2) in Japan. Seasonal variations of Japanese rivers are quite regular due to the rather well-defined rainy season there. The ground soil is also rarely dry. It is, therefore, reasonable to expect that most of the cyclical variation of the riverflow data can be explained by that of the rainfall data if only the transformation from the latter to the former is adequately modelled. As a result, we may treat the latter (denoted by Y_t) as an instrumental time series and fit a TARSO model to the former (denoted by X_t, after a logarithmic transformation to the base 10).

The following model is fitted, using only record of the first 280 days.

$$X_t = \begin{cases} 0{\cdot}0185 + 0{\cdot}9992X_{t-1} + 0.0065Y_{t-1} - 0{\cdot}1519X_{t-2} - 0{\cdot}0017Y_{t-2} \\ \qquad + 0{\cdot}1236X_{t-3} - 0{\cdot}0004Y_{t-3} - 0{\cdot}0295X_{t-4} - 0{\cdot}0014Y_{t-4} \\ \qquad + 0{\cdot}0065X_{t-5} + \varepsilon_t^{(1)} \quad \text{if } Y_{t-1} \leqslant 4{\cdot}6000, \\ 0{\cdot}1281 + 0{\cdot}5044X_{t-1} + 0{\cdot}0146Y_{t-1} + 0{\cdot}2767X_{t-2} + 0{\cdot}0014Y_{t-2} + \varepsilon_t^{(2)} \\ \qquad \text{if } Y_{t-1} > 4{\cdot}6000, \end{cases} \tag{9.5}$$

where $\operatorname{var}\varepsilon_t^{(1)} = 0{\cdot}0012$, $\operatorname{var}\varepsilon_t^{(2)} = 0{\cdot}0173$ (pooled variance $= 0{\cdot}0047$).

Based on this fitted model (9.5), we have obtained one-step-ahead predictions of the next 86 days, and Fig. 16 represents an 18 per cent reduction in the RMSE when compared with the linear model. We would suggest that the TARSO models could be useful for the purpose of synthetic hydrology. However, a practically more important problem is the modelling of the rainfall, which so far seems to have bedevilled time series analysts! The solution of this difficult problem will pave the way for a long-range forecasting of floods.

10. SOME DISCUSSION

Through our practical experiences in applying the threshold models to real data, we are led to believe that this new class of models offers exciting potential in the analysis of cyclical data and opens up new vistas. However, much work remains to be done and we would just mention a few areas.

Following the same idea as in Ozaki and Tong (1975), we can partition the time axis suitably so as to arrive at a class of locally stationary TAR models. For example, the rainfall–riverflow relationship may change in an obvious way between the summer seasons and the winter seasons for some rivers. We have some encouraging results in a non-stationary TARSO modelling of the River Cam data, which will be reported elsewhere.

We are certainly conscious of the possible shortcomings in using the minimum AIC method in our model identification. We have made it clear in our proposed procedure, and we emphasize once again, that this method is not the only tool we have used, although our experiences have led us to believe that it can give us good service, provided we use it sensibly. For example, we have been particularly cautious when the minimum AIC method selects a model whose parametric dimensionality is near to the maximum possible dimension entertained. (See, for example, Shimizu, 1978.) It seems that the latest Bayesian extension of the minimum AIC method developed by Akaike (1979) holds out the possibility of a more sophisticated procedure. Briefly, we may treat $\exp(-\frac{1}{2}\text{AIC}(k))$ as the "likelihood" of the kth order model from which we may obtain the posterior distribution over the class of models under consideration, prior being some reasonably simple distribution, say proportional to $(k+1)^{-1}$. A Bayesian model may then be obtained by averaging the class of models under consideration with respect to the posterior distribution.

Of course, in principle there is no difficulty in extending our TAR by including the moving average terms, obtaining a TARMA. We have as yet insufficient practical experience in the identification of a TARMA, the main difficulty being the computer time consideration. Another possibly useful direction of extension is to allow the $\mathbf{A}^{(j)}$'s and $\mathbf{B}^{(j)}$'s of (4.1) to be functions of $\mathbf{X}_{n-1}$, which includes the *piece-wise polynomial* approximation. (See also Tong, 1980a.)

Finally, in the case of linear models, the notion of a *state* has been fully developed and is identified with a set of *observable* basis vectors of the *predictor space* of Akaike (1974). This fundamental notion gives a precise mathematical meaning to the information reduction process expressed by the linear ARMA model under the *only* assumption of finiteness of the dimension of the predictor space. Such a notion is lacking in the non-linear case. In this respect, the TAR (or the TARMA) models, as well as all other known classes of non-linear time series models, must be regarded at present as *ad hoc* (Akaike, private communication). We would argue that the formulation of this fundamental notion will be a most challenging and urgent problem for the next stage of development in non-linear time series modelling. Towards this end, it seems that a topological approach might offer some insight. Now, let $\mathscr{X}$ denote a separable metric space generated by $\hat{X}_1, \hat{X}_2, \ldots$, the metric being the mean square norm. Here

$$\hat{X}_i = E[X_i \mid X_0, X_{-1}, \ldots], \quad i = 0, 1, 2, \ldots.$$

We call $\mathscr{X}$ the *general predictor space.* (If the $\hat{X}_i$'s are linear in $X_0, X_{-1}, \ldots$, then this general predictor space reduces to the predictor space of Akaike.) We now follow the rigorous definition of the *dimension* of a separable metric space given by Menger and Urysohn. (See, for example, Hurewicz and Wallman, 1941, p. 24.) Now, a fundamental theorem in *dimension theory* (*op. cit.* p. 52) shows that if $\mathscr{X}$ has dimension $n(<\infty)$ then among the totality of continuous real-valued functions defined over $\mathscr{X}$, there is a set of $2n+1$ (but not any fewer) functions $\xi_1, \xi_2, \ldots, \xi_{2n+1}$ (the co-ordinate functions), which form a *basis*, in the sense that every continuous real-valued function f defined on $\mathscr{X}$ is expressible in the form

$$f = g(\xi_1, \ldots, \xi_{2n+1}),$$

where g is a continuous function of $2n+1$ variables. We may identify $\boldsymbol{\xi} = (\xi_1, \xi_2, \ldots, \xi_{2n+1})$ as a *state vector*, which seems to offer possibilities of further developments towards a fuller understanding of the structural aspects of non-linear time series models.

ACKNOWLEDGEMENTS

We are most grateful to Dr H. Akaike, Mr N. Komura, Mr T. Ozaki and Professor M. B. Priestley for their assistance, discussions and criticisms during the formative years of the present work. The two referees' very careful scrutinies of the paper and helpful suggestions have led to a considerably improved version.

Our thanks also to Mr P. K. Wong of UMIST for supplying us with Fig. 1.

Dr Tong's research was supported in part by a grant from the Science Research Council of the United Kingdom.

REFERENCES

AIZERMAN, M. A. (1963). *Theory of Automatic Control.* New York : Pergamon.

AKAIKE, H. (1973), Information theory and an extension of the maximum likelihood principal. *2nd Int. Symp. on Inf. Theory* (B. N. Petrov and F. Csaki, pp. 267–281. Budapest : Akademiai Kiado.

—— Stochastic theory of minimal realisations. *I.E.E.E. Trans. Auto. Control*, **AC-19**, 667–674.

—— (1978). On the likelihood of a time series model. *The Statistician*, **27**, 217–235.

—— (1979). A Bayesian extension of the minimum AIC procedure of autoregressive model fitting. *Biometrika*, **66**, 237–242.

BARTLETT, M. S. (1966). *Stochastic Processes*, 2nd ed. Cambridge : Cambridge University Press.

BHANSALI, R. (1979). A mixed spectrum analysis of the lynx data. *J. R. Statist. Soc.* A, **142**, 199–209.

BOX, G. E. P. and JENKINS, G. M. (1970). *Time Series Analysis, Forecasting and Control.* San Francisco : Holden-Day.

BULMER, M. G. (1974). A statistical analysis of the 10 year cycle in Canada. *J. Anim. Ecol.*, **43**, 701–718.

—— (1975). Phase relations in the 10 year cycle. *J. Anim. Ecol.*, **44**, 609–621.

CAMPBELL, M. J. and WALKER, A. M. (1977). A survey of statistical work on the MacKenzie River series of annual Canadian lynx trappings for the years 1821–1934, and a new analysis. *J. R. Statist. Soc.* A, **140**, 411–431; Discussion 448–468.

CHAN, W. Y. T. and WALLIS, K. F. (1978). Multiple time series modelling: another look at the mink–muskrat interaction. *Appl. Statist.*, **27**, 168–175.
FUJISHIGE, S. and SAWARAGI, Y. (1974). Optimal estimation for continuous systems with jump process. *I.E.E.E. Trans. on Auto. Control*, **AC-19**, 225–228.
GOLUB, G. (1965). Numerical methods for solving linear least square problems. *Numerische Mathematik*, **7**, 206–216.
HAGGAN, V. and OZAKI, T. (1980). Amplitude-dependent AR model fitting for non-linear random vibrations. *Biometrika*, **67**,
HAKEN, H. (1978). *Synergetics: An Introduction*, 2nd ed. Heidelberg: Springer.
HUREWICZ, W. and WALLMAN, H. (1941). *Dimension Theory*. New Jersey: Princeton University Press.
JACOBS, P. A. and LEWIS, P. A. W. (1978). Discrete time series generated by mixtures—I: correlational and runs properties. *J. R. Statist. Soc.* B, **40**, 94–105.
JENKINS, G. M. (1975). The interaction between the muskrat and mink cycles in North Canada. In *Proc. of the 8th Int. Biometric Conference* (Constança, Romania, August 1974) (L. C. A. Corsten and T. Postelnicu, eds).
JONES, D. A. (1978). Non-linear autoregressive processes. *Proc. Roy. Soc. London.* A, **360**, 71–95.
JONES, J. W. (1914). *Fur-Farming in Canada*, 2nd ed. Ottawa: Commission of Conservation.
KLIMKO, L. A. and NELSON, P. I. (1978). On conditional least squares estimation for stochastic processes. *Ann. Statist.*, **6**, 629–642.
LAWRANCE, A. J. and KOTTEGODA, N. T. (1977). Stochastic modelling of riverflow time series (with Discussion). *J. R. Statist. Soc.* A, **140**, 1–47.
LI, T.-Y. and YORKE, J. A. (1975). Period three implies chaos. *Amer. Math. Monthly*, **82**, 988–992.
MACLULICH, D. A. (1937). *Fluctuations in the Number of the Varying Hare (Lepus americanus)*. University of Toronto Studies No. 43, Biol. series. Toronto: University of Toronto Press.
MAY, R. M. (1976). Simple mathematical models with very complicated dynamics. *Nature*, **261**, No. 5560, 459–467.
MINORSKY, N. (1962). *Non-linear Oscillations*. New York: Van Nostrand.
OZAKI, T. (1980). Non-linear time series models for non-linear random vibrations. *J. Appl. Prob.* (to appear).
OZAKI, T. and TONG, H. (1975). On fitting of non-stationary autoregressive models in time series analysis. In *Proc. 8th Hawaii Int. Conf. on System Sciences*, pp. 225–226. North Hollywood: Western Periodicals.
PRIESTLEY, M. B. (1965). Evolutionary spectra and non-stationary processes", *J. R. Statist. Soc.* B, **27**, 204–237.
—— (1977). Discussion of papers by Campbell *et al. J. R. Statist. Soc.* A, **140**, 448–450.
PRIESTLEY, M. B. and TONG, H. (1973). On the analysis of bivariate non-stationary processes. *J. R. Statist. Soc.* B, **35**, 153–166, 179–188.
RISHEL, R. (1975). Control of systems with jump Markov disturbances. *I.E.E.E. Trans. Auto. Control*, **AC-20**, 241–244.
ROBINSON, V. G. and SWORDER, D. D. (1974). A computational algorithm for design of regulators for linear jump parameter systems. *I.E.E.E. Trans. on Auto. Control*, **AC-19**, 47–49.
SHIMIZU, R. (1978). Entropy maximisation principle and selection of the order of an autoregressive Gaussian process. *Ann. Inst. Stat. Maths.*, **30**, 263–270.
SUGAWARA, M. (1952). On the method of deriving the daily discharge of the River Koza from the daily precipitation (in Japanese). *Res. Memo of Inst. of Stat. Maths., Tokyo*, Vol. **8**, No. 10.
—— (1962). On the analysis of run-off structure about several Japanese rivers. *Jap. J. Geophy.*, **2**, 1–76.
TODINI, E. and WALLIS, J. R. (1977). Using CL for daily or longer period rainfall—run-off modelling. In *Math. Models for Surface Water Hydrology* (C. Cirane, L. Marine and D. Wallis, eds). London: Wiley.
TONG, H. (1977a). Some comments on the Canadian lynx data. *J. R. Statist. Soc.*, A, **140**, 432–436, 448–468.
—— (1977b). Discussion of a paper by A. J. Lawrance and N. T. Kottegoda. *J. R. Statist. Soc.* A, **140**, 34–35.
—— (1978). On a threshold model. In *Pattern Recognition and Signal Processing* (C. H. Chen, ed.). The Netherlands: Sijthoff and Noordhoff.
—— (1980a). A view on non-linear time series model building. *Time Series* (O. D. Anderson, ed.). Amsterdam: North-Holland.
TUNNICLIFFE-WILSON, G. (1977). Discussion on papers by Campbell *et al. J. R. Statist. Soc.* A, **140**, 455–456.
TWEEDIE, R. L. (1975). Sufficient conditions for ergodicity and recurrence of Markov chains on a general state space. *Stochastic Processes and Their Applications*, **3**, 383–403.
WALTMAN, P. and BUTZ, E. (1977). A threshold model of antigen–antibody dynamics. *J. Theor. Biol.*, **65**, 499–512.
YULE, G. U. (1927). On the method of investigating periodicities in distributed series with special reference to Wolfer's sunspot numbers. *Phil. Trans. Roy. Soc. London*, A, **226**, 267–298.

DISCUSSION OF THE PAPER BY DR TONG AND MS LIM

Dr C. CHATFIELD (Bath University): I would like to congratulate the authors on making a substantial contribution to non-linear time-series modelling. I particularly welcome the fact that the paper combines new theoretical work with a number of practical examples using real data.

The authors are certainly right in suggesting that the time is ripe to look at alternatives to linear time-series models. The newcomer to non-linear models would do well to start by reading Granger and Andersen (1978) and Priestley (1978). These references and Subba Rao (1979) introduce an alternative class of models called *bilinear* models. The TAR models have the useful properties that they are locally linear and that they admit limit cycles, but in some other respects I find bilinear models more appealing. I hope that it will not be too long before bilinear and TAR models can be compared on real data. In particular I hope the authors can tell us how their model for the sunspot series compares with the bilinear models fitted by Granger and Andersen and by Subba Rao.

The first general question that might be asked in respect of non-linear modelling is: "How can we tell if a given time series is non-linear?" or "How can we decide if it is worth trying to fit a non-linear model?" The answer does not appear to be easy. In particular it is no use fitting a linear model, carrying out the usual diagnostic checks (such as looking at the autocorrelation function of the residuals), and hoping that these will indicate non-linearity because they won't. The tests, which are based on second-order properties, are designed to see if the "best" linear model has been fitted and not to indicate non-linearity. Indeed Granger and Andersen (1978) have shown that one can find a linear model and a bilinear model with the same second-order properties, and so they suggest looking at the second-order properties of $\{X_t^2\}$ as well as $\{X_t\}$ in order to distinguish between a linear and a bilinear model. More generally one might look at moments of $\{X_t\}$ which are higher than second-order, and the bispectrum is one possibility. In their examples, the authors have tried a TAR model because, for one reason or another, the "best" linear model was felt to be inadequate. For example, in the lynx data, the time "going up" systematically exceeds the time "coming down". What other features should we be looking for? Can the authors suggest a more general tactic for detecting non-linearity?

Let me now turn to the requirements listed by the authors in the introduction. Firstly they say that statistical identification should not entail excessive computation. Reading Section 8, I formed the impression that the computational problems are very much harder than those in both the linear and bilinear cases, so that it is not clear if the first requirement is satisfied. I would like to ask the authors how much more computing time is typically required to fit a TAR model.

Another sensible requirement proposed by the authors is that the overall prediction performance should be an improvement upon the linear model. Here I must confess to being a little disappointed. The reduction in RMSE is only 10 per cent for the lynx data, though 18 per cent for the riverflow data. Would the authors give us similar comparisons for the other two examples? Would the improvement be more substantial if predictions were compared for more than one step ahead? The extra complexity of fitting TAR models can, of course, only be justified by a substantial reduction in RMSE and/or by additional insight into the process mechanism.

Despite my queries and suggestions for future research, which are inevitable in any good read paper, I would like to conclude by saying how much I have enjoyed today's paper, and I have great pleasure in proposing the vote of thanks.

Dr G. Tunnicliffe Wilson (University of Lancaster): This paper strikes a welcome balance between theory and applications, but I confess to being more impressed by the latter. Following a tradition of empirical modelling, the authors have recognized features in the data which are not explained by linear models and have sought model extensions that *are* successful in representing these features.

Fundamentally, they use different linear models for different parts of the data, and I admire their ingenuity in demonstrating by simulation examples that TARs have sufficient potential for their task. The success so far demonstrated in practical applications suggests strongly that these models "approximate to the truth". Perhaps this success needs explaining as much from a data-analytic viewpoint as by investigating the theoretical properties of the models. For example in linear modelling a stable autoregression is *ensured* by almost all fitting processes. Is there a similar law to ensure the stability and cyclical properties of fitted TARS? I would like to know more about the failures which the authors have decently buried.

My main concern upon reading the theoretical part of the paper, was that according to their own arguments the authors should have used thresholds in all the predicting variables of their autoregressions. How, therefore, have they managed to achieve success using a threshold in one variable only? Most of the applications are to series with strong cycles so the predicting variables do not wander over the whole of their possible range, but are effectively confined to a closed orbit in this space. The choice of a single threshold in one lagged variable is effectively a means of defining the two parts of this orbit, or equivalently of the cycle, over which different linear predictors may most profitably be constructed. I would suggest that if more thresholds were used, then the choice of threshold variable would not be so critical, but the direction in which a threshold was crossed would become important.

I would expect worthwhile improvements to follow from attempts to better define the state of the system producing the cycle. A second order system is likely, so that a "level" and "slope" measurement should adequately represent the state. I believe that classical time series operations such as smoothing to remove noise, and filtering to correct for trends and low frequency modulating effects could be useful in extracting these measurements. This approach recognises that stochastic effects may enter in many different ways, and whilst in linear models all the components may be gathered into one ARIMA model with no loss of information, for non linear models it may be best to decompose the series so as to extract the basic cycle. This cycle should be predictable using a non-linear function of the two state variables only—possibly linearized at different points of the cycle. Forecasts of the original series could then be resynthesized from the components.

With their emphasis on producing a simple prediction formula, TARS may be failing to exploit the evident structure of many cyclical series.

The models which have been presented to us this evening may have to be refined in many ways, but a good start to empirical non-linear modelling has been made and the authors should be congratulated for their perseverance with TARS. I have much pleasure in seconding the vote of thanks.

The vote of thanks was passed by acclamation.

Dr R. J. BHANSALI (Department of Computational and Statistical Science, University of Liverpool): I would like to extend my congratulations to the authors on an interesting paper. Although considerable work on the development of the sampling properties of the identification methods proposed in Section 8 still needs to be done, the threshold autoregressive models appear to offer novel possibilities for the modelling of practical time series.

Apart from the applications to biological and other physical time series discussed in the paper, I might mention commodity price series as a possible class of Economic time series where applications of these models may be useful, in particular for describing the Cob-web phenomena—that is, cycles arising because of the interaction between price and production of agricultural commodities. The inadequacy of the Random Walk hypothesis (Labys and Granger, 1970) for the modelling of monthly Cocoa price series, 1949–73, is discussed by Beenstock and Bhansali (1980), who have suggested that within the class of linear autoregressive models, a second-order model provides a better fit to the changes of Cocoa prices. However, over the forecasting period of July 1974–July 1977, the second-order model provides only a modest improvement in the predictions of the future cocoa prices. The need for fitting a non-linear model is indicated by an examination of the residuals obtained after fitting the second order model. These are found not to be approximated by the Normal distribution, though the Laplace distribution provides a better fit.

I was also interested to note the authors' rather pragmatic attitude towards the usefulness of Akaike's information criterion for the identification of time series models. This pragmatic attitude appears to be in marked contrast to the almost religious attitude adopted earlier by Dr Tong in his analysis of the lynx data.

Dr M. G. BULMER (Oxford University): I should like to comment on the biological interpretation of the lynx and mink–muskrat data discussed in Section 9. The authors suggest that the lynx cycle is driven by a predator–prey interaction between the lynx and the snowshoe hare. There is good biological evidence that this is not the case. The hare cycle almost certainly drives the lynx cycle, but direct assessment of the impact of lynx predation on hare populations shows that it is too weak to be capable of causing the hare cycle. It has been suggested that the hare cycle is due to a plant–herbivore interaction (the hare being the predator and its plant food the prey). This situation might have been inferred from the periodogram of the lynx, which should be symmetrical about its peak value (which it is not) if the lynx–hare interaction drives the cycle, whereas it will exhibit a red shift (as observed) if the hare drives the lynx (Bulmer, 1978).

For the mink–muskrat data the authors fit a model which has a limit cycle of period 5 years. All previous authors have agreed that both mink and muskrat have a periodicity of about $9\frac{1}{2}$ years, the same as the lynx. A possible (though rather speculative) explanation is based on the facts that horned owls eat both hares and mink, and that mink eat muskrat. Thus the hare cycle drives an owl cycle, which drives a mink cycle, which drives a muskrat cycle. The observed phase lags are consistent with this explanation, mink being in phase with hares and muskrat two years earlier.

In conclusion, I must admit that I am rather doubtful of the gain in understanding which is likely to result from fitting the type of model developed in this paper. I would give higher priority to the fourth of the five requirements proposed in the Introduction.

Dr E. KHABIE-ZEITOUNE (North East London Polytechnic): It is my pleasure to congratulate the authors for a most stimulating paper. The non-stationary threshold models raise some challenging problems.

I would like to put forward the thesis that the juggling with the AIC criterion in this paper might one day be thought of as a preliminary identification/estimation method, only paving the way towards a fully fledged maximum likelihood estimation, applicable to a class of, say, non-linear SETARMA models.

The authors mention that the Levinson–Durbin procedure is not available for non-Toeplitzian block covariance matrices. Perhaps I should state here that this procedure has been generalized to deal with the inversion of "Toeplitzian" and "non-Toeplitzian" block covariance matrix Γ under mild conditions, with the computation of "generalized partial autocorrelations" (unpublished paper). This generalization leads to some very interesting results: If $\{\mathbf{X}_{t_1}, \ldots, \mathbf{X}_{t_n}\}$ is a set of random p-vectors with covariance matrix $\mathbf{\Gamma} = (\gamma_{i,j})$, $\gamma_{i,j}$ being the covariance matrix of (X_{t_i}, X_{t_j}), then one can compute p^2-matrix coefficients $\alpha_{n,h} = \alpha_{n,h}(\{\gamma_{i,j}\})$, dependent on $\gamma_{i,j}$'s, such that the set of random vectors $\mathbf{Y}_1, \ldots, \mathbf{Y}_n$, defined by

$$\mathbf{X}_{t_i} = \sum_{j=1}^{i} \alpha_{i,j} \mathbf{Y}_j \tag{1}$$

is uncorrelated, and such that

$$\mathbf{X}_{(n)}^{\mathrm{T}} \mathbf{\Gamma}^{-1} \mathbf{X}_{(n)} = \sum_{i=1}^{n} \mathbf{Y}_i^{\mathrm{T}} \mathbf{Y}_i \tag{2}$$

with

$$\mathbf{X}_{(n)}^{\mathrm{T}} = (\mathbf{X}_{t_1}^{\mathrm{T}}, \ldots, \mathbf{X}_{t_n}^{\mathrm{T}}).$$

A computationally feasible methodology for exact maximum constrained likelihood estimation of model parameters can now be put forward for a number of models of stochastic processes, both stationary or otherwise, in a unified approach. This methodology, which can be embedded into a computer program, requires small memory storage. It isolates correlational properties from model properties: given a model with parameters to be estimated, compute autocovariances of, and their first- and second-order derivatives with respect to, model parameters; then compute "partial autocovariances", then $\alpha_{i,j}$'s and hence $\mathbf{Y}_j$'s from (1); further, under some probability density assumption $f(y)$ for $\mathbf{Y}_j$'s (an additional assumption of independence might be made here, not needed in the Gaussian case) compute the "exact" likelihood of, and derivatives with respect to, model parameters; finally maximize locally using constrained non-linear optimization or Newton–Raphson routines (the necessary Kuhn–Tucker conditions may be written down).

The above method may be successfully applied to the following models:

(i) Stationary time series
- ARMA (no problem for starting values of autocovariances recursions);
- Random phase (nonlinear: $X_t = A\cos(\omega t + \varepsilon)$)
- DARMA of Jacobs and Lewis;
- RARMA (unpublished; $\Leftrightarrow$ ARMA with random orders and coefficients, all coefficients independent apart from AR ones which can be dependent, similarly MA coefficients).

(ii) Non-stationary processes
- ARMA/RARMA (problems there: more unknown parameters than data values);
- Processes with independent increments;
- SETARMA???

The method will be illustrated by reference to a SETAR(1) model, the problem being that of the computation of the autocovariances from the model parameters.

Consider the following threshold model: $X_t = U_t X_{t-1} + \varepsilon_t$, $E(X_t) = 0$, ε_t being white noise, with the random variable

$$U_t = \sum_{l=1}^{L} \phi(l)\,\mathbf{1}\{X_{t-d} \in R_l\},$$

$\mathbf{1}\{\cdot\}$ being the indicator function of the event $\{\cdot\}$. This model can be written

$$X_t = \phi^{(l)} X_{t-1} + \varepsilon_t \quad \text{if } X_{t-d} \in R_l.$$

If $\text{Prob}(X_{t-d} \in R_l)$ is independent of t, then the difficulty I am going to mention will not arise. However, when it cannot be assumed that $\text{Prob}(X_{t-d} \in R_l)$ is constant with respect to t, then after some algebraic manipulation, one can show that

$$\gamma_{t,t+k} = E(X_t X_{t+k}) = \sum_{l=1}^{L} \phi^{(l)} \gamma^{(l)}_{t,t+k-1}\, \pi^{(l)}_t, \tag{3}$$

where

$$\gamma^{(l)}_{t,t+k-1} = \text{cov}\,[(X_t, X_{t+k-1}) \,|\, \text{under model } l],$$

and

$$\pi^{(l)}_t = \text{Prob}\,(\mathbf{1}\{X_{t-d} \in R_l\} = 1) = \text{Prob}\,(\mathbf{1}\{\textstyle\sum_{h=1}^{t-d} \alpha_{t-d,h} Y_h \in R_l\} = 1).$$

If one assumes further that $Y_1, \ldots, Y_n$ are independent, then

$$\pi^{(l)}_t = \int \cdots \int_{\{\sum_{h=1}^{t-d} \alpha_{t-d,h} Y_h \in R_l\}} f(y_1) \ldots f(y_n)\, dy_1 \ldots dy_n$$

This results in a high order non-linear system of equations (3) to be solved in order to obtain $\mathbf{\Gamma}$.

If each of the l AR(1) models, $l = 1,\dots, L$ is stationary, then $\gamma_{t,t+k-1}^{(l)} \equiv \gamma_{k-1}^{(l)}$ can be computed without difficulty. Further, under the Gaussian assumption, X_t is then strictly stationary and hence $\pi_t^{(l)} = \pi_t$, independent of t. System (3) then shows that the $\gamma_{t,t+k} \equiv \gamma_k$ does not depend on t, and hence the SETAR(1) process is stationary. Moreover, the exact likelihood can be computed without difficulty. Now the likelihood maximisation will assign the same values to the parameters $\pi^{(l)}$, even if one considers another partition $R_1' \cup \dots \cup R_L'$, such that $\text{Prob}(X_{t-d} \in R) \equiv \text{Prob}(X_{t-d} \in R_l')$. $l = 1,\dots, L$ Hence no information on the choice of the partition is provided there by likelihood. The AIC criterion is irrelevant as both partitions have the same number of parameters. In this respect, the authors' split of the data into subsets, though heuristic, is invaluable for identifying the preferred partition. The generalization of these ideas will be presented elsewhere.

I have difficulty in interpreting the event {the observation $x_{t-d} \in R_l$}. If it means that $X_{t-d} \in R_l$ conditional upon information I_{t-d} available prior to $(t-d)$, and also information from $(t-d+1)$ to $(t-1)$, then $\pi_t^{(l)}$ depends on t and the difficulty remains.

Professor M. B. PRIESTLEY (University of Manchester Institute of Science and Technology): Tonight's paper is one of a group which have appeared recently on non-linear time series models, and which I feel represent a significant advance in the methology of time series. We now have several classes of "tractable" non-linear models (e.g. bilinear, threshold autoregressive and exponential autoregressive) which have been shown to be capable of providing good fits to a wide variety of data, and which possess more interesting structural properties than the conventional linear models. The basic idea underlying the TAR models is that, when we abandon linear models, we should look first at models which are "locally" linear. However, in this context the term "local" does not refer to a neighbourhood of a particular time point–rather, it refers to a particular region of the "state space" of the process. (The former notion is related to "non-stationarity", rather than "non-linearity", and there is an interesting form of duality between these two concepts') For the AR(k) process,

$$X_t + a_1 X_{t-1} + \dots + a_k X_{t-k} = \varepsilon_t,$$

the evolution of the process is determined by $(X_{t-1}, X_{t-2} \dots, X_{t-k})$ (together with the future ε_t's), these k quantities acting as "initial conditions" in determining the solution of the above difference equation from time t onwards. Consequently, the "state" of the process at time t is represented by the k-dimensional vector, $\mathbf{x}_t^{\mathrm{T}} = (X_t, X_{t-1}, \dots, X_{t-k+1})$, and the most general form of locally linear AR (k) model would be one in which the coefficients were all functions of $\mathbf{x}_{t-1}$, i.e. would take the form

$$X_t + a_1(\mathbf{x}_{t-1}) X_{t-1} + \dots + a_k(\mathbf{x}_{t-1}) X_{t-k} = \varepsilon_t. \qquad (*)$$

We may refer to this as a general "*state-dependent model*". Although this type of model can be put into the form of the authors' equation (5.4), their "piecewise linearization" approach would involve the partition of the k-dimensional space, R^k, into a multitude of "small" regions in each of which the coefficients $(a_1, \dots, a_k)$ were assumed to take constant values. Such an approach would be quite horrendous from a computational point of view, and the authors' way round this difficulty is to assume that the coefficients depend on only *one component* of $\mathbf{x}_{t-1}$, namely X_{t-d} (d being some specified integer, $1 \leqslant d \leqslant k$).

There is, however, an alternative way of dealing with general state-dependent models, which I will now indicate very briefly. The simplest form of functional dependence of the coefficients on the state-vector is that in which each a_i is a linear function of $\mathbf{x}_{t-1}$, i.e.

$$a_i(\mathbf{x}_{t-1}) = a_i^{(0)} + \mathbf{x}_{t-1}^{\mathrm{T}} \boldsymbol{\beta}_i, \quad \text{say.}$$

This assumption is quite restrictive, but we may relax it by allowing the "gradients", $\boldsymbol{\beta}_i$, to be themselves state-dependent, so that the (a_i) are then only locally linear functions of $\mathbf{x}_{t-1}$. If we do this we are then faced with the problem of specifying the functional form of the $(\boldsymbol{\beta}_i)$, but we can obviate this difficulty by simply letting the $\boldsymbol{\beta}_i$ "wander" over time, i.e. we allow $\boldsymbol{\beta}_i = \boldsymbol{\beta}_i^{(t)}$ to depend purely on the time parameter. The basic idea now is to let the $(\boldsymbol{\beta}_i^{(t)})$ wander in the form of a *random walk*, i.e. to set $\boldsymbol{\beta}_i^{(t)} = \boldsymbol{\beta}_i^{(t-1)} + \mathbf{v}_t$, where the $(\mathbf{v}_t)$ are independent zero mean random variables with variance matrix, $\boldsymbol{\Sigma}_v$, say. The estimation procedure then determines, for each t, those values of $(\boldsymbol{\beta}_i^{(t)})$ which, roughly speaking, minimise the discrepancy between X_{t+1} and its predictor, $\tilde{X}_{t+1}$, computed from the model. The estimation procedure is thus based on a sequential type of algorithm, similar in nature to the *Kalman filter* algorithm, and it leads to coefficients (a_i) which are "locally optimal" in the sense that they provide the best "local predictor" for the next observation. (The "smoothness" of the (a_i) as functions of $\mathbf{x}_{t-1}$ is controlled by the ratio of $\|\boldsymbol{\Sigma}_v\|$ to σ_ε^2.) Once we have determined suitable values of the (a_i) over a range of time points we can plot these as functions of the corresponding state vectors, and then, using some form of multidimensional smoothing (e.g. via "splines" or the "kernel" method), we can build up a graphical picture of the functional form of the (a_i). Thus, for a TAR model the (a_i) should appear as "ridges" of step-functions, depending only on one component of $\mathbf{x}_{t-1}$.

The general state-dependent model (*) includes, as special cases, the TAR and exponential autoregressive models, and, by adding moving average terms, it can also accommodate bilinear models (see Priestley, 1979).

As far as threshold models are concerned, the authors have given a convincing demonstration of their applicability to a wide range of data, and their modelling fitting expertise is certainly most impressive. As the authors show, these models can give rise to some fascinating features (such as limit cycles and jump phenomena), and they will, I am sure, stimulate much interest in this new and rapidly growing area of time series analysis.

Dr B. W. SILVERMAN (University of Bath): It would be interesting to know whether any connections can be made between the threshold models discussed tonight and the ideas of catastrophe theory, which might well give rise to models with piecewise behaviour of the kind described. Certainly the electric relay can be viewed in these terms. Models based on catastrophe theory would be attractive from the point of view of the authors' criteria (iv) and (v), while any relations with the authors' methodology would help with the fitting of catastrophe-theoretic ideas to real data.

Mr E. J. GODOLPHIN (Royal Holloway College): I would like to join the other discussants in congratulating the authors on an interesting and thought-provoking paper.

I have two questions to ask the authors, the first of which is about the R_js defined in Section 4, which seem to be best regarded either as random variables or possibly as deterministic but unknown quantities to be derived from the available data. Am I right in thinking that these quantities are likely to be considerably more important to the specification of the model than even the various sets of autoregressive parameters themselves? If this is so, I wonder if the authors could say a little more about the properties of their threshold estimates beyond the comments made in Section 8? For example, in one of the authors' examples the number 4·6 appears in equation (9.5); but how useful an estimate is that?

Secondly, in Section 8 the authors also refer to the eventual forecast function which they adopt for specification purposes. Have the authors considered obtaining a functional form for the eventual forecast function for variable lead times? I am thinking, for example, about results which would parallel a paper of my own (Godolphin, 1975) which deals with the non-stationary linear case, including seasonal models. If it were possible to compare the different kinds of functional forms for these eventual forecast functions with those for the linear models, this might provide an interesting way of exploring the manner in which the authors have succeeded in generalizing the linear case.

Dr. D. A. JONES (Institute of Hydrology, Crowmarsh Gifford, Wallingford, Oxon OX10 8BB): Recently a number of wide classes of non-linear time-series models have been proposed: autoregressive models, a class of which are discussed by the authors, and bilinear models (Granger and Andersen, 1978; Granger, 1978; Priestley, 1978). An example of the use of a smooth non-linear autoregressive structure, as opposed to one which is sectionally linear, appears in O'Connell and Jones (1979). These developments make it appropriate to question whether time-series models are necessary.

A "model" here means a complete probabilistic description of a series (apart from certain parameters). The answer will be different depending on the purpose of the data analysis. I am mainly thinking of problems where forecasts are to be constructed. Follow-up questions concern whether the model properties that are used in practice can be replaced by methods which are not model-dependent, and whether checks of the complete structure of models are actually available. A question which possibly encompasses these is whether linear *models* are actually used at present. It can be argued that relatively little use is made of linear models, as opposed to linear forecasts, on the basis that standard techniques involving ARMA structures are concerned essentially with forecasts rather than models. A question of a different character is whether discrete-time models are realistic. Should not all real processes be thought of as evolving continuously in time, even if at a rather basic level?

Much of time-series analysis is directed towards constructing forecasts. In this situation a forecasting rule can be fitted directly, rather than fitting a model. A class of possible rules for estimating y_t from $y_{t-1}, y_{t-2}, \ldots$ is first defined in terms of a number of parameters θ: let $\hat{y}_t(\theta)$ be the value of the forecast of y_t and let $\hat{Y}_t(\theta)$ be used when y_{t-1}, y_{t-2} are treated as random variables. Given observations $(y_1, \ldots, y_T)$ on a random process, the rule is fitted as follows:

(a) choose a loss function: squared-error loss is used here for convenience,

(b) for any θ define

$$s_T(\theta) = T^{-1} \sum_{t=1}^{T} (y_t - \hat{y}_t(\theta))^2,$$

(c) find $\hat{\theta}_T$ such that $s_T(\hat{\theta}_T) \leqslant s_T(\theta)$ for all θ,

(d) use $\hat{y}_t(\hat{\theta}_T)$ to forecast y_t, and $s_T(\hat{\theta}_T)$ as the estimated mean square error of the forecast.

Under weak conditions on the forecast rule and the process $\{Y_t\}$,

$$\lim_{T\to\infty} s_T(\theta) = \sigma(\theta) = E[\{Y_t - \hat{Y}_t(\theta)\}^2]$$

and $\hat{\theta}_T \to \theta^0$, where $\sigma(\theta^0) \leqslant \sigma(\theta)$ for all θ. Thus the best rule out of the chosen class is found. Note that the forecast need not be a one-step-ahead forecast, and that no explicit model needs to be assumed. The above is just a fit *via* minimization of the sum of squares of forecast errors, as used in the authors' paper.

The authors of this paper have shown that, for various data sets, certain non-linear forecasts are better than linear forecasts. It is a very large step to go from these forecasts to writing down a model in terms of impulses which are both independent and Gaussian, as seems to be implied in the paper. It is only too easy to interpret a fitted forecast $\hat{x}_t = \alpha x_{t-1}^2$, say, as meaning that a model $x_t = \alpha x_{t-1}^2 + \varepsilon_t$ ($\varepsilon_t \sim$ i.i.d. $N(0, \sigma^2)$) is being used, but this should always be avoided. A quadratic forecasting rule, as used by Cox (1977), is a perfectly valid choice, whereas the supposedly corresponding model would usually be rejected out of hand.

Dr H. AKAIKE (The Institute of Statistical Mathematics, Tokyo): Strictly speaking, every real system is non-linear and non-stationary. Thus, when Dr Tong and Ms Lim try to generalize their TAR model it inevitably leads to the blurring of the nature of their model.

The authors note that the era of practical non-linear time series modelling is long overdue. Actually the modelling of each particular non-linearity was the key to our success in implementing computer controls of cement rotary kilns and supercritical thermal power plants (Otomo, Nakagawa and Akaike, 1972; Nakamura and Akaike, 1979). Our experience with these systems suggest the importance of identifying a variable which characterizes the dynamics of a system. In Sugawara's tank model this variable is generated with the aid of an imaginary system of underground reservoirs. In these examples it was the analysis of the physical characteristic of each system that led to the choice of a particular "conditioning variable".

The variable J_n of TAR is an example of such a "conditioning variable", but since we are not told how practically to identify the variable we must resort to the examples. The examples of Canadian lynx and sunspot data show that the conditioning variables characterize the beginnings of the downward and upward paths of one cycle of oscillation. This observation clarifies why the model does not work well with the Mink–muskrat data, where the periodicity is not so clear. In the case of the Kanna riverflow data, again the input series is used to identify the upward and downward paths of the riverflow.

If these simple observations can capture the essence of these examples, what is the use of the elaborate generalization of TAR?

Professor D. R. Cox (Imperial College, London): The authors' account of non-linear models, and in particular threshold autoregressive models, is very valuable. It is interesting and important to see the kinds of qualitative behaviour that simple systems of this kind can produce. I am, however, extremely uneasy at the analysis of the data in Section 9. For instance, I can see that it is interesting to show that (9.1) has limit cycles, but are the authors claiming that fitting 14 (or really more) parameters in this way to 114 observations tells us anything about what is "really happening"? Is the mechanical use of AIC, or any other criterion, a good idea: perhaps there are much simpler models that give nearly as good a value of AIC? Once the need for an irreversible process is clear, the possibilities are so rich that in the absence of strong guidance from theory, graphical or other preliminary analysis to establish the approximate form of dependencies present seems very desirable. Finally, in terms of prediction, how does the authors' model compare with the much simpler, although explosive, model I reported in the discussion of Campbell and Walker (1977)?

Professor K. W. HIPEL (University of Waterloo) and Professor A. I. MCLEOD (University of Western Ontario): We fully concur with Dr Tong and Ms Lim when they state that "The new era of practical non-linear time series modelling is, without doubt, long overdue." The authors should be commended for not only describing the theory for a new class of non-linear time series models but also for presenting procedures for model identification and efficient estimation of the model parameters.

When modelling hydrologic time series that are measured at short periods of time such as hourly or daily time intervals, the fitted stochastic models must take into account unique non-linear properties of the data that are caused by complex physical processes. For example, when precipitation falls on a river basin this causes the flows at a given location in a river to increase while after the precipitation has ceased the flows return to their former levels. This ascension–recession behaviour of the hydrograph of flow *versus* time makes the modelling of daily and hourly flows an arduous task. Comprehensive appraisals regarding research in stochastic hydrology have stressed the need for flexibility in modelling this type of phenomenon (see, for example, Lawrance and Kottegoda (1977), Kibler and Hipel (1979) and Hipel and McLeod (1980)) and recently some new non-linear models have been examined. Some of these models include the non-linear autoregressive model of O'Connell and Jones (1979) which is based upon the theoretical work of Jones (1978), the model of Yakowitz (1973) which is similar to the model examined by O'Connell and Jones (1979), and also the non-parametric Markov model of Yakowitz (1979). In addition, the bilinear model described by Granger and Andersen (1978) and also Priestley (1978) may be useful in hydrology. It may be instructive to compare the models in today's paper to some of the aforementioned models in order to judge which models would be most appropriate to use in practice.

Some hydrologists are also concerned with the types of stochastic models that are employed to model data that are available at longer time periods such as weekly, monthly or yearly intervals. This is because the correlation structure for river flows that are relatively high may be different from the correlation pattern for low flows. Possible physical reasons for this behaviour include the manner in which the carrying capacity of the river channel varies, depending upon the level of the river, and the way in which different water table levels can affect the base flow of a river. The non-linear models of today's paper may prove to be effective for modelling phenomena of this type. Although short memory ARMA models have been shown to provide a statistical explanation for the Hurst phenomenon when modelling yearly geophysical time series (Hipel and McLeod, 1978), perhaps it may be worthwhile to determine if non-linear models provide a significant improvement over linear models when modelling annual data.

When modelling annual sunspot numbers from 1700 to 1960, McLeod *et al.* (1977) found that after taking a square root transformation of the data, the most appropriate ARMA model to fit to the transformed series is a constrained AR(9) model with the autoregressive parameters from lags three to eight left out of the model. In (9.2) and (9.3), Dr Tong and Ms Lim present their piecewise linear autoregressive models for modelling specified portions of the sunspot number series. Would a data transformation and perhaps omitting some of the less significant parameters from their models, help to lessen the values of the AIC?

Professor MITUAKI HUZII (Tokyo Institute of Technology, Dept. of Information Sciences): This paper gives us new ideas and methods for modelling non-linear systems. I think it will be an interesting problem to investigate the statistical properties of the process defined by (4.1) or (4.2). The reason is as follows:

(i) The likelihood function of the observations depends on the condition $\{X_{n-d} \in R_j; j = 1, 2, ..., l\}$. So, if we intend to examine the statistical properties of the maximum likelihood estimates of the unknown parameters, we have to know the properties of the process.

(ii) When $\{R_j\}$ or $\{\gamma_0, \gamma_1, ..., \gamma_l\}$ are unknown, we have to give a method for estimating these values. For this, the statistical properties of the process will be needed.

Dr I. T. JOLLIFFE (University of Kent at Canterbury): I have four questions on this interesting and useful paper.

The first concerns forecasts for more than one step ahead. It is stressed in Section 8 that the *eventual forecast function* (eff) is a plot of one-step-ahead predictions. However, with cyclical data it will often be required to forecast several time periods (at least one full cycle) into the future. How can such forecasts best be made?

A related point is that in practice different cycles in the same series will often have very different amplitudes and different cycle lengths; the sunspot data supply a good example, since amplitudes vary widely and the average cycle length was shorter in the first half of this century than in earlier periods. Can such variability be captured or, even better, forecast using the authors' models?

Thirdly, the authors' real data series are assumed to be stationary for all or part (sunspot data) of their length. Do the authors have any suggestions for dealing with non-stationary series?

My final question concerns the impressive range of behaviour exhibited by the examples of the threshold models given in Tables 3–7. How difficult was it to construct these examples, and are there other types of behaviour for which the authors failed to find models within their class?

Mr T. OZAKI (The Institute of Statistical Mathematics, Tokyo): This paper is of particular interest to me as it is closely related with some of my own work. I have the following three comments:

First, the set of threshold AR models is not general enough to include linear AR models, if (iii) of Definition 5.1 is assumed. If the proposed set of non-linear models and their identification method were appropriately defined we would expect to get a linear model for a linear process and a stable non-linear model for a stable non-linear process. My experience suggests that the linear threshold AR model often fails to satisfy this requirement (see Ozaki (1979a)).

Secondly, although the authors stress the importance of limit cycles in non-linear time series modelling, they did not give clear explanation of its mechanism for the threshold AR models. Extensive discussions of the limit cycle of non-linear time series models are given in Ozaki (1980), Haggan and Ozaki (1979) and Ozaki (1979c), based on the van der Pol equation, where the "shift back to center" property, which makes the process stationary ergodic in the sense of Tweedie (1975), was realized by making the instantaneous characteristic roots of the model state-dependent.

Thirdly, I would like to mention that the *jump resonance* and *amplitude–frequency dependency*, which are well known in relation to the *Duffing equation*, are also discussed extensively in Ozaki and Oda (1977), Haggan and Ozaki (1979) and Ozaki (1979b). If the authors would try to develop an analysis of the mechanism of the phenomena, these papers would be useful references. In these papers the "amplitude-dependent restoring property" of the Duffing equation is realized by the amplitude-dependence of the arguments of the instantaneous characteristic roots of the non-linear time series model.

Professor P. M. ROBINSON (University of Surrey): The threshold models are an interesting and important class, particularly in view of the connections that Tong and Lim have established with the theory of non-linear vibrations. Once the decision is made to forgo the great simplicity of the linear model, however, one is confronted with an embarrassingly wide choice of non-linear ones, even though many have yet to come under close scrutiny. This has recently led me to investigate a nonparametric approach to non-linear time series analysis. Let the conditional distribution of X_n given $X_{n-1}, X_{n-2}, \ldots$, have expectation $f(X_{n-1})$, so that one has the general NLAR (1) (equation (5.1)). Instead of assuming a specific form for $f(x)$ we could estimate $f(x)$, for any given x, by

$$f(x) = \left(\sum_{n=1}^{N} w_{nN}(x) \right)^{-1} \sum_{n=1}^{N} w_{nN}(x) X_n. \tag{1}$$

The non-negative weight functions $w_{nN}(x)$ can depend on $X_1, \ldots, X_N$, and will generally give greatest weight to n-values for which X_{n-1} is close to x; one possibility is $w_{nN}(x) = 1$, $|x - X_{n-1}| \leqslant c$; $= 0$, $|x - X_{n-1}| > c$. The estimator (1) can be shown to minimize a certain loss function; alternative loss functions will produce, for example, better robustness properties. Non-parametric regression estimators such as (1) have been considered previously, but in the case of independent, non-time series, observations; a recent reference is Stone (1977). Note that we could use the same type of approach to estimate other features of the distribution of X_n conditional on X_{n-1}. Extension is also possible to the general NLAR (p), for $p > 1$. Examination of the estimated $f(x)$ could provide a test of linearity or suggest a class of non-linear model for subsequent parametric analysis. If Tong and Lim's threshold model is selected, the non-parametric estimate may be of some use in its identification and estimation.

Dr T. SUBBA RAO (University of Manchester Institute of Science and Technology): The class of threshold autoregressive models proposed by the authors is definitely very useful, but sometimes other types of non-linear models fitted to the real time series may lead to models with fewer parameters. One class of such models—bilinear time series models—has recently been extensively studied by Granger and Andersen (1978) and Subba Rao (1979). The advantage of this model is that it is possible to obtain theoretical expressions for the moments, spectra, etc., whereas such things are not always possible in quite a number of non-linear models. Using the estimation methods described by Subba Rao (1979), bilinear models are fitted by Mr M. M. Gabr and myself to sunspot data and the Canadian lynx data (the details of which will be reported elsewhere).

Sunspot data

The first 246 observations are used for fitting, and the fitted model is

$$X_t - 1{\cdot}209X_{t-1} + 0{\cdot}502X_{t-2} - 0{\cdot}173X_{t-9}$$
$$= 5{\cdot}891 - 0{\cdot}0098X_{t-2}e_{t-1} + 0{\cdot}0103X_{t-8}e_{t-1} - 0{\cdot}0048X_{t-8}e_{t-3}$$
$$+ 0{\cdot}0016X_{t-3}e_{t-2} + 0{\cdot}0014X_{t-4}e_{t-7} + e_t.$$

The mean sum of squares of residuals is 141·18 and the AIC value is 1186·2.

The one-step-ahead predicted values from the bilinear model together with the true values are given in Table D1.

TABLE D1

t	247	248	249	250	251	252	253	254	255	256
True values	92·6	151·6	136·3	134·7	83·9	69·4	31·5	13·9	4·4	38·0
Predicted values	77·9	130·0	149·8	119·8	86·2	51·4	38·9	18·8	3·3	25·7

Since the authors have not actually tabulated their predicted values, it is not possible to compare the performance.

Canadian lynx data

We now consider the Canadian lynx data, discussed by the authors in Section 9. The data is logarithmically transformed. The bilinear model is fitted to the first hundred observations, and the fitted model is

$$X_t - 0{\cdot}8845X_{t-1} + 0{\cdot}1699X_{t-2} + 0{\cdot}1271X_{t-4} - 0{\cdot}5514X_{t-10} + 0{\cdot}5280X_{t-11}$$
$$= 1{\cdot}117 - 0{\cdot}1653X_{t-8}e_{t-10} - 0{\cdot}0970X_{t-5}e_{t-8} + 0{\cdot}0922X_{t-1}e_{t-1} + e_t.$$

This model has nine parameters and the mean sum of squares of residuals is 0·0329 and the AIC is −283·577 which are considerably less than the values obtained by the authors. The one-step-ahead predictors for the next fourteen observations together with the true values are given in Table D2.

TABLE D2

t	101	102	103	104	105	106	107
True values	2·360	2·601	3·054	3·386	3·553	2·468	3·187
Predicted values	2·442	2·756	2·897	3·135	3·411	3·512	2·922
t	108	109	110	111	112	113	114
True values	2·723	2·686	2·821	3·000	3·201	3·424	3·531
Predicted values	2·706	2·583	2·844	2·966	3·159	3·299	3·415

Since the models we are fitting are non-linear models, it is interesting to fit the models to the *original* data rather than the transformed data. It is known that the predictors obtained for the original data from the transformed data are highly biased. Now the question is why the authors did not consider fitting the threshold model to the original Canadian lynx data. The bilinear model fitted to the first 100 observations for the *original* Canadian lynx data is

$$\begin{aligned} X_t - 0{\cdot}941X_{t-1} + 0{\cdot}455X_{t-2} + 0{\cdot}102X_{t-5} - 0{\cdot}238X_{t-9} &= 612{\cdot}68 - 0{\cdot}00256X_{t-8}\,e_{t-10} - 0{\cdot}000375X_{t-1}\,e_{t-7} \\ &\quad + 0{\cdot}000164X_{t-2}\,e_{t-5} + 0{\cdot}000142X_{t-7}\,e_{t-2} \\ &\quad + 0{\cdot}000049X_{t-5}\,e_{t-4} + e_t. \end{aligned}$$

The mean sum of squares of residuals is 439075·2 and AIC = 1189·3.

The one-step-ahead predictors, together with their true values, are given in Table D3. Except for two values, the predictors are reasonably good.

TABLE D3

t	101	102	103	104	105	106	107
True values	229·0	399·0	1132·0	2432·0	3574·0	2945·0	1537·0
Predicted values	127·0	953·9	1147·2	2295·1	2931·4	2775·2	1653·3

t	108	109	110	111	112	113	114
True values	529·0	485·0	662·0	1000·0	1590·0	2657·0	3396·0
Predicted values	561·8	391·6	620·3	775·4	1474·9	2187·2	3252·5

It may be pointed out here that the test proposed by Subba Rao and Gabr (1981) has shown that the log (Canadian lynx data) is linear although it is not Gaussian.

An alternative non-linear model which may be useful for representing "cyclical data" is

$$Y(t) = \alpha \cos\{\theta t + \phi + \beta X(t)\}$$

where α, θ and β are constants, ϕ is a uniform random variable and $\{X(t)\}$ is a strictly stationary process. (ϕ and $\{X(t)\}$ are independent.) This model is used for representing frequency modulated signals. If $\{X(t)\}$ is Gaussian with correlation coefficient $\gamma(t)$, then it can be shown (Hannan, 1970, p. 85) that the auto-covariance function of $Y(t)$ is

$$E\{Y(t)\,Y(t+s)\} = \tfrac{1}{2}\alpha^2\, e^{-\beta^2\{1-\gamma(s)\}} \cos(\theta s)$$

which shows that the covariance function is a harmonic function with decreasing amplitude as the lag s increases. The sample autocovariance functions of both the sunspot data and the Canadian lynx data are harmonic functions with decreasing amplitudes suggesting that the above model may also be very appropriate. The statistical analysis of this model is under investigation.

The AUTHORS replied later, in writing, as follows.

Dr Silverman's suggestion is most exciting, although we did speculate upon some "catastrophic" connection in the original version of the paper! A fuller exploration of the TAR-modelling/catastrophe relation is now available in Tong (1980b) and here we give only a *brief* indication as follows. The most famous catastrophe is the so-called cusp-catastrophe, which is characterized by the five *qualitative* features of *bimodality*, *inaccessibility*, *hysteresis* (*limit cycle*), *sudden jumps* and *divergence*. (See, for example, Zeeman, 1977.) It seems that the lynx data exhibit these features (Tong, 1980b) and that our SETAR model is really an execution of the cusp-catastrophic paradigm. (See Fig. D1.) Table D5 shows that the fitted SETAR (2; 8, 3) model of (9.1) has captured some of the probabilistic structure of the lynx data (cf. Table D4). The *bimodality* of some of the fitted conditional distributions is particularly interesting. The "crater"

TABLE D4

Bivariate histogram of log lynx data

X(N) / X(N−2)	0·2	−0·4	0·6	−0·8	1·0	−1·2	1·4	−1·6	1·8	−2·0	2·2	−2·4	2·6	−2·8	3·0	−3·2	3·4	−3·6	3·8	−4·0	4·2	−4·4	4·6	−4·8 5·0	Total
5·0–	0	0	0	0	0	0	0	0	0	0	0	0	0	0	0	0	0	0	0	0	0	0	0	0	0
4·8–	0	0	0	0	0	0	0	0	0	0	0	0	0	0	0	0	0	0	0	0	0	0	0	0	0
4·6–	0	0	0	0	0	0	0	0	0	0	0	0	0	0	0	0	0	0	0	0	0	0	0	0	0
4·4–	0	0	0	0	0	0	0	0	0	0	0	0	0	0	0	0	0	0	0	0	0	0	0	0	0
4·2–	0	0	0	0	0	0	0	0	0	0	0	0	0	0	0	0	0	0	0	0	0	0	0	0	0
4·0–	0	0	0	0	0	0	0	0	0	0	0	0	0	1	0	1	1	0	0	0	0	0	0	0	3
3·8–	0	0	0	0	0	0	0	0	0	0	0	2	2	0	0	0	1	0	0	0	0	0	0	0	5
3·6–	0	0	0	0	0	0	0	0	2	1	0	2	2	2	1	3	3	3	1	0	0	0	0	0	20
3·4–	0	0	0	0	0	0	0	0	1	1	0	2	0	2	1	1	4	1	1	0	0	0	0	0	14
3·2–	0	0	0	0	0	0	0	0	0	0	0	1	1	0	0	0	5	0	1	0	0	0	0	0	8
3·0–	0	0	0	0	0	0	0	0	1	0	2	1	1	0	0	2	6	0	0	0	0	0	0	0	13
2·8–	0	0	0	0	0	0	0	1	0	1	1	0	0	3	1	3	2	0	0	0	0	0	0	0	12
2·6–	0	0	0	0	0	0	1	0	0	0	3	2	1	2	2	3	0	1	0	0	0	0	0	0	15
2·4–	0	0	0	0	0	0	0	0	0	0	0	0	3	2	3	1	0	0	0	0	0	0	0	0	9
2·2–	0	0	0	0	0	0	0	0	1	0	0	1	1	1	0	0	0	0	0	0	0	0	0	0	4
2·0–	0	0	0	0	0	0	0	1	0	1	1	1	1	0	0	0	0	0	0	0	0	0	0	0	5
1·8–	0	0	0	0	0	0	0	0	0	0	2	1	0	0	0	0	0	0	0	0	0	0	0	0	3
1·6–	0	0	0	0	0	0	0	1	0	0	0	0	0	0	0	0	0	0	0	0	0	0	0	0	1
1·4–	0	0	0	0	0	0	0	0	0	0	0	0	0	0	0	0	0	0	0	0	0	0	0	0	0
1·2–	0	0	0	0	0	0	0	0	0	0	0	0	0	0	0	0	0	0	0	0	0	0	0	0	0
1·0–	0	0	0	0	0	0	0	0	0	0	0	0	0	0	0	0	0	0	0	0	0	0	0	0	0
0·8–	0	0	0	0	0	0	0	0	0	0	0	0	0	0	0	0	0	0	0	0	0	0	0	0	0
0·6–	0	0	0	0	0	0	0	0	0	0	0	0	0	0	0	0	0	0	0	0	0	0	0	0	0
0·4–	0	0	0	0	0	0	0	0	0	0	0	0	0	0	0	0	0	0	0	0	0	0	0	0	0
0·2–																									
Total	0	0	0	0	0	0	1	3	5	4	9	13	12	13	8	14	22	5	3	0	0	0	0	0	112

TABLE D5

Bivariate histogram from fitted SETAR$(2;\ 8,\ 3)$, $d = 2$

X(N) (N − 2)	0·2	–0·4	0·6	–0·8	1·0	–1·2	1·4	–1·6	1·8	–2·0	2·2	–2·4	2·6	–2·8	3·0	–3·2	3·4	–3·6	3·8	–4·0	4·2	–4·4	4·6	–4·8 5·0	*Total*
5·0–	0	0	0	0	0	0	0	0	0	0	0	0	0	0	0	0	0	0	0	0	0	0	0	0	0
4·8–	0	0	0	0	0	0	0	0	0	1	0	1	0	0	0	0	0	0	0	0	0	0	0	0	2
4·6–	0	0	0	0	0	0	0	1	0	1	0	1	0	0	0	0	0	1	1	0	0	0	0	0	5
4·4–	0	0	0	0	0	0	2	1	0	4	4	3	4	4	2	3	3	1	0	0	0	0	0	0	31
4·2–	0	0	1	0	0	1	0	5	9	8	8	15	11	20	10	15	6	3	1	1	0	0	0	0	114
4·0–	0	0	1	0	1	2	4	7	13	20	20	28	33	42	41	36	32	15	7	2	1	1	0	0	306
3·8–	0	0	0	3	2	3	9	19	14	32	34	50	78	66	68	68	68	49	28	13	4	0	1	0	609
3·6–	0	0	1	1	5	6	14	18	22	40	58	66	102	104	106	127	102	88	55	32	9	2	0	0	958
3·4–	0	1	0	2	7	8	14	17	32	52	62	80	104	162	159	177	169	131	67	28	11	1	1	0	1285
3·2–	1	0	0	1	1	1	5	14	28	34	44	65	117	142	212	242	196	127	86	31	5	1	0	0	1353
3·0–	0	0	0	0	2	4	10	11	23	43	46	70	113	142	226	243	187	117	40	5	1	0	0	0	1283
2·8–	0	0	0	0	0	5	9	16	31	38	60	77	126	190	218	182	112	47	15	1	0	0	0	0	1127
2·6–	0	0	1	1	3	6	10	23	22	26	46	79	142	162	131	122	50	23	5	1	0	0	0	0	853
2·4–	0	0	1	1	2	10	8	8	15	43	66	97	119	110	101	45	19	6	1	0	0	0	0	0	652
2·2–	0	0	3	3	7	3	5	14	28	59	70	100	85	75	51	15	10	1	0	0	0	0	0	0	529
2·0–	0	0	0	2	0	4	4	16	35	54	65	50	53	36	14	7	4	0	0	0	0	0	0	0	344
1·8–	0	1	0	0	1	5	12	23	33	39	40	36	22	19	10	4	0	0	0	0	0	0	0	0	245
1·6–	0	1	0	0	3	4	20	22	19	22	14	19	14	7	1	1	0	0	0	0	0	0	0	0	147
1·4–	0	0	0	0	2	6	10	12	15	11	8	9	2	2	2	0	0	0	0	0	0	0	0	0	79
1·2–	0	0	0	1	3	8	5	11	5	1	4	5	1	0	0	0	0	0	0	0	0	0	0	0	44
1·0–	0	0	1	0	3	2	3	5	0	0	1	0	0	0	0	0	0	0	0	0	0	0	0	0	15
0·8–	0	1	2	0	1	1	2	1	0	0	2	2	0	0	0	0	0	0	0	0	0	0	0	0	12
0·6–	0	0	0	0	1	0	1	1	0	1	0	0	0	0	0	0	0	0	0	0	0	0	0	0	4
0·4– 0·2–	0	0	1	0	0	0	0	0	0	0	0	0	0	0	0	0	0	0	0	0	0	0	0	0	1
Total	1	4	12	15	44	79	147	245	344	529	652	853	1126	1283	1352	1287	958	609	306	114	31	5	2	0	9998

† Based on 10 000 point simulations.

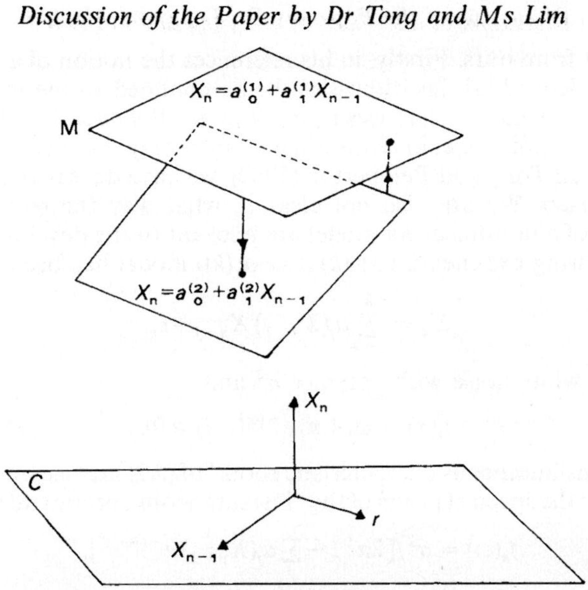

FIG. D1. Catastrophe paradigm for SETAR(2; 1, 1).

shape of the fitted bivariate distribution is certainly related to the concept of *perturbed limit cycles.* We have found simulation exercises with decreasing noise variance quite instructive. Not only would the connection with catastrophe theory *elevate* TAR *modelling above the ad hoc status,* it could also suggest ways of refining our models. In fact, for the lynx data, different threshold values may be used depending on whether the data are going up or down. This idea is also related to Dr Tunnicliffe-Wilson's suggestion of incorporating a "slope" in TAR modelling. As a result of some preliminary investigation, the following more refined SETAϕ is fitted to the first 100 lynx observations (logarithmically transformed):

$$X_n = \begin{cases} 0\cdot5382+1\cdot0602X_{n-1}-0\cdot2547X_{n-2}+0\cdot1598X_{n-3}-0\cdot3626X_{n-4} \\ \quad +0\cdot2100X_{n-5}-0\cdot2201X_{n-6}+0\cdot2753X_{n-7}-0\cdot0264X_{n-8}+\varepsilon_n^{(1)} \\ \quad \text{if } \{X_{n-1}-X_{n-2}\geqslant 0 \text{ and } X_{n-2}\leqslant 3\cdot4\} \\ \quad \text{or } \{X_{n-1}-X_{n-2}\leqslant 0 \text{ and } X_{n-2}\leqslant 3\cdot3\}, \\ 0\cdot6354+1\cdot6359X_{n-1}-1\cdot1985X_{n-2}+0\cdot3032X_{n-3}+\varepsilon_n^{(2)} \\ \quad \text{if } \{X_{n-1}-X_{n-2}\geqslant 0 \text{ and } X_{n-2}>3\cdot4\} \\ \quad \text{or } \{X_{n-1}-X_{n-2}\leqslant 0 \text{ and } X_{n-2}>3\cdot3\}, \end{cases} \qquad \text{(A1)}$$

where $\operatorname{var}\varepsilon_n^{(1)} = 0\cdot0322$, $\operatorname{var}\varepsilon_n^{(2)} = 0\cdot0537$. (The pooled mean sum of squares of residuals = 0·0383.) This new model seems to give encouraging results when compared with the SETAR model reported in Tong (1980a). More details about this refined SETAR class will be reported by Lim and Tong elsewhere. The catastrophe paradigm has highlighted the significance of the *piecewise* aspect of our approach and relegated the actual choice of the class of submodels to a secondary position. Thus, as we have indicated in Tong (1980a), threshold polynomial AR models, threshold bilinear models and threshold exponential AR models, etc. are all ripe for exploitation. A connection between TAR modelling and catastrophe theory implies a connection between TAR models and non-linear vibrations, because the latter are often most elegantly explained in the language of catastrophe theory. Viewed in this light, our demonstrations in Section 6 are really quite trivial and natural, although we are indeed most delighted with the generally favourable reception of them. Mr Ozaki's approach to random vibrations is different from ours. Firstly, in his references the notion of a *limit cycle in discrete time* has never been defined, all his discussion being confined to the *classical* (*deterministic*) *continuous time* case, typified by references to the Van der Pol equation. We must confess that our earlier discussion of limit cycles in terms of a predator–prey system (e.g. Tong, 1978) suffered from a similar defect. In Tong and Pemberton (1980), we have drawn attention to the possible pitfalls of this approach. We are also not clear in what way the so-called "instantaneous characteristic roots" of a non-linear AR model are relevant to the development of TAR models. Specifically, the following exponential AR(k) (EXPAR(k)) model has been proposed:

$$X_n = \sum_{j=1}^{k} a_j(X_{n-1})\, X_{n-j} + \varepsilon_n, \tag{A2}$$

where ε_n is the usual white noise with variance σ^2 and

$$a_j(x) = \phi_j + \pi_j\, e^{-\gamma x^2} \quad (\gamma \geqslant 0). \tag{A3}$$

Now, the so-called "instantaneous characteristic roots" of (A2) are just another way of looking at the "frequencies" of the spectral peaks of the "instantaneous spectral density function" for X_n

$$f_n(\omega) = \sigma^2 / [2\pi \,|\, 1 - \sum_j a_j(X_{n-1})\, e^{-ij\omega} \,|^2]. \tag{A4}$$

These "frequencies" are functions of the "amplitude" X_{n-1}^2 (*a random variable*). Whilst we sympathize with the use of this kind of *physical consideration* to deliver a new class of non-linear time series models, we prefer not to treat these "frequencies" as the most dominant *physical notion*. For, if (A2) defines a stationary ergodic stochastic process, it has a *unique spectrum* and therefore *unique peak frequencies* (*or a unique proper frequency when* $k = 2$) *for all n*. Then, what are the amplitude-dependent frequencies? Incidentally, Atkinson and Caughey (1968a, b) have discussed continuous time first-order SETAR with emphasis on the spectrum. We would also prefer retaining our original terminology of SETAR to Mr Ozaki's own preference of "linear threshold AR".

Dr Tunnicliffe-Wilson and Mr Ozaki have raised the important question of stability. In fact, there are at least two types of stability, one systematic with the innovation absent and the other stochastic with the innovation present. Let us consider the former case now. In the linear case, local stability implies global stability. More specifically, if a linear model is stable over the dynamical range, say S, of the observations, then its extrapolation beyond S would not cause any problem. However, in the non-linear case, this is not and indeed cannot be expected to be so. There is simply no information contained in the data about the (non-linear) behaviour beyond S. Engineers have long recognized this point and the famous saturation system is a fine example. Our Definition 5.1 and Theorem 5.1 is a modest attempt to formalize this recognition for our purpose of model building and is useful not only for TAR models but also for EXPAR, bilinear (BL) and polynomial AR (PAR) models, etc. Naturally, the final product is necessarily a *threshold* model and whether it is a good model or not is a different matter. Using this artifact, we can also avoid what seems to us a rather unnatural distinction between a model based forecast and a non-model based forecast discussed by Dr Jones. For, the parameters involved in his forecast rule inevitably define a model (albeit not necessarily unique) whether we like it or not. Professor Cox's lynx *model* can be thus stabilized. Of course, if a model is already globally stable then there is no need for invoking the stabilization. Thus, Mr Ozaki has really missed the point here. We are also puzzled by the second part of his first comment because the *sole* SETAR model quoted in Ozaki's research memo (1979a) is at variance with the original SETAR model fitted by us, which is now in print (Tong, 1980a). Incidentally, we would just mention that the EXPAR model for the lynx data reported in Haggan and Ozaki (1980b, p. 67) appears to be explosive in our simulation studies *when the innovation is present*. The fact that some of the "characteristic roots" of the purely ϕ-operator are virtually on the unit circle might account for this stochastic instability. (Coefficients are correct to two decimal places as reported in their paper). Yet, our stabilization could still have come to its rescue. It might be wise to re-examine the appropriateness of fitting a *symmetric* process, such as EXPAR (a fact which he has correctly recognized), to *asymmetric* data such as the lynx.

Many discussants have commented on our identification and fitting of SETAR models. The general view seems to be favourable with the significant exception of Professor Cox. It is a little amusing to see that he finds our use of AIC *mechanical* while, at the other extreme, Dr Bhansali has gone almost as far as labelling one of us a heretic! Perhaps Professor Cox's concern about the number of parameters might be alleviated if some of the intermediate terms are suppressed as in subset AR models. (See, for example, Tong, 1977a.) It is gratifying to note that all the three classes of non-linear time series models mentioned in Professor Priestley's contribution have relied on AIC for their *practical* identification. Of course, this does not imply that AIC is the only tool for this job. In fact, we would concur with Professor Cox and Dr Tunnicliffe-Wilson on the importance of graphical analysis and other classical time series operations. We do have some experience in the use of univariate and bivariate histograms, simulation studies, scatter diagrams and sample regression functions etc. as aids for our model identification and detecting non-linearity, some of which have been indicated in Tong (1980b) and more details will be reported by Lim and Tong elsewhere. We would even argue that, since a *non-linear* time series model is a transformation of the probability distributions of the (unobservable) input white noise process to those of the (observable) output process, it is not so much the residual sum of squares (RSS) or the number of parameters (p), but rather the general shape of the distributions (univariate, bivariate, trivariate, etc.) which is of paramount importance. In practice, we would caution against attaching too much significance to RSS and p as means of comparing *different classes* of models because of the almost inevitably different optimization algorithms and the often different number of effective observations. After all, the most important purpose of fitting a *non-linear* model is to gain a *better* understanding of the probabilistic structure underlying the data. At this point, we should appreciate the greater insight gained from moving away from *linear Gaussian* models. We are also interested in Dr Khabie-Zeitoune's suggestion of the feasibility of performing exact maximum likelihood computation using small memory storage.

Whilst Professor Priestley seems to be striving for greater flexibility, Dr Akaike seems to be urging us to go in quite a different direction. Rather than blurring the essential feature of our model, we feel that the introduction of the indicator variable J_n is similar to the facility provided by a modern vari-focal lens in photography (from where?) which enables us to focus on the more interesting aspects of reality. We should have made it clearer that the definition of J_n need not always be restricted to the few specific suggestions we have given. *It can be much more general.* Catastrophe consideration has suggested model (A1). The threshold model of neuron firing (see, for example, Brillinger and Segundo, 1979) suggests a J_n dependent on $\Sigma_{j=1}^{p} a_j X_{n-j}$ for some p and some a_j's. In a private communication, Dr G. Gudmundsson has suggested a particularly interesting case for hydrological application in which the precipitation is alternately in the form of rain and snow. Professors Hipel and McLeod have also given us valuable suggestions for hydrological applications. The several engineering references quoted in our paper suggest a J_n which follows a Markov chain. In economic applications, Chien and Chan (1979) and Dr Bhansali's discussion might also lead to some useful suggestions. Professor Priestley's striving for greater flexibility is undoubtedly interesting and we look forward to seeing some real applications. We suspect that Professor Robinson's non-parametric regression is not unrelated to one aspect of Professor Priestley's approach. This type of exercise is more appealing to us than Dr Chatfield's suggestion of comparing the different forecasts, because we know only too well the kind of tangle the latter might lead to even in the linear case. However, to satisfy Dr Chatfield's expressed wish (and no doubt that of many others) we give Tables D6, D7 and D8, although we must caution against any general inference from them. For the purpose of discrimination, we think that it is also important to know more about the different types of distributions to which the different classes of non-linear models can give rise. Dr Jones' work (1978) would be very valuable here. A Ph.D. student at UMIST, Mr J. Pemberton, has obtained

TABLE D6

One-step-ahead predictions of log lynx (fitting period 1821–1920)

Year	Real Data	Predicted values											
		(1) SETAR	(2) AR(12)	(3) *Moran*	(4) *Moran*-BL	(5) *C-W*1	(6) *C-W*2	(7) *C-W*3	(8) *Subset* BL	(9) *Cox*	(10) EXPAR	(11) TPAR	(12) (A1)
1921	2·3598	2·3109	2·455	2·4504	2·5059	2·5511	2·4895	2·4262	2·442	3·3514			2·3415
1922	2·6010	2·8770	2·807	2·8099	2·8369	2·8745	2·8768	2·9018	2·756	2·6910			2·8051
1923	3·0539	2·9106	2·899	2·8974	2·9589	2·9412	3·0180	3·0945	2·897	2·8823			3·0012
1924	3·3860	3·3703	3·231	3·3495	3·3003	3·1803	3·2790	3·2942	3·135	3·3626			3·4171
1025	3·5532	3·5875	3·388	3·4676	3·4578	3·2164	3·3053	3·2368	3·411	3·5282	3·4223	3·4191	3·3691
1026	3·4676	3·4261	3·332	3·4465	3·4226	3·1788	3·2145	3·1633	3·512	3·4662	3·4299	3·4074	3·5093
1927	3·1867	3·0936	3·007	3·1966	3·1907	3·0345	3·0009	3·0425	2·922	3·1365	2·7499	2·7771	3·0762
1928	2·7235	2·7706	2·688	2·8666	2·8694	2·8719	2·7789	2·8519	2·706	2·8033	2·5991	2·5928	2·7699
1929	2·6857	2·4217	2·428	2·4307	2·4715	2·6442	2·5343	2·5311	2·583	2·4470	2·5011	2·4626	2·5724
1930	2·8209	2·7644	2·765	2·7357	2·6585	2·9063	2·8005	2·7258	2·844	2·7925	2·5536	2·5572	2·7168
1931	3·0000	2·9397	2·984	2·9554	2·9336	3·1378	3·0968	3·0607	2·966	2·9959	2·8832	2·8777	2·9222
1932	3·2014	3·2462	3·217	3·1036	3·0913	3·2538	3·3013	3·3571	3·159	3·1493	3·2280	3·2096	3·2367
1933	3·4244	3·3701	3·365	3·2490	3·2217	3·2840	3·3999	3·4653	3·299	3·2985	3·2802	3·2933	3·3765
1934	3·5310	3·4468	3·503	3·4077	3·3598	3·2851	3·4181	3·3966	3·415	3·4532	3·3080	3·3389	3·5602
MSE		0·0144	0·018	0·0168	0·0204	0·0371	0·0232	0·0297	0·018	0·0093	0·0415	0·0395	0·0093
Var*	0·3287	0·3897	0·3632	0·3151					0·1932				0·2961

* Var = Var X_n. For the real data, it is based on the fitting period. Other entries are based on simulated samples of 10 000 points from the fitted models, with transients removed. The same convention is adopted in tables D7 and D8.

some results in this respect. Partly as an answer to Dr Chatfield's first question we would suggest that qualitative analyses such as bimodality, skewness, etc., are vital for deciding whether a linear Gaussian model is adequate or not. Incidentally, Rosenblatt (1979) might be relevant for Professor Robinson's examination of the estimated non-parametric regression as a test of linearity.

Dr Bhansali, Mr Godolphin and Professor Huzii are right in suggesting the need for a more thorough study of the sampling properties of the parameter estimates. We will report some results elsewhere. There can be no definite answer to Mr Godolphin's first question. For the lynx data the answer is affirmative and for the Kanna data the answer is negative. The number 4.6 in equation (9.5) merely indicates the state of very low rainfall. As for his second question, which is related to ones raised by Dr Chatfield and Dr Jolliffe, our answer is no, but we are currently

---

(1) See Tong (1980a, p. 54).
(2) See Tong (1977, p. 466).
(3) See Tong (1977, p. 468) for reference of P. A. P. Moran's AR(2) with fitting period 1821–1934.
(4) See Tong (1977, p. 454, under item Dr T. Subba Rao). Fitting period is 1821–1934.
(5), (6) and (7). See M. J. Campbell and A. M. Walker (1977, especially equations (4.13) and (4.19) and p. 462). See also Tong (1980a). Fitting periods are all 1821–1934.
(8) See Dr T. Subba Rao's contribution to the discussion of this paper.
(9) See Tong (1977, p. 453, under item Professor D. R. Cox) with regression coefficients 0·345173, 1·099411, 0·120404, 0·116176, −0·383841 obtained by us.
(10) See unpublished M.Sc. dissertation by Mr M. C. Wong (1980), University of Manchester, who gave the following parameter estimates:

i	1	2	3	4	5	
$\hat{\phi}_i$	1·097	−0·444	0·376	−0·330	0·257	
$\hat{\pi}_i$	−0·265	−0·151	−0·525	−0·091	−0·021	
i	6	7	8	9	10	11
$\hat{\phi}_i$	−0·247	0·245	−0·232	0·197	0·290	−0·341
$\hat{\pi}_i$	0·140	−0·601	0·293	0·151	−0·069	

$\exp\{-2{\cdot}45X_{n-i}^2\}$ is associated with $\hat{\pi}_i$, $i = 1, 2, ..., 10$. His fitting period is 1821–1924 and he has followed Haggan and Ozaki (1980b) for notation.

(11) See above dissertation. He has followed Ozaki (1979a) for notation. $T = 1{\cdot}02$.

i	1	2	3	4	5	
$\hat{\phi}_i$	0·822	−0·593	−0·188	−0·417	0·219	
$\hat{\pi}_{ii}$	0·236	0·160	0·549	0·083	0·027	
i	6	7	8	9	10	11
$\hat{\phi}_i$	−0·083	−0·366	0·094	0·374	0·229	−0·329
$\hat{\pi}_i$	−0·130	0·556	−0·301	−0·176	0·050	

(12) See equation (A1) of our reply.

working on his suggestion. Many-step-ahead predictions are non-trivial for the non-linear case. Whether a 10 per cent reduction in RMSE is considered substantial in the case of one-step-ahead predictions should presumably depend on the relative inadequacy of the linear models and the relative importance of the forecasts.

TABLE D7
One-step-ahead predictions of sunspot numbers (fitting period 1770–1869)

		Predicted values		
Year	*Real data*	(1) AR(2)	(2) BL	(3) SETAR
1870	139·0	92·664	94·240	95·308
1871	111·2	158·681	157·054	133·366
1872	101·6	71·551	109·128	79·982
1873	66·2	78·193	76·328	82·609
1874	44·7	34·765	39·099	44.418
1875	17·0	30·005	33·503	32·384
1876	11·3	6·249	7·262	6·884
1877	12·4	18·376	20·562	9·377
1878	3·4	24·110	23·818	31·721
1879	6·0	10·481	3·589	1·922
1880	32·3	20·765	20·968	18·998
1881	54·1	56·342	59·799	62·342
1882	59·7	68·466	69·462	61·432
1883	64·7	60·079	57·866	46·870
1884	63·5	61·832	63·699	53·870
1885	52·2	58·623	58·976	52·468
1886	25·4	42·667	41·630	39·907
1887	13·1	12·737	6·936	13·756
1888	6·8	14·800	17·236	8·754
1889	6·3	14·814	13·282	23·483
Var	1385·2	1541·5	2107·5	1275·4
MSE for 20 point predictions		346·6	293·4	267·6
MSE for 10 point predictions		622·6	507·5	422·1

Key to Table D7

(1) $X_n = 14{\cdot}70 + 1{\cdot}425X_{n-1} - 0{\cdot}731X_{n-3} + \varepsilon_n$, $\operatorname{var} \varepsilon_n = 228$. See Granger and Andersen (1978, p. 86).

(2) $X_n = 14{\cdot}70 + 1{\cdot}425X_{n-1} - 0{\cdot}731X_{n-2} + \varepsilon_n$, where $\varepsilon_n = -0{\cdot}0222\varepsilon_{n-2}\eta_{n-1} + 0{\cdot}202\varepsilon_{n-1} + \eta_n$, $\operatorname{var} \eta_n = 197$. See Granger and Andersen (1978, p. 86).

(3)
$$X_n = \begin{cases} 5{\cdot}2659 + 1{\cdot}8891X_{n-1} - 1{\cdot}5289X_{n-2} + 0{\cdot}3039X_{n-3} \\ \qquad + 0{\cdot}3387X_{n-4} + \varepsilon_n^{(1)} \quad \text{if } X_{n-3} < 36{\cdot}6, \\ 0{\cdot}3900 + 1{\cdot}1366X_{n-1} - 0{\cdot}3645X_{n-2} + 0{\cdot}0524X_{n-3} + \varepsilon_n^{(2)} \\ \quad \text{if } X_{n-3} > 36{\cdot}6, \end{cases}$$

where $\operatorname{var} \varepsilon_n^{(1)} = 154{\cdot}88$, $\operatorname{var} \varepsilon_n^{(2)} = 94{\cdot}00$ (pooled variance $= 121{\cdot}73$).

As for computation time, we can only blame ourselves for giving such a detailed description of our identification procedure in Section 8, which has undoubtedly given Dr Chatfield the wrong impression. In fact, it has taken our CDC 7600 computer twelve seconds for the complete SETAR identification, as described in Section 8, of the lynx data. In a private communication, Dr Tunnicliffe-Wilson has indicated the feasibility of using GENSTAT for fitting TAR models which should make TAR modelling more readily available.

TABLE D8

One-step-ahead predictions of sunspot numbers (fitting period from 1700 to 1920 for model (1) and from 1700 to 1945 for all others)

Year	Data	Predicted values (1) SETAR	(2) BL(3, 4)	(3) BL	(4) EXPAR	(5) TPAR
1921	26·1	29·182				
1922	14·2	10·236				
1923	5·8	5·294				
1924	16·7	15·728				
1925	44·3	39·363				
1926	63·0	69·650				
1927	69·0	72·138				
1928	77·8	75·358				
1929	64·9	67·323				
1930	35·7	55·433				
1931	21·2	23·508				
1932	11·1	16·952				
1933	5·7	22·958				
1934	8·7	17·707				
1935	36·1	24·472				
1936	79·7	64·559				
1937	114·4	106·468				
1938	109·6	121·579				
1939	88·8	86·124				
1940	67·8	67·788				
1941	47·5	42·865				
1942	30·6	25·737				
1943	16·3	15·686				
1944	9·6	10·921				
1945	32·2	22·277				
1946	92·6	64·050	61·099	77·9	59·4681	58·8084
1947	151·6	133·464	126·314	130·0	127·7654	127·8114
1948	136·3	169·018	135·483	149·8	123·5977	127·2716
1949	134·7	110·580	85·781	119·8	98·7405	96·2994
1950	83·9	97·849	64·032	86·2	97·8016	96·7381
1951	69·4	57·312	30·662	51·4	67·7415	68·0585
1952	31·5	33·665	67·742	38·9	63·0849	63·0153
1953	13·9	21·137	58·286	18·8	7·5928	9·3160
1954	4·4	2·767	37·504	3·3	5·9210	6·2002
1955	38·0	19·766	0·095	25·7	11·3457	11·4411
MSE*		148·205	1173·744	164·75	506·640	515·334

The prediction period 1921–1955 consists of 3 fairly representative cycles of different amplitudes.

(1) See equation (9.2) of this paper. Var X_n = 1340·3 (c.f. 1168·9 of the observed.)
(2) See Subba Rao (1979). Var $X_n = 1 \times 10^{58}$ (c.f. 1155·1 of the observed.)
(3) See Dr T. Subba Rao's discussion of this paper. Var X_n = 1059·2 (c.f. 1155·1 of the observed.)
(4) See note (10) of Table A3. $\hat{\gamma}$ = 0·000168.

$\hat{\phi}_i$	0·789	−0·170	−0·053	0·166	−0·034	−0·078	0·113
$\hat{\pi}_i$	0·802	−0·402	−0·252	−0·120	−0·182	0·273	

(5) See note (11) of Table A3. $\hat{T}$ = 96·1.

$\hat{\phi}$	1·717	−0·655	−0·318	0·010	−0·246	0·240	0·109
$\hat{\pi}_i$	−0·008	0·004	0·002	0·001	0·002	−0·003	

* A linear AR (10) fitted to 1700–1945 has MSE = 482·0.

Finally, we come to the analysis of the real data in Section 9. Dr Bulmer has clarified some of the doubts we did have regarding the lynx-hare hypothesis. However, he has not stated what model he would use for the lynx. Plainly he cannot retain his AR plus harmonic component model. Our current view is that empirical evidence seems to support a cusp-catastrophe model in which the amount of food in the present year is the one control parameter and the population density of recent years the other. Our SETAR model may then be regarded as a statistical expression of this cusp-catastrophe model. (See our discussion in the first paragraph and for more details see Tong, 1980b.) Dr Subba Rao has fitted bilinear models to the lynx data, for both the log transformed data and the original data. Our simulation studies suggest that BL models give *skew, unimodal* bivariate distributions. (Gaussian white noise is assumed throughout.) It seems clear that it is the linear AR part of the BL models which "explains" the cyclicity of the data; the bilinear terms probably account for the skewness of the probability distributions. We conjecture that the non-existence of limit cycles of BL models (see, for example, Brockett, 1977) implies that a BL process has, under general conditions, unimodal joint distributions. Despite these remarks, it is noteworthy that Dr Subba Rao has apparently succeeded in making BL modelling a practical proposition. His *subset* BL models represent an important step in this direction because a *full* BL model usually consists of too many parameters for efficient computation. Now, regarding his point about transformation, besides making the usual Gaussian assumption of the white noise more plausible, a logarithmic transformation might also have some stabilising effect. (See, for example, Rosenblatt, 1971, p. 164.) In fact, our simulation studies suggest that his BL model for the *original* lynx data tends to have a rather wide dynamical range, with a substantial proportion on the negative side extending beyond −20 000. Our simulated sample of 10 000 data has a mean 1450 and a variance $1{\cdot}4\times10^{10}$ which may be compared with observed values of 1528 and $2{\cdot}662\times10^{7}$ respectively.

Dr Bulmer seems to have overlooked the fact that our analysis of the mink–muskrat data is for the period of 1767–1849 and a *first differencing* operation is applied to both the log transformed mink data and the log transformed muskrat data. The observed limit cycle of period five years is probably due to the high-pass filtering property of a differencing operation. The following TARSC model is now fitted to the 1848–1909 data, which were used by Jenkins (1975) and Chan and Wallis (1978). (We had some difficulty in obtaining these data previously.) Let $X_n = \ln$ (number of mink in year $1847+n$), $Y_n = \Delta \ln$ (number of muskrats in year $1848+n$).

$$Xn = \begin{cases} 8{\cdot}1624+0{\cdot}3437X_{n-1}+0{\cdot}4510Y_{n-1}+0{\cdot}0696X_{n-2}-0{\cdot}0713Y_{n-2} \\ \quad -0{\cdot}4119X_{n-3}+0{\cdot}5353Y_{n-3}+0{\cdot}2228X_{n-4}+\eta_n^{(1)} \\ \quad \text{if } Y_{n-5}\leqslant -0{\cdot}0443, \\ 5{\cdot}4058+0{\cdot}5266X_{n-1}+0{\cdot}4653Y_{n-1}+0{\cdot}3631X_{n-2}-0{\cdot}2820Y_{n-2} \\ \quad -0{\cdot}2207X_{n-3}+0{\cdot}2009Y_{n-3}-0{\cdot}1585X_{n-4}+\eta_n^{(2)} \\ \quad \text{if } Y_{n-5}> -0{\cdot}0443, \end{cases}$$

where $\operatorname{var}(\eta_n^{(1)}) = 0{\cdot}0369$, $\operatorname{var}(\eta_n^{(2)}) = 0{\cdot}0234$ pooled variance $= 0{\cdot}0282$),

$$Y_n = \begin{cases} 2{\cdot}9045-0{\cdot}0212Y_{n-1}-0{\cdot}6994X_{n-1}-0{\cdot}3047Y_{n-2}+0{\cdot}4254X_{n-2} \\ \quad +0{\cdot}0485Y_{n-3}+\varepsilon_n^{(1)} \quad \text{if } X_{n-5}\leqslant 10{\cdot}9616, \\ 5{\cdot}8527+0{\cdot}3032Y_{n-1}-0{\cdot}5387X_{n-1}-0{\cdot}1289Y_{n-2}+\varepsilon_n^{(2)} \\ \quad \text{if } X_{n-5}>10{\cdot}9616, \end{cases}$$

where $\operatorname{var}(\varepsilon_n^{(1)}) = 0{\cdot}0385$, $\operatorname{var}(\varepsilon_n^{(2)}) = 0{\cdot}0841$ (pooled variance $= 0{\cdot}0589$). This fitted model has a 10 year period limit cycle with six ascension years and four descension years for the mink, and four ascension years, three descension years and then two ascension years and one descension year for the muskrat.

Given the limited data length and noisiness of the data, our TARSC model seems reasonably successful. It also seems to be one of the very few real examples of bivariate non-linear time series models.

We must admit that we are a little disappointed with the results of all the non-linear time series models, including SETAR, BL, EXPAR and TPAR, which have been fitted to the sunspot numbers. The very large number of sunspot numbers near the minimum is the main source of difficulty. The other source of difficulty is the well-known inhomogeneity of the data. One feature has come to light during our simulation studies which concerns the full BL (3, 4) model reported in Subba Rao (1979). The AR operator there has one pair of complex roots in the unstable region ($\hat{a}_3$ should read $-0{\cdot}27$).

REFERENCES IN THE DISCUSSION

ATKINSON, J. D. and CAUGHEY, T. K. (1968a). Spectral density of piecewise linear first order systems excited by white noise. *Int. J. Non-linear Mech.*, **3**, 137–156.

—— 1968b). First order piecewise linear systems with random parametric excitation. *Int. J. Non-linear Mech.*, **3**, 399–411.

BEENSTOCK, M. and BHANSALI, R. J. (1980). Analysis of cocoa price series by autoregressive model fitting techniques. *J. Agric. Econ.*, **31**, 237–242.

BRILLINGER, D. R. and SEGUNDO, J. P. (1979). Empirical examination of the threshold model of neutron firing. *Biol. Cybern.*, **35**, 213–220.

BROCKETT, R. W. (1977). Convergence of Volterra series of infinite intervals and bilinear approximations. In *Non-linear Systems and Applications* (V. Lakshmikanthan, ed.), pp. 39–46. New York: Academic Press.

BULMER, M. G. (1978). The statistical analysis of the ten year cycle. In *Time Series and Ecological Processes* (H. H. Shugart, ed.), pp. 141–153. Philadelphia: SIAM. (SIAM-SIMS Conference No. 5.)

CHIEN, M. J. and CHAN, L. (1979). Non-linear input-output model with piecewise affine coefficients. *J. Econ. Theory*, **21**, 389–410.

COX, D. R. (1977). Discussion on papers by Campbell *et al. J. R. Statist. Soc.* A, **140**, 453–454.

GODOLPHIN, E. J. (1975). A direct basic form for predictors of autoregressive integrated moving average processes. *Biometrika*, **62**, 483–496.

GRANGER, C. W. J. (1978). New classes of time series models. *The Statistician*, **27**, 237–253.

GRANGER, C. W. J. and ANDERSEN, A. P. (1978). *Introduction to Bilinear Time Series Models.* Gottingen: Vandenhoeck and Ruprecht.

HAGGAN, V. and OZAKI, T. (1980b). Amplitude-dependent exponential AR model fitting for non-linear random vibrations. In *Time Series* (O. D. Anderson, ed.), pp. 57–71. Amsterdam: North Holland.

HANNAN, E. J. (1970). *Multiple Time Series.* New York and London: Wiley.

HIPEL, K. W. and MCLEOD, A. I. (1978). Preservation of the rescaled adjusted range. 2, simulation studies using Box–Jenkins models. *Water Resources Research*, **14**(3), 509–516.

HIPEL, K. W. and MCLEOD, A. I. (1979). Perspectives in stochastic hydrology. In *Time Series* (O. D. Anderson, ed.), pp. 73–102. Amsterdam: North-Holland.

KIBLER, D. F. and HIPEL, K. W. (1979). Surface water hydrology. *Rev. Geophys. and Space Phys.*, **17**(6), 1186–1209.

LABYS, W. C. and GRANGER, C. W. J. (1970). *Speculation, Hedging and Commodity Price Forecasts.* D. C. Heath and Co., Lexington, Mass.

MCLEOD, A. I., HIPEL, K. W. and LENNOX, W. C. (1977). Advances in Box–Jenkins modelling, 2. Applications. *Water Resources Research*, **13**(3), 577–586.

NAKAMURA, H. and AKAIKE, H. (1979). Use of statistical identification for optimal control of a supercritical thermal power plant. In *Identification and System Parameter Estimation* (R. Isermann, ed.). Oxford: Pergamon.

O'CONNELL, P. E. and JONES, D. A. (1979). Some experience with the development of models for the stochastic simulation of daily flows. In *Inputs for Risk Analysis in Water Systems* (E. A. McBean, K. W. Hipel and T. E. Unny, eds), pp. 287–314. Fort Collins, Colorado: Water Resources Publications.

OTOMO, T., NAKAGAWA, T. and AKAIKE, H. (1972). Statistical approach to computer control of cement rotary kilns. *Automatica*, **8**, 35–48.

OZAKI, T. and ODA, H. (1977). Non-linear time series model identifications by Akaike's Information Criterion. In *Information and Systems* (B. Dubuisson ed.). Oxford: Pergamon.

OZAKI, T. (1979a). Non-linear threshold AR models for non-linear random vibrations, Research Memo. No. 157, Institute of Statistical Mathematics, Tokyo. (To appear in *J. of Appl. Prob.*)

—— (1979b). Statistical analysis of Duffing process through non-linear time series models, Research Memo. No. 151, Institute of Statistical Mathematics, Tokyo. (To appear in *J. of Appl. Mechanics.*)

—— (1979c). Statistical analysis of perturbed limit cycle processes through non-linear time series models, Research Memo. No. 158, Institute of Statistical Mathematics, Tokyo. (Also submitted for publication.)

PRIESTLEY, M. B. (1978). Non-linear models in time series analysis. *The Statistician*, **27**, 159–176
—— (1979). On a general class of non-linear time series models. *Bull. Int. Statist.*, **42**, to appear.
ROSENBLATT, M. (1979). *Markov Processes: Structural and Asymptotic Behaviour.* Berlin: Springer -Verlag.
—— (1977). Linearity and non-linearity in time series: prediction. *Bull. Int. Statist. Inst.*, **42**, to appear.
STONE, C. J. (1977). Consistent non-parametric regression (with Discussion). *Ann. Statist.*, **5**, 595–645.
SUBBA RAO, T. (1979). On the theory of bilinear time series models—II. Technical Report No. 121, Department of Mathematics, UMIST.
SUBBA RAO, T. and GABR, M. M. (1981). A test for linearity of stationary time series. Submitted to *Appl. Statist.*
TONG, H. (1980b). Catastrophe theory and threshold autoregressive modelling. Technical Report No. 125, Dept. of Mathematics, UMIST. (Abstract in *Resumé des Communications*, Journées de Statistique, Toulouse, 19–22 May 1980.)
TONG, H. and PEMBERTON, J. (1980). On stability and limit cycles of non-linear autoregression in discrete time. *Cahiers du CERO*, Bruxelles, **22**, No. 2, 137–148.
YAKOWITZ, S. J. (1973). A stochastic model for daily river flows in an arid region. *Water Resources Research*, **9**(5), 1271–1285.
—— (1979). A non-parametric Markov model for daily river flow. *Water Resources Research*, **15**(5), 1035–1043.
ZEEMAN, E. C. (1977). *Catastrophe Theory: Selected Papers 1927–1977.* Mass.: Addison-Wesley.

Review of the Paper by Howell Tong and K. S. Lim: "Threshold Autoregression, Limit Cycles and Cyclical Data (with Discussion)"

H. Z. AN

Academy of Mathematics & Systems Science
Chinese Academy of Sciences, Beijing 100080, China
E-mail: ahz@amt.ac.cn

It gives me great pleasure to write this review on Howell's paper, "Threshold autoregression, limit cycles and cyclical data--with discussion" (1980, with K.S.Lim). I think this paper is one of the most important papers by Howell, in which he has made substantial contributions to nonlinear time series analysis. His novel ideas have had great impacts on the subsequent development in this field.

Two decades have passed since this paper was published; the benefits of hindsight are such that reviews on it now must be quite different from first impressions in the days in the 1980s. Especially for me, before I read this paper I attended a seminar course held in 1981 in Beijing, in which Howell spent one month introducing his audience to the TAR (Threshold Autoregressive) model. After this initiation, I not only read the above seminal paper but also followed his many papers and books. Since my early exposure to the threshold models, I have been working mainly on nonlinear time series analysis, especially the TAR models. There is no doubt that Howell's work, especially the above paper, which contained many original ideas, has had the greatest influence on my research. Now I would like to review this paper based on my understanding and perception.

1. Model Switching

Starting with the linear autoregressive (AR) models first introduced by Yule (1927), it is then natural to write a nonlinear autoregressive model in the following form

$$x_t = \varphi(x_{t-1}, x_{t-2}, \ldots, x_{t-p}) + \varepsilon_t, \tag{1}$$

where $\varphi(\ldots)$ is a function from R^p to R^1, $\{\varepsilon_t\}$ is an i.i.d. series with zero mean, p is an integer, the order of the model. However, for one reason or another we may be interested in some special cases of the function $\varphi(\ldots)$ for the study and use of model (1). In fact, in the past two decades or so we have seen several cases of nonlinear time series in the literature, for example, the fractional autoregressive models (Jones, 1965), the random coefficient autoregressive models (e.g. Andel, 1976), the class of the bilinear models (e.g. Granger and Andersen, 1978), and others. Each one has its own merits and limitations. It is worth mentioning that the TAR model enjoys the property of being capable of switching between several linear sub-models, which has much practical value in many fields, including for example hydrology, atmospheric science, economy and finance. In fact, this merit has been widely recognized in the literature, the latest including Wu and Chen (2007).

2. Density of the Threshold Autoregressive Functions

As mentioned above, the TAR model is only a special case of the general nonlinear autoregressive models. However, piecewise linear (TAR) functions are dense in the class of the measurable functions from R^p to R^1. This property may offer potential values for theoretical exploration and practical applications. The following experience impressed me tremendously. In 1990, I studied several papers on the test of linearity for time series in the literature, and found that most of the testing methods were based on testing linearity against some specific class of nonlinear models. That is, in these papers, their alternative hypothesis H_1 referred to some specific class of nonlinear models. The authors of these papers suggested that their methods could be used to test linearity approximately against the general case of nonlinear models, and showed some simulation results in their papers. Chan and Tong (1986) also proposed their testing method by taking H_1 as TAR models. Their test turned out to enjoy generally better properties than other test developed up to then by comparison.

3. Partly Linear Parametric Form

Although the full non-parametric form of model (1) is more general, it is typically too difficult to fit with real data because of computational problems, as well as the curse of dimensionality. Consequently the non-parametric model defined by (1) is still not widely used in practice. Even finite parametric versions of model (1) are not always easy to fit, for example, the fractional autoregressive models. Against this background, the TAR models have shown another remarkable advantage, namely partly linear parametric form. As we know that, given known threshold and delay that, with no loss of generality, are taken to be respectively 0 and 1 below, the autoregressive function of the TAR model is linear in the parameters. Thus the procedure for fitting TAR models with known threshold and delay is much the same as fitting linear AR models. For example, let us consider the following simple TAR model

$$x_t = \begin{cases} \alpha_0 + \alpha_1 x_{t-1} + \varepsilon_t, & if \quad x_{t-1} < 0, \\ \beta_0 + \beta_1 x_{t-1} + \varepsilon_t, & if \quad x_{t-1} \geq 0. \end{cases} \tag{2}$$

It could be rewritten as the following form

$$\begin{aligned} x_t &= \alpha_0 I(x_{t-1} < 0) + \alpha_1 x_{t-1} I(x_{t-1} < 0) + \beta_0 I(x_{t-1} \geq 0) + \beta_1 x_{t-1} I(x_{t-1} \geq 0) + \varepsilon_t \\ &= a_1 f_1(x_{t-1}) + a_2 f_2(x_{t-1}) + a_3 f_3(x_{t-1}) + a_4 f_4(x_{t-1}) + \varepsilon_t, \end{aligned} \tag{3}$$

where $I(...)$ denotes the indicator function, the parameter vector

$$a = (a_1, a_2, a_3, a_4)^T = (\alpha_0, \alpha_1, \beta_0, \beta_1)^T,$$

and the known functions

$$f_1(x_{t-1}) = I(x_{t-1} < 0),\ f_2(x_{t-1}) = x_{t-1} I(x_{t-1} < 0),\ f_3(x_{t-1}) = I(x_{t-1} \geq 0),\ f_4(x_{t-1}) = x_{t-1} I(x_{t-1} \geq 0).$$

Model (3) is in a typical linear regression form. For a given data set it is easy to fit such a model. Of course the theory of the statistical inference for the case of unknown threshold and delay is far from being straightforward. In the general case of unknown threshold and delay, the indicator function $I(x_{t-1} \geq 0)$ has to be replaced by $I(x_{t-d} \geq r)$ where r is the threshold and d is the delay. In this case the threshold parameter and the delay parameter are not in a linear form.

4. SETAR Models and Conditional Variance

To continue to discuss the above example, we set

$$\varphi(x_{t-1}) = a_1 f_1(x_{t-1}) + a_2 f_2(x_{t-1}) + a_3 f_3(x_{t-1}) + a_4 f_4(x_{t-1}),$$

so model (2) takes the form of equation (1). In fact, model (1) is known as an additive-noise model in the literature. However, Tong and Lim (1980) proposed the following form

$$x_t = \begin{cases} \alpha_0 + \alpha_1 x_{t-1} + \varepsilon_{1t}, & if \quad x_{t-1} < 0 \\ \beta_0 + \beta_1 x_{t-1} + \varepsilon_{2t}, & if \quad x_{t-1} \geq 0 \end{cases}, \tag{4}$$

where $\{\varepsilon_{1t}\}$ and $\{\varepsilon_{2t}\}$ are two independent i.i.d. noises with zero means. By the same arguments as used in model (3), model (4) can be written as

$$\begin{aligned} x_t &= \varphi(x_{t-1}) + \varepsilon_{1t} I(x_{t-1} < 0) + \varepsilon_{2t} I(x_{t-1} \geq 0) \\ &= \varphi(x_{t-1}) + e_t \end{aligned} \tag{5}$$

and

$$e_t = \varepsilon_{1t} I(x_{t-1} < 0) + \varepsilon_{2t} I(x_{t-1} \geq 0). \tag{6}$$

Although model (5) takes the form of model (1), the noise series $\{e_t\}$ is not i.i.d. series, but a martingale difference series (e.g. Tong, 1990). Because model (5) is a dynamical system (e.g. Tong, 1990), Tong referred to the above TAR model as a self-exciting threshold autoregressive (SETAR) models in Tong and Lim (1980). In fact, system (5) is driven by two independent noises, which is different from system (3). On the other hand, if a stationary time series $\{x_t\}$ satisfies model (5), the conditional variance of the series $\{x_t\}$ is given by

$$Var\{x_t \mid x_{t-1}, x_{t-2}, ...\} = \sigma_1^2 I(x_{t-1} < 0) + \sigma_2^2 I(x_{t-1} \geq 0), \tag{7}$$

where $\sigma_1^2 = E\varepsilon_{1t}^2$, $\sigma_2^2 = E\varepsilon_{2t}^2$. Letting

$$S^2(x_{t-1}) = \sigma_1^2 I(x_{t-1} < 0) + \sigma_2^2 I(x_{t-1} \geq 0), \tag{8}$$

we may define another model by the following equation

$$x_t = \varphi(x_{t-1}) + S(x_{t-1})\varepsilon_t, \tag{9}$$

where $\{\varepsilon_t\}$ is same as in model (1). In particular, if $\{\varepsilon_{1t}\}$, $\{\varepsilon_{2t}\}$ and $\{\varepsilon_t\}$ are i.i.d. normal noises with zero means, the structure of model (9) is the same as model (5). The generalization of model (9) is called the autoregressive model with changing conditional variances (e.g. Chen and Chen, 2000). Changing conditional variances is another stylistic property of the TAR models, which is similar to the ARCH models proposed by Engle (1982).

5. Ergodicity and Stationary Solutions

For the TAR models as well as the general class of nonlinear time series models like model (1), we may be interested in conditions ensuring the stationarity and the ergodicity of the model, because this kind of results is of importance in the study of the statistical inference of the model. Tong has mentioned Jones' (1978) paper in Section 5 of the seminal paper, although no further detail was given beyond referring to Tweedie (1975) concerning the ergodicity of a Markov chain. In fact, in the last two decades many authors have worked on the stationarity and the ergodicity of nonlinear time series, for example, Chan and Tong (1985), Tjostheim (1990), An and Huang (1996), Chen and Chen (2000) and many others. Many results are available in the literature, and remarkably Tweedie (1975) has, as foreseen by Howell, played an important role in almost every one of them.

References

1. An, H.Z. and Huang, F.C. (1996). The geometric ergodicity of nonlinear autoregressive models, *Statistica Sinica,* 6: 943-956.
2. Andel, J. (1976).Autoregressive series with random parameters. *Math. Op. Stat.,* 7, 735-741.
3. Chen, M. and Chen, G. (2000). Geometric ergodicity of nonlinear autoregressive model with changing conditional variances. *The Canadian J. Statist.,* 28(3), 605-613.
4. Chan, K.S. and Tong, H. (1985). On the use of the deterministic Lyapunov function for the ergodicity of stochastic difference equations. *Adv. Appl. Probab.,* 17, 666-678.
5. Chan, K.S. and Tong, H. (1986). On test for nonlinearity in time series analysis. *J. Forecasting* 5, 217-228.
6. Engle, R.F. (1982). Autoregressive conditional heteroscedasticity with estimates of the variance of U.K. inflation. *Econometrica*, 50, 987-1008.
7. Granger, C.W.J. and Andersen, A.P. (1978). *An introduction to bilinear time series.* Academic Press, New York.
8. Jones, D.A. (1978). Non-linear autoregressive processes. Proc. *Roy. Soc. London. A,* 360, 71-95.
9. Jones, R.H. (1965).An experiment in non-linear prediction. *J.Appl.Meterol.,* 4, 701-705.
10. Tjøstheim, D. (1990). Nonlinear time series and Markov chains, Adv. *Appl. Probab.,* 22, 587-611.
11. Tong, H. (1990). *Non-linear Time Series.* Oxford Science Publications.

12. Tong, H. and Lim, K.S. (1980).: Threshold autoregression, limit cycles and cyclical data--with discussion. *J. Roy. Statist. Soc., B*, vol. 42, 245-292.
13. Tweedie, R.L. (1975). Sufficient conditions for ergodicity and recurrence of Markov chain on a general state space. *Stochastic Processes Appl.,* 3, 385-403.
14. Wu, S. and Chen, R. (2007). Threshold variable determination and threshold variable driven switching autoregressive models, *Statistica Sinica,* 17(1), 242-264.
15. Yule, G.U. (1927). On a method of investigating periodicities in disturbed series with special reference to Wolfer's sunspot numbers. *Philos. Trans. R. Soc.,* **A226**, 267-298.

Reflections on Threshold Autoregression

PETER J. BROCKWELL

Colorado State University, Statistics Department
Fort Collins, Colorado 80523-1877, USA
E-mail: pjbrock@stat.colostate.edu

In 1980, the discussion paper, "Threshold Autoregression, Limit Cycles and Cyclical Data" (Tong and Lim (1980)) was presented to the Royal Statistical Society. In this article we review the contents of that paper and the impact of the paper on the study of nonlinear time series in the subsequent twenty nine years.

For the past thirty years the subject of time series analysis has been evolving at a remarkable rate, driven by the need to account for the complex behaviour of many of the time series encountered in practice and to develop forecasting methods superior to the best linear forecasts which dominated time series analysis for many years.

According to the Wold decomposition, every weakly stationary time series $\{X_n, n = 0, \pm 1, \ldots\}$ with mean μ and such that $\{X_n - \mu\}$ has zero deterministic component, can be represented as

$$X_n = \mu + \sum_{j=0}^{\infty} \psi_j \varepsilon_{n-j}, \tag{1}$$

where $\{\varepsilon_n\}$ is a sequence of uncorrelated, zero-mean random variables with constant variance such that ε_n is a function of $\{X_t, t \leq n\}$ and $\sum \psi_j^2 < \infty$. If $\{X_n\}$ is a Gaussian series then the sequence $\{\varepsilon_n\}$ is also Gaussian and hence iid (independent and identically distributed). More generally the process $\{X_n\}$ is said to be *linear* if it has a representation of the form (0.1) with $\{\varepsilon_n\}$ iid but not necessarily Gaussian. In particular, the class of causal invertible ARMA processes $\{X_n\}$ driven by iid noise $\{\varepsilon_n\}$ can be written in the form (0.1) and constitutes a parsimoniously parameterized subfamily of the class of linear models. For the fitting of ARMA models to observed data, a large body of techniques for model-selection, estimation and forecasting has been developed over the years and applied successfully to a range of observed time series.

However in spite of the wide applicability and utility of linear time series models, by the nineteen seventies it had long been clear that there were many features of empirical time series in ecology, hydrology, finance and other fields which could not possibly be explained within the established framework of linear models. Asymmetry in the rates of increase and decrease of observed sample-paths, bursts of high variance (or volatility), and the apparent existence of limit cycles were just a few of the observed phenomena which indicated the need to move beyond linear models.

Such phenomena, together with others such as amplitude-frequency dependency and jump resonance, were well understood in the theory of nonlinear vibrations and accounted for by a variety of well-known non-linear differential equation models. It was therefore a natural step to search for a convenient family of stochastic models in discrete time to account for the corresponding behaviour of observed time series.

A major problem in the selection of an appropriate nonlinear model for a given time series is the vast array of possibilities. A very large family of non-linear models is obtained from (0.1) by replacing the sum on the right by an arbitrary nonlinear measurable function of $\{\varepsilon_t, t \leq n\}$ to obtain

$$X_n = f(\varepsilon_n, \varepsilon_{n-1}, \ldots). \tag{2}$$

If f is sufficiently smooth, then we can also replace the right-hand side of (0.2) by a mean-square convergent *Volterra expansion*,

$$X_n = \mu + \sum_{j\geq 0} \psi_j \varepsilon_{n-j} + \sum_{j,k\geq 0} \psi_{jk} \varepsilon_{n-j}\varepsilon_{n-k} + \sum_{j,k,l\geq 0} \psi_{jkl} \varepsilon_{n-j}\varepsilon_{n-k}\varepsilon_{n-l} + \ldots. \tag{3}$$

Although the Volterra expansion provided a convenient sequence of polynomial approximations to a very general class of nonlinear series, for the purpose of selecting and fitting nonlinear models in practice, there remained a significant challenge, namely to find a parametric family of nonlinear models which could play a role somewhat analogous to that of ARMA processes in linear time series analysis. A family of models was required which is capable not only of generating sample-paths with the desired nonlinear characteristics, but for which statistical identification and estimation is computationally feasible and for which prediction is both feasible and superior to linear prediction. In addition one might hope to obtain from the fitted model some insight into the underlying mechanism generating the data. The family should possess a degree of generality in application, with the potential for extension to the analysis of multivariate data. These were the stated goals of the paper of Tong and Lim (1980), henceforth referred to as TL. This was the first systematic statistical investigation of the properties and applications of threshold autoregressive models proposed in Tong (1978).

The study of nonlinear time series and the related study of their moments of order higher than two, has a long history (see e.g. Wiener (1958), Shiryaev (1960), Brillinger and Rosenblatt (1967)), and there have been numerous applications of time series models tailored to deal with specific sources of non-linearity as, for example, in Otomo et al. (1972), Nakamura and Akaike (1979) and O'Connell and Jones (1979). However the systematic study of parametric families of nonlinear models did not gather momentum until the late nineteen seventies, when a variety of useful and relatively tractable families began to appear, all motivated by essentially the same goals as those outlined in the previous paragraph. Besides the threshold model, other examples were the bilinear model of Granger and Anderson (1978) (see also Subba-Rao and Gabr (1984)), the exponential autoregressive model of Ozaki and Oda (1978), the state-dependent model of Priestley (1980), the random-coefficient autoregression model of Nicholls and Quinn (1982), and the ARCH and GARCH models of Engle (1982) and Bollerslev (1986) respectively. An excellent discussion of these models can be found in Chapter 3 of Tong (1990). Although the ARCH and GARCH models, specifically designed to capture the volatility clustering observed widely in financial data, have had the greatest impact in econometrics, the threshold models have been highly influential in a broader range of applications.

The *general threshold autoregression* (TAR) was defined in TL to be a sequence $\{\mathbf{X}_n; J_n\}$, where J_n is a random variable (measurable with respect to the σ-algebra of events generated by the information available up to time $n-1$) taking values in some finite set $\{1, 2, \ldots, m\}$ and $\{\mathbf{X}_n\}$ is a k-dimensional time series satisfying,

$$\mathbf{X}_n = \mathbf{B}^{(J_n)}\mathbf{X}_n + \mathbf{A}^{(J_n)}\mathbf{X}_{n-1} + \boldsymbol{\varepsilon}_n^{(J_n)} + \mathbf{C}^{(J_n)}, \tag{4}$$

where, for each fixed j, $\mathbf{A}^{(j)}$ and $\mathbf{B}^{(j)}$ are $k \times k$ (non-random) matrix coefficients, $\mathbf{C}^{(j)}$ is a $k \times 1$ vector of constants, and $\boldsymbol{\varepsilon}_n^{(j)}$ is a k-dimensional strict white noise sequence of independent random vectors with a diagonal covariance matrix. It was also assumed that $\{\boldsymbol{\varepsilon}_n^{(j)}\}$ and $\{\boldsymbol{\varepsilon}_n^{(j')}\}$ are independent for $j \neq j'$.

Within this general framework, three processes of particular interest were identified, the *self-exciting threshold autoregressive process* SETAR, the *open loop threshold autoregressive system*, TARSO, and the *closed loop autoregressive system*, TARSC.

If $\{R^{(j)}, j = 1, \ldots, m\}$ is a partition of $\mathbb{R}^k$ and if, for each $j \in \{1, \ldots, m\}$,

$$\mathbf{A}^{(j)} = \begin{bmatrix} a_1^{(j)} & a_2^{(j)} & \cdots & a_{k-1}^{(j)} & a_k^{(j)} \\ 1 & 0 & \cdots & 0 & 0 \\ 0 & 1 & \ddots & 0 & 0 \\ \vdots & \vdots & \ddots & \ddots & \vdots \\ 0 & 0 & \cdots & 1 & 0 \end{bmatrix}, \quad \boldsymbol{\varepsilon}_n^{(j)} = \begin{bmatrix} \varepsilon_n^{(j)} \\ 0 \\ \vdots \\ 0 \\ 0 \end{bmatrix}, \quad \mathbf{C}^{(j)} = \begin{bmatrix} a_0^{(j)} \\ 0 \\ \vdots \\ 0 \\ 0 \end{bmatrix},$$

$\mathbf{B}^{(j)} = \mathbf{0}$, $\mathbf{X}_n = (X_n, X_{n-1}, \ldots, X_{n-k+1})'$, $J_n = \sum_{j=1}^m j I_{R^{(j)}}(\mathbf{X}_{n-1})$ and $\{\varepsilon_n^{(j)}\}$ is a strict white noise sequence, then the equations (0.4) are equivalent to the system of m linear autoregressions,

$$X_n = a_0^{(j)} + \sum_{i=1}^{k} a_i^{(j)} X_{n-i} + \varepsilon_n^{(j)} \;\text{ if } J_n = j;\; j = 1, \ldots, m, \tag{5}$$

and, since J_n is a function of $\mathbf{X}_{n-1}$, the univariate process $\{X_n\}$ is said to be a *self-exciting* process. If each indicator function $I_{R^{(j)}}(\mathbf{x})$ has the form $I_{R_j}(x_d)$ where x_d is the dth component of $\mathbf{x}$, and if $R_j = (r_{j-1}, r_j], j = 1, \ldots m-1$, and $R_m = (r_{m-1}, \infty)$, where $-\infty = r_0 < r_1 < \ldots < r_{m-1} < \infty$, then the relevant equation for X_n in (0.5) depends only on the value of X_{n-d} and the process $\{X_n\}$ is called a SETAR process with *delay* d and *thresholds*, $r_1, \ldots, r_{m-1}$.

A TARSO system (X_n, Y_n) consists of an observable output series $\{X_n\}$ and an observable input series $\{Y_n\}$ related by equations of the form,

$$X_n = a_0^{(j)} + \sum_{i=1}^{k_j} a_i^{(j)} X_{n-j} + \sum_{i=0}^{k_j'} b_i^{(j)} Y_{n-i} + \varepsilon_n^{(j)},\; Y_{n-d} \in R_j;\; j = 1, \ldots, m, \tag{6}$$

where each sequence $\{\varepsilon_n^{(j)}\}$ is strict white noise with zero mean and finite variance, and the sequences $\{\{Y_n\}, \{\varepsilon_n^{(j)}\}, j = 1, \ldots, m\}$, are independent. As before $\{R_j, j = 1, \ldots m\}$ is a partition of $\mathbb{R}$ into intervals.

If (X_n, Y_n) and (Y_n, X_n) both satisfy equations of the form (0.6) and if all of the white noise sequences are independent, then $\{X_n, Y_n\}$ is called a *closed-loop threshold autoregressive system*, or TARSC,

In TL the authors demonstrated by means of examples the ability of the TAR family to exhibit jump resonance, amplitude-frequency dependency, limit cycles, subharmonics and superharmonics. They also proposed estimation and identification procedures based primarily on Gaussian likelihood and the AIC criterion and used them to fit SETAR models to the logged Canadian lynx series and the annual sunspot series, a TARSC model to the differenced mink-muskrat series (after taking logarithms), and a TARSO model to the logged Kanna daily riverflow and rainfall data. The models fitted to the first three of these models exhibited limit cycles and the first and last models showed considerable improvement in forecast mean squared errors over corresponding linear models. Possible physical interpretations of the models were also given and many indications of directions for future research, including the development of generalizations to threshold ARMA and bilinear models, investigation of the statistical properties of the parameter estimators and the development of multi-step optimal forecasting methods.

In the years since its publication TL has been particularly influential in the promotion of research, not only into the questions raised by the paper itself, many of which were resolved in Tong (1983), but to the development of nonlinear time series modelling generally. The *threshold principle*, or the idea in (0.5) of partitioning the range $\mathbb{R}^k$ of the state vector $\mathbf{X}_{n-1}$ into sets, on each of which $\mathbf{X}_n$ is determined by a linear autoregression, is a very natural way to approximate an autoregressive relationship which is nonlinear but approximately linear on the subsets of the partition. The idea is closely related to the idea of approximating non-stationary time series by series which are stationary over small time intervals (see Ozaki and Tong (1975) and Priestley (1988)). Because such a piecewise linear approximation allows the approximation of such a large class of nonlinear models, the SETAR, TARSO and TARSC models have been used with great success in a many fields of application. These include epidemiology, hydrology, astrophysics, oceanography, population dynamics and finance. For details of some of the financial applications see Tsay (2005).

The principle of thresholding lends itself to broad modification and generalization, giving rise to many more processes of interest than the three specific models emphasized in TL. For example the partition of $\mathbb{R}^k$ need not depend only on the dth component of the state vector as in the SETAR model (see e.g. Boucher and Cline (2007)) and the discontinuities in the conditional mean which are implicit in the SETAR model can be avoided as in the STAR model of Chan and Tong (1986). The threshold GARCH model of Glosten et al. (1993) is another application of thresholding, designed to account for leverage effects in financial data. Ccntinuous-time SETARMA models have been studied by Stramer et al. (1996) and applied to the analysis of financial data in Brockwell and Williams (1997). Further extensions and generalizations of the thresholding principle now abound in the literature.

Associated with threshold models, and with the other non-linear models mentioned earlier, are many interesting and fundamental probabilistic questions related to ergodicity and the existence of stationary versions. For the self-exciting threshold AR(1) process with delay 1, necessary and sufficient conditions for the existence of a stationary version and properties of the least squares estimators of the coefficients when the thresholds are assumed known were established by Chan et al. (1985). For the same process with a single threshold and arbitrary delay, necessary and sufficient conditions for ergodicity were obtained by Chen and Tsay (1991). In general however such questions are far from fully resolved and remain active areas of research. The selection and estimation of threshold models for given data sets also remains a challenging problem. Systematic approaches are proposed in the papers of Tsay (1989, 1998) and an overview of more recent developments is contained in the book of Fan and Yao (2005).

The paper TL played an important role in drawing the attention of probabilists and statisticians to the need for and the potential benefits to be derived from nonlinear models. It also led to the two books Tong (1983, 1990), the first of which contains computer programs for implementing the model-fitting described in the paper, and the second of which provides an overview of nonlinear time series modelling, highlighting the use of the dynamical systems approach to the analysis of these models and the use of Markovian methods and Lyapunov functions for their analysis. The dynamical system framework has now become fundamental in tackling many of the problems associated with non-Gaussian and non-linear models in time series analysis. In conjunction with Markov chain Monte Carlo and particle filtering methods, it has permitted in principle the analysis of extremely complex non-linear problems. The development of computationally efficient and rapidly convergent algorithms however is still a challenging area. The theoretical analysis of ergodicity for many important classes of widely used non-linear models is also an area in which much remains to be done.

Acknowledgment I am indebted to the National Science Foundation for support of this work under the grant DMS-0744058.

References

1. T. Bollerslev, Generalized autoregressive conditional heteroskedasticity, *J. of Econometrics.* **31**, 307-327, (1986).
2. T. R. Boucher and D. B. H. Cline, Stability of cyclic threshold and threshold-like autoregressive time series models, *Stat. Sinica.* **15** (1), 43-62, (2001).
3. D. R. Brillinger and M. Rosenblatt, Asymptotic theory of kth order spectra, In *Spectral Analysis of Time Series.* Ed. B. Harris, 153-188. (Wiley, New York, 1967).
4. P. J. Brockwell and R. J. Williams, On the existence and application of continuous-time autoregressions of order two, *Adv. Appl. Prob..* **29**, 205-227, (1997).
5. K. S. Chan, Joseph D, Petrucelli, H. Tong and Samuel Woolford, A multiple-threshold AR(1) model, *J. Appl. Probability.* **22**, 267-279, (1985).
6. K. S. Chan and H. Tong, On estimating thresholds in autoregressive models, *J. Time Series Analysis.* **7**, 179-190, (1986).
7. K. S. Chan and H. Tong, *Chaos: A Statistical Perspective.* (Springer-Verlag, New York, (2001)).
8. R. Chen and R. S. Tsay, On the ergodicity of TAR(1) processes. *Ann. Appl. Probability.* **1**, 813-634, (1991).
9. R. F. Engle, Autoregressive conditional heteroscedasticity with estimates of the variance of United Kingdom inflations, *Econometrica.* **50**, 987-1007, (1982).
10. J. Fan and Q. Yao, *Nonlinear Time Series: Nonparametric and Parametric Methods.* (Springer-Verlag, New York, (2005)).
11. L. R. Glosten, R. Jagannathan and D. E. Runkle, On the relation between the expected value and the volatility of nominal excess return on stocks, *J. Finance.* **48**, 1779-1801, (1993).
12. C. W. J. Granger and A. P. Andersen, *An Introduction to Bilinear Time Series Models.* (Vanderhoeck and Reprecht, Göttingen, 1978).
13. H. Nakamura and H. Akaike, Use of statistical identification for optimal control of a supercritical thermal power plant. In *Identification and System Parameter Estimation.* Ed. R. Isermann, (Pergamon, Oxford, 1979).
14. D. F. Nicholls and B. G. Quinn, *Random Coefficient Autoregressive Models: An Introduction.* (Springer Lecture Notes in Statistics, **11**, (1982)).
15. P. E. O'Connell and D. A. Jones, Some experience with the development of models for the stochastic simulation of daily flows. In *Inputs for Risk Analysis in Water Systems.* Eds. E. A.

McBean, K. W. Hipel and T. E. Unny, 287-314, (Colorado Water resources Publications, Fort Collins, Colorado, 1979).

16. T. Otomo, T. Nakagawa and H. Akaike, Statistical approach to computer control of cement rotary kilns, *Automatica.* **8**, 35-48, (1972).
17. T. Ozaki and H. Oda, Nonlinear time series model identification by Akaike's Information Criterion. In *Information and Systems.* Ed. B. Dubuisson, 83-91. (Pergamon, Oxford, 1978).
18. T. Ozaki and H. Tong, On the fitting of non-stationary autoregressive models in time series analysis. In *Proceedings of the 8th Hawaii International Conference on System Sciences.* 224-226. (Western Periodical Co., Hawaii, 1975).
19. M. B. Priestley, State-dependent models: a general approach to nonlinear time series analysis, *J. Time Series Analysis.* **1**, 47-71, (1980).
20. M. B. Priestley, *Nonlinear and Nonstationary Time Series Analysis.* (Academic Press, London, 1988). *J. Finance.* **48**, 1779-1801, (1993).
21. T. Subba-Rao and M. M. Gabr, *An Introduction to Bispectral Analysis and Bilinear Time Series Models*, (Springer Lecture Notes in Statistics, **24**, (1984)).
22. A. N. Shiryaev, Some problems in the spectral theory of higher-order moments I. *Theory Prob. Appl..* **5**, 265-284, (1960).
23. H. Tong, On a threshold model. In *Pattern Recognition and Signal Processing.* Ed. C. H. Chen (Sijthoff and Noordhoff, The Netherlands, 1978). 224-226. (Western Periodical Co., Hawaii, 1975).
24. H. Tong and K. S. Lim, Threshold autoregression, limit cycles and cyclical data (with discussion), *J. Roy. Stat. Soc. B.* **42** (3), 245-292, (1980).
25. H. Tong, *Threshold Models in Nonlinear Time Series Analysis.* (Springer Lecture Notes in Statistics, **21**, (1983)).
26. H. Tong, *Nonlinear Time Series: A Dynamical Systems Approach.* (Oxford University Press, Oxford, (1990)).
27. R. S. Tsay, *Analysis of Financial Time Series*, 2nd edition. (Wiley-Interscience, Hoboken, New Jersey, (2005)).
28. R. S. Tsay, Testing and modeling threshold autoregressive processes, *J. American Statistical Association.* **84**, 231-240, (1989).
29. R. S. Tsay, Testing and modeling multivariate threshold models, *J. American Statistical Association.* **93**, 1188-1202, (1998).
30. N. Wiener, *Nonlinear Problems in Random Theory.* (MIT Press, Cambridge, Massachusetts, 1958).

Threshold Autoregression: Its Seed Corn, Meeting the Market Test, and Two of Its Spillover Effects

THOMAS B. FOMBY

Department of Economics, Southern Methodist University
3300 Dyer Street,301M, Umphrey Lee Center, Dallas, TX 75275-0496, USA
E-mail: tfomby@smu.edu

The Tong and Lim (1980) paper is shown to be a seminal paper in the statistical literature by examining the depth, breadth, and durability of its citation and subject area counts as tabulated from the ISI Web of Knowledge citation database. Two progenitor threshold models related to the SETAR model, Threshold Cointegration and Threshold GARCH, are presented, along with their citation and subject area counts, to illustrate two of the spillover effects generated by the Tong and Lim paper.

1. Introduction

"The most important function of a bibliographic entry is to help the reader obtain a copy of the cited work"

Daniel J. Bernstein, American Mathematician and Computer Scientist[1]

Leave it to an expert to define basic concepts in the field of Library Science. Take, for example, the definition of a "seminal" paper. Consider the following two definitions provided by Library Scientists, the first being quite succinct while the second is more expansive:

"A seminal paper is a kind of 'classic' in a broad meaning of the term. It is a paper which has served as a model for other papers, which first has presented an influential view of theory."
– Professor Birger Hjorland, Professor of Library and Information Science, Royal School of Library and Information Science, Copenhagen, Denmark. (2006): http://www.db.dk/bh.

"The model begins with a new theory published in a research paper. If the scholarly community comes to accept the validity of the new theory this paper is considered a seminal paper. This seminal paper influences the scholarly community's thinking and ultimately, the body of knowledge. The seminal paper stimulates the writing of other scholarly papers. Last, the novel thinking, expressed in the seminal paper and subsequent scholarly papers, is organized into new patterns of thinking which can be recorded in subject heading schemes and then

[1] http://www.brainyquote.com/quotes/quotes/d/danieljbe391195.html.

applied to the subject indexing of newly published scholarly papers." – L.P. Lussky (2004, 4-5).

I will argue here that, as it relates to literature citations, a seminal paper is one that has **depth**, **breadth**, **durability**, and causes **spillover effects** in terms of giving rise to new research models and methods in many disciplines. The paper I am reviewing here, Tong and Lim (1980), "Threshold Autoregression, Limit Cycles and Cyclical Data," Journal of the Royal Statistical Society, Series B, 245 – 292, is one such paper.[2] As noted above, seminal papers often lead to new patterns of thinking and, for Library Scientists, new subject heading schemes and new subject indexes. Alas, the Tong and Lim paper caused librarians more work: "Threshold Autoregression" for a new subject index and a proliferation of related subject indexes starting with the adjective "threshold"! Work and more work for the Library Scientists! Although the Library Scientists might be complaining, I am not because I, along with many others, have definitely benefited from this new subject index. You will see this, at least in part, in further discussion below.

The subheading to the title of this paper refers to the **seed corn**, **market test**, and **spillover effects** associated with the Tong and Lim (1980) paper, hereafter referred to simply as TL. These terms will be discussed as we go down through the TL bibliographical "family tree" represented by Figure 1 below. At the top is the "grandparent stage" and is represented by the "seed corn" of data sets analyzed by Tong (1977a, b, 1978) and later by Tong and Lim (1980). Of particular germinating effect was the Canadian Lynx data commented on by Tong (1977b).[3] The "parent stage" resides with TL and their paper. It has definitely met the economists' "market test". Finally, two of the "spillover children" of TL and their implied progeny are represented in the stages below. TL has been a prolific parent! This is one population explosion that even the Rev. Thomas Malthus would not have been concerned about!

The outline of this paper is as follows. In Section 2 we will discuss the role that the "seed corn" data sets played in the development of the threshold autoregression model. In the following section we will examine the **depth**, **breadth, and durability** of the TL paper as it has affected the literature, not only in the statistics field, but in many other fields as well. The depth, breadth, and durability of citations are surely dimensions that Library Scientists would agree are useful indicators of the degree of "seminality"[4] of a paper. In Section 4 we will briefly discuss two of TL's progeny, Threshold Cointegration and Threshold GARCH, to demonstrate how universally the idea of

---

[2] For an excellent autobiographical sketch of "The Birth of the Threshold Time Series Model" see Tong (2007). Of particular interest are Tong's discussions of his philosophy of "the inseparability of theory and practice" (p. 9) and his "divide and rule" (divide and conquer) approach to making progress on difficult problems in statistical theory (p. 10).

[3] Interestingly, a Google search for the phrase "Canadian Lynx Data" yielded 79,900 hits. For researchers who are interested in upping their citation counts, creating interesting and challenging data sets would certainly appear to be one way do so! Being interested in the analysis of challenging data sets myself, I would say all the more power to such individuals!

[4] Seminality – n. "The quality or state of being seminal" according to freedictionary.com. Microsoft Word's spell check doesn't think that seminality is a seminal word however! Let's do a citation count on it ten years from now!

FIGURE 1
FAMILY TREE OF
THRESHOLD AUTOREGRESSION

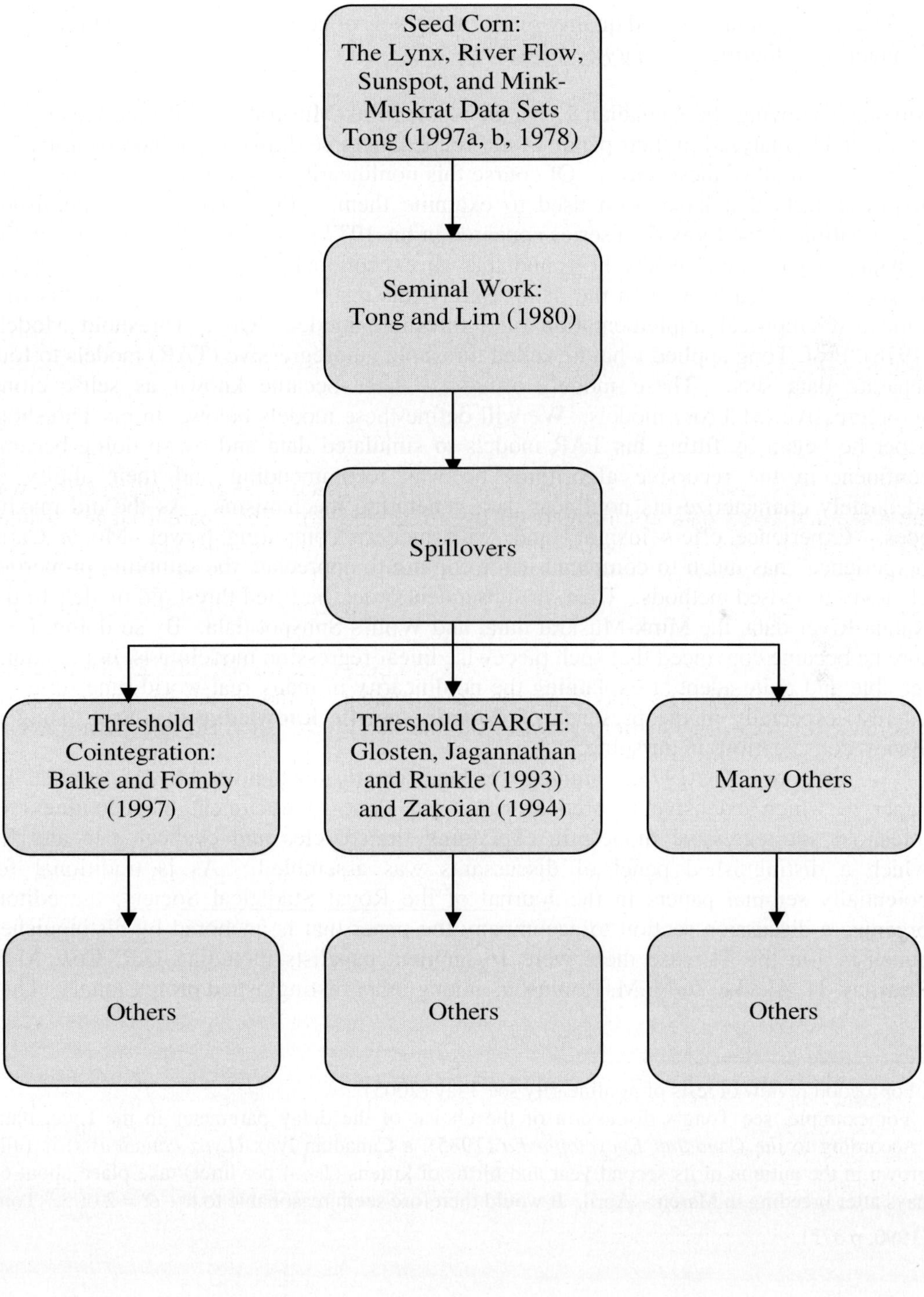

threshold effects has been applied and continues to be applied. Finally, the paper concludes with a brief discussion of the "signature" of a seminal paper and how we need more of them for the advancement of the sciences.

2. The Seed Corn

Seed Corn – noun. "Good quality seeds (as kernels of corn) that are reserved for planting." thefreedictionary.com

Anyone reviewing the Canadian Lynx, Sunspot, Mink-Muskrat, and Kanna River flow data that TL analyzed in their paper would immediately be drawn to the nonlinearity that is apparent in all of these series. Of course this nonlinearity is substantiated in the many nonlinear tests that have been used to examine them.[5] Prof. Tong's first published investigation of the Lynx data series appeared in his 1977 paper "Some Comments on the Canadian Lynx Data" where he found that an exceptionally high order autoregressive model was needed to explain the data. Surely this result set him thinking. In his first published empirical implementation of a threshold model, "On a Threshold Model" (1978), Prof. Tong applied what he called threshold autoregressive (TAR) models to four separate data sets. These models of course later became known as self-exciting autoregressive (SETAR) models. We will define these models below. In his Threshold paper he began by fitting his TAR models to simulated data and by so doing became confident in the recursive algorithms he was recommending and their ability to adequately characterize his nonlinear data generating mechanisms. As the old maxim goes, "Experience offers insight" and, with modern computing power, Monte Carlo "experience" has much to commend it for coming to appreciate the sampling properties of newly proposed methods. Then, in subsequent order, he fitted threshold models to the Kanna River data, the Mink-Muskrat data, and Wolf's Sunspot data. By so doing, I am sure he became convinced that such piecewise linear regression modeling is, in fact, quite feasible and quite adept at explaining the nonlinearity in many real-world time series of interest, especially in the presence of **domain-specific knowledge** that Prof. Tong is always conscientious in pursuing.[6]

Of course this 1978 beginning was significantly substantiated by the seminal TL paper in which extensive theoretical work was done to document the usefulness of threshold autoregressive models in explaining limit cycles and cyclical data and for which a distinguished panel of discussants was assembled. As is traditional for potentially seminal papers in the Journal of the Royal Statistical Society, the editors organize a discussion section to accompany the paper that is anchored by distinguished panelists. In the TL case there were 16 eminent panelists including D.R. Cox, M.B. Priestley, H. Akaike, and P.M. Robinson, among other distinguished professionals. They

---

[5] For a good review of tests of nonlinearity see Tsay (2005).

[6] For example, see Tong's discussion of the choice of the delay parameter in the Lynx data. "According to *the Canadian Encyclopaedia* (1985), a Canadian lynx (*Lynx canadensis*) is fully grown in the autumn of its second year and births of kittens (1 – 4 per litter) take place about 63 days after breeding in March – April. It would therefore seem reasonable to try $\alpha = 2$ or 3." Tong (1990, p.377).

offered 12 pages of comments which in turn gave rise to 10 and ½ pages of return comments by Tong and Lim! Dr. C. Chatfield commented, "I would like to congratulate the authors on making a substantial contribution to non-linear time-series modeling. I particularly welcome the fact that the paper **combines** new theoretical work with a number of practical examples, using **real data**." (My emphasis) From their comments, the other panelists totally agreed with Prof. Chatfield and saw the TL threshold methods as being ahead of their time. Mr. E.J. Godolphin congratulated the authors on "an interesting and thought-provoking paper," the "thought-provoking" part being, in my mind, a very strong "leading" indicator of a soon to be seminal paper.[7]

In summary, one must say that the Lynx et.al. "seed corn" data never led Prof. Tong astray but rather sharpened his eye for incisive analysis and modeling building. As commented by Peter Brockwell in the 2007 special volume on "Threshold Models and New Developments in Time Series" in Statistica Sinica, Vol. 17, pp. 3-4:

> "Motivated by physical considerations and sample-path indications of nonlinearity such as irreversibility, bursts of outlying observations, and the existence of limit cycles, Howell showed excellent judgment in specifying and developing the theory and applications for the class of threshold models. The applications given in his original lecture notes[8] and those that have been made by him and others in the ensuing twenty years cover a remarkable range of research areas including electronics, ecology, hydrology, medical research, astrophysics and finance."

With this lead-in let's look to the depth, breadth, and durability of the citations that TL has generated since it was written.

3. Meeting the Market Test

To put matters straight and to allow a comparison with two of TL's progenitors below, a k-regime, self-exciting TAR (SETAR) model with threshold variable x_{t-d} is defined by the piecewise regression equation

$$x_t = \phi_0^{(j)} + \phi_1^{(j)} x_{t-1} + \cdots + \phi_p^{(j)} x_{t-p} + a_t^{(j)}, \text{ if } \gamma_{j-1} \le x_{t-d} < \gamma_j \qquad (1)$$

where $j = 1,2\cdots,k$, k and d are positive integers, the γ_j are real numbers such that $-\infty = \gamma_0 < \gamma_1 < \cdots < \gamma_k = \infty$, the superscript (j) is used to signify the regime, and

[7] Forgive me for what I view as a light-hearted aside. One of the panelists commented that "... the authors' pragmatic attitude towards the usefulness of Akaike's information criterion for identifying time series models" stood in marked contrast to "the almost religious attitude adopted earlier by Dr. Tong in his analysis of the lynx data." Now I would ask those of you who have had the good fortune to attend one of Prof. Tong's seminars, "Have you ever seen him present his research with religious fervor?" To the positive I can say that I have and often! Passion in the analysis of data is one thing, passion in a seminar and in one's research is quite another matter. The literature has definitely benefited from Prof. Tong's passion for deep and substantive research!

[8] Tong (1983).

$\{a_t^{(j)}\}$is an iid sequence with mean 0 and variance σ_j^2 and are mutually independent across regimes. It is understood that one or more of the coefficients of the piecewise autoregressions are different across the regimes or otherwise the model could be simplified by combining some of the regimes.

When judging the desirability of a product, economists often invoke the so-called "market test" of the product. In reference to a product that has made continual profits for a sustained period of time, economists would say that the product has passed the "market test" and thus is a worthy product, given that consumers evidently want it or, if not, they wouldn't continue to buy it at prices that sustain its production and profitability.[9] It is the same way with a professional article. One would hope that it would make the "market test" in that it would be frequently cited and the citations would continue for a substantial period of time. A plus, of course, would be if the article generated interest not only in one's own field but also in other fields of academic endeavor. As we will show below, all of these conditions have been met in the case of TL.

In examining the **depth**, **breath**, and **durability** of the effects of the TL article on the literature, we use the ISI Web of Knowledge (hereafter WOK) database (1975 – present) to search for the citations of the TL article and other articles discussed below.[10] We conducted this search on 2/15/08. WOK provides coverage of the Science Citation Index Expanded (1975 – present), Social Science Citation Index (1975 – present), and Arts & Humanities Citation Index (1975 – present). Not only can one search for the number of citations of an article by year but one can also break out the total citation count by subject areas across the above three citation indices e.g. statistics & probability, economics, business & finance, social sciences – mathematical methods, political science, computer science, planning & development, international relations, urban studies, environmental studies, etc. According to WOK (as of 2/15/08) the TL article has generated a total of 273 citations in the various professional journals covered by WOK. This compares to an **average** (per article) all-time citation count of 19.81 (72, 37, 35, 20, 10, 9, 9, 7, 7, 7, and 5) for the other 11 articles that appeared in the same issue of JRSS (Series B, 1980, Vol. 42, No. 3) that the TL article appeared in. So the TL article definitely had **depth** with respect to the **effect that it has had on the literature in total**.

In examining the **breadth** of the effect of the TL article on the literature we note that the all-time **number of subject areas that have been affected** by the TL article is 50 (!), according to WOK. This compares to an average (per article) all-time subject area count of 9.54 (29, 15, 13, 11, 8, 7, 6, 5, 4, 4, and 3) for the other 11 articles that appeared in the same issue of JRSS. So defined, the TL article has substantial breadth of effect as well.

To emphasize these points of **depth** and **breadth** of effect of the TL article, one only has to examine the Box plots of the total number of citations and subject areas impacted by the TL article versus its JRSS companion papers. The TL citation count is significant positive outlier in the sense that it lies well above the upper "whiskers" in the

---

[9] Of course one might take exception to addictive products like cigarettes, heroin, and cocaine. But, otherwise, the rule has merit as a proxy for measuring the desirability of products.

[10] One should remember that the citation research reported here is for only **one paper** in the extensive vitae of Prof. Tong.

Figure 2
Box Plots of
Citation and Subject Area Counts
of TL and <u>JRSS</u> Companion Articles

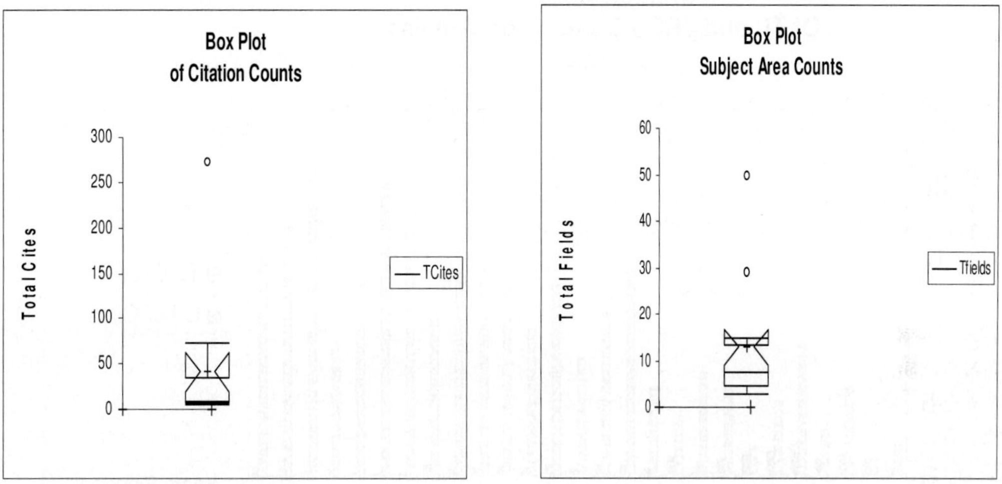

Box plots. The same comment applies to the subject area counts generated by the TL paper. The TL subject area count is a very strong positive outlier. These Box plots are reported above in Figure 2.[11]

As a measure of the **durability** of an article's impact on the literature, one only needs to look at the **"duration" of the citations of the article in the literature**. What is the incidence of the citations of the article year-by-year? Does the citation count immediately become less and less through time and quickly die away? Do the citations grow over time and eventually reach a peak and then die away and, if so, how long does it take for the peak to be reached? Or do the citations simply continue to grow in effect through time? In fact the duration of the TL article's effects seem to behave more like the latter duration – its effects have yet to display a strong downward trend. This stands in contrast to the duration effect of the per-paper average of the citations of the <u>JRSS</u> companion papers which apparently reached a peak after 5 years and has fallen away in count since. The duration plots of the citations for the TL article and the average of its <u>JRSS</u> companion articles are depicted in Figure 3 below.

---

[11] The main box of the Box plot extends from the first quartile, Q1, of the data to the third quartile, Q3, of the data. The notch in the box is at the mean of the data. The upper "whisker" (Upper extreme horizontal line) is equal to the maximum value in the data set that is just below the cutoff = Q3 + 1.5(Q3 – Q1). The lower "whisker" is equal to the minimum value in the data set which is just above the cutoff = Q1 - 1.5(Q3 – Q1). All values that are less then the lower whisker or greater than the upper whisker are treated as the outliers in the Box plot.

Figure 3
Durability of Citations

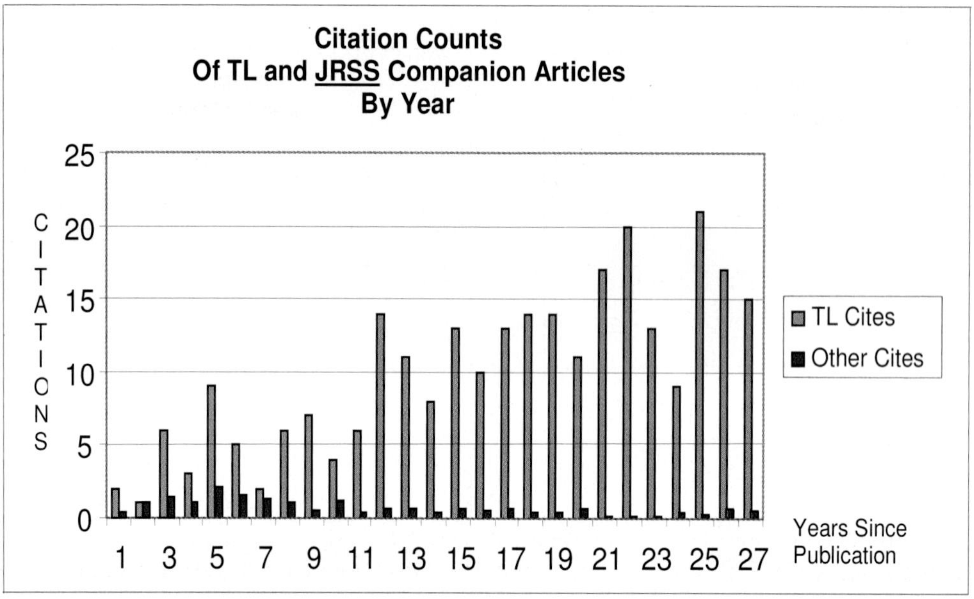

In summary it appears that the TL article has met the "market" test of a seminal article. It has exhibited depth, breadth, and durability in its impact on the literature. Now let us consider the **"spillover" impact** that the TL article has had on two progenitors – Threshold Cointegration and Threshold GARCH, among many other progenitors of the TL article.

4. Two Spillover Effects, Among Many, of the TL Paper

4.1. Threshold Cointegration

Since the publication of the TL paper, several threshold-type models have appeared in the literature. One of these models is the Threshold Cointegration model of Balke and Fomby (1997). This model was suggested to the authors through having previously read the TL article. Of course, as economists Balke and I were familiar with the cointegration concept of Engle and Granger (1987). However, we had been thinking for sometime about the intermittency of shocks to economic systems and how they should be statistically modeled. See, for example, Balke and Fomby (1991a, b, 1994). This naturally led us to the question, "Is cointegration always continually active among time series? Might there be an "on-off" cointegration based upon whether a certain threshold had been surpassed or not? This naturally falls into the genre of models suggested by the TL article.

As a beginning for notation, let $\mathbf{y}_t = (y_{t1}, y_{t2,} \cdots, y_{tK})'$ denote a K-dimensional vector of time series, each of which is I(1) and let $\Delta\mathbf{y}_t = (\Delta y_{t1}, \Delta y_{t2,} \cdots, \Delta y_{tK})'$ denote the K-dimensional vector of first differences of these variables. Furthermore let us assume that there is one cointegrating relationship between the $y_{t1}, y_{t2,} \cdots, y_{tK}$ time series and let this cointegrating relationship be represented by

$$\beta_1 y_{t1} + \beta_2 y_{t2} + \cdots + \beta_k y_{tK} = z_t \tag{2}$$

where z_t is a stationary random variable with possibly a non-zero mean. The Error Correction Model (ECM) of Engle and Granger (1987) can then be written as

$$\Delta\mathbf{y}_t = \boldsymbol{\alpha} + \sum_{i=1}^{p} \boldsymbol{\Phi}_i \Delta\mathbf{y}_{t-i} + \boldsymbol{\theta} z_{t-1} + \mathbf{a}_t \tag{3}$$

where $\boldsymbol{\alpha}$ is a K x 1 vector of intercept coefficients, the $\boldsymbol{\Phi}_i$ represent K x K matrices of lag coefficients associated with the lagged vectors $\Delta\mathbf{y}_{t-i}$, $\boldsymbol{\theta}$ is a K x 1 vector of error correction coefficients, z_{t-1} is the so-called scalar error-correction term, and $\mathbf{a}_t$ is a K x 1 vector of white noise error terms, the error terms having possibly different variances.

Now a piece-wise, threshold version of the ECM of (3) has been postulated by Balke and Fomby (1997). We call this model the Threshold Cointegration model. It is of the form

$$\Delta\mathbf{y}_t = \begin{cases} \boldsymbol{\alpha}_{(1)} + \sum_{i=1}^{p} \boldsymbol{\Phi}_i^{(1)} \Delta\mathbf{y}_{t-i} + \boldsymbol{\theta}_{(1)} z_{t-1} + \mathbf{a}_t^{(1)} & , z_{t-1} \le \gamma_1 \\ \boldsymbol{\alpha}_{(2)} + \sum_{i=1}^{p} \boldsymbol{\Phi}_i^{(2)} \Delta\mathbf{y}_{t-i} + \boldsymbol{\theta}_{(2)} z_{t-1} + \mathbf{a}_t^{(2)} & , \gamma_1 < z_{t-1} \le \gamma_2 \\ \boldsymbol{\alpha}_{(3)} + \sum_{i=1}^{p} \boldsymbol{\Phi}_i^{(3)} \Delta\mathbf{y}_{t-i} + \boldsymbol{\theta}_{(3)} z_{t-1} + \mathbf{a}_t^{(3)} & , \gamma_2 < z_{t-1} \end{cases} \tag{4}$$

Again we assume a single cointegrating relationship as in (2) but this time we have a three-regime Error Correction Model. Each regime has its own intercept vector, $\boldsymbol{\alpha}_{(j)}$, j = 1, 2, and 3, its own autoregressive coefficient matrices, $\boldsymbol{\Phi}_i^{(j)}$, j = 1, 2, and 3, its own error correction coefficient vectors, $\boldsymbol{\theta}_{(j)}$, j = 1, 2, and 3, and its own white noise error vectors, $\mathbf{a}_t^{(j)}$, j = 1, 2, and 3. If, as we postulate, the cointegrating relationship is "on" in the outer regimes (j = 1, 3) but is "off" in the middle regime (j = 2), we have the "outer" error correction coefficient vectors, $\boldsymbol{\theta}_{(1)}$ and $\boldsymbol{\theta}_{(3)}$, being non-zero while the "middle" error correction coefficient vector, $\boldsymbol{\theta}_{(2)}$, is equal to zero.

Model (4) is naturally suggested in the context of markets subject to arbitrage like the futures markets and foreign exchange markets where the costs of transactions render the trading within the cost band (the middle regime) unprofitable but not so in the outer regimes. Thus, the cost of arbitrage causes the cointegrating relationships of arbitrage markets to be of an "on-off" variety. In fact the cost of transactions in specific markets can be used to set initial estimates for the threshold boundaries γ_1, γ_2, and γ_3. See, for example, Lo and Zivot (2001) for an extensive review of applications of Threshold Cointegration in markets susceptible to arbitrage and incurring consequential transactions costs. Model (4), which one can view as a "discrete" threshold model, has been nicely extended to a smooth transition Threshold Cointegration model as in Michael, Nobay, and Peel (1997) and Sarantis (1999).[12] Also, Tsay (1998) has extensively discussed the testing and modeling of the Threshold Cointegration model.

Obviously the Threshold Cointegration model (4) is a direct progenitor (descendant) of the TL Self-exciting Threshold Autoregressive (SETAR) model. It is a piecewise linear (Error Correction) model - a modification of the "continually operating" Error Correction Model of Engle and Granger (1987) which was one of the citations of the 2003 Nobel Prize in Economics.

4.2. Threshold GARCH

There is another way that the TL threshold concept has influenced research in a Nobel Prize winning area. In 2003, in addition to citing the concept of cointegration, the Nobel Prize Committee cited the research of Engle (1982) on the univariate volatility model called Autoregressive Conditional Heteroscedasticity (ARCH). This model specifies that the variance of a time series y_t can be expressed in the form

$$y_t = \mathbf{x}_t'\boldsymbol{\beta} + \varepsilon_t \tag{5}$$

where

$$\varepsilon_t = a_t \sqrt{\alpha_0 + \alpha_1 \varepsilon_{t-1}^2 + \alpha_2 \varepsilon_{t-2}^2 + \cdots + \alpha_q \varepsilon_{t-q}^2} \tag{6}$$

and a_t is assumed to be distributed as a standard normal random variable. For ease of exposition we assume a simple linear regression form for the mean of $y_t, \mu_t = \mathbf{x}_t'\boldsymbol{\beta}$, although the linearity is not a necessity. In finance, y_t is often a return to an asset, that is $y_t = r_t$, and the mean is assumed to be constant or determined as a simple autoregressive model like $\mu_t = E(r_t) = \beta_0 + \beta_1 r_{t-1}$. Given (5) and (6) we can write the variance (volatility) of returns in this so-called ARCH(q) model as

$$\sigma_t^2 = \alpha_0 + \alpha_1 \varepsilon_{t-1}^2 + \alpha_2 \varepsilon_{t-2}^2 + \cdots + \alpha_q \varepsilon_{t-q}^2 \tag{7}$$

Equation (7) is made estimable by replacing σ_t^2 with the squared OLS residuals $\hat{\varepsilon}_t^2$ obtained from (5) and using $\hat{\varepsilon}_{t-1}^2, \hat{\varepsilon}_{t-2}^2, \cdots, \hat{\varepsilon}_{t-q}^2$ as the proxies for $\varepsilon_{t-1}^2, \varepsilon_{t-2}^2, \cdots, \varepsilon_{t-q}^2$. Estimation and tests of hypotheses can then proceed straight-forwardly given the

[12] Interestingly, Professors Tong and K.S. Chan have a "parent" paper of the smooth transition Threshold Cointegration model, namely Chan and Tong (1986).

estimable form of (7). Subsequently, Bollerslev (1986) extended the ARCH(q) model to the GARCH(p,q) model that is written as

$$\sigma_t^2 = \alpha_0 + \delta_1\sigma_{t-1}^2 + \delta_2\sigma_{t-2}^2 + \cdots + \delta_p\sigma_{t-p}^2 + \alpha_1\varepsilon_{t-1}^2 + \alpha_2\varepsilon_{t-2}^2 + \cdots + \alpha_q\varepsilon_{t-q}^2 \tag{8}$$

In an attempt to explain the leverage effect[13] often seen in financial analysis, Glosten, Jagannathan, and Runke (1993) and Zakoian (1994) extended the GARCH(p,q) model to the so-called Threshold GARCH (TGARCH) model written as

$$\sigma_t^2 = \alpha_0 + \sum_{i=1}^{q}(\alpha_i + \gamma_i N_{t-i})\varepsilon_{t-i}^2 + \sum_{j=1}^{p}\delta_j\sigma_{t-j}^2 \tag{9}$$

where N_{t-i} is an indicator function for **negative** ε_{t-i}, that is,

$$N_{t-i} = \begin{cases} 1 & if\ \varepsilon_{t-i} < 0 \\ 0 & if\ \varepsilon_{t-i} \geq 0 \end{cases}. \tag{10}$$

This specification uses zero as the threshold to separate the two volatility regimes but other threshold values can be searched for. If the leverage effect is present in a return equation ($y_t = r_t$), one would expect the estimated parameters of the model to be such that the volatility implied by the model would be greater in the negative error regime than in the positive error regime. In addition to the ARCH, GARCH, and TGARCH models, several additional volatility models have been devised for describing volatility of time series variables around their means. For a good survey of these other models, see Tsay (2005).

4.3. Spillover Citation Results for Two Progenitors of the TL Paper: Threshold Cointegration and Threshold GARCH

To examine the **spillover impact** of the TL Threshold Autoregressive model, we again consulted the ISI Web of Knowledge and obtained the all-time citation counts of the aforementioned "threshold" papers. The citation counts and number of subject areas (fields) encountered for these papers as well as the TL paper are depicted in the bar chart of Figure 4. The TL (1980) paper is labeled TL, the Balke and Fomby (1997) paper is labeled BF, the Zakoian (1994) paper is labeled Z, while the Glosten, Jagannathan, and Runkle (1993) paper is labeled GJR. Thus, in addition to the 273 cites and 50 fields for the TL article, we have 140 cites and 18 fields for the Balke and Fomby article, 105 cites and 23 fields for the Zakoian article, and finally 418 cites and 38 fields for the Glosten, Jagannathan, and Runkle article. Noticeably the TL article has the greatest breath of all of the threshold papers.

In summary, the TL article seems to have generated a worthy number of spillover effects in just the two progenitors of Threshold Cointegration and Threshold GARCH.

[13] The leverage effect occurs when the volatility of an asset is greater when there is a price drop than when there is a price increase. With a price drop there is a greater likelihood of "default" in some sense than when there is a price increase. Therefore negative returns imply a greater uncertainty in this respect.

Figure 4

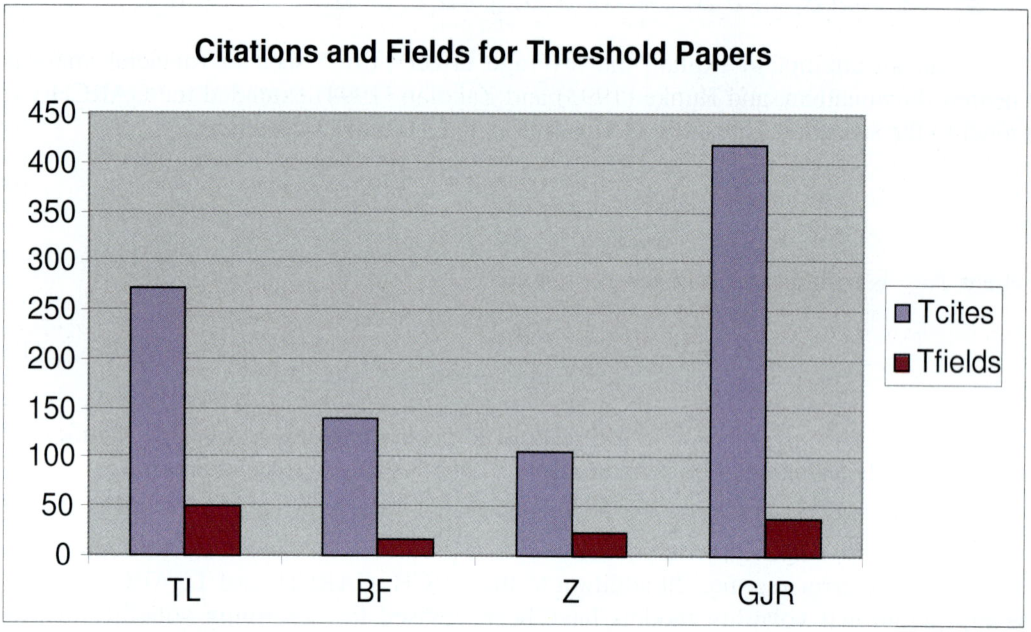

Surely, many other spillover effects will eventually be registered in other threshold models like the Threshold Multivariate GARCH model of Audrino and Trojani (2004).

5. The "Signature" of a Seminal Paper

What is the "signature" of a seminal research paper in the academic literature? We have argued here that the signature of a seminal paper is one that has substantial impact on the literature in terms of **depth**, **breadth**, **durability,** and **spillovers**. **Depth** can be thought of as the number of citations of the paper in the literature **in total**. **Breadth** can be thought of as the **number of subject areas** in which the paper is cited. **Durability** can be thought of as the **time profile of the citations** to the paper, irrespective of subject area. That is, are the citations quickly dissipating or long-lasting? The **spillover impact** of a paper can be viewed as the cumulative citation counts of papers that derived their own seminal ideas from the "parent" paper and thereby generated many citations of their own. For the spillover impact of a parent paper to be substantial, there should be a substantial number of progenitor articles that themselves have strong citation counts, each having substantial depth, breadth, durability, and spillover impacts.

Given the ISI Web of Knowledge citation and subject area counts of the Tong and Lim (1980) paper and two of its progenitors, it is certainly the case that TL has thoroughly met these criteria. It has met the "market test" for a seminal paper. Thanks go to Prof. Tong for pushing the frontiers of time series analysis light years forward and doing so with real-world problems in mind.

In the future we will need more seminal papers so that the sciences can make even greater progress in the area of empirical-based decision-making. As an economist I applaud those professional societies, professional journals, universities, and public and private organizations that have established incentive systems to recognize and reward producers of seminal research. We, as a society, will surely benefit by supporting and adding to those incentive systems.

Acknowledgments

I would like to thank Sarah Haight and Toni Nolen of the SMU University Library and Diane Shepelwich of the University of Texas at Arlington for their patience in instructing me on how to do this citation research on WOK and my research assistant, Juan Wang, for helping me refine the graphs in this paper. I also would like to express my appreciation for the helpful comments that I received from my colleague, Nathan Balke, and the referees of this paper. Any errors that may remain, however, are the sole responsibility of the author.

References

1. Audrino, F. and F. Trojani (2004), "A General Multivariate Threshold GARCH Model with Dynamic Conditional Correlations," Dec. 2004 Working paper in Institute of Finance, University of Lugano, in Lugano, Switzerland.

2. Balke, N. and T. Fomby (1991a), "Shifting Trends, Segmented Trends, and Infrequent Permanent Shocks," Journal of Monetary Economics, 28, 61 – 85.

3. Balke, N. and T. Fomby (1991b), "Infrequent Permanent Shocks and the Finite-Sample Performance of Unit Roots," Economic Letters, 36, 269 – 273.

4. Balke, N. and T. Fomby (1994), "Large Shocks, Small Shocks, and Economic Fluctuations: Outliers in Macroeconomic Time Series," Journal of Applied Econometrics, 9, 181 – 200.

5. Balke, N. and T. Fomby (1997), "Threshold Cointegration," International Economic Review, 38, No. 3, 627 – 645.

6. Bollerslev, T. (1986), "Generalized Autoregressive Conditional Heteroskedasticity," Journal of Econometrics, 31, 307 – 327.

7. Brockwell, P. (2007), "Beyond Linear Time Series," Statistica Sinica, 17, no. 1, 3–5.

8. Chan, K.S. and H. Tong (1986), "On Estimating Thresholds in Autoregressive Models," Journal of Time Series Analysis, 7, 179 – 190.

9. Engle, R. F. (1982), "Autoregressive Conditional Heteroscedasticity with Estimates of the Variance of United Kingdom Inflations," Econometrica, 50, 987 – 1007.

10. Engle, R.F. and C.W.J. Granger (1987), "Co-Integration and Error Correction: Representation, Estimation, and Testing," Econometrica, 55, No. 2, 251–276.

11. Glosten, L.R., R. Jagannathan, and D.E. Runkle (1993), "On the Relation Between the Expected Value and the Volatility of Nominal Excess Return on Stocks," Journal of Finance, 48, 1779 – 1801.

12. Hjorland, B. (2006). Quote from Prof. Hjorland's website at http://www.db.dk/bh.

13. ISI Web of Knowledge (2008), Thomson Reuters, Inc., UK (version 4.2)

14. Lo, M.C. and E. Zivot (2001), "Threshold Cointegration and Nonlinear Adjustment of the Law of One Price," Macroeconomic Dynamics, 5, 533 – 576.

15. Lussky, J.P. (2004), "Bibliometric Patterns in an Historical Medical Index: Using the Newly Digitized Index Catalogue of the Library of the Surgeon General's Office, United States Army." Thesis, College of Information Science and Technology, Drexel University. Full text available at http://dspace.library.drexel.edu/retrieve/3815/Lussky_Joan.pdf

16. Michael, P., R.A. Nobay, and D.A. Peel (1997), "Transactions Costs and Nonlinear Adjustment in Real Exchange Rates: An Empirical Investigation," Journal of Political Economy, 105, 862 – 879.

17. Sarantis, N. (1999), "Modeling Non-Linearities in Real Effective Exchange Rates," Journal of International Money and Finance, 18, 27 – 45.

18. Tong, H. (1977a), "Discussion of a Paper by A.J. Lawrance and N.T. Kottegoda," Journal of the Royal Statistical Society, Series A, 34-35.

19. Tong, H. (1977b), "Some Comments on the Canadian Lynx Data – with Discussion, Journal of the Royal Statistical Society, Series A, 432-435, 448-468.

20. Tong, H. (1978), "On a Threshold Model," in Pattern Recognition and Signal Processing (ed. C.H. Chen), Sijthoff and Noordhoff, Amsterdam.

21. Tong, H. and K.S. Lim (1980), "Threshold Autoregression, Limit Cycles and Cyclical Data," Journal of the Royal Statistical Society, Series B (with discussion), 245 – 292.

22. Tong, H. (1983), Threshold Models in Non-linear Time Series Analysis. Lecture Notes in Statistics, No. 21. Springer, Heidelberg.

23. Tong, H. (1990), Non-linear Time Series: A Dynamical System Approach, Oxford Statistical Science Series, Clarendon Press, Oxford, UK.

24. Tong, H. (2007), "Birth of the Threshold Time Series Model," Statistica Sinica, 17, no. 1, 8-14.

25. Tsay, R. (1998), "Testing and Modeling Multivariate Threshold Models," Journal of the American Statistical Association, 93, 1188 – 1202.

26. Tsay, R.S. (2005), Analysis of Financial Time Series, second edition, Wiley-Interscience, New York.

27. Zakoian, J.M. (1994), "Threshold Heteroscedastic Models," Journal of Economic Dynamics and Control, 18, 931 – 955.

The SETAR Model of Tong and Lim and Advances in Computation*

JOHN GEWEKE
Center for the Study of Choice
University of Technology Sydney,
Ultimo NSW 2007, Australia
and
Department of Finance, University of Colorado
Boulder, CO 80309, USA
E-mail: John.Geweke@uts.edu.au

This discussion revisits Tong and Lim's seminal 1980 paper on the SETAR model in the context of advances in computation since that time. Using the Canadian lynx data set from that paper, it compares exact maximum likelihood estimates with those in the original paper. It illustrates the application of Bayesian MCMC methods, developed in the intervening years, to this model and data set. It shows that SETAR is a limiting case of mixture of experts models and studies the application of one variant of those models to the lynx data set. The application is successful, despite the small size of the data set and the complexity of the model. Predicive likelihood ratios favor Tong and Lim's original model.

Tong and Lim (1980), hereafter TL, provided the first complete exposition of the self-exciting threshold autoregressive (SETAR) model. TL addressed many of the population properties of a time series that follows a SETAR model, while at the same time attending to the practical issues encountered in applying SETAR given the technology available at the time. Attention to both features was critical to its subsequent influence and successful applications.

Since 1980 there have been enormous advances in the field of practical nonlinear time series modeling in which Tong and Lim were pioneers. The dramatic progress in computing since that time has been a key factor in this evolution. Progress in the intervening years has moved back the frontiers of nonlinear time series modeling in ways that could scarcely have been anticipated thirty years ago when the research for TL began. This discussion focuses on some aspects of these advances as they relate to SETAR, centered around one of the illustrations in TL. It begins by reviewing the findings reported in TL (Section 1), before moving on to a Bayesian treatment of SETAR (Section 2), which became possible only around 1990. Section 3 shows that SETAR is a limiting case of a much broader class known as mixture of experts models introduced in the neural computation literature in the mid-1990's, and illustrates that these models can be applied in at least one of the illustrations taken up in TL. The findings provide support for the TL specification, but also show that mixture of experts models can be applied using much smaller data sets than is typically the case.

---

*This comment was written while the author was Professor of Economics and Statistics at the University of Iowa. Partial financial support for the work was provided by National Science Foundation Grant SBR-0720547.

1. The SETAR of TL and Its Application

The SETAR in TL pertained to a univariate time series $\{y_t\}$ with a single threshold,

$$y_t = \beta_{10} + \sum_{j=1}^{k_1} \beta_{1j} y_{t-s} + \varepsilon_t, \quad \varepsilon_t \overset{iid}{\sim} N\left(0, \sigma_1^2\right) \text{ if } y_{t-d} \le r, \tag{1}$$

$$y_t = \beta_{20} + \sum_{j=1}^{k_2} \beta_{2j} y_{t-s} + \varepsilon_t, \quad \varepsilon_t \overset{iid}{\sim} N\left(0, \sigma_2^2\right) \text{ if } y_{t-d} > r.$$

This model is widely applied today, as is the generalization in which there are three or more regimes defined by the value of y_{t-d} and two or more thresholds r_j. This discussion will be confined to the two-regime case (1), as was TL. Conditional on d, k_1, k_2 and r the likelihood function is conventional, but absent the conditioning it is nonstandard. Conditional on k_1, k_2 and d finding the maximum likelihood estimate of r is today a straightforward problem, but this was not the case in 1980. Conditional on this estimate, TL selected d, k_1 and k_2 using the Akaike Information Criterion (AIC), introduced shortly before the publication of TL.

TL contained four applications of (1), of which two – one using Canadian lynx data and one using sunspot numbers – provided canonical data sets for subsequent nonlinear time series analysis (e.g. Lewis and Stevens (1991); Stenseth et al. (1998)). This discussion utilizes the lynx data, which pertain to the years 1821 through 1934 (114 observations) and are displayed in Figure 1. The time series is dominated by a cycle of nine to ten years, with an asymmetry – more rapid declines and slower recoveries – that today we recognize as characteristic of a two-regime SETAR time series. TL applied their methods to these data, taking as the range of the variable y_t in (1) all of the observations but the first ten, those being withheld to allow for ten lags, the maximum number considered; thus the sample size employed in TL was 104, with the range of the dependent variable being the years 1831 through 1934. TL reported the choices $k_1 = 8$, $k_2 = 3$, $d = 2$, based on AIC, together with the estimate $r = 3.1163$. Conditional on these choices, the data set used for this discussion reproduces the estimates of the parameters β_{ji} and σ_j^2 reported in TL.

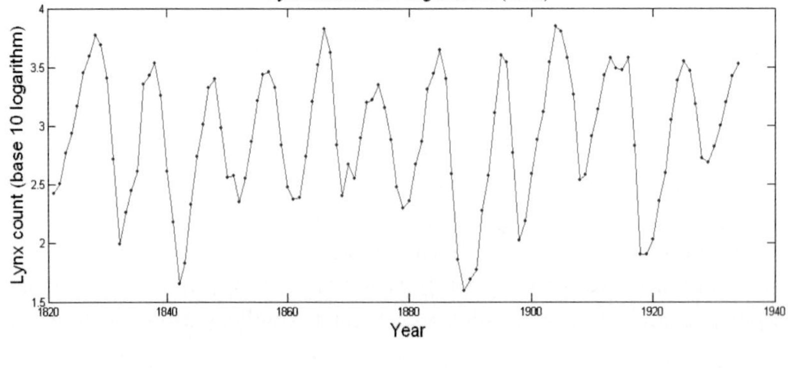

Fig. 1. Canadian lynx data used in TL.

It is today well understood that the essential nonstandard feature of the SETAR likelihood function arises from the break at $y_{t-d} = r$. For any combination of k_1, k_2 and d, concentration of the likelihood function in r leads to a step function, with the steps occurring at the values of y_{t-d}. Figure 2 provides the concentrated log-likelihood function for the choices $k_1 = 8$, $k_2 = 3$, $d = 2$ of TL. It is constructed from maximum likelihood estimates of the parameters β_{ji} and σ_j^2 at 1600 points equally spaced between the extreme values shown in the figure. These computations are trivial today, but that was not the case in the late 1970's when the research underlying TL was carried out. As explained in TL, only five alternative threshold values were explored. The vertical line in Figure 2 indicates the value $r = 3.1163$ chosen in TL. The true maximum of the likelihood function occurs on the interval $(3.3101, 3.3261)$, which is separated by some 12 observations y_{t-d} from the value selected in TL.

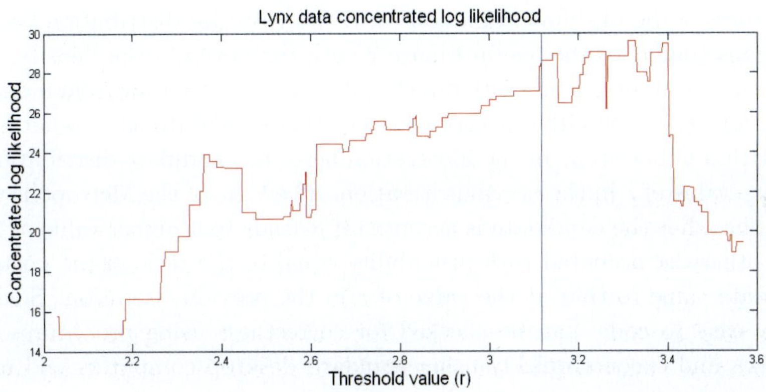

Fig. 2. Evaluation of the log likelihood function for $k_1 = 8$, $k_2 = 3$, $d = 2$, after maximizing in α_j, β_j and σ_j^2. The vertical line indicates the estimate of r in TL.

The near-continuous representation of the log-likelihood function in Figure 2 illustrates severe asymmetry over the range relevant for inference. The function drops precipitously less than 0.1 units to the right of the maximum, by nearly 8 log-likelihood points, to about 22. By contrast there are plateaus exceeding this value beyond 1.0 units to the left of the maximum. Keeping in mind that each "step" in the concentrated log-likelihood function corresponds to an observation, it is clear that there is substantial uncertainty about the threshold r, even before turning to formal methods for characterizing this uncertainty, which is the next step in this discussion.

2. A Bayesian treatment of SETAR

Bayesian methods implemented with Markov chain Monte Carlo (MCMC) algorithms are well suited to the application of SETAR for at least two reasons. First, the unconventional nature of the likelihood function illustrated in Figure 2 and studied by Chan causes no difficulties for Bayesian inference in principle. Second, MCMC algorithms known as Gibbs samplers are well suited to situations in which likelihood functions are conventional but for a single parameter or a small group of parameters, as is the case with the SETAR likelihood function. Consequently Bayesian inference is straightforward in practice as well as in principle, and there is a large Bayesian SETAR literature (e.g. Geweke and Terui (1993), Sorensen et al. (1995), Koop and Potter (1999)).

To illustrate how these methods characterize the threshold uncertainty evident in Figure 2, complete the SETAR model with the independent prior specifications

$$\beta_{1i} \sim N(0,1) \quad (i = 0,\ldots,8)\,; \quad \beta_{2i} \sim N(0,1) \quad (i = 0,\ldots,3)\,;$$
$$0.2/\sigma_j^2 \sim \chi^2(5) \quad (j = 1,2)\,; \quad r \sim N(3.5,1)\,.$$

The results that follow are not particularly sensitive to this choice of prior distribution. The Gibbs sampling algorithm draws parameters in successive groups based on the form of the posterior distribution. Conditional on all other parameters, the joint distribution of the coefficients $\beta_{10},\ldots,\beta_{18},\beta_{20},\ldots,\beta_{23}$ is multivariate normal and the parameters σ_1^2 and σ_2^2 are each each inverted gamma.

The posterior distribution of r conditional on all other parameters inherits the unconventional features of the likelihood function, with the posterior distribution being a product of step functions similar to the one in Figure 2 with the normal prior density. Rather than sample from the unconditional density directly, it is possible to draw from a different candidate distribution, and then either accept or reject that candidate using what is known as a Metropolis within Gibbs step. In the illustration here the candidate distribution is uniform centered at the value of r in the previous iteration. The logic of the Metropolis within Gibbs step then implies that the candidate is accepted if it leads to a higher value of the posterior density, and otherwise accepted with probability equal to the ratio of the posterior density at the candidate value to that at the value of r in the previous iteration. Such algorithms are relatively easy to code, can be checked for correctness using procedures described in Geweke (2004), and execute quickly using standard desktop computers and mathematical applications software.

Figure 3 provides some information about this algorithm and the posterior distribution, based on 10^5 iterations of the MCMC algorithm just described. The upper left panel shows the value of r at 1000 equally spaced iterations – that is, each adjacent pair is separated by 100 iterations. While MCMC algorithms in general produce autocorrelated samples from the posterior distribution there is no evidence of autocorrelation at intervals of 100 iterations, and indeed formal measurements confirm (near) absence of serial correlation at even closer intervals. To achieve this mixing in the MCMC sample, it is important that the uniform distribution used in the Metropolis within Gibbs step for r have sufficiently great support. In the algorithm used the length of the support is 0.25. If the support is much smaller – for example, 0.04 – then steps between the analogues in the posterior density of the higher plateaus shown in Figure 2 are improbable in the Markov chain, implying that the number of iterations required for an adequate representation of the posterior distribution can be large – perhaps impractically so. This is a key technical point in implementing MCMC in this and other SETAR models, and it is essential to coping with the unconventional likelihood function in this approach.

The upper right panel of Figure 3 represents the posterior distribution of the threshold parameter r using a histogram, the posterior probability that r is in each bin being proportional to the height shown. The correspondence between this panel and Figure 2 is not exact because the latter figure is based on the likelihood function maximized in parameters other than r, whereas the former is based on the posterior density integrated in parameters other than r. Nevertheless the broad correspondence is clear. The posterior distribution, like the concentrated log-likelihood function, is sharply truncated on the right, has a long left tail, and exhibits multiple modes. Notice that there are also very thin tails of the distribution

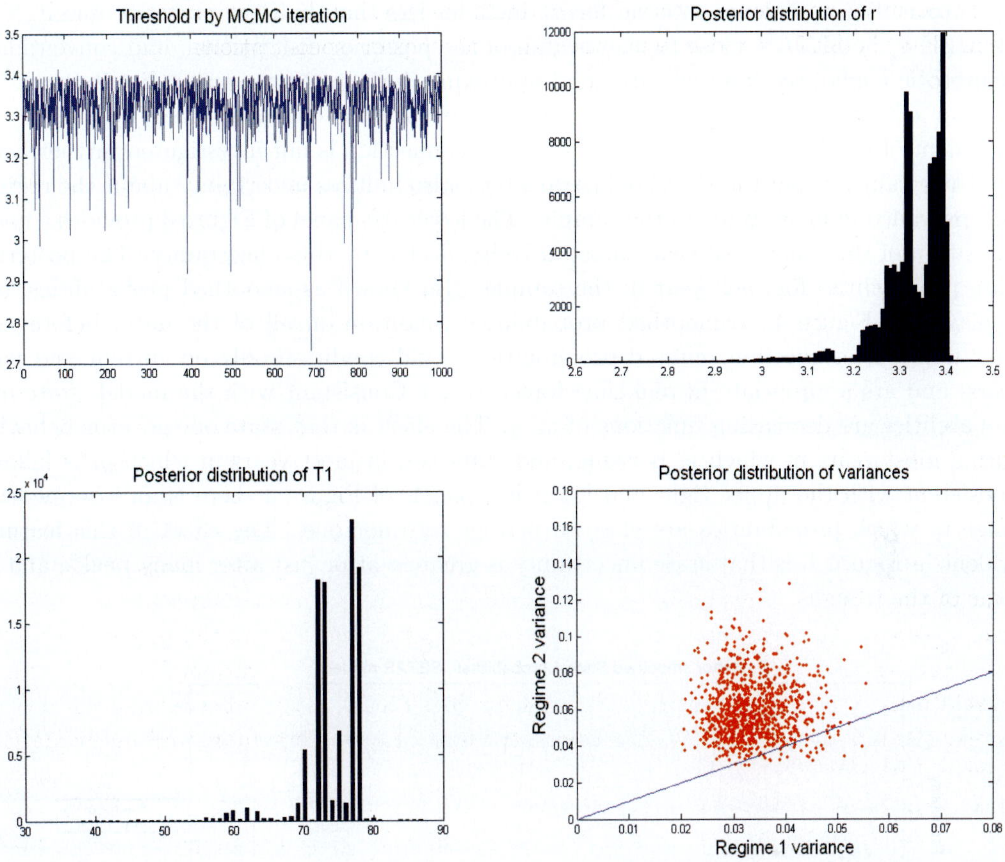

Fig. 3. Some aspects of the MCMC posterior simulator (upper left) and the posterior distribution (other panels).

that extend over almost the entire range of the horizontal axis. The posterior distribution of r collapses about the (pseudo) true value at rate $1/T$, for the same reasons conveyed in the analysis of Chan (1993) for the maximum likelihood estimator $\widehat{r}$ of r. The results in Figure 3 indicate that there is, none the less, substantial uncertainty about r. This in turn suggests cautious interpretation of any analysis (like TL, but unlike the posterior distribution) that proceeds as if $r = \widehat{r}$.

The posterior distribution of r implies a posterior distribution of the number of observations in each regime. The lower left panel of Figure 3 depicts the posterior distribution of the number of observations in the first regime, that is, the number of observations for which $y_{t-d} \leq r$; since r is uncertain, this number is uncertain. The posterior distribution places the number of observations in the first regime between 69 and 78 with high probability, but there is non–negligible probability that the number could be substantially lower. It is almost certain that the majority of observations are in the first regime.

The lower right panel of Figure 3 plots 1,000 values of σ_1^2 (horizontal axis) and σ_2^2 (vertical axis) drawn at equally spaced intervals of 100 iterations in the MCMC simulation. The maximum likelihood estimate $\widehat{\sigma}_1^2$ in TL, which conditions on the value $r = 3.1163$, is 0.0255, well to the left of most of the posterior mass of this parameter; a conventional asymptotic 95% confidence interval for σ_1^2 includes less than half the posterior mass. The TL MLE $\widehat{\sigma}_2^2 = 0.0516$ is close to the median of the posterior distribution, and conventional asymptotic confidence intervals are good approximations to the posterior distribution. To a major extent these discrepancies in the two approaches to analysis may be due to the treatment of $r = 3.1163$ as fixed in TL, but that connection is not investigated further here.

Uncertainty about the threshold parameter r also induces uncertainty about the regime that prevailed in many years in the sample. The lower left panel of Figure 3 provides a good indication of the number of years substantively affected by this uncertainty. The posterior state probabilities for each year in the sample, also known as smoothed probabilities, are displayed in Figure 4. (Smoothed probabilities condition on all of the data, before and after the year in question; filtered probabilities would condition only on current and past years, and are appropriate in real-time forecasting.) Consistent with the model, state one probabilities are decreasing functions of y_{t-2}. The effect is that state one governs behavior during most years in which y_t is rising, and state two in most years in which y_t is falling. Consistent with the upper right and lower left panels of Figure 3 there is an intermediate range in which probabilities are close to neither zero nor one. The effect of this feature, evident in Figure 4, is that state uncertainty is greatest at or just after many peaks, and at some of the troughs.

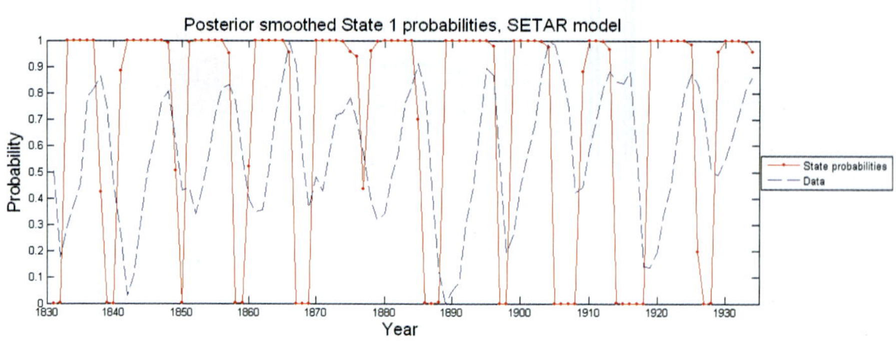

Fig. 4. Posterior probabilities of the event $y_{t-2} < r$ for each year in the sample.

3. Generalizing SETAR

Advances in computation inspired significant progress in nonlinear modeling in the 1990's. An important component of this research is the study of conditional distributions without imposing assumptions like linearity in the mean, normality in the distribution, and so on. Distribution mixtures have played a significant role. Jordan and Jacobs (1994) introduced mixture of experts models. These models are characterized by two or more latent states. In the case of a continuous variable of interest y and a vector of conditioning random variables $\mathbf{x}$, the distribution conditional on $\mathbf{x}$ and the latent state is the familiar normal linear regression model. For each state there is a unique set of parameters and thus

$$y \mid (\mathbf{x}, s = j) \sim N\left(\beta_j' \mathbf{x}, \sigma_j^2\right) \quad (j = 1, \ldots, m). \tag{2}$$

The latent states, in turn, have probabilities that are also affected by the vector of covariates $\mathbf{x}$. In much of the substantial literature that has built on Jordan and Jacobs (1994) the relationship between s and $\mathbf{x}$ is specified by a multinomial logit model. Jiang and Tanner (1999) contains important results on the ability of these models to approximate arbitrarily well any conditional distribution in the exponential family.

Geweke and Keane (2007) introduced a variant of these models, called the smoothly mixing regression (SMR) model, in which the link between $\mathbf{x}$ and s is given by a multinomial probit model. This specification is particularly advantageous for a Bayesian approach using MCMC. In that approach the covariates entering this model need not be the same as the ones entering the components (2). Thus,

$$\mathbf{w} \mid \mathbf{z} \sim N\left(\mathbf{\Gamma z}, \mathbf{I}_m\right), \quad s = \arg\max\left(\mathbf{w}\right). \tag{3}$$

The vectors $\mathbf{x}$ and $\mathbf{z}$ are subsets of a more extensive list of possible covariates. In both this approach as well as those based more directly on Jordan and Jacobs (1994) the richness of the model comes from the fact that given $\mathbf{x}$, $\mathbf{z}$ and the model parameters the value of s is not known with certainty, and therefore via (2) the distribution is a mixture of normal distributions. For example, because σ_j^2 varies with j, the model can display conditional heteroscedasticity; but it can also display conditional skewness, a uni- or multimodal distribution depending on the value of $\mathbf{z}$, and so on.

To draw the link between the SMR and SETAR models, let $y = y_t$, let $\mathbf{x} = (1, y_{t-1}, \ldots, y_{t-8})'$, and let $\mathbf{z} = (1, y_{t-d})'$. With only two regimes the matrix $\mathbf{\Gamma}$ in (3) is 2×2. If

$$\Gamma = \begin{bmatrix} cr & -c \\ -cr & c \end{bmatrix}, \tag{4}$$

where r is the threshold and c is an arbitrary positive constant, then as $c \to \infty$ SMR is equivalent to SETAR. Thus SMR may be interpreted as setting up a "soft" transition between regimes with probabilities changing most rapidly at $y_{t-2} = r$, whereas SETAR imposts a "hard" transition in which probabilities shift discontinuously from zero to one. (The transition probabilities, so described, condition on parameter values. If the parameters are unknown then SETAR also leads to "soft" transitions through the posterior distribution of r as illustrated in Figure 4. But in the posterior distributions, one would expect to find greater ambiguity about states at particular times in SMR than in SETAR.)

In (2) the covariate vector is the same in each state. In general there are strong arguments for imposing this restriction due to identification issues for states; for further discussion see Geweke and Keane (2006) and Geweke (2007). In the case of the threshold model with the Canadian lynx data, however, the identification of states is so clear that these issues do not arise. Therefore, in the interest of achieving the greatest comparability with SETAR, the application of SMR in this discussion imposes $\beta_{24} = \ldots = \beta_{28} = 0$.

The applications of SMR in Geweke and Keane (2007) used samples of over 2,000 observations and models with two covariates. Applications of mixture of experts models, more generally, similarly use models with few covariates and large samples. I am unaware of any applications of such models with so many covariates and so few observations as this one. Thus the application is also an experiment in examining the limits of complex models used in conjunction with small data sets. The applications here use the same prior distributions as in the previous section for SETAR for the parameters β_{ji} and σ_j^2; the prior distributions of the parameters γ_j are independent $N(0, 100)$, the large standard deviation reflecting the limiting instances of SMR approaching SETAR. The results reported here for the application of SETAR to the lynx data utilize 10^5 iterations of the MCMC algorithm described in Geweke and Keane (2006). The initial values for the MCMC recursions are taken from the SETAR posterior distribution of the parameters β_{ji}, σ_j^2 and r, with $c = 20$ in (4). Without well-chosen initializations, the algorithm does not discover the set of SMR model parameters that, as described next, makes it similar to SETAR. Once in that neighborhood, however, the algorithm does not stray far and the sequence of MCMC draws mixes well, similar to the mixing in the SETAR MCMC algorithm portrayed in the upper left panel of Figure 3.

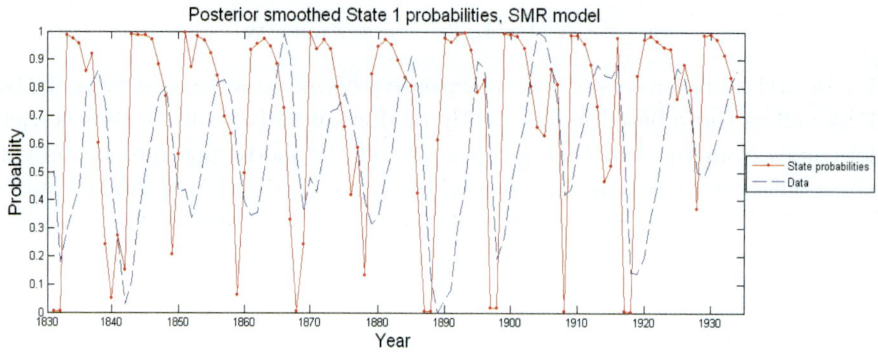

Fig. 5. Posterior probabilities of the event $s_t = 1$ for each year in the sample.

The SMR posterior distribution implies posterior state probabilities just as the SETAR posterior distribution does. These posterior probabilities are displayed in Figure 5, which is organized in the same way as Figure 4. As anticipated, smoothed state probabilities display less tendency to be very close to zero or one than they did in SETAR. However the patterns in the two figures are strikingly similar, and it is clear that the interpretation of the dynamics in the two models is similar as described earlier in this section.

Which model is preferred? In a Bayesian approach a standard answer to this question would be given by the Bayes factor

$$\frac{p(\mathbf{y}\mid SMR)}{p(\mathbf{y}\mid SETAR)} = \prod_{t=1}^{T} \frac{p(y_t \mid y_1, \ldots, y_{t-1}, SMR)}{p(y_t \mid y_1, \ldots, y_{t-1}, SETAR)},$$

where $T = 114$ is the sample size. Each of the T terms on the right is a predictive Bayes factor, with term t providing the multiplicative updating factor for the Bayes factor due to the observation y_t. The corresponding additive decomposition of the log Bayes factor is

$$\sum_{t=1}^{T} \left[\log p(y_t \mid y_1, \ldots, y_{t-1}, SMR) - \log p(y_t \mid y_1, \ldots, y_{t-1}, SETAR)\right]. \tag{5}$$

Reliable numerical approximations of the Bayes factor are difficult for these models, and in any event Bayes factors can be very sensitive to prior distributions. On the other hand, except for small values of t, the predictive Bayes factor is relatively insensitive to the prior distribution. Its approximation is straightforward as the average of the one-step-ahead predictive density averaged over the parameter values drawn from the posterior distribution.

The results of these computations are shown in Figure 6, which displays the log predictive Bayes factors for the last 30 observations in the sample. Half are positive and half are

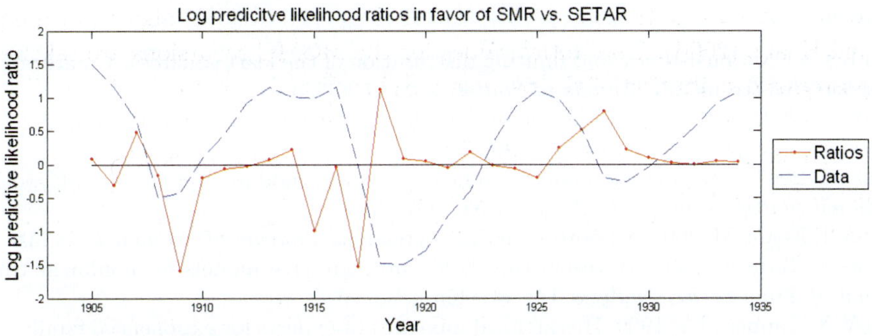

Fig. 6. Log predictive Bayes factors $\log p\,(y_t \mid y_1, \ldots, y_{t-1}, SMR) - \log p\,(y_t \mid y_1, \ldots, y_{t-1}, SETAR)$ for the last 30 observations in the sample.

negative, and the log Bayes factors to not appear to have any systematic association with the lynx cycle. The sum of these terms is the log predictive Bays factor for the last thirty observations jointly,

$$\log \frac{p\,(y_{T-29}, \ldots, y_T \mid y_1, \ldots, y_{T-30}, SMR)}{p\,(y_{T-29}, \ldots, y_T \mid y_1, \ldots, y_{T-30}, SETAR)} = -1.1192.$$

Thus the predictive Bayes factor favors SETAR over SMR by about 3:1.

4. Conclusion

Advances in computation have greatly increased the scope of practical nonlinear time series modeling, a field in which TL was an important pioneering achievement. The method of estimation in TL was maximum likelihood. As illustrated in Section 1, it is now straightforward to compute exact maximum likelihood estimates, which were not reported in TL. The concentrated likelihood function (Figure 2) underscores the difficulty in interpreting these estimates using conventional asymptotic distribution theory. Advances in computing since TL make exact Bayesian inference no more difficult – and, arguably, simpler – than was approximate maximum likelihood estimation when TL was written. Section 2 illustrated Bayesian inference for the same model and data set, showing how MCMC methods can be used to study posterior distributions of cyclical behavior in the context of the SETAR model.

In historical context, TL may be interpreted as formulating a nonlinear model for time series with one eye on the population properties of the model and the other eye on both the promise and limits of computation. Such considerations remain important in advances in nonlinear time series and in modelling more generally. Section 3 took up one recent innovation, the smoothly mixing regression model of Geweke and Keane (2007). It showed that their approach provides a foundation for flexible nonlinear time series models, and that TL is a limiting case of mixture of experts models. That section illustrated MCMC-based Bayesian inference for the small lynx data set in TL, suggesting that it may be possible to use these methods even in small sample sizes. At least for the lynx data, formal model comparison favors the SETAR of TL, but not overwhelmingly so. This underscores the insights of TL into nonlinear time series, while at the same time suggesting that smoothly mixing regression models or mixture of experts models may also prove useful in nonlinear time series modelling.

References

1. Chan KS. 1993. Consistency and limiting distribution of the least squares estimator of a threshold autoregressive model. *Annals of Statistics* **21**: 520-533.
2. Geweke J. 2004. Getting it right: Joint distribution tests of posterior simulators. *Journal of the American Statistical Association* **99**: 799-804.
3. Geweke J. 2007. Interpretation and inference in mixture models: Simple MCMC works. *Computational Statistics and Data Analysis* **51**: 3529-3550.
4. Geweke J, Keane M. 2007. Smoothly mixing regressions. *Journal of Econometrics***138**: 252-291.
5. Geweke J, Terui N. 1993. Bayesian threshold autoregressive models for nonlinear time series. *Journal of Time Series Analysis* **14**: 441-455.
6. Jiang WX, Tanner MA. 1999. Hierarchical mixtures-of-experts for exponential family regression models: Approximation and maximum likelihood estimation. *The Annals of Statistics* **27**: 987-1011.
7. Jordan MI, Jacobs RA. 1994. Hierarchical mixtures of experts and the EM algorithm. *Neural Computation* **6**: 181-214.
8. Koop G, Potter SM. 1999. Dynamic asymmetries in US unemployment. *Journal of Business and Economic Staistics* **18**: 298-312.
9. Lewis PAW, Stevens JG. 1991. Nonlinear modeling of time series using multivariate adaptive regression splines. *Journal of the American Statistical Association* **86**: 864-877.
10. Sorensen DA, Andersen S, Gianola D, Korsgaard I. 1995. Bayesian inference in threshold models using Gibbs sampling. *Genetics, Selection, Evolution* **27**: 229-249.
11. Stenseth NC, Falck W, Chan KS, Bjornstad ON, O'Donoghue M, Tong H, Boonstra R, Boutin S, Krebs CJ, Yoccoz NG. 1998. From patterns to processes: Phase and density dependencies in the Canadian lync cycle. *Proceedings of the National Academy of Sciences* **95**: 15430-15435.
12. Tong H, Lim KS. 1980. Threshold autoregression, limit cycles and cyclical data. *Journal of the Royal Statistical Society Series B* **42**: 245-292.

The Threshold Approach in Volatility Modelling

W. K. LI

University of Hong Kong, Department of Statistics and Actuarial Science
Pokfulam Road, Hong Kong, P. R. China
E-mail: hrntlwk@hku.hk

Extensions of Tong's threshold approach to other fields of statistics abound. Among these, the application of the threshold approach to model volatility changes in financial time series has been particularly noteworthy. This paper aims to give a brief survey on this vast and important development since the birth of the threshold autoregression models.

1. Introduction

The modelling of volatility is an important problem in many financial applications such as option pricing and the computation of the value-of-risk. It is a well established stylized fact that volatility in a bull market is somewhat less than the volatility in a bear one. Hence, asymmetry is exhibited in the volatility of many financial time series and this provides ample of room for employing and extending Tong's threshold model to this area. In fact, this development has been predicted in Tong (1990) where the so-called second-generation models are suggested. One of these is the SETAR-ARCH model

$$X_t = f(X_{t-1}, \cdots, X_{t-k}) + a_t, \tag{1}$$

where SETAR stands for self-excited threshold autoregression, ARCH stands for autoregressive conditional heteroscedasticity (Engle, 1982), $f(\cdot)$ is piece-wise linear, a_t is i.i.d. with mean 0 and conditional variance h_t given by

$$h_t = \alpha_0 + \sum_{i=1}^{m} \alpha_i a_{t-i}^2, \tag{2}$$

where $\alpha_0 > 0$ and $\alpha_i \geq 0$, $i = 1, \ldots, m$.

The above idea was first applied to the daily closing Hong Kong Hang Seng Index (HSI) by Li and Lam (1995), with $f(\cdot)$ satisfying the skeleton of a two-regime threshold autoregressive model. An interesting finding is that during the study period the autoregressive parameters are essentially positive when $X_{t-1} \geq 0$ while they are negative otherwise. This suggests not only that returns of the HSI has a larger chance of being positive but also that simply based on the second order analysis the so-called efficient market hypothesis is hard to reject because of the cancelation of signs in the parameters in the nonlinear $f(\cdot)$. The threshold idea clearly opens up a lot of possibilities in areas such as financial data analysis. The next section discusses a major extension of the threshold idea – ARCH types models with a threshold structure. In section 3, some other types of threshold models for volatility that require Bayesian treatment are discussed.

2. ARCH Models with a Threshold Structure

A more general specification than the SETAR-ARCH model and one that is closer to the original spirit of Tong (1978) and Tong & Lim (1980) is to allow the conditional variance to depend also on the market situation and hence admitting a threshold structure. This was taken up by Li and Li (1996). Let $\mathcal{F}_{t-1}$ be the σ-field generated by the random variables $\{a_{t-i}|i = 1, 2, \ldots\}$. For each t, given information $\mathcal{F}_{t-1}$, a_t is a normally distributed random variable, with mean zero and conditional variance, $E(a_t^2|\mathcal{F}_{t-1}) = h_t$, where $E(\cdot|\mathcal{F}_{t-1})$ denotes conditional expectation given $\mathcal{F}_{t-1}$. A time series $\{X_t\}$ is a Double-Threshold Autoregressive Conditional Heteroscedastic (DTARCH) process, if it satisfies,

$$\begin{aligned} X_t &= \Phi_0^{(j)} + \sum_{i=1}^{p_j} \Phi_i^{(j)} X_{t-i} + a_t, \quad r_{j-1} < X_{t-d} \leq r_j, \\ h_t &= \alpha_0^{(j)} + \sum_{r=1}^{q_j} \alpha_r^{(j)} a_{t-r}^2, \end{aligned} \tag{3}$$

where $j = 1, 2, \ldots, m$ and $d \geq 1$ is the delay parameter. The threshold parameters r_j satisfy $-\infty = r_0 < r_1 < r_2 \cdots < r_m = \infty$. Note that it is straightforward to allow for different delays and different sets of threshold parameters for the mean and variance. Further, the threshold effect might not be present in both the mean and the variance.

A full inference procedure was developed in Li and Li (1996). In particular, an iteravely weighted least squares scheme was proposed for maximum likelihood estimation and diagnostic checking procedures were derived for checking model adequacy.

The DTARCH model has been applied to the daily closing data of HSI and the results suggest that asymmetry could be present in the mean and the variance specification. Moreover, the down market regime usually has a larger conditional variance than that of the up market one which is consistent with observations in the financial market. The DTARCH model provides an easily understood alternative to the EGARCH model of Nelson (1991) and is akin to the floor and ceiling model of Pesaran and Potter (1997).

An alternative threshold ARCH model has been suggested by Rabemananjara and Zakoian (1993) where the ARCH-like structure is defined for the conditional standard deviation and the parameters take on different values depending on whether the corresponding X_{t-i} is positive or not. However, in their paper no threshold structure is considered for the conditional mean.

To facilitate statistical inference for the threshold structure the Chan and Tong likelihood ratio test (Chan and Tong, 1990) for testing for the presence of threshold autoregressive structure was extended to include the presence of ARCH feature in Wong and Li (1997). Extension to include the presence of a threshold ARCH feature was considered in Wong and Li (2000). The score test is considered in these two papers which requires only estimation under the null of no threshold structure. In general, in the presence of ARCH the empirical sizes of the Chan and Tong test were found to be much greater than the nominal sizes. Some critical values of the new tests are tabulated in these two papers.

A variant of the DTARCH model using a_{t-d} as the threshold indicator variable for h_t instead of X_{t-d} was considered in Liu, Li and Li (1997). An observable time series $\{X_t\}$ is a DTARCH model of order $(\ell_1, p_1, \ldots, p_{\ell_1}, \ell_2, q_1, \ldots, q_{\ell_2})$ if it satisfies

$$X_t = \sum_{j=1}^{\ell_1} \left[\phi_0^{(j)} + \sum_{i=1}^{p_j} \phi_i^{(j)} X_{t-i} \right] 1_{(X_{t-d} \in R_j)} + a_t, \tag{4}$$

$$\mathrm{Var}(a_t|\mathcal{F}_{t-1}) = h_t = \sum_{k=1}^{\ell_2} \left[\sigma_0^{(k)} + \sum_{r=1}^{q_k} \sum_{s=1}^{q_k} \sigma_{rs}^{(k)} a_{t-r} a_{t-s} \right] 1_{(a_{t-d'} \in R'_k)}, \tag{5}$$

where $E(a_t|\mathcal{F}_{t-1}) = 0$,

$$\sigma_0^{(k)} > 0, \quad \mathcal{F}_{t-1} = \sigma(X_{t-1}, \ldots,),$$

$$R_j = (r_{j-1}, r_j],\ j = 1, \ldots, \ell_1,\ R'_k = (r'_{k-1}, r'_k],\ k = 1, \ldots, \ell_2,\ \{R_j\} \text{ and } \{R'_k\}$$

constitute two different partitions of the real line, $\Sigma^{(k)} = [\sigma_{rs}^{(k)}]$ is nonnegative definite and d and d' are nonnegative integers. Ling (1999) considered an extension of DTARCH model to include a threshold autoregressive moving average (ARMA) conditional mean and also a threshold generalized ARCH (GARCH) specification (Bollerslev, 1986) and established the conditions for strict stationarity and finiteness of moments.

Direct generalizations to cover the generalized ARCH setting has also been considered by Brooks (2001). Here, the h_t specification in (3) is replaced by,

$$h_t = \alpha_0^{(j)} + \sum_{r=1}^{q_j} \alpha_r^{(j)} a_{t-r}^2 + \sum_{k=1}^{p_j} \beta_k^{(j)} h_{t-k}. \tag{6}$$

Smooth transition double threshold model was studied in Lee and Li (2000). Unlike Chan and Tong (1986) a logistic transition function was used instead of the Gaussian cumulative distribution function. While the theory in Chan and Tong (1986) applies to all distributions the logistic function makes the technical details a bit easier. However, estimation of the parameters in the transition function seems to require a larger sample size than that of the classical threshold case.

3. Bayesian Inference for Threshold Volatility Models

An alternative to the ARCH models is the stochastic volatility (SV) model proposed by Taylor (1982),

$$\begin{aligned} X_t &= \psi_0 + \psi_1 X_{t-1} + a_t, \\ a_t &= u_t \sqrt{h_t}, \quad u_t \sim N(0,1), \\ \log h_{t+1} &= \alpha + \phi \log h_t + \eta_t, \ , \quad \eta_t \sim N(0, \sigma^2), \end{aligned} \tag{7}$$

where u_t and η_t are independent white noise. Estimation of (7) can be based on the expectation maximization algorithm as in Harvey and Shephard (1993) resulting in quasi-maximum likelihood estimates. However, estimation of such models can be done more efficiently using a Bayesian approach by adopting the Markov chain Monte Carlo (MCMC) method (Jacquier, Polson and Rossi, 1994). So, Li and Lam (2002) considered a threshold stochastic volatility model (THSV) as follows:
Define a set of Bernoulli random variables s_t by

$$s_t = \begin{cases} 0 & \text{if} \quad X_{t-1} < 0, \\ 1 & \text{if} \quad X_{t-1} \geq 0. \end{cases} \tag{8}$$

The THSV model is then given by the following:

$$\begin{aligned} X_t &= \psi_{0s_t} + \psi_{1s_t} X_{t-1} + a_t \\ a_t &= \sqrt{h_t} u_t \quad u_t \sim N(0,1) \\ \log h_{t+1} &= \alpha_{s_{t+1}} + \phi_{s_{t+1}} \log h_t + \eta_t \ , \quad \eta_t \sim N(0, \sigma^2) \end{aligned} \tag{9}$$

where u_t and η_t are stochastically independent. At time $t-1$, when there is an unexpected drop in price due to the presence of bad news, $X_{t-1} < 0$ and $s_t = 0$. In contrast, if there is good news at time $t-1$, then $X_{t-1} \geq 0$ and $s_t = 1$. Therefore, the value of s_t is determined by the sign of X_{t-1}. In the THSV model, the parameters ψ_0, ψ_1, α and ϕ switch between the two regimes corresponding to the rise and fall in asset prices. In the symmetric case, the two sets of parameters are identical. In particular, if $\phi_0 = \phi_1$, then $\alpha_0 \geq \alpha_1$ implies that variance is higher when the past return is negative than when it is positive. In the general model, ϕ_0 can be different from ϕ_1. Here ϕ_{s_t} measures the effect of the previous variance on the current variance. If ϕ_0 is greater than ϕ_1, the previous variance will have greater impact on the present variance after falls in price than after rises in price. In this hypothetical situation, it is expected that it will take longer for the bad news contained in the previous variance to be digested by the market. This kind of asymmetry has not been considered in the stochastic volatility literature.

Bayesian methodology has also been developed for the family of threshold GARCH models. So, Chen and Chen (2005) proposed a threshold nonlinearity test to distinguish GARCH and threshold GARCH models by adopting the reversible jump Markov chain Monte Carlo method to calculate the posterior probabilities of the two competing models. Chen and So (2006) considered a DTARCH model with regime indicator given by a weighted average of auxiliary variables where estimation is done by the MCMC method.

Fractionally integrated autoregressive moving average (ARFIMA) models (Hosking, 1981) has been popular with time series exhibiting the so-called long memory property where the autocorrelation function decays hypobolically rather than exponentially. Chen and Yu (2005) considered ARFIMA models with conditional variance modelled by a threshold GARCH specification. Estimation is again done by the MCMC approach. To study causality between stock returns in different countries Chen, Chiang and So (2003) considered the following double TAR-GARCH model:

$$R_t^i = \begin{cases} \phi_0^{(1)} + \phi_1^{(1)} R_{t-1}^i + \psi_1^{(1)} R_{t+m-d}^j + a_t, & \text{if } R_{t+m-d}^j \leq r, \\ \phi_0^{(2)} + \phi_1^{(2)} R_{t-1}^i + \psi_1^{(2)} R_{t+m-d}^j + a_t, & \text{if } R_{t+m-d}^j > r, \end{cases} \tag{10}$$

$$h_t = \begin{cases} \alpha_0^{(1)} + \alpha_1^{(1)} a_{t-1}^2 + \beta_1^{(1)} h_{t-1}, & \text{if } R_{t+m-d}^j \leq r, \\ \alpha_0^{(2)} + \alpha_1^{(2)} a_{t-1}^2 + \beta_1^{(2)} h_{t-1}, & \text{if } R_{t+m-d}^j > r, \end{cases} \tag{11}$$

where i and j are country indices, R_t^i is the return of the i-th market index at time t and m is the time difference between the i and j markets. Note that R_{t+m-d}^j is exogenous to R_t^i. All the above models are but a small sample of the many models for volatility that make use of the threshold approach of Tong & Lim (1980).

Conclusion

It can be seen from the previous sections that the threshold approach has found very fruitful applications in the modelling of the (conditional) variance or volatility process. In particular, the idea has been successfully employed in the field of financial time series. Because of the limitation of space we will not discuss the many other possibilities in detail. These possibilities include the threshold autoregressive conditional duration models of Zhang, Russell and Tsay (2001) which is useful in the study of high frequency financial data. Multivariate threshold GARCH models with dynamic correlation has been studied by Kwan, Li and Ng (2005) and threshold direct value-at-risk models by Jin, Li and Yu (2004). As stressed by Tong and Lim (1980) in their rejoinder, the regime indicator variable can be defined in quite a general way, for example, it can follow a Markov chain process. This is an idea of great foresight and in fact this predates and foretells the use of hidden Markov models which has been so widely used in econometrics (Hamilton, 1989). For volatility modelling, Markov switching ARCH model has been considered by Cai (1994) and Hamilton and Susmel (1994). Markov switching stochastic volatility model was considered by So, Lam and Li (1998). The comment also covers the mixture models in volatility, as discussed in Wong and Li (2001), Zhang, Li and Yuen (2006) and Gray (1996), which have received some attention in the literature recently. The idea of using thresholds to approximate nonlinearity is clearly one of those few great ideas in science that is simple yet encompassing. More novel and influential use of the idea could just be around the corner.

Acknowledgment

The author would like to thank two referees for their helpful comments and the Hong Kong Research Grant Council grant HKU 7036/06P for partial support.

References

1. Bollerslev, T. (1986). Generalized autoregressive conditional heteroskedasticity. *Journal of Econometrics* **31**, pp. 307–327.
2. Brooks, C. (2001). A double-threshold GARCH model for the French Franc/ Deutschmark exchange rate. *Journal of Forecasting* **20**, pp. 135–143.
3. Cai, J. (1994). A Markov model of switching-regime ARCH. *Journal of Business & Economic Statistics* **12**, 309–316.
4. Chan, K. S. and Tong, H. (1986). On estimating thresholds in autoregressive models. *Journal of Time Series Analysis* **7**, pp. 179–194.
5. Chan, K. S. and Tong, H. (1990). On likelihood ratio tests for threshold autoregression. *Journal of Royal Statistical Society* **B52**, pp. 469–476.
6. Chen, C. W. S., Chiang, T. C. and So, M. K. P. (2003). Asymmetrical reaction to US stock-return news: evidence from major stock markets based on a double-threshold model. *Journal of Economics and Business* **55**, pp. 487–502.
7. Chen, C. W. S. and So, M. K. P. (2006). On a threshold heteroscedastic model. *International Journal of Forecasting*, **22**, 73–89.
8. Chen, C. W. S. and Yu, T. H. K. (2005). Long-term dependence with asymmetric conditional heteroscedasticity in stock returns. *Physica A* **353**, pp. 413–424.
9. Engle, R. F. (1982). Autoregressive conditional heteroskedasticity with estimates of the variance of U.K. inflation. *Econometrica* **50**, pp. 987–1008.
10. Gray, S. F. (1996). Modeling the conditional distribution of interest rates as a regime-switching process. *Journal of Financial Economics* **42**, pp. 27–62.
11. Hamilton, J. D. (1989). A new approach to the economic analysis of nonstationary time series and the business cycle. *Econometrica* **57**, pp. 357–384.
12. Hamilton, J. D. and Susmel, R. (1994). Autoregressive conditional heteroscedasticity and changes in regime. *Journal of Econometrics* **64**, pp. 307–333.
13. Harvey, A. C. and Shephard, N. (1993). Estimation and testing of stochastic variance models. *STICERD Econometrics Discussion Paper* LSE.
14. Hosking, J. R. M. (1981). Fractional differencing, *Biometrika* **68**, pp. 165–176.
15. Jacquier, E., Polson, N. G. and Rossi, P. E. (1994). Bayesian analysis of stochastic volatility models. *Journal of Business and Economic Statistics* **12**, pp. 371–389.

16. Jin, S., Li, W. K. and Yu, P. L. H. (2004). On some models for value-at-risk. To appear in *Econometric Reviews.*
17. Kwan, K. C., Li, W. K. and Ng, K. (2005). A multivariate threshold varying conditional correlation model. To appear in *Econometric Reviews.*
18. Lee, Y. N. and Li, W. K. (2000). On smooth transition double threshold models. In *Statistics and Finance: An Interface*, 205–225. Editors W. S. Chan, Li, W. K. and Tong, H., Imperial College Press.
19. Li, C. W. and Li, W. K. (1996). On a double threshold autoregressive heteroskedastic autoregressive time series model. *Journal of Applied Econometrics* **11**, pp. 253–274.
20. Li, W. K. and Lam, K. (1995). Modelling asymmetry in stock returns by a threshold ARCH model. *The Statistician* **44**, pp. 333–341.
21. Ling, S. (1999). On the probabilistic properties of a double threshold ARMA conditional heteroskedastic model. *Journal of Applied Probability* **36**, pp. 688–705.
22. Liu, J., Li, W. K. and Li, C. W. (1997). On a threshold autoregressive with conditional heteroscedastic variances. *Journal of Statistical Planning and Inferences* **62**, pp. 279–300.
23. Nelson, D. B. (1991). Conditional heteroskedasticity in asset returns: a new approach. *Econometrica* **59**, pp. 347–370.
24. Pesaran, H. and Potter, S.M. (1997). A floor and ceiling modeling of US output. *Journal of Economic Dynamics and Control* **21**, pp. 661–695.
25. Rabemananjara, R. and Zakoian, J. M. (1993). Threshold ARCH models and asymmetries in volatility. *Journal of Applied Econometrics* **8**, pp. 31–49.
26. So, M. K., Chen, C. W. S. and Chen, M. T. (2005). A Bayesian threshold nonlinearity test for financial time series. *Journal of Forecasting* **24**, pp. 61–75.
27. So, M. K. P., Lam, K. and Li, W. K. (1998). A stochastic volatility model with Markov switching. *Journal of Business and Economic Statistics* **16**, pp. 244–253.
28. So, M. K. P., Li, W. K. and Lam, K. (2002). A threshold stochastic volatility model. *Journal of Forecasting* **21**, pp. 473–500.
29. Taylor, S. J. (1982). Financial returns modelled by the product of two stochastic processes, a study of daily sugar prices 1961–79. In *Time Series Analysis: Theory and Practice 1*, 203–226, Anderson OD (ed.). North-Holland, Amsterdam.
30. Tong, H. (1978). On a threshold model, in C. H. Chen (ed.), *Pattern Recognition and Signal Processing*, 575–586, Sijthoff and Noordhoff, Amsterdam.
31. Tong, H. (1990). *Nonlinear Time Series: A Dynamical Systems Approach.* Oxford University Press: Oxford.
32. Tong, H. and Lim, K. S. (1980). Threshold autoregression, limit cycles and cyclical data (with discussion). *Journal of the Royal Statistical Society B* **42**, pp. 245–292.
33. Wong, A. C. S. and Li, W. K. (2001). On a mixture of autoregressive conditional heteroscedastic model. *Journal of the American Statistical Association* **96**, pp. 982–995.
34. Wong, C. S. and Li, W. K. (1997). Testing for threshold autoregression with conditional heteroscedasticity. *Biometrika* **84**, pp. 407–418.
35. Wong, C. S. and Li, W. K. (2000). Testing for double threshold autoregressive conditional heteroscedastic model. *Statistica Sinica* **10**, pp. 173–189.
36. Zhang, M. Y., Russell, J. R. and Tsay, R. S. (2001). A nonlinear autoregressive conditional duration model with application to financial transaction data. *Journal of Econometrics* **104**, pp. 179–207.
37. Zhang, Z., Li, W. K. Li and Yuen, K. C. (2006). On a mixture GARCH time series model. *Journal of Time Series Analysis* **27**, pp. 577–597.

Dependence and Nonlinearity

MURRAY ROSENBLATT
University of California, San Diego
Department of Mathematics
9500 Gilman Drive #0112, La Jolla, CA 92093, USA
mrosenbl@math.ucsd.edu

The autoregressive moving average (ARMA) sequences have been used as time series models for a long time. It was clear that these linear models could not exhibit effects characteristic of data gathered on various wildlife populations and in their paper Tong and Lim 1980 proposed using threshold autoregressive schemes as a class of nonlinear models that might capture some of these effects. An ARMA sequence satisfies a system of equations

$$\sum_{j=0}^{p} a_j x_{n-j} = \sum_{k=0}^{q} b_k \varepsilon_{n-k} \tag{1}$$

with the ε_n's independent, identically distributed random variables. If $E\varepsilon_n^2 < \infty$, $E\varepsilon_n = 0$ and the polynomial $a(z)$ has its roots outside the unit disc, there is a unique stationary solution to the system of equations. Let $\overline{X}_{n-1} = (X_{n-1}, \ldots, X_{n-k})$ and let $\overline{R} = R_1 \cup \cdots \cup R_s$ be a partition of s-dimensional space. Assume that $a_1(\overline{X}), \ldots, a_k(\overline{X})$ are constant on each set of the partition but may take on different values on different sets of the partition. A system of the form

$$X_n - \sum_{j=1}^{k} a_j(\overline{X}_{n-j}) = \varepsilon_n \tag{2}$$

with the ε_n's iid is a simple example of a threshold autoregressive scheme. The system (2) provides the framework of a Markov process as does (1) but the process is now a nonlinear rather than a linear process. Questions of existence of a Markov process as a solution of the system, whether it is stationary, geometrically ergodic, etc. naturally occur. One should also note that the coefficients of the scheme are discontinuous. Aspects of the theory of Markov processes (see Meyn and Tweedie 1993) have been used to establish existence of solutions and examine properties of the solutions. The paper of Tong and Lim stimulated much interest and research on threshold autoregressive models.

It is worthwhile to consider recent work of Wu which is set in the context of earlier research. In 1958 Wiener posed the question as to when a stationary sequence $\{X_n\}$ can be represented as a function of a one-sided sequence $\{\xi_n, \xi_{n-1}, \cdots\}$ of iid random variables and its shifts. It is clear that the random variables $\{\xi_n\}$ could be taken as uniformly distributed on $[0, 1]$. Let $\mathcal{B}_n = \mathcal{B}\{X_j,\ j \leq n\}$ be the σ-field generated by the random variables $X_j,\ j \leq n$. A process with such a one-sided representation must be purely nondeterministic in the sense that the σ-field $\mathcal{B}_{-\infty} = \bigcap_n \mathcal{B}_n$ must be the trivial σ-field consisting of the empty set and the whole probability space (see Rosenblatt 1971). It is still an open question as to whether the process being purely nondeterministic is a sufficient condition for such a representation. We shall later on comment on reversibility, a topic that Tong has found interesting. Univariate stationary Gaussian processes which motivated much early research in times series analysis are reversible, i.e. the probability structure of the process is the same as the probability structure of the process with time reversed. Most stationary processes do not have this property of reversibility which is also the case for many natural phenomena, e.g. hysteresis in magnetization. One can already see aspects of this here. There are many purely nondeterministic processes which are not purely nondeterministic with time reversed. A simpler classical example is given by the first order autoregressive scheme

$$X_n = \frac{1}{2} X_{n-1} + \varepsilon_n \tag{3}$$

with the ε_n's iid binary random variables with $P(\epsilon_n = 0) = P(\epsilon_n = 1) = \frac{1}{2}$. The example is noted in Rosenblatt 1964 and was pointed out to me by B. Jamison.

Wu (2005), Shao and Wu (2007) consider a stationary process $X_n = G(\ldots, \varepsilon_{n-1}, \varepsilon_n)$ with the ε_n's iid. Let (ε'_n) be an independent iid copy of (ε_n). Set $X'_n = G(\ldots, \varepsilon'_{-1}, \varepsilon'_0, \varepsilon_1, \ldots, \varepsilon_n)$, a coupled counterpart of X_n. Wu called X_n $GMC(\alpha)$, $\alpha > 0$, if there are G and ρ, $0 < \rho(\alpha) < 1$, $C > 0$ such that for all integers $n > 0$

$$E(|X'_n - X_n|^\alpha) \leq C\rho^n.$$

Under appropriate conditions a class of threshold models can be shown to be $GMC(\alpha)$ for some $\alpha > 0$. A condition like this is often useful in dealing with nonlinear models and proving limit theorems for such processes.

It is clearly of interest to determine when a discrete parameter ARMA scheme can be interpolated so as to get a continuous time ARMA scheme written formally as

$$\sum_{k=0}^{\gamma} \alpha_k x^{(k)}(t) = \sum_{j=0}^{s} \beta_j \varepsilon^{(j)}(t) \tag{4}$$

with $\varepsilon(t)$ a process of independent increments and $x^{(k)}(t)$, $\varepsilon^{(j)}(t)$ the formal k^{th} and j^{th} derivatives of $x(t)$, $\varepsilon(t)$. Such an interpolation cannot be carried out generally. But characterizing when it can be carried out has not been fully resolved. Consistent with this, attention has been drawn to continuous time threshold autoregressive processes. A first order threshold autoregressive model of a simple character leads one to an equation of the form

$$dx(t) = b(x_t)dt + \sigma(x_t)dB(t) \tag{5}$$

with $b(y) = a_j y + c_j \sigma_j$, $\sigma(y) = \sigma_j$ with $j = 1, 2$ according as $y > 0$ or $y \leq 0$. Here let us assume that $B(t)$ is the Brownian motion process. The equation (5) then has the character of a stochastic diffusion equation but with possibly discontinuous coefficients. Call a measurable function f locally integrable $f \in L'_{loc}(d)$ at a point d if there is $\delta > 0$ such that

$$\int_{d-\delta}^{d+\delta} |f(x)| dx < \infty$$

f is said to be locally integrable on a set D, $f \in L'_{loc}(D)$, if $f \in L'_{loc}(d)$ for each point $d \in D$. Under the assumption $(1+|b|)/\sigma^2 \in L'_{loc}(R)$ with R the real line, Cherny and Engelbert 2005 determine conditions for existence of a global solution of (5). They also examine the case of an isolated point d with $(1+|b|)/\sigma^2 \notin L'_{loc}(d)$ which they call a singular point. Stramer, Tweedie, and Brockwell (1996) and Brockwell and Williams (1997) have considered related questions for threshold autoregressive processes of 1st and 2nd order.

Even though Gaussian univariate stationary sequences are reversible, this is generally not the case for stationary vector-valued Gaussian sequences. A simple example is given by the 2 vector Gaussian sequence $X_n = (X_n^{(1)}, X_n^{(2)})$ with $X_n^{(1)} = Y_n$, $X_n^{(2)} = Y_{n+1}$ and Y_n a sequence of independent, identically distributed normal random variables with mean 0 and variance 1. A necessary and sufficient condition for reversibility for vector-valued Gaussian stationary sequences is that the covariance matrices be symmetric. In the usual definition of a complex multivariate normal vector X with covariance matrix $\sum$, the covariance matrix of (ReX, ImX) is

$$\frac{1}{2}\begin{pmatrix} \mathrm{Re}\sum & -\mathrm{Im}\sum \\ \mathrm{Im}\sum & \mathrm{Re}\sum \end{pmatrix}. \tag{6}$$

If $X(t) = (X^{(1)}(t), X^{(2)}(t))$ is a 2 vector-valued weakly stationary process, $EX(t) = 0$, then

$$X^{(i)}(t) = \int_0^\pi \cos t\lambda dz_1^{(i)}(\lambda) + \int_0^\pi \sin t\lambda dz_2^{(i)}(\lambda),$$
$$Edz_i^{(k)}(\lambda)dz_j^{(k)}(\mu) = 2\delta_{i,j}\delta_{\lambda,\mu}dG_{i,j}(\lambda),$$

and

$$\begin{aligned} Edz_i^{(1)}(\lambda)dz_i^{(2)}(\mu) &= 2\delta_{\lambda,\mu}\mathrm{Re}dG_{1,2}(\lambda) \\ Edz_1^{(1)}(\lambda)dz_2^{(2)}(\mu) &= -Edz_2^{(1)}(\lambda)dz_1(\mu) \\ &= 2\delta_{\lambda,\mu}\ \mathrm{Im}\ dG_{1,2}(\lambda) \end{aligned}$$

with $i, j, k = 1, 2$ as well as $\lambda, \mu \geq 0$. Here $z(\lambda) = (z_1(\lambda), z_2(\lambda))$ is the random spectral function of $X(t)$ and $G(\lambda) = (G_{i,j}(\lambda), i, j = 1, 2)$ the matrix-valued spectral distribution function of the process. If $X(t)$ is a complex multivariate normal stationary process in the sense that every finite collection of $X(t)$ random variables is complex normal in the sense of (6), then

$$\begin{aligned} G_{11}(\lambda) &= G_{22}(\lambda) \\ \mathrm{Re}\ G_{1,2}(\lambda) &\equiv 0. \end{aligned}$$

If we consider a strictly stationary complex-valued process $X(t) = X^{(1)}(t) + iX^{(2)}(t)$ with real and imaginary parts $X^{(1)}(t)$ and $X^{(2)}(t)$ jointly normal, $G(\lambda)$ the spectral distribution function of $(X^{(1)}(t), X^{(2)}(t))$, then the spectral distribution function of $X(t)$ is given by

$$F(\lambda) = G_{11}(\lambda) + G_{2,2}(\lambda) + 2\ \mathrm{Im}\ G_{1,2}(\lambda)$$

(it is understood that $dG(\lambda) = \overline{dG(-\lambda)}$). If the process is reversible

$$\text{cov}\ (X^{(1)}(t), X^{(2)}(\tau)) = \text{cov}\ (X^{(1)}(-t), X^{(2)}(-\tau))$$

implying that $\text{Im}\ G_{1,2}(\lambda) = 0$ and hence that the mass of F is symmetric about zero. The converse also holds in this case. Notice that a complex stationary process (satisfying (6)) needn't be reversible. Of course, these are simple remarks concerning their reversibility in complexified form. Tong and colleagues discuss reversibility for vector-valued ARMA schemes in their paper of 2006.

In Rosenblatt (1961) a simple example was given of a stationary non Gaussian sequence whose normalized partial sums converged in distribution to a nonnormal distribution. This was an example of a process with long-range dependence. Let y_k be the stationary Gaussian sequence with mean zero and covariance sequence $r_k = (1+k^2)^{-\gamma}$ with $0 < \gamma < \frac{1}{4}$. Set $x_k = y_k^2 - 1$. The sequence x_k is stationary with covariance function $r_k = 2(1+k^2)^{-2\gamma}$. One can show that $n^{-1+2\gamma} \sum_{k=1}^{n} x_k$ has the non Gaussian limiting distribution with characteristic function

$$\exp\left\{ \sum_{k=2}^{\infty} (2it)^k c_k / k \right\}$$

where

$$c_k = \int_0^1 \cdots \int_0^1 |x_1 - x_2|^{-2\gamma} |x_2 - x_3|^{-2\gamma} \ldots |x_k - x_1|^{-2\gamma} dx_1 \ldots dx_k.$$

The processes can be viewed as long–range dependent in the sense that their covariance functions are not finitely summable in k. In 1979 Dobrushin and Major, and Taqqu in separate papers obtained extensive results on limiting nonnormal distributions for nonnormal nonlinear stationary functions of normal variables. Since then there has been extensive research in theory and applications on the distinction between processes with long and short range dependence (see Doukhan et al 2003).

This is intended as a brief set of remarks on some work related to that of Howell Tong on dependence and nonlinearity, research that was stimulated and motivated by his many researches in this area or related to his interests.

Addendum

In the paper "A comment on a conjecture of N.Wiener" (2009) Stat. and Prob. Letters 79, 347–348, I have shown that having a stationary sequence purely nondeterministic is not sufficient generally for a one-sided representation in terms of iid random variables.

References

1. Brockwell, P. and Williams, R. "On the existence and application of continuous-time threshold autoregressions of order two" Adv. in Appl. Probab. 29 (1997) 205–227
2. Chan, K.S., Ho, L., and Tong, H. "A note on time-reversibility of multivariate linear processes" Biometrika (2006) 93, 221–227
3. Cherny, A. and Engelbert, H. *Singular Stochastic Differential equations*, Springer 2005
4. Dobrushin, R. and Major, P. "Non-central limit theorems for non-linear functionals of Gaussian fields" Z. Wahrsch. Verw. Gebiet (1979) 50, 27–52

5. Doukhan, P., Oppenheim, G., and Taqqu, M. (Editors). *Theory and applications of long–range dependence*, Birkhauser 2003
6. Meyn, S. and Tweedie, R. *Markov chains and stochastic stability*, Springer 1993
7. Rosenblatt, M. "Independence and dependence" Proc. Fourth Berkeley Symp. Math. Statist. (1961) 431–443 Univ. Calif. Press
8. Rosenblatt, M., "Some nonlinear problems arising in the study of random processes" Radio Sci. J. Research, NBS (USNC–URSI) 68D (1964), 933–936
9. Rosenblatt, M. *Markov processes, structure and asymptotic behaviour*, Springer 1971
10. Shao, X. and Wu, W. "Asymptotic spectral theory for non-linear time series" Ann. Statist. (2007)
11. Stramer, O., Tweedie, R. and Brockwell, P. "Existence and stability of continuous time threshold ARMA processes" Statist. Sinica 6 (1996), 715–732
12. Tong, H. *Non-linear time series*, Oxford 1990
13. Tong, H. and Lim, K. "Threshold autoregression, limit cycles and cyclical data" J.R. Statist. Soc. B (1980) 42, 245–292
14. Taqqu, M. "Convergence of integrated processes of arbitrary Hermite rank." Z. Wahrsch Verw. Gebiete (1979) 50, 53–83
15. Wiener, N., *Nonlinear problems in random theory*, John Wiley 1958
16. Wu, W., "Nonlinear system theory: another look at dependence " Proc. Natl. Acad. Sci. USA 102 (2005), no. 40, 14150–14154

The Threshold Approach: An Appreciation

RUEY S. TSAY

Booth School of Business, University of Chicago
5807 S. Woodlawn Avenue, Chicago, IL 60637, USA
E-mail: ruey.tsay@chicagobooth.edu

1. Introduction

It is my honor to congratulate Professor Howell Tong for exceeding yet another important threshold in his life. This is particularly significant, for his research productivity does not show any sign of aging. I am also deeply privileged to have the opportunity to comment on his fundamental and path-breaking article *Threshold Autoregression, Limit Cycles and Cyclical Data* (joint with Lim and published in JRSSB, 1980, with discussion), which recently earned him the Guy medal in silver from the Royal Statistical Society.

2. Comment

Many nonlinear time series models, new and old, have been proposed in the literature; see, for instance, the models discussed in Tong (1990). However, it is the threshold approach of Tong (1978) that has surpassed the key threshold of being one of the oldest and widely used nonlinear models. The threshold model popularized by Tong and Lim (1980) has found many successful applications in diverse fields, including hydrology, ecology, economics, finance, and public health. I was fortunate enough to meet Howell in the 80s and to have the opportunity to study his threshold approach in my early academic career. A quarter of century later, I still find myself thrilled by the idea and presentation of this seminal paper.

An influential paper often consists of several important ingredients. First, it develops a new methodology or proposes new statistical methods for solving practical problems. Examples include the proportional regression model of Cox (1972), the longitudinal data analysis of Liang and Zeger (1986), and the Markov chain Monte Carlos (MCMC) methods of Gelfand and Smith (1990). The threshold model to time series is the Cox regression to survival analysis. Tong and Lim (1980) develops a methodology that enables time series analysts to employ simple nonlinear models in describing observed phenomena. Most, if not all, real-world time series are indeed nonlinear. Second, the article must be simple, easy to understand and is based on sound statistical reasoning. Using simple threshold models throughout the article, Tong and Lim (1980) has provided ample theoretical reasons to support the threshold models. Their simple threshold models, which possess limit cycles and jump resonance, are insightful and informative. Third, it must contain multiple examples of good applications. Via rigorous and penetrating analysis of the lynx data, sunspot numbers, mink-muskrat data, and Kanna riverflow and rainfall series, Tong and Lim (1980) has successfully demonstrated the wide applicability of the threshold model. Fourth, it must be presented in such a way that attracts reader's attention and imagination. The use of jump resonance and limit cycle in Tong and Lim (1980) is illuminating and can be an eye opener for those who are used to seeing theoretical derivations in a statistical article. Finally, the article is often incomplete and invites criticisms and further study. Although Tong and Lim made it clear in every opportunity they could that their use of the Akaike information criterion (AIC) to select a threshold model is only preliminary and, indeed, they made heavy use of other statistical tools in their real data analysis, the article still invites several discussants to raise concerns about model identification. These concerns are well justified, and history has proved that both the authors and discussants are correct. The AIC remains a useful identification tool in threshold modeling, but significant advances have been made to ease the difficulty in specifying a threshold model in time series analysis. I was lucky enough to make some minor contributions in this area; see Tsay (1989). Many properties of threshold models, even for the subclass of self-exciting threshold autoregressive models, remain unknown, e.g., the ergodicity of general self-exciting threshold autoregressive models. However, much important progress concerning threshold models has been made over the past 25 years. See, for example, the ergodicity condition of Petruccelli and Woolford (1984) and Chen and Tsay (1991), the limiting properties of least squares estimates of Chan (1993) and Chan and Tsay (1998), among many others. From a Bayesian perspective, MCMC methods have been successfully used in analysis of threshold autoregressive models. See, for example, Geweke and Terui (1993) and Koop and Potter (1999). It is fair to say that more methodological advances and interesting applications of the threshold models would continue to appear in the future.

3. The Simple Idea that Works

Piecewise linear models or local linear models are commonly used in statistics. They existed well before the threshold models were proposed. What distinguishes threshold models from other piecewise linear models is the introduction of the threshold variable. The linear approximation to the true model is then achieved in the threshold space. The fact that water turns into ice at $32^o F$ is just one of many real-life examples of how threshold works. It is then not surprising to see wide acceptance of threshold models in so many diverse fields. Consider, for example, the application in finance. The idea of threshold fits nicely with the concept of no arbitrage, which plays an important role in asset pricing. Consider the prices of a product in two different cities. Assume that the tariffs on the product are fixed and finite at the two cities. Then, the two prices must be the same except for the transportation cost and the difference in taxes. Let x be the price difference and y be the sum of transportation cost and difference in taxes. If $x > y$, then one can buy the product at the cheaper city and sells it at the more expensive one for a net profit of $x - y > 0$. This profitable strategy is referred to as an arbitrage opportunity in finance and is exploited by market participants if it exists. When mutliple market participants try to capitalize the arbitrage profit, the competition would force the value of x to decrease and, hence, close out the arbitrary opportunity. Therefore, the market force would force the prices of the product to be sufficiently close between the two cities, implying that there is no arbitrary opportunity. Consequently, if we were to study the dynamic structure of the prices at the two cities, the price difference would play an important role as a price equalizer with the sum of transportation cost and difference in taxes being the cut-off point. In statistical term, the price difference would become the threshold variable and the sum of transportation cost and difference in taxes would be the threshold. If the prices are unit-root nonstationary, then they must have threshold co-integration in the sense that the prices can behave as a random walk individually when the price difference is small, but the market force (or arbitrary opportunity) would not allow the prices to deviate substantially from each other. For more information on threshold co-integration and arbitrage in finance, readers are referred to Tsay (2005, Section 8.7) and the references therein.

This simple and useful concept to capture the market dynamics is the beauty of the threshold approach. After so many years, I continue to admire the authors for proposing such a simple model that can elegantly describe the essence of many real-world phenomena. The article of Tong and Lim (1980) may not be the favorite of some time series analysts, but it certainly exceeds the threshold of being an influential and fundamental paper in nonlinear time series analysis. It is my great pleasure to add my vote of thanks to Howell.

References

1. Chan, K. S. (1993). Consistency and limiting distribution of the least squares estimator of a threshold autoregressive model. *The Annals of Statistics*, 21, 520-533.
2. Chan, K. S. and Tsay, R. S. (1998). Limiting properties of the conditional least squares estimator of a continuous TAR model. *Biometrika*, 85, 413-426.
3. Chen, R. and Tsay, R. S. (1991). On the ergodicity of TAR(1) processes. *Annals of the Applied Probability*, 1, 613-634.
4. Cox, D. R. (1972). Regression models and life tables (with discussion). *Journal of the Royal Statistical Society, Series B*, 34, 187-220.
5. Gelfand, A. E. and Smith, A. F. M. (1990). Sampling-based approaches to calculating marginal densities. *Journal of the American Statistical Association*, 85, 398-409.
6. Geweke, J. and Terui, N. (1993). Bayesian threshold autoregressive models for nonlinear time series. *Journal of Time Series Analysis*, 14, 441-455.
7. Koop, G. and Potter, S. M. (1999). Dynamic asymmetries in US unemployment. *Journal of Business and Economic Statistics*, 18, 298-312.
8. Liang, K. Y. and Zeger, S. L. (1986). Longitudinal data analysis using generalized linear models. *Biometrika*, 73, 13-22.
9. Petruccelli, J. and Woolford, S. W. (1984). A threshold AR(1) model. *Journal of Applied Probability*, 21, 270-286.
10. Tong, H. (1978). On a threshold model. In *Pattern Recognition and Signal Processing*, ed. C.H. Chen. The Netherlands: Sijthoff and Noordhoff.
11. Tong, H. (1990). *Non-linear Time Series: A Dynamical System Approach.* Oxford University Press, Oxford, U.K..
12. Tsay, R. S. (1989). Testing and modeling threshold autoregressive processes. *Journal of the American Statistical Association*, 84, 231-240.
13. Tsay, R. S. (2005). *Analysis of Financial Time Series*, 2nd Edition, Wiley: Hoboken, New Jersey.

With young researchers at the Chinese University of Hong Kong
in 1983 (Kung-Sik Chan, Ka Sing Man, HT, Man Cheung Hau)

First National Time Series Meeting in Beijing, 1984
(Front row, to the right of HT: George Tiao, Maurice Priestly, Shou-Ren Wang,
Manny Parzen,?, Ritei Shibata,
? Ogata,?, Hong-zhi An,?,?,?,?,?, Zhao-Guo Chen)

With Maurice Priestley, Carol Parzen,
and Manny Parzen in Beijing, 1984

Rudolph Kalman visiting the Chinese University of Hong Kong after receiving the 1985 Kyoto Prize

With fellow UK delegates at the First Bernoulli World Congress in Tashkent, 1986. (Geoffrey Grimmett, Richard Smith, Terry Lyons, HT, Henry Wynn)

With Akiva Yaglom and his wife at the foot of the Tian Shan Mountain on theTashkent side, 1986.

J. R. Statist. Soc. B (1992)
54, *No.* 2, *pp.* 427–449

On Consistent Nonparametric Order Determination and Chaos

By B. CHENG and H. TONG†

Chinese Academy of Sciences, Beijing, China — *University of Kent, Canterbury, UK*

[*Read before* The Royal Statistical Society *at a meeting on Chaos organized by the* Research Section *on Wednesday, October 16th, 1991*]

SUMMARY

We give a brief introduction to deterministic chaos and a link between chaotic deterministic models and stochastic time series models. We argue that it is often natural to determine the embedding dimension in a noisy environment first in any systematic study of chaos. Setting the stochastic models within the framework of non-linear autoregression, we introduce the notion of a generalized partial autocorrelation and an order. We approach the estimation of the embedding dimension via order determination of an unknown non-linear autoregression by cross-validation, and give justification by proving its consistency under global boundedness. As a by-product, we provide a theoretical justification of the final prediction error approach of Auestad and Tjøstheim. Some illustrations based on the Hénon map and several real data sets are given. The bias of the residual sum of squares as essentially a noise variance estimator is quantified.

Keywords: ATTRACTORS; BANDWIDTH; BIAS CORRECTION; BINARY SHIFT MAP; CANADIAN LYNX; CHAOS; CONSISTENCY; CROSS-VALIDATION; DIMENSION; DOUBLE WINDOWS; EMBEDDING DIMENSION; EPIDEMICS; EXPERIMENTAL DATA; FINAL PREDICTION ERROR; FRACTALS; GENERALIZED PARTIAL AUTOCORRELATION FUNCTION; GLOBAL BOUNDEDNESS; HÉNON MAP; KERNEL ESTIMATION; LIMIT CYCLES; LIMIT POINTS; LOCAL INSTABILITY; LYAPUNOV EXPONENT; MAP RECONSTRUCTION; MEASLES; NON-LINEAR AUTOREGRESSION; ORDER DETERMINATION; ORDER OF NON-LINEAR AUTOREGRESSION; PREDICTIVE RESIDUALS; RESIDUAL SUM OF SQUARES; SKELETON; U-STATISTICS; WOLF'S SUNSPOT NUMBERS

1. INTRODUCTION

A non-linear dynamicist would often be interested in 'low dimensional' attractors in a dissipative dynamical system, because their existence permits drastic reduction of the complexity of the system at least in qualitative terms. In the past, limit points and limit cycles were the attractors of central importance of research and they have very simple and low 'dimensions', such as 0 and 1, according to any reasonable definition of dimension. They represent respectively a long run steady state and a long run periodic state. We shall not address the complicated issue of dimension but merely mention that there are numerous and not always equivalent definitions, namely the Hausdorff dimension, the Kolmogorov capacity, the correlation dimension, the Lyapunov dimension, the information dimension, etc. (See, for example, Farmer *et al.* (1983).) Some of these will be addressed by other speakers at this meeting. Henceforth the dimension of an attractor will be understood to be one of the above-mentioned varieties. Since the 1970s much more exotic attractors have become the central objects of intense interest. These typically have dimensions other than

† *Address for correspondence*: Institute of Mathematics and Statistics, University of Kent, Cornwallis Building, Canterbury, Kent, CT2 7NF, UK.

 0035-9246/92/54427 $2.00

0 or 1 but they are generally still of a low number, say D. The time plots of these exotic attractors tend to have the appearance of realizations of a stochastic process, at least to human eyes. They constitute the *raison d'être* of the new discipline called *chaos*.

To generate chaos we need two basic ingredients, namely *global boundedness* and *local instability*. We may explain them by using the *binary shift map*:

$$X_{t+1}=2X_t \pmod 1, \qquad t=0, 1, 2, \ldots, \tag{1.1}$$

where $X_0 \in [0, 1]$. Clearly $0 \leqslant X_t \leqslant 1$ for all $t \geqslant 0$ and this is what we mean by global boundedness. The map (1.1) admits an elementary solution (such instances are rare):

$$X_t=2^t X_0 \pmod 1, \qquad t=1, 2, \ldots . \tag{1.2}$$

Write X_0 in the binary expansion as

$$X_0=0.k_1k_2k_3k_4 \ldots \tag{1.3}$$

where $k_i \in \{0, 1\}$, each i. Since each application of map (1.1) removes the foremost binary digit, two initial values, say X_0 and X_0', which are very close together with their respective binary expansions agreeing to the first p positions will, after p iterations, lead to vastly different consequents, i.e. map (1.1) is locally unstable or equivalently *sensitive to initial conditions*. A standard way to quantify the sensitive dependence of $G:\mathbf{R}\rightarrow\mathbf{R}$ on an initial condition, say $X_0 \in \mathbf{R}$, is to evaluate the so-called *Lyapunov exponent* at X_0, $\lambda(X_0)$, where

$$\lambda(X_0)=\lim_{t\rightarrow\infty}\left\{t^{-1}\ln\left|\frac{\mathrm{d}}{\mathrm{d}X}\,G^{(t)}(X_0)\right|\right\}$$

and $G^{(t)}$ denotes the t-fold composition of G. Positive $\lambda(X_0)$ confirms sensitive dependence of G on X_0 in that

$$\Delta_t \approx \exp\{t\lambda(X_0)\}\Delta_0,$$

where Δ_t denotes the discrepancy between two iterations of G at time t if their respective initial values are X_0 and $X_0+\Delta_0$. For the binary shift map, $\lambda(X_0)=\ln 2>0$, trivially and for all $X_0 \in (0, 1)$. (If we replace ln by $\log_2$, then $\lambda(X_0)=1$, which means that one bit of information is lost on every application of G.) The concept of the Lyapunov exponent can be generalized to the case $G:\mathbf{R}^k\rightarrow\mathbf{R}^k$. (See, for example, Nychka *et al.* (1992).)

Let d be a positive integer. A fairly common dynamical system takes the form

$$X_{t+d}=G(X_{t+d-1}, \ldots, X_t), \qquad t=0, 1, \ldots, \tag{1.4}$$

where $\{X_t\}$ is a discrete time univariate time series and $G:\mathbf{R}^d\rightarrow\mathbf{R}$ is usually assumed to be a well-behaved function. In the dynamical system literature, this formulation is often associated with Takens's (1981) theorem, which, roughly, states that corresponding to a continuous time multidimensional deterministic dynamical system which has an attractor of dimension D there exist a positive d and a $G:\mathbf{R}^d\rightarrow\mathbf{R}^1$ such that system (1.4) holds and that the $(d+1)$-dimensional vectors $(X_{t+d}, \ldots, X_t)$ are themselves attracted towards an equivalent attractor in $\mathbf{R}^{d+1}$. The positive integer d is called an *embedding dimension*. (The vector $(X_{t+d}, \ldots, X_t)$ and its generalization is sometimes called a vector with *delay co-ordinates*, which, together with an associated plot called a *delay map*, was apparently introduced into the dynamical system literature in about 1980, initially without reference to the much earlier work of U. Yule in the time series literature. A similar construction, now called the directed scatterplot, was independently used by one of us in 1980 in non-linear time series analysis. See, for example, Tong (1990a), pages 216–217, for a brief account of the history.) In a practical situation where a univariate time series, say $\{X_t\}$, is given, it seems natural to search for d first and then to search for a low dimensional attractor of dimension say $D'(\leqslant d)$. Indeed this was the approach adopted by one of us (see, for example, Tong and Lim (1980) and Tong (1990a)). As commented by Sugihara and May (1990), it is rather curious that the usual procedure in the non-linear dynamical literature goes in the reverse direction, i.e. D is determined before d. (We are nearly tempted to replace D by cart and d by horse!) They have also suggested that the usual procedure would result in the loss of information in a time series of *finite* duration (Sugihara and May, 1990). In a more general situation, $\{X_t\}$ may be just some experimental data and need not even be a univariate component time series of an underlying multidimensional dynamical system. In this case, it would then seem even more compelling to determine d before D.

We argue that, in practice, observations rarely evolve according to system (1.4), which, among other things, demands absolute accuracy in measurements. Following the terminology of Tong (1990a), system (1.4) is called a *skeleton*. Let ξ_t denote the *state vector* $(X_{t+d-1}, \ldots, X_t)$. Equation (1.4) then defines a trajectory (i.e. $\xi_0, \xi_1, \ldots$) in the *state space* $\mathbf{R}^d$. Associated with each ξ_i is a Dirac δ-measure. Thus, a natural way to incorporate stochastic perturbation in system (1.4) is to enlarge the above trajectory to a trajectory (say $\mu_0, \mu_1, \ldots$) in the space of probability measures on $\mathbf{R}^d$. Here, μ_t stands for $\mu_t(\mu_0)$ and denotes the probability measure of ξ_t given that ξ_0 has the probability measure μ_0. Let us now impose a *Markovian* assumption such that $\{\xi_t\}$ follows a Markov chain on $\mathbf{R}^d$ and that $\mu_{s+t}(\mu_0)=\mu_s(\mu_t(\mu_0))$. It is reasonable to represent the Markov chain $\{\xi_t\}$ on $\mathbf{R}^d$ in the form

$$\xi_{t+1}=\phi(\xi_t, e_{t+1}), \tag{1.5}$$

where $\phi:\mathbf{R}^{2d}\rightarrow\mathbf{R}^d$, $\{e_t\}$ is a sequence of independent and identically distributed d-dimensional random vectors and e_t is independent of ξ_s, $s<t$. (See, for example, Eckmann and Ruelle (1985), Tong (1990a), p. 97, and the references therein.) In this paper, we consider a special case of equation (1.5), namely

$$\xi_{t+1}=\phi(\xi_t, \mathbf{0})+e_{t+1}, \tag{1.6}$$

where the only possible non-zero entry of e_t is in its first component, say ϵ_{t+d}. Effectively, we consider *perturbing skeleton (1.4) with additive stochastic noise* to obtain the *stochastic model*

$$Z_{t+d}=F(Z_{t+d-1}, \ldots, Z_t)+\epsilon_{t+d}, \qquad t=0, 1, \ldots, \tag{1.7}$$

where, to compensate partly for the loss of generality in going from equation (1.5) to equation (1.6), we relax ϵ_t to be a sequence of martingale differences representing *dynamic noise*. We further assume that the distribution of ϵ_t has bounded support with variance σ^2. A similar idea has been proposed by Farmer and Sidorowich (1990), equation (3), and Eubank and Farmer (1990). This way of incorporating stochastic noise is consistent with our approach to non-linear time series including chaotic time series, as explained in Tong (1990a). Here we use Z and X to denote the observable and the unobservable respectively. Our discussion has been heuristic.

However, it can be shown that under an appropriate mode of convergence of $F(x_1, \ldots, x_d)$, or more precisely $F^{(\sigma)}(x_1, \ldots, x_d)$, to $G(x_1, \ldots, x_d)$ as $\sigma^2 \to 0$, a unique stationary marginal distribution of Z_t exists and converges weakly to the stationary marginal distribution of X_t (appropriately defined) as $\sigma^2 \to 0$ (Chan and Tong, 1991). In this sense, the deterministic model (1.4) is 'embedded' in the stochastic model (1.7).

We maintain that models such as equation (1.4) are mathematical fictions whereas models such as equation (1.7) are closer to reality. As discussed elsewhere (Tong, 1990b, c), we take the view that ϵ_t is unavoidable and *must not be ignored* even though $\{\epsilon_t\}$ might correspond to the trajectory of a high dimensional attractor. The crux of the matter is that as long as the skeleton of model (1.7) explains a substantial proportion of the variation of Z_t the search for low dimensional attractors is a meaningful generalized signal extraction exercise.

2. CROSS-VALIDATORY ESTIMATE

We therefore take as our model

$$Z_t = F(Z_{t-1}, \ldots, Z_{t-d}) + \epsilon_t, \tag{2.1}$$

where F is *unknown* and $\{\epsilon_t\}$ is a sequence of martingale differences with variance σ^2. Assume that $\{Z_t\}$ is a strictly stationary univariate time series with finite variance and absolutely continuous distribution. Our objective is to determine d from the observations $(Z_1, \ldots, Z_N)$ assuming that there is no 'redundancy' among $Z_{t-1}, \ldots, Z_{t-d}$. (We shall make this notion precise in definition 2 later.)

We shall use kernel estimation for autoregression. (See, for example, Robinson (1983).) Let P denote the class of non-negative even functions $k: \mathbf{R}^1 \to \mathbf{R}^1$, satisfying

$$\int_{\mathbf{R}^1} k(u)\, \mathrm{d}u = 1,$$

$$\int_{\mathbf{R}^1} |u|\, k(u)\, \mathrm{d}u < \infty.$$

Now, for $k \in P$ and $u = (u_1, \ldots, u_l) \in \mathbf{R}^l$, we define

$$K_l(u) = \prod_{i=1}^{l} k(u_i). \tag{2.2}$$

(For emphasis, we sometimes attach an l to functions of K_l.) It is not essential for our results to have K in the form of a 'product' kernel. Denote the row vector $(Z_{t-d}, \ldots, Z_{t-1})$ by Y_t. Let f (or more precisely f_d) denote the density of Y_t. Let $B(\ )$ denote the bandwidth. For $y \in \mathbf{R}^d$ and $t \geqslant i \geqslant r$, t and r being positive integers, define

$$\hat{f}_{t,\setminus i}(y) = \frac{1}{(t-r)\, B(t)^d} \sum_{\substack{s=r \\ s \neq i}}^{t} K_d\left\{\frac{y - Y_s}{B(t)}\right\}, \tag{2.3}$$

$$\hat{F}_{t,\setminus i}(y) = \frac{1}{(t-r)\, B(t)^d} \sum_{\substack{s=r \\ s \neq i}}^{t} Z_s K_d\left\{\frac{y - Y_s}{B(t)}\right\} \hat{f}_{t,\setminus i}(y)^{-1}.$$

If these summations include the terms when $s = i$ as well, then we drop $\setminus i$ in the suffices of $\hat{f}$ and $\hat{F}$ to give $\hat{f}_t(y)$ and $\hat{F}_t(y)$ respectively. (Obviously the divisor $t-r$ will have to be altered to $t-r+1$ then.) Note that $\hat{F}$ remains invariant with respect to the transformation $y \to ay+b$, $Y \to aY+b$ and $B(t) \to a\,B(t)$. In effect, $\hat{f}_t$ is a kernel estimate of f using observations $Y_r, Y_{r+1}, \ldots, Y_t$ and $\hat{f}_{t,\setminus i}$ is the same but with the observation Y_i left out. Similarly, $\hat{F}_t$ is a kernel estimate of $E[Z_t | Z_{t-1}, \ldots, Z_{t-d}]$ using observations $Z_{r-d}, \ldots, Z_t$ and $\hat{F}_{t,\setminus i}$ is the same but with the summand $Z_i K_d(y-Y_i)/B(t)$ left out.

We now describe the cross-validatory (CV) approach which is a modification of that outlined in Tong (1990a, b). Let Σ denote the summation over t from r to N. Let

$$\mathrm{CV}(d) = (N-r+1)^{-1} \sum \{Z_t - \hat{F}_{N,\setminus t}(Y_t)\}^2\, W(Y_t), \tag{2.4}$$

and

$$\mathrm{RSS}(d) = (N-r+1)^{-1} \sum \{Z_t - \hat{F}_N(Y_t)\}^2\, W(Y_t), \tag{2.5}$$

where W is a suitably chosen non-negative weight function and r is so chosen that the edge effects are minimal. For bounded time series, we may simply set W identically equal to 1. (We shall adopt this choice in all our examples later.) The general idea is that we replace Z_t by its conditional mean given its recent past where the conditional mean function is estimated by using the kernel method and *with Z_t deleted*. It is a fact that a leave-one-out approach penalizes the complexity of the model. (See, for example, Stone (1974).) We then evaluate a weighted sum of squares of the 'residuals' and minimize CV(d) with respect to d over a prefixed range, say $\{1, 2, \ldots, L\}$. The minimizer $\hat{d}$ is the CV estimate of d.

To justify the use of $\hat{d}$ fully, we would need to investigate its sampling properties, at least for large samples. We summarize our results in the form of the following theorems. A complete proof of the main result, theorem 2, is given in Appendix A but proofs of the other theorems are given in Cheng and Tong (1991).

Theorem 1. Under conditions (a)–(o) which are listed in Appendix A

$$\mathrm{CV}(d) = \mathrm{RSS}(d)\{1 + 2\alpha\gamma\rho^d/N + o_p(\rho^d/N)\}, \tag{2.6}$$

where $\alpha = K_d^{1/d}(0)$, $\gamma = \int W(x)\,\mathrm{d}x / \int W(x) f(x)\,\mathrm{d}x$ and $\rho = B(N)^{-1}$.

Essentially, equation (2.6) expresses CV(d) as an adjusted RSS(d). We recall that for a linear $F(Y_t)$, say $\theta' Y_t$, F may be estimated by replacing θ by its least squares estimate, say $\hat{\theta}_N$, or $\hat{\theta}_{N,\setminus t}$ if Y_t is left out. In this case, it is well known (Kavalieris, 1989) that the effect of leaving one out is an adjustment of the residual from $Z_t - \hat{\theta}_N' Y_t$ to $(Z_t - \hat{\theta}_N' Y_t)/\{1 - a_{tt}(d)\}$, where $a_{tt}(d)$ is the tth diagonal element of the 'hat' matrix, which is commonly then approximated by N^{-1}trace(hat matrix) = d/N. (We shall refer to any CV method using this approximation as an *approximate CV method*.) Much of the tedious algebra in the proof of theorem 1 involves the extensive use of results in U-statistics for weakly dependent data from Denker and Keller (1983) to evaluate $\hat{F}_N(Y_t)$ and $\hat{F}_{N,\setminus t}(Y_t)$ to obtain an appropriate adjustment. Naturally, properties of the kernel have a role to play here. For N sufficiently large such that $B(N) << 1$, the 'penalty term' $2\alpha\gamma\rho^d/N \to \infty$ as $d \to \infty$. Thus, in a practical situation, increasing d will decrease the 'residual sum of squares' (i.e. 'badness of fit')—to be quantified in theorem 3, but at the expense of incurring a higher penalty due to the complexity of the fitted model. This is in the same spirit of modern model selection criteria such as Akaike's information criterion (AIC) (see, for example, Tong (1990a)). For the (parametric) case of linear autoregression, the penalty term for AIC is $2d/N$. In our case, the price that we have to pay for not knowing the autoregression function F is quantified by α due to the functional form of the kernel and ρ due to the bandwidth of the kernel. Note also that equation (2.6) may be considered a generalization of a similar result obtained by Härdle *et al.* (1988), equation (2.5), for the regression problem.

In his analysis of AIC for linear autoregressive models, Shibata (1976) used a fundamental characterization of the partial autocorrelation function for linear time series due to Ramsey (1974), p. 1299. We shall imitate some of his arguments to analyse our $\hat{d}$. For this we need to generalize the partial autocorrelation function so that it is also applicable to non-linear time series. Let $\sigma^2(0)=\mathrm{var}(Z_t)$ and, for $d\geqslant 1$,

$$\sigma^2(d)=E[Z_t-E(Z_t|Z_{t-1},\ldots,Z_{t-d})]^2=\min_g\{E[Z_t-g(Z_{t-1},\ldots,Z_{t-d})]^2\},$$

for any Borel function g. Clearly $\sigma^2(d)\leqslant\sigma^2(d-1)$.

Definition 1. $\phi(d)=\{1-\sigma^2(d)/\sigma^2(d-1)\}^{1/2}$ for $d\geqslant 1$ is called the *generalized partial autocorrelation function*. Note that $\phi(d)$ is invariant with respect to scale and location transformations of Z_t.

Definition 2. A strictly stationary time series $\{Z_t\}$ is said to be a *non-linear autoregression of order* d_0 if $\phi(d_0)\neq 0$ and $\phi(d)=0$ for $d>d_0$. Note that d_0 is invariant with respect to a one-to-one transformation of Z_t. This definition makes precise the notion of no 'redundancy' among $Z_{t-1},\ldots,Z_{t-d_0}$ if d_0 is the true order.

In the next theorem, let d_0 denote the true order, i.e. the 'non-redundant' value of d in model (2.1). For $1\leqslant d<L$, let A be an arbitrary compact set in $\mathbf{R}^L$ and B_d be its projection in $\mathbf{R}^d$. We specify the weight function W for each d, W_d, by identifying it with the indicator function of the set B_d.

Theorem 2. Under the conditions of theorem 1:

(a) for $0\leqslant d<d_0$,

$$\lim_{N\to\infty}\{P(\hat{d}=d)\}=0;$$

(b) for $d_0<d\leqslant L$,

$$\limsup_{N\to\infty}\{P(\hat{d}=d)\}\leqslant 1-\liminf_{N\to\infty}P\bigcap_t\{(Z_{t-1},\ldots,Z_{t-d})\in B_d\}.$$

Note that $P\{(Z_{t-1},\ldots,Z_{t-d})\in B_d\}$ may be approximated by $\int_{B_d}\hat{f}_N(x)\,dx$. From the proof of theorem 2 it follows immediately that, for $d_1>d_2\geqslant d_0$,

$$\overline{\lim_{A\to\mathbf{R}^d}}\ \overline{\lim_{N\to\infty}}\ [P\{\mathrm{CV}(d_1)\leqslant\mathrm{CV}(d_2)\}]=0. \qquad (2.7)$$

It follows from theorem 2 that, if $\{Z_t\}$ is a *bounded* time series, then the CV estimator $\hat{d}$ is a *consistent* estimator of d_0. This result is particularly relevant in view of the global boundedness of chaos. This result may be surprising at first sight because it is well known that methods such as the final prediction error (FPE) and

AIC are asymptotically equivalent to the approximate CV method for the parametric case of linear autoregressive (AR) models (Kavalieris, 1989), and they give *inconsistent* estimators of d_0 for these models. The basic difference between the CV estimator for the nonparametric non-linear case and the above methods (to be denoted by $\widetilde{\mathrm{CV}}$) for the parametric linear case is that, on ignoring the factor $1/N$, the penalty term for the former follows a 'power law' dependent on N whereas that for the latter follows a 'linear law' independent of N. This has the consequence, as shown in the proof of theorem 2 in Appendix A, that for the nonparametric case

$$N B(N)^{d_1}\{\mathrm{CV}(d_1)-\mathrm{CV}(d_0)\}=c(d_1)+o_p(1), \tag{2.8}$$

where $c(d_1)>0$, $d_1>d_0$. In contrast, for the parametric linear AR case with $d_1>d_0$ we have from Kavalieris (1989), theorem 2, and Hannan and Quinn (1979), equation (7),

$$N\{\widetilde{\mathrm{CV}}(d_1)-\widetilde{\mathrm{CV}}(d_0)\}=-2(d_1-d_0)\sigma^2\ln(\ln N)+2(d_1-d_0)\sigma^2+o_p(1). \tag{2.9}$$

Recall that to restore positivity to the right-hand side of equation (2.9) (modulo $o_p(1)$) Hannan and Quinn (1979) replaced the penalty term $2d/N$ by $2cd\ln(\ln N)/N$ $(c>1)$.

3. FINAL PREDICTION ERROR APPROACH

Auestad and Tjøstheim (1990) have considered the use of the notion of FPE due to Akaike (1970) to construct criteria for the determination of d. They have acknowledged that their arguments are fairly rough. By demonstrating the asymptotic equivalence of the FPE approach and our CV approach, we shall give a theoretical justification for the former. Before doing so, we give theorem 3, which is of independent interest and underpins the various FPE criteria proposed by Auestad and Tjøstheim. Define

$$\epsilon_t^{(d)}=Z_t-E[Z_t|Z_{t-1},\ldots,Z_{t-d}]$$

and

$$\sigma_N^2(d)=(N-r+1)^{-1}\sum(\epsilon_t^{(d)})^2 W_d(Y_t).$$

To simplify the notation, we sometimes omit reference to d if the context is clear.

Theorem 3. Under the conditions of theorem 1,

$$\mathrm{RSS}(d)=\sigma_N^2(d)\{1-(2\alpha-\beta)\gamma\rho^d/N+o_p(\rho^d/N)\}, \tag{3.1}$$

where $\beta=\int K_d(u)^{2/d}\,du$. (We sometimes write α, β and γ as $\alpha(d)$, $\beta(d)$ and $\gamma(d)$ respectively for emphasis.)

Auestad and Tjøstheim (1990) have conjectured a downward asymptotic bias of $\mathrm{RSS}(d)$ as an estimator of a quantity closely related to $E(\sigma_N^2)$, the correction of which is crucial in their construction of the FPE-type criteria. Theorem 3 gives an explicit expression for the bias, which is similar to but not identical with the conjectured values.

Let us consider an expression similar to that in equation (2.4):

$$(N-r+1)^{-1}\sum\{Z_t-\tilde{F}_N(Y_t)\}^2\,W(Y_t), \tag{3.2}$$

where $\tilde{F}_N$ is now obtained from an *independent* copy of $(Z_1, Z_2, \ldots, Z_N)$. Effectively $\tilde{F}_N$ is a 'leave-all-out' estimate of F! The replacement by an (imaginary) independent copy originates from Akaike (1970). It is not difficult to imitate the proof of theorem 3 to show that under the conditions of theorem 1 the expectation of expression (3.2) is

$$E(\sigma_N^2)\{1+\beta\gamma\rho^d/N+o(1)\}. \tag{3.3}$$

Substituting $E\{\sigma_N^2(d)\}$ by its unbiased estimator as given in theorem 3, we obtain the criterion

$$\mathrm{FPE}(d)=\mathrm{RSS}(d)(1+N^{-1}\beta\gamma\rho^d)/\{1-N^{-1}(2\alpha-\beta)\gamma\rho^d\}. \tag{3.4}$$

The minimizer $\tilde{d}$ of FPE(d), over a prefixed range $\{1, 2, \ldots, L\}$, is called an *FPE-type estimator of d*. It is obvious that under the conditions of theorem 1

$$\mathrm{FPE}(d)=\mathrm{CV}(d)+o_p(\rho^d/N), \tag{3.5}$$

with a similar relation holding for $\hat{d}$ and $\tilde{d}$. We have therefore provided a justification for the use of FPE(d).

Now, there is an interesting spin-off from equation (3.1). If we choose a kernel such that $k(0)=0$, then $\alpha=0$. In this case, RSS(d) = CV(d) to $o_p(\rho^d/N)$ and has an asymptotic bias which is *not* negative but positive. Indeed, $E\{\mathrm{RSS}(d)\}$ is now equal to expression (3.3), which forms the basis of FPE-type criteria for order determination. Unlike the FPE, *neither RSS(d) nor CV(d) invokes the (Akaike) assumption of independent copies*. In effect, any kernel which is such that $k(0)=0$ performs the role of 'leaving one out'. Such a kernel may be realized by imitating the 'double-window' technique in the analysis of mixed spectra. (See, for example, Priestley (1981).) For example, we may take as our kernel k any standard bimodal symmetric probability density function with compact support which satisfies the Lipschitz condition (g) (Appendix A) and has a zero antimode at the origin. However, as we do not have much practical experience in the use of bimodal kernels, we leave the use of RSS(d) with such kernels to future investigation.

4. EXAMPLES

4.1. *Example 1*

We clothe a Hénon map with dynamic noise to obtain

$$Z_t=1-1.4Z_{t-1}^2+0.3Z_{t-2}+\epsilon_t, \tag{4.1}$$

where ϵ_t is uniformly distributed on $(-0.01, 0.01)$. It is a long-standing conjecture that the skeleton corresponding to equation (4.1) exhibits 'chaotic' behaviour, with its Hausdorff dimension reckoned to be approximately 1.26. Fig. 1 gives the undirected scatterplot of Z_{t-1} *versus* Z_t based on 500 (standardized) simulated data from equation (4.1). (By standardization we mean a division by the sample standard deviation.) Fig. 2 gives the minimum CV estimate $\hat{d}$ of the order d against the bandwidth $B(N)$, which shows that there is a sizable interval of $B(N)$ values which produce $\hat{d}=2$, the true order. Fig. 3 and many others like it lead us to suggest the use of the *data-driven bandwidth*, i.e. one that is obtained by minimizing CV($\hat{d}$) with

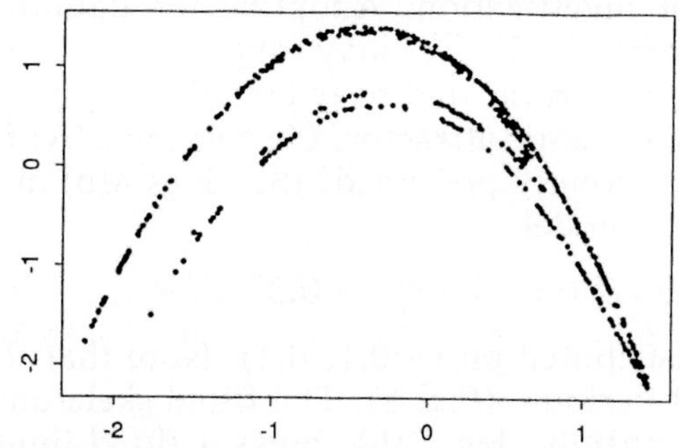

Fig. 1. Undirected scatterplot of 500 standardized observations from model (4.1), (Z_1, Z_2) lying in the basin of attraction (standard deviation (for original data), 0.7171)

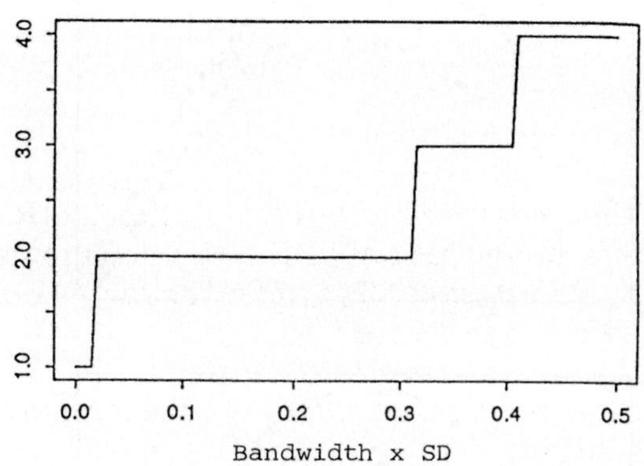

Fig. 2. $\hat{d}$ *versus* $B(N) \times$ standard deviation for the 500 standardized observations from model (4.1): henceforth, unless otherwise stated, we use the kernel $k(x) = 0.5 \exp(-|x|/\sqrt{2}) \sin(|x|/\sqrt{2} + \pi/4)$ (Silverman, 1985) (although this kernel violates the assumption of non-negativity, it enhances numerical stability in our experience; the proofs of theorems 1 and 3 may be modified to accommodate this violation at the expense of increasing their length)

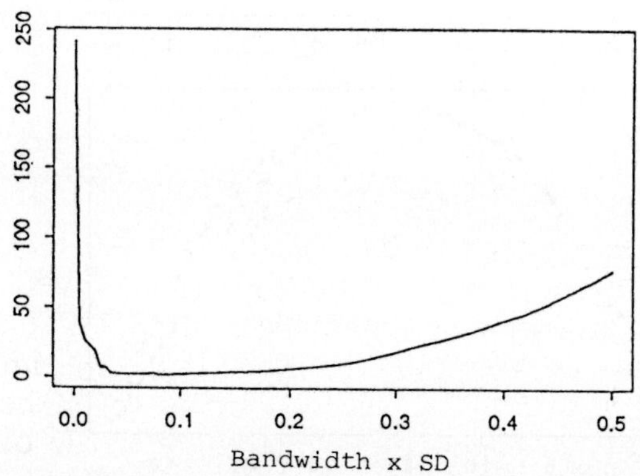

Fig. 3. CV($\hat{d}$) *versus* $B(N) \times$ standard deviation for the 500 standardized observations from model (4.1) (global minimum at $B(N) \times$ standard deviation, 0.0801)

respect to $B(N)$. However, we hasten to add that this suggestion falls outside theorems 1–3 but deserves further investigation. Adopting the data-driven bandwidth (i.e. $B(N)=0.0801/0.7171$) and $\hat{d}=2$, we may obtain the fitted skeleton through $X_t=\hat{F}_N(X_{t-1}, X_{t-2})$. Fig. 4(a) is the undirected scatterplot of 500 points of the fitted skeleton, which suggests a chaotic attractor. On varying $B(N)$ from 0.05 to 0.20, the chaotic appearance is apparently preserved. (See Figs 4(b) and 4(c).) Figs 5–7 are analogous figures for the model

$$Z_t=1-1.3Z_{t-1}^2+0.3Z_{t-2}+\epsilon_t, \tag{4.2}$$

where ϵ_t is uniformly distributed on $(-0.1, 0.1)$. Note that the skeleton of model (4.2) has a limit cycle of period 7 (Fig. 8). The fitted skeletons (Figs 7(a) and 7(c)) resemble Fig. 8 quite faithfully. Fig. 7(b) shows a fitted limit cycle of period 21, representing three almost overlapping cycles each of period 7. The different dynamical

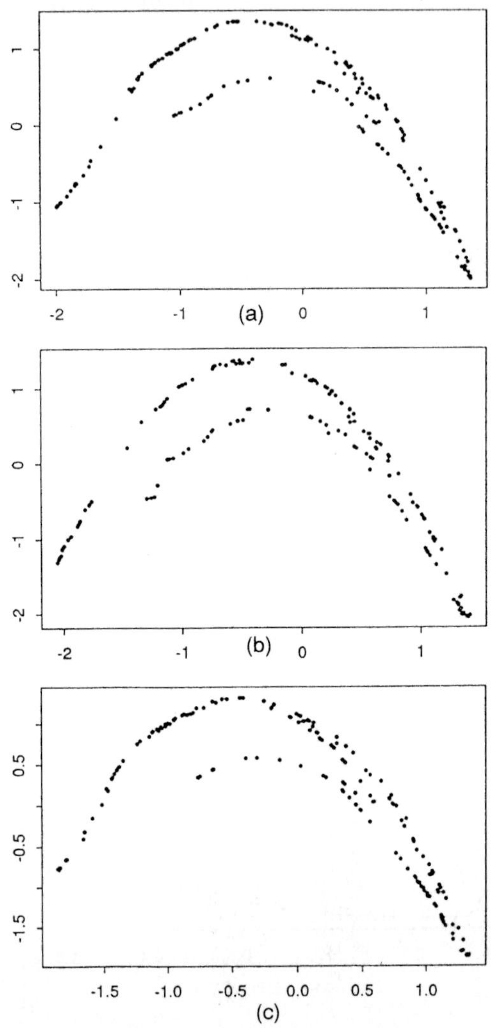

Fig. 4. Undirected scatterplot of 500 points of the skeleton fitted to standardized data generated by model (4.1): (a) $B(N)=0.1117$ (global minimum); (b) $B(N)=0.05$; (c) $B(N)=0.20$

behaviours of the skeletons of models (4.1) and (4.2) do not seem to affect the CV determination of d.

4.2. *Example 2*

Applying our CV approach to the 'bench-mark' Canadian lynx data (listed in Tong (1990a)) on a natural logarithm scale, we obtain $\hat{d}=3$ with $B(N)$ for the standardized data varying between 0.31 and 0.52 and with the minimum among these at $B(N)=0.3917$. The Gaussian kernel is used and $N=114$. (Recall that we set $W(\ )\equiv 1$ for all our examples.) This estimate of d is quite close to those of the low order models reported in the literature (see, for example, Tong (1990a)), which are mostly of order 2. We have here two tuning parameters: d and $B(N)$. With $\hat{d}=3$ and $B(N)=0.3917$ we can obtain its skeleton as in example 1. The skeleton clearly shows a connected curve as its attractor (Fig. 9), thus lending further support to the limit cycle skeletons of several fitted parametric non-linear AR models summarized in Tong (1990a), section 7.2. The time plot corresponding to Fig. 9 exhibits a period of approximately 9.2 years, in reasonable agreement with the period of approximately 9.5 years for the observed data. We have varied $B(N)$ in the neighbourhood of 0.3917 (from 0.36 to 0.60) and have obtained a similar attractor to Fig. 9 in each

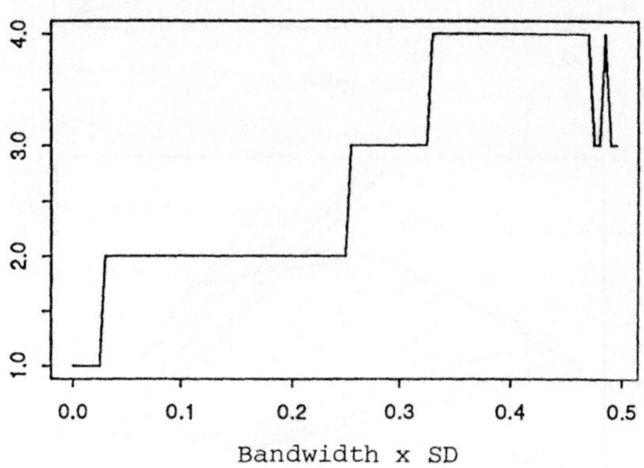

Fig. 5. $\hat{d}$ *versus* $B(N)\times$standard deviation for the 500 standardized observations for model (4.2) (standard deviation (for original data), 0.6898)

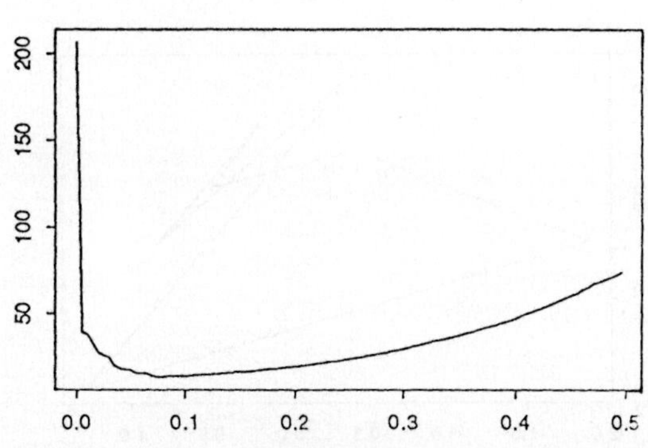

Fig. 6. CV($\hat{d}$) *versus* $B(N)\times$standard deviation for the 500 standardized observations from model (4.2) (global minimum at $B(N)\times$standard deviation, 0.0751)

case. However, here and later, we have noticed that the size of the neighbourhood tends to depend on the initial values. The CV(3) value is 18.4% for the standardized data, which may be interpreted as an approximate ratio of the 'variance of the noise' to the 'variance of the data' by virtue of theorems 1 and 3. This value may be compared with the ratio of 15.2% for the five-parameter quadratic AR model of order 2 initiated by Cox and recorded on p. 410 of Tong (1990a). To gain further insights, it is worth exploring the possibility of a substantive approach in the sense of Cox (see, for example, Tong (1990c)) by clothing a population model, e.g. the Oster–Ipaktchi model (see equation (7.3) of Tong (1990a)), which incorporates a delayed regulation time. Tong (1990a), p. 377, has suggested that it might be reasonable to experiment with

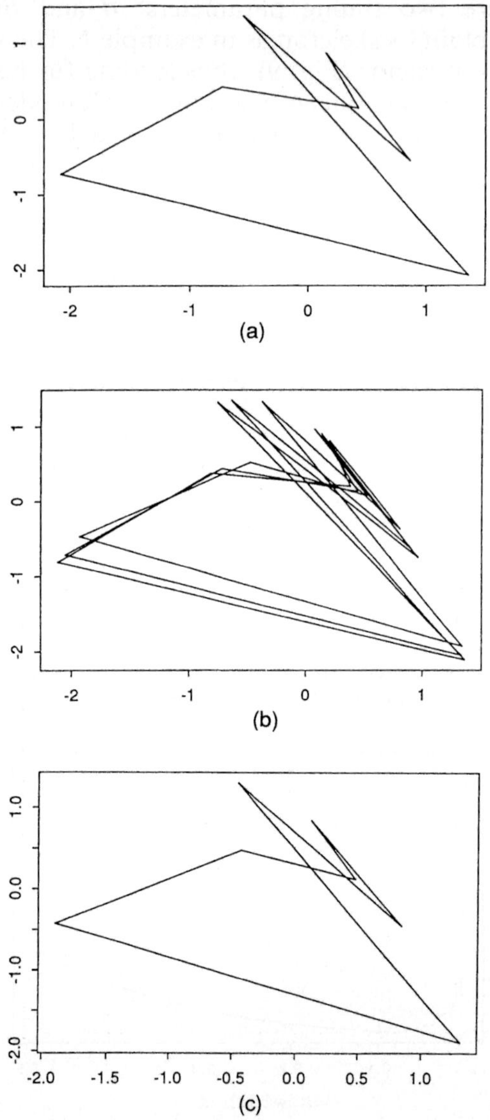

Fig. 7. Directed scatterplot of 500 points of the skeleton fitted to standardized data generated by model (4.2): (a) $B(N)=0.1089$ (global minimum); (b) $B(N)=0.08$; (c) $B(N)=0.20$

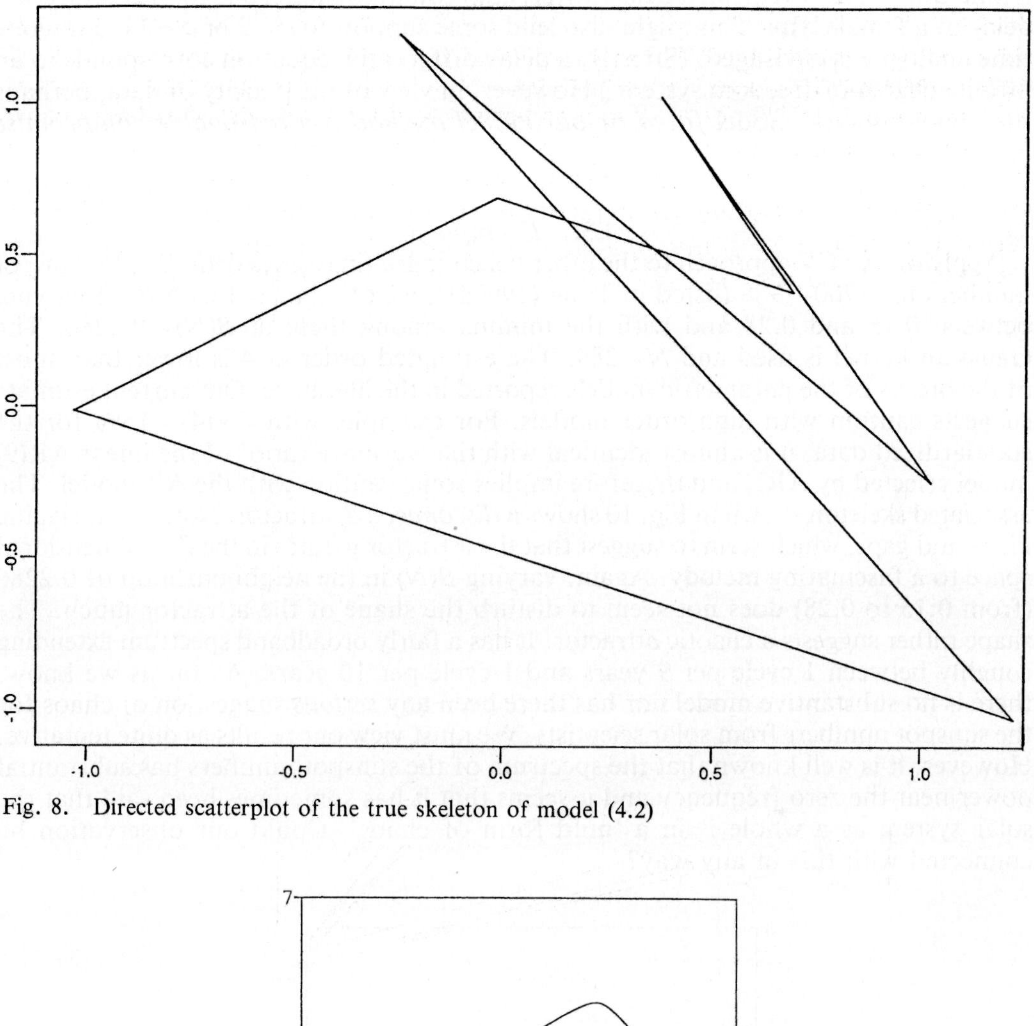

Fig. 8. Directed scatterplot of the true skeleton of model (4.2)

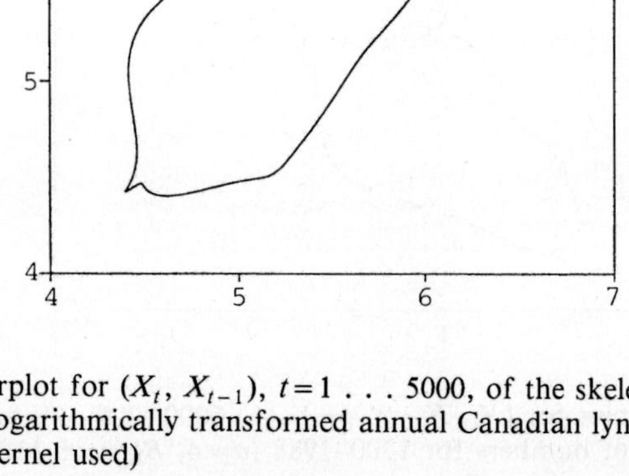

Fig. 9. Undirected scatterplot for (X_t, X_{t-1}), $t=1 \ldots 5000$, of the skeleton of the nonparametric AR model for the base e logarithmically transformed annual Canadian lynx data (1821–1933) ($\hat{d}=3$, $B(N)=0.3917$, Gaussian kernel used)

a time delay of 2–3 years in a delay differential equation model due to the maturation delay of a female lynx. This might also lend some support to $d=2$ or $d=3$ if a discrete time analogue is envisaged. (Strictly, a delay differential equation corresponds to an infinite degree-of-freedom system.) However, in view of the paucity of data, perhaps the nonparametric model *fitted by our kernel method is a reasonable compromise in the circumstances.*

4.3. *Example 3*

Applying our CV approach to the other bench-mark time series data, Wolf's sunspot numbers for 1700–1988 (listed in Tong (1990a)), we obtain $\hat{d}=4$ with $B(N)$ varying between 0.15 and 0.28 and with the minima among these at $B(N)=0.2286$. The Gaussian kernel is used and $N=289$. The estimated order at 4 is lower than most of the orders of the parametric models reported in the literature. Our current estimate suggests caution with high order models. For example, with CV(4) = 15% for the standardized data, it is almost identical with the 'variance ratio' of the linear AR(9) model selected by AIC, and therefore implies some caution with the AR model. The associated skeleton shown in Fig. 10 shows a *disconnected* attractor. Note the intriguing kinks and gaps, which seem to suggest that the attractor gyrates in the five-dimensional space to a fascinating melody. Again, varying $B(N)$ in the neighbourhood of 0.2286 (from 0.15 to 0.28) does not seem to disturb the shape of the attractor much. The shape rather suggests a chaotic attractor. It has a fairly broadband spectrum extending roughly between 1 cycle per 9 years and 1 cycle per 10 years. As far as we know, there is no substantive model nor has there been any *serious* suggestion of chaos for the sunspot numbers from solar scientists. We must view our results as quite tentative. However, it is well known that the spectrum of the sunspot numbers has substantial power near the zero frequency and it seems that it has sometimes been said that the solar system as a whole is in a 'mild form of chaos'. Could our observation be connected with this in any way?

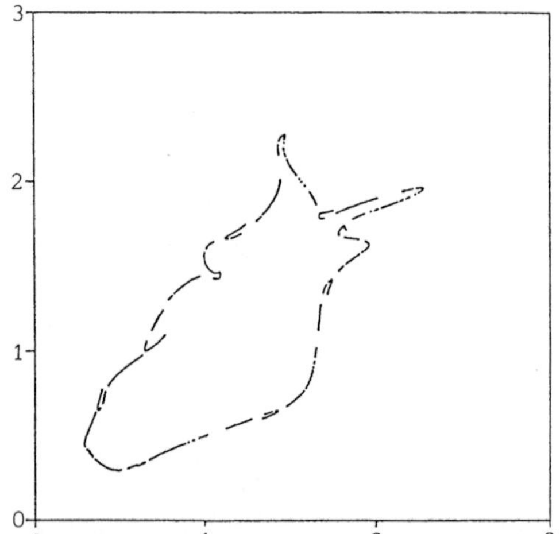

Fig. 10. Undirected scatterplot for (X_t, X_{t-1}), $t=1 \ldots 5000$ of the skeleton of the nonparametric AR model for Wolf's sunspot numbers for 1700–1988 ($\hat{d}=4$, $B(N)=0.2286$, Gaussian kernel used)

4.4. *Example 4*

Applying our CV method to the monthly New York measles data (1928–63) analysed by Sugihara and May (1990) gives a local minimum at $d=6$ *or* $d=7$ over a fairly wide range of bandwidth choices (0.03–0.64). Fig. 11 gives CV(d) *versus* d at $B(N)=0.0327$ (for the standardized measles data). Sugihara and May (1990) have suggested an optimal embedding dimension of between 5 and 7 inclusively for the data. Since CV(7) = 21.6% in Fig. 11, it is also comparable with their measure of the goodness of fit and we may *tentatively* adopt $\hat{d}=7$. The skeleton of the fitted nonparametric autoregressive model (Fig. 12) shows a limit cycle of period 97, which may be compared with Fig. 13 of the data. It seems that the limit cycle twists and turns many times in a fairly high (i.e. 7) dimensional embedding space. The time plot (Fig. 14) mimics the almost biennial interepidemic oscillations of the real data (Fig. 15). Varying the bandwidth in the neighbourhood of 0.0327 (from 0.020 to 0.038) does not appear to destroy the limit cycle but alters its period. In fact, no chaos has been detected even when we vary the bandwidth beyond this neighbourhood.

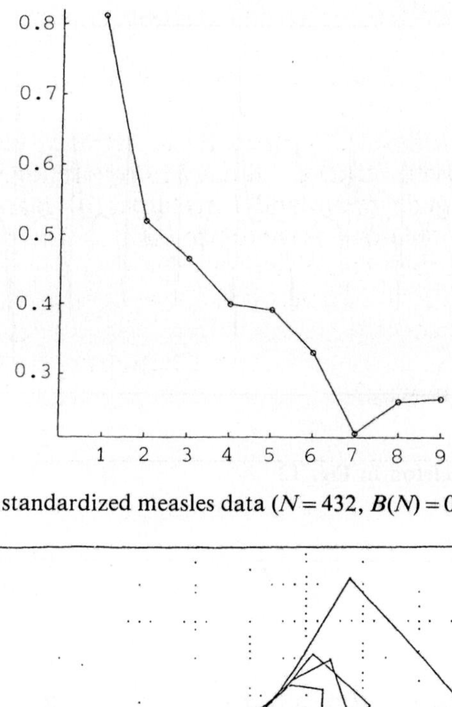

Fig. 11. CV(d) *versus* d for standardized measles data ($N=432$, $B(N)=0.0327$, Gaussian kernel used)

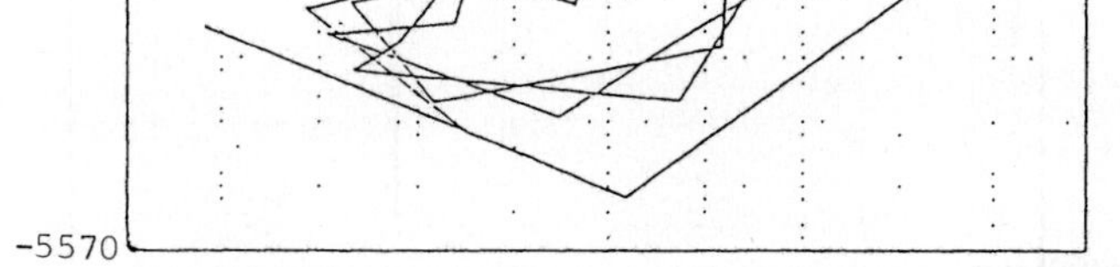

Fig. 12. Limit cycle (period 97) skeleton of the nonparametric AR model for the measles data ($\hat{d}=7$, $B(N)=0.0327$)

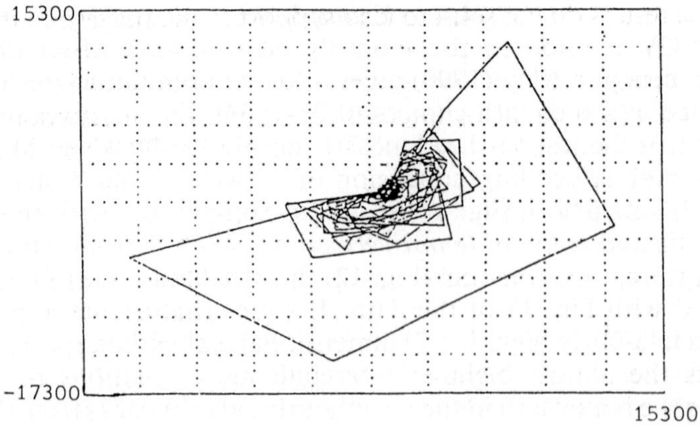

Fig. 13. Directed scatterplot of the New York monthly measles data (1928–1963)

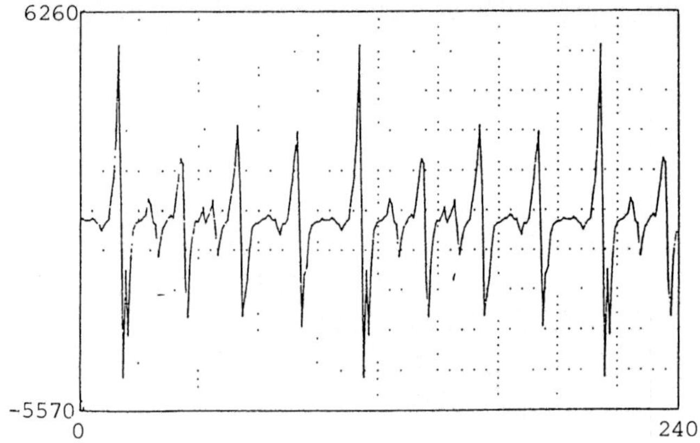

Fig. 14. Time plot of the skeleton in Fig. 12

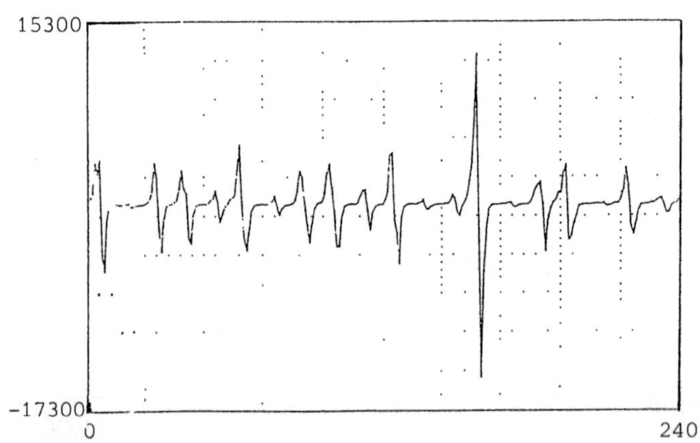

Fig. 15. Time plot of the measles data (first 240 points)

Although these results do not seem to lend support to the suggestion that the measles data may be 'shadowed' by a chaotic skeleton (see Sugihara and May (1990)), they must *not* be taken as being conclusive. For, we have restricted our investigation to the Gaussian kernel only (this implicitly imposes constraints on the functional form for the skeleton), and we find that CV(d), $d \geqslant 10$, falls below CV(7). Therefore the choice of $\hat{d}=7$ may not be optimal in the CV context. It is also a little too close to the maximum order (9) for comfort. (See Tong (1990a), p. 289). However, we would be reluctant to rely on CV(d) for d larger than 9 or 10 unless we have a substantially larger sample.

It is perhaps relevant to report that, using the STAR3 personal computer package accompanying Tong (1990a), we have fitted a four-regime self-exciting threshold autoregressive (SETAR) (4; 1, 4, 2, 8) model (notation as in Tong (1990a)) with delay parameter 3 and thresholds parameters $\{-450, 0, 450\}$ to the 1928–68 data and have obtained (variance of noise)/(variance of data) $\approx 22\%$. Diagnostics suggest that this model is moderately adequate if due allowance is made for the very sharp spikes in the data. This four-regime model does admit a seemingly chaotic skeleton (Fig. 16). Interestingly, if a different initial point is used, the same model admits instead a limit cycle of period 24 (Fig. 17). We have so far not discovered any other attractors for the model.

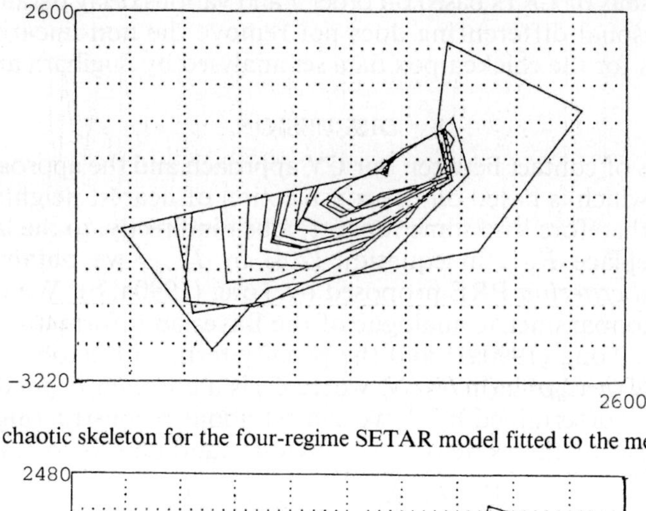

Fig. 16. Possible chaotic skeleton for the four-regime SETAR model fitted to the measles data (1928–68)

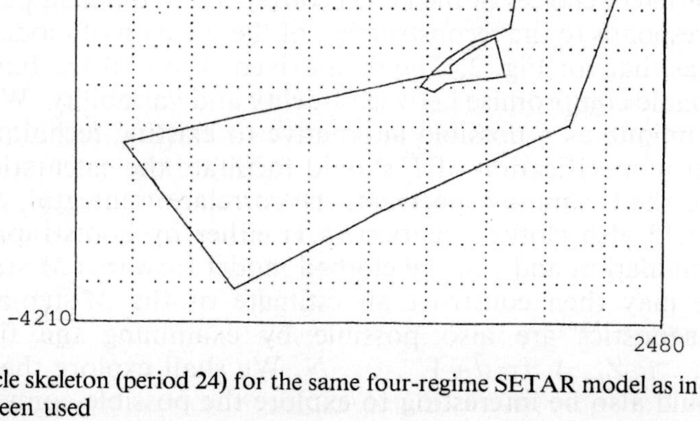

Fig. 17. Limit cycle skeleton (period 24) for the same four-regime SETAR model as in Fig. 16; a different initial point has been used

TABLE 1
LRT statistic for the New York monthly measles data for 1928–63†

Delay	1	2	3	4	5	6	7
LRT	181.29	268.61	293.25	377.15	443.82	734.26	358.25

†$\hat{d}=7$, $N=432$ (the 0.1% point is 34.14, where the threshold parameter is searched over the interquartile range of the data (Chan, 1991)).

We may argue that, given the noise level of the data and the sample size, we are not sure whether there is sufficient information in the data to enable us to come to a definitive answer to the key question: 'Is there clear evidence of a low dimensional chaotic skeleton extractable from the noisy measles data?'. In this respect, it is unclear to us whether the users of techniques such as correlation dimension, Lyapunov exponents, etc., on these data (references in Sugihara and May (1990)) have addressed and quantified the effects of *dynamic* noise adequately. However, what is quite clear is that the data display overwhelming non-linearity as revealed by the likelihood ratio test (LRT) statistics of Chan and Tong, the cumulative sum test of Petruccelli and Davies and Tsay's test. (See chapter 5 of Tong (1990a) for details of these tests.) Table 1 lists the results of LRTs based on order 7 and various delay parameters. We have observed that seasonal differencing does not remove the non-linearity of this data set whereas it does for the chicken-pox data set analysed by Sugihara and May (1990).

5. DISCUSSION

There are points of contact between our CV approach and the approach of Sugihara and May (1990), which is based on the construction of nearest neighbouring convex hulls. Our results therefore lend some support, albeit indirectly, to the latter approach. Further, if we replace $\hat{F}_{N,\setminus t}$ in equation (2.4) by $\hat{F}_{t-1}$, we obtain the so-called *predictive residual criterion* PRE proposed by Tong (1990a, b). We conjecture that PRE will be the nonparametric analogue of the Bayesian information criterion BIC (see, for example, Tong (1990a)) and the penalty term will involve a factor of the form $C_d\rho^d \ln \rho^d/N$ or $C_d\rho^d \ln(\ln N)/N$, where C_d is a constant depending on d only.

With the order d determined by $\hat{d}$, we can set about reconstructing or estimating the functional form of the skeleton. An obvious candidate is to start with

$$X_t=\hat{F}_N(X_{t-1}, \ldots, X_{t-\hat{d}}). \tag{5.1}$$

Fundamentally, different choices of the kernel correspond to different parameterizations of $\hat{F}$. Fig. 18 corresponds to the reconstruction of the skeleton of model (4.1) by using the same kernel as that for Fig. 2. The data-driven choice of the bandwidth seems to strike a reasonable compromise between fidelity and variability. We are currently studying this technique as a possible alternative to existing techniques (e.g. Mees (1989)). An explicit specification of $\hat{F}$ should facilitate the calculation of intrinsic quantities such as the Lyapunov spectrum, the correlation integral, etc., at least in principle. We could also clothe equation (5.1) either by bootstrapping the fitted residuals or by simulation and run the clothed model forwards M steps. Repeating this B times we may then construct an estimate of the M-step-ahead forecast distribution. Diagnostics are also possible by examining the fitted residuals $Z_t-\hat{F}_{N,\setminus t}(Z_{t-1}, \ldots, Z_{t-\hat{d}})$, $t=\hat{d}+1, \ldots, N$. We shall explore these possibilities elsewhere. It would also be interesting to explore the possible connection between

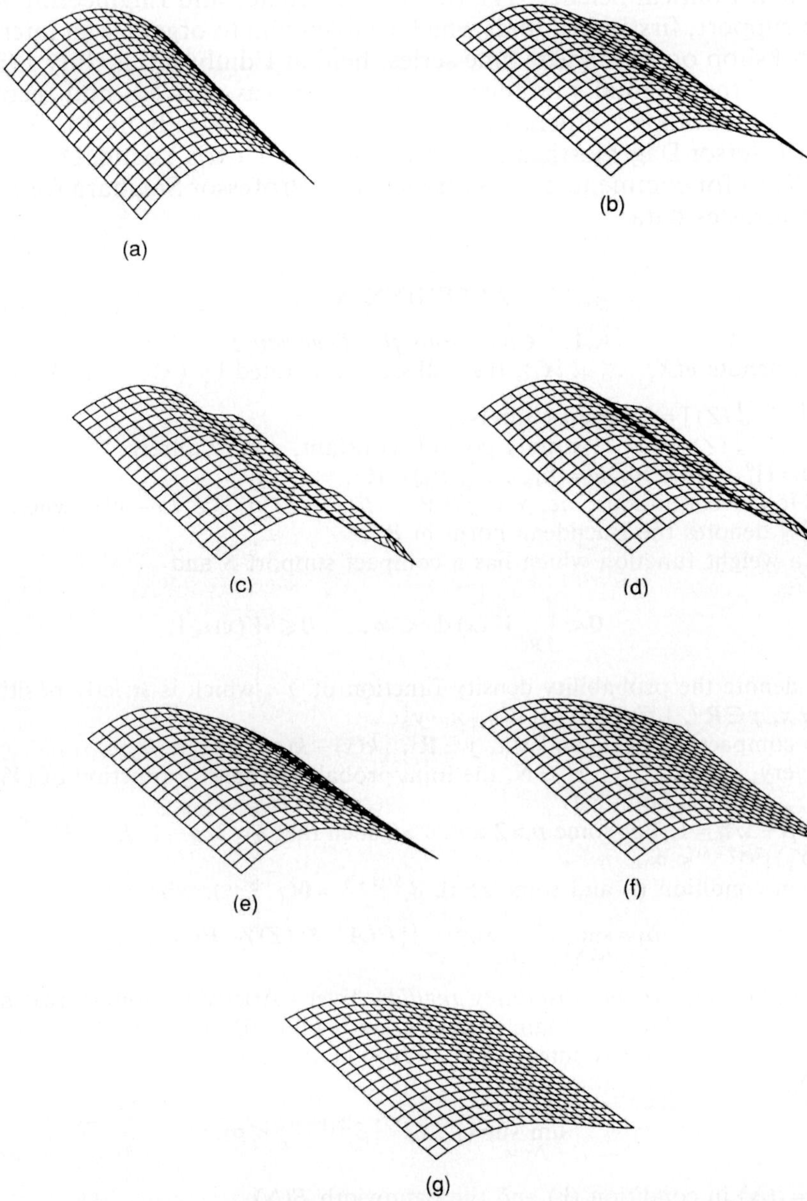

Fig. 18. Reconstruction of the skeleton of model (4.1), $\hat{d}=2$: (a) true skeleton; (b) $B(N)=0.1117$ (global minimum CV choice); (c) $B(N)=0.04$; (d) $B(N)=0.08$; (e) $B(N)=0.20$; (f) $B(N)=0.30$; (g) $B(N)=0.40$

the data-driven estimate of the bandwidth $B(N)$ and the (smoothing) parameter k of Casdagli (1992) in their interpretation.

ACKNOWLEDGEMENTS

BC thanks the Royal Society (UK) for financial support and Professor P. M. Robinson for his kindness and guidance during his visit to the London School of

Economics and Political Science. HT thanks the Science and Engineering Research Council for support, firstly for funds which enabled him to organize an international research workshop on non-linear time series, held at Edinburgh in July 1989, from which much of the original stimulus for this paper was derived, and secondly for funds from their Complex Stochastic Systems initiative. We thank Professor A. J. Lawrance, Professor Dag Tjøstheim, Dr A. E. Sorour, Dr Iris Yeung, Dr Eryl Bassett and Dr W. K. Li for comments and assistance, and Professor Sugihara for providing us with the measles data.

APPENDIX A

A.1. *Conditions for Theorem 1*

Let $\mathscr{F}_s^t(X)$ denote $\sigma(X_s, \ldots, X_t)$, the σ-algebra generated by $(X_s, \ldots, X_t)$.

(a) $E[\epsilon_t \mid \mathscr{F}_{-\infty}^{t-1}(Z)] = 0$, almost surely.
(b) $E[\epsilon_t^2 \mid \mathscr{F}_{-\infty}^{t-1}(Z)] = \sigma^2$, a strictly positive constant, almost surely.
(c) $K_d(u) = \Pi_{i=1}^d k(u_i)$ for $u = (u_1, \ldots, u_d) \in \mathbf{R}^d$.
(d) F is Hölder continuous, i.e. $\forall x, y \in \mathbf{R}^d$, $|F(x) - F(y)| \leqslant c_1 \|x-y\|^\mu$, where $0 < \mu \leqslant 1$ and $\| \;\|$ denotes the Euclidean norm in $\mathbf{R}^d$.
(e) W is a weight function which has a compact support S and

$$0 < \int_{\mathbf{R}^d} W(x)\,\mathrm{d}x < \infty, \qquad 0 \leqslant W(x) \leqslant 1.$$

(f) Let f denote the probability density function of Y_t, which is strictly positive on S, and $\forall x, y \in \mathbf{R}^d$, $|f(x) - f(y)| \leqslant c_2 \|x-y\|$.
(g) k has compact support, and $\forall x, y \in \mathbf{R}^1$, $|k(x) - k(y)| \leqslant c_3 \|x-y\|$.
(h) For every, $t, s, \tau, t', s', \tau' \in \mathbf{N}$, the joint probability density function of $(Y_t, Y_s, Y_\tau, Y_{t'}, Y_{s'}, Y_{\tau'})$ is bounded.
(i) Let $1/p + 1/q = 1$. For some $p > 2$ and $\delta > 0$ such that $\delta < 2/q - 1$, $E|\epsilon_s|^{2p(1+\delta)} < \infty$ and $E|F(Y_1)|^{2p(1+\delta)} < \infty$.
(j) For δ in condition (i) and some $\epsilon > 0$, $\beta_j^{\delta/(1+\delta)} = \mathbf{0}(j^{-2+\epsilon})$, where

$$\beta_j = \sup_{i \in \mathbf{N}} \left(E\left[\sup_{A \in \mathscr{F}_{i+j}^\infty(Z)} \{ |P(A \mid \mathscr{F}_1^i(Z)) - P(A)| \} \right] \right).$$

($\{Z_t\}$ is then said to be *absolutely regular*. Every strictly stationary real aperiodic Harris recurrent Markov chain is absolutely regular (Bradley, 1986).)
(k) Let $j = j(N)$ be a positive integer and $i = i(N)$ be the largest positive integer such that $2ij \leqslant N$,

$$\limsup_{N \to \infty} (1 + 6e^{1/2} \beta_j^{1/(1+i)})^i < \infty.$$

(l) For $i = i(N)$ in condition (k) and the bandwidth $B(N)$,

$$\limsup_{N \to \infty} \{ i(N)\, B(N)^d \} < \infty.$$

(m) $N B(N)^{2d} \to \infty$ as $N \to \infty$.
(n) For μ in assumption (d) $N B(N)^{2d+2\mu} \to 0$ as $N \to \infty$.
(o) For q, δ and ϵ in conditions (i) and (j), $N^\epsilon B(N)^{-2d+\theta} \to 0$ as $N \to \infty$, where $\theta = 4d/(q + q\delta)$.

Some explanation of these conditions is in order. Conditions (a)–(d) are self-explanatory. Condition (e) is the introduction of a weight function W, the purpose of which is to overcome the 'infinite integration problem' in asymptotic expansion encountered by Auestad and Tjøstheim (1990). Conditions (f), (g), (i), (m) and (n) are standard conditions in nonparametric inference. Condition (h) is a mild condition, which will be useful when we use a mixing inequality. Condition (j) is a very mild condition, which is weaker than geometric absolute regularity. Conditions (k) and (l) were given by Roussas (1988). They may be replaced by other assumptions on the mixing coefficient β, if other methods are used to show the almost sure convergence of $\hat{f}_N$ and $\hat{F}_N$. Condition (o) is necessary for proposition 2 of Denker and Keller (1983). Note that conditions (j) and (o) do not contradict each other.

A.2. *Proof of Theorem 2*

To prove part (a) of theorem 2, we normalize $\sigma_N^2(d)$ to

$$\tilde{\sigma}_N^2(d) = \sigma_N^2(d) \Big/ \int_{\mathbf{R}^d} W_d(x) f_d(x)\,\mathrm{d}x. \tag{A.1}$$

Then by theorem 3 and an ergodic theorem,

$$\mathrm{RSS}(d) = \tilde{\sigma}_N^2(d) \int_{\mathbf{R}^d} W_d(x) f_d(x)\,\mathrm{d}x + o_p(1) \tag{A.2}$$

and

$$\tilde{\sigma}_N^2(d) = \sigma^2(d) = o_p(1).$$

We 'estimate' $\phi(d)$ by

$$\phi_N(d) = \{1 - \tilde{\sigma}_N^2(d)/\tilde{\sigma}_N^2(d-1)\}^{1/2}. \tag{A.3}$$

Then

$$\tilde{\sigma}_N^2(d) = \tilde{\sigma}_N^2(0) \prod_{i=1}^{d} \{1 - \phi_N^2(i)\}, \tag{A.4}$$

and by an ergodic theorem

$$\lim_{N\to\infty} \phi_N(d_0) = \phi(d_0), \qquad \text{almost surely.} \tag{A.5}$$

Now, for $0 \leqslant d < d_0$,

$$\frac{\mathrm{RSS}(d)}{\mathrm{RSS}(d_0)} = \frac{\int W_d(x) f_d(x)\,\mathrm{d}x}{\int W_{d_0}(x) f_{d_0}(x)\,\mathrm{d}x}\,\frac{\tilde{\sigma}_N^2(d)}{\tilde{\sigma}_N^2(d_0)}$$

$$\geqslant \frac{\tilde{\sigma}_N^2(d)}{\tilde{\sigma}_N^2(d_0)} \sim \prod_{i=d+1}^{d_0} \{1 - \phi_N^2(i)\}^{-1} \geqslant \{1 - \phi_N^2(d_0)\}^{-1} > 1. \tag{A.6}$$

For any $\delta > 0$ $\exists$ an integer $M > 0$ such that for any $N \geqslant M$

$$\frac{1 + 2\,\alpha(d_0)\,\gamma(d_0)\rho^{d_0}/N}{1 + 2\,\alpha(d)\,\gamma(d)\rho^{d}/N} < 1 + \delta. \tag{A.7}$$

Finally, using inequalities (A.6) and (A.7) and theorem 1, we have for any $\epsilon > 0$, $\exists$ δ such that

$$P(\hat{d} = d) = P\{\mathrm{CV}(d) \leqslant \mathrm{CV}(d'),\ 1 \leqslant d' \leqslant L\} \leqslant P\{\mathrm{CV}(d) \leqslant \mathrm{CV}(d_0)\}$$

$$= P\left\{\frac{\mathrm{RSS}(d)}{\mathrm{RSS}(d_0)} \leqslant \frac{1 + 2\alpha\,(d_0)\,\gamma(d_0)\rho^{d_0}/N}{1 + 2\alpha\,(d)\,\gamma(d)\rho^{d}/N}\right\} \leqslant P\left\{\frac{\mathrm{RSS}(d)}{\mathrm{RSS}(d_0)} < 1 + \delta\right\} < \epsilon,$$

which completes the proof of part (a). For part (b), let $d_1 > d_2 \geqslant d_0$. Note that $\epsilon_t^{(d_1)} = \epsilon_t^{(d_2)} = \epsilon_t^{(d_0)}$, and

$$P\{\mathrm{CV}(d_1) - \mathrm{CV}(d_2) \leqslant 0\} = P[\{\mathrm{CV}(d_1) - \mathrm{CV}(d_2) \leqslant 0\} \cap \bigcap_t \{(Z_{t-1}, \ldots, Z_{t-d_1}) \in B_{d_1}\}]$$
$$+ P[\{\mathrm{CV}(d_1) - \mathrm{CV}(d_2) \leqslant 0\} \cap \bigcap_t \{(Z_{t-1}, \ldots, Z_{t-d_1}) \notin B_{d_1}\}]. \qquad (A.8)$$

Note that $(Z_{t-1}, \ldots, Z_{t-d_1}) \in B_{d_1} \Rightarrow (Z_{t-1}, \ldots, Z_{t-d_2}) \in B_{d_2}$ for each t, from which it follows that $\sigma_N^2(d_1) = \sigma_N^2(d_2)$. Therefore on the set $\cap_t \{(Z_{t-1}, \ldots, Z_{t-d_1}) \in B_{d_1}\}$ we have $\sigma_N^2(d_1) = \sigma_N^2(d_2) = \sigma^2(d_0)$ and $1/N B(N)^{d_2} = o\{1/N B(N)^{d_1}\}$.

Hence, on $\cap_t \{(Z_{t-1}, \ldots, Z_{t-d_1}) \in B_{d_1}\}$,

$$N B(N)^{d_1} \{\mathrm{CV}(d_1) - \mathrm{CV}(d_2)\} = \beta(d_1)\, \gamma(d_1)\, \sigma^2(d_0) + o_p(1),$$

i.e.

$$\limsup_{N\to\infty} (P[\{\mathrm{CV}(d_1) - \mathrm{CV}(d_2) \leqslant 0\} \cap \bigcap_t \{(Z_{t-1}, \ldots, Z_{t-d_1}) \in B_{d_1}\}]) = 0.$$

Thus

$$\limsup_{N\to\infty} [P\{\mathrm{CV}(d_1) - \mathrm{CV}(d_2) \leqslant 0\} \leqslant P \bigcap_t \{(Z_{t-1}, \ldots, Z_{t-d_1}) \notin B_{d_1}\}]$$
$$= 1 - \liminf_{N\to\infty} P \bigcap_t \{(Z_{t-1}, \ldots, Z_{t-d_1}) \in B_{d_1}\}. \qquad \square$$

REFERENCES

Akaike, H. (1970) Statistical prediction identification. *Ann. Inst. Statist. Math.*, **22**, 203–217.

Auestad, B. and Tjøstheim, D. (1990) Identification of nonlinear time series: first order characterisation and order determination. *Biometrika*, **77**, 669–688.

Bradley, R. (1986) Basic properties of strong mixing conditions. In *Dependence in Probability and Statistics* (eds E. Eberlein and M. S. Taqqu). Boston: Birkhauser.

Casdagli, M. (1992) Chaos and deterministic *versus* stochastic non-linear modelling. *J. R. Statist. Soc.* B, **54**, 303–328.

Chan, K. S. (1991) Percentage points of likelihood ratio tests for threshold autoregression. *J. R. Statist. Soc.* B, **53**, 691–696.

Chan, K. S. and Tong, H. (1991) On embedding a deterministic map in nonlinear autoregression. To be published.

Cheng, B. and Tong, H. (1991) On residual sums of squares in non-parametric autoregression. *Technical Report*. Institute of Mathematics and Statistics, University of Kent, Canterbury.

Denker, M. and Keller, G. (1983) On *U*-statistics and von Mises' statistics for weakly dependent processes. *Z. Wahrsch. Ver. Geb.*, **64**, 505–522.

Eckmann, J.-P. and Ruelle, D. (1985) Ergodic theory of chaos and strange attractors. *Rev. Mod. Phys.*, **57**, 617–656.

Eubank, S. and Farmer, J. D. (1990) An introduction to chaos and randomness. *Technical Report LA-UR90-1874*, sect. 3.3. Los Alamos National Laboratory, Los Alamos.

Farmer, J. D., Ott, E. and Yorke, J. A. (1983) The dimension of chaotic attractors. *Physica* D, **7**, 153–180.

Farmer, J. D. and Sidorowich, J. J. (1990) Optimal shadowing and noise reduction. *Physica* D, **47**, 373–392.

Hannan, E. J. and Quinn, B. G. (1979) The determination of the order of an autoregression. *J. R. Statist. Soc.* B, **41**, 190–195.

Härdle, W., Hall, P. and Marron, J. S. (1988) How far are automatically chosen regression smoothing parameters from their optimum? *J. Am. Statist. Ass.*, **83**, 86–101.

Kavalieris, L. (1989) The estimation of the order of an autoregression using recursive residual and cross-validation. *J. Time Ser. Anal.*, **10**, 271–282.

Mees, A. L. (1989) Modelling complex systems. *Technical Report*. Mathematics Department, University of Western Australia, Perth.

Nychka, D., Ellner, S., McCaffrey, D. and Gallant, A. R. (1992) Finding chaos in noisy systems. *J. R. Statist. Soc.* B, **54**, 399–426.
Priestley, M. B. (1981) *Spectral Analysis and Time Series*, vol. 1. London: Academic Press.
Ramsey, F. L. (1974) Characterization of the partial autocorrelation function. *Ann. Statist.*, **2**, 1296–1301.
Robinson, P. M. (1983) Non-parametric estimation for time series models. *J. Time Ser. Anal.*, **4**, 185–208.
Roussas, G. G. (1988) Nonparametric estimation in mixing sequences of random variables. *J. Statist. Planng Inf.*, **15**, 135–149.
Shibata, R. (1976) Selection of the order of an autoregressive model by Akaike information criterion. *Biometrika*, **63**, 117–126.
Silverman, B. W. (1985) Some aspects of the spline smoothing approach to non-parametric regression curve fitting (with discussion). *J. R. Statist. Soc.* B, **36**, 1–52.
Stone, M. (1974) Cross-validatory choice and assessment of statistical predictions (with discussion). *J. R. Statist. Soc.* B, **36**, 111–147.
Sugihara, G. and May, R. M. (1990) Nonlinear forecasting as a way of distinguishing chaos from measurement error in time series. *Nature*, **344**, 734–741.
Takens, F. (1981) Dynamical systems and turbulence. *Lect. Notes Math.*, **898**.
Tong, H. (1990a) *Non-linear Time Series: a Dynamical System Approach*. Oxford: Oxford University Press.
——(1990b) Discussion on Chance or chaos? (by M. S. Bartlett). *J. R. Statist. Soc.* A, **153**, 330–332.
——(1990c) Contrasting aspects of non-linear time series analysis. In *Proc. Time Series Meet. Institute of Mathematics and Applications, Minneapolis*. New York: Springer. To be published.
Tong, H. and Lim, K. S. (1980) Threshold autoregression, limit cycles and cyclical data (with discussion). *J. R. Statist. Soc.* B, **42**, 245–292.

Recent Developments on Semiparametric Regression Model Selection

JITI GAO
School of Economics, The University of Adelaide
Adelaide SA 5005, Australia
E-mail: jiti.gao@adelaide.edu.au

This paper provides a survey of the recent development on model selection in semiparametric time series regression. In order to avoid using more data for model validation, this review briefly discusses a two–step model selection procedure, in which we extend the conventional nonparametric CV1 method proposed in Cheng and Tong (1992) to deal with both the optimum subset selection and optimum bandwidth choice. The main ideas and methodology of this review are based on the unpublished paper by Gao and Tong (2005).

1. Introduction

Consider a nonlinear time series model of the form

$$\xi_t = F(\xi_{t-1}, \cdots, \xi_{t-d}) + \epsilon_t, \tag{1}$$

where F is unknown and $\{\epsilon_t\}$ is a sequence of martingale differences with finite variance σ^2. Assume that $\{\xi_t\}$ is a strictly stationary univariate time series with finite variance and absolutely continuous distribution. The paper by Cheng and Tong (1992) proposes estimating both F and d using a nonparametric leave–one–out cross–validation (CV1) method. Asymptotic theory is established and illustrated using simulated and real–data examples.

Since the publication of this paper, such ideas have been extended and studied extensively. For example, Cheng and Tong (1993) discuss the simultaneous selection of both d and the bandwidth involved in the nonparametric kernel method. The paper by Cheng and Tong (1994) shows that the form of $F(\cdot)$ need not be estimated accurately while the main interest is in estimating d. Vieu (1994) and Yao and Tong (1994) then independently extend such ideas to nonparametrically choose an optimum subset of $\zeta_t = (\xi_{t-1}, \cdots, \xi_{t-d})^\tau$. Both the asymptotic properties and practical implementation of such selection procedures have been discussed by the authors. Other closely related studies in the field of nonparametric regression model selection include Auestad and Tjøstheim (1990), Tjøstheim and Auestad (1994a, 1994b), Vieu (1995), Yang (1999), Tschernig and Yang (2000), Vieu (2002), and Yang and Tschernig (2002).

Since the optimum value of d may still be greater than three, model (1) may still suffer from the "curse of dimensionality". To partially solve such a problem, Gao, Anh and Wolff (2001) propose using a semiparametric variable selection method to choose optimum subsets U_t and X_t of ζ_t such that

$$P\left(F(\xi_{t-1}, \cdots, \xi_{t-d}) = U_t^\tau \beta + \phi(X_t)\right) = 1, \tag{2}$$

where $U_t = (U_{t1}, \cdots, U_{tp})^\tau$, $X_t = (X_{t1}, \cdots, X_{tq})^\tau$, the vector of unknown parameters β and the unknown function $\phi(\cdot)$ are both identifiable and estimatable semiparametrically.

In the statistics literature, such an issue has been treated as a model selection problem. In the econometrics literature, this problem has instead been regarded as a model specification issue of testing $H_0: \ P\left(F(\xi_{t-1},\cdots,\xi_{t-d}) = U_t^\tau \beta + \phi(X_t)\right) = 1$. The main similarity is that both methods focus on finding the true form of the conditional mean function. The main difference is that unlike the model specification method, the variable selection method may be able to treat each of the components of ζ_t equally without assuming that $\{U_t\}$ is a subset of the parametric components. As a result, the variable selection method will be more expensive computationally than the model specification method.

In Gao, Anh and Wolff (2001), the authors extend the nonparametric cross–validation selection method to a semiparametric setting and propose using a semiparametric kernel cross–validation method to estimate both β and $\phi(\cdot)$ in (2) and then establish a semiparametric selection criterion. The authors show that if the true model is a semiparametric model of the form $Y_t = U_t^\tau \beta + \phi(X_t) + \epsilon_t$, the semiparametric kernel selection method can asymptotically find it as well as an optimum bandwidth parameter involved in the kernel method.

This kind of selection is considered as the first step to choose an optimum semiparametric model. Assume that the true model is a semiparametric model of the form

$$Y_t = U_t^\tau \beta + \phi(X_t) + \epsilon_t. \tag{3}$$

The second step is to determine whether both U_t and X_t are of the smallest dimension.

When the dimensionality of X_t is smaller than three, model (3) can be used to overcome the dimensionality problem. Otherwise, model (3) itself may still suffer from the curse of dimensionality. Thus, before using model (3) we need to determine whether both the parametric and nonparametric components are of the smallest possible dimension. To achieve this objective, Gao and Tong (2004) develop a simultaneous semiparametric leave–more–out cross–validation selection method for the optimum choice of both U_t and X_t. As observed in the simulations in Section 3.1 of Gao and Tong (2004), the number of observations used to fit the model is, however, quite small (with $T_c = 69$ in Section 3.1 for the semiparametric case), while the number of observations used to validate the proposed method is relatively large ($T_v = 219$, respectively). This may impede the implementation of the nonparametric–based selection method in practice because nonparametric estimation theory requires $T_c \to \infty$.

In order to avoid using more data for model validation, this review briefly discusses a two–step model selection procedure, in which we extend the conventional nonparametric CV1 method proposed in Cheng and Tong (1992) to deal with both the optimum subset selection and optimum bandwidth choice. The main ideas and methodology of this review are based on the unpublished paper by Gao and Tong (2005). Following an earlier version of Gao and Tong (2005), Avramidis (2005) also proposes a different two–step selection criterion for the optimum choice of U_t and X_t.

2. Cross–Validation Model Selection in Semiparametric Regression

Let $A_0 = \{1, 2, \ldots, p\}$, $D_q = \{1, 2, \ldots, q\}$, $\mathcal{A}$ denote all nonempty subsets of A_0 and $\mathcal{D}$ denote all nonempty subsets of D_q. For any subset $A \in \mathcal{A}$, U_{tA} is defined as a column vector consisting of $\{U_{ti}, i \in A\}$. For any subset $D \in \mathcal{D}$, X_{tD} is defined as a column vector consisting of $\{X_{ti}, i \in D\}$. Throughout this paper, $B \subseteq C$ means that B can be the maximum subset C, and $B \subset C$ means that B cannot attain the maximum subset C. We use $d_E = |E|$ to denote the cardinality of a set E.

In this paper, we assume that there is a unique pair (A_*, D_*) with $A_* \in \mathcal{A}$ and $D_* \in \mathcal{D}$ such that there is a true and compact version of model (1.1) defined by

$$Y_t = U_{tA_*}^\tau \beta_* + \phi_*(X_{tD_*}) + \epsilon_t^*, \tag{4}$$

where β_* is a vector of unknown parameters, $\phi_*(\cdot)$ is an unknown function over $I\!R^{|D_*|}$, and $\epsilon_t^* = Y_t - E[Y_t|U_t, X_t]$.

Detailed conditions for the existence and the uniqueness of A_* and D_* are discussed in Sections 2.1 and 2.2 below. This review then considers the following cases:

- Case I: If the linear component of model (3) is already compact but the nonparametric component is not compact, we then take $A_* = A_0 = \{1, 2, \ldots, p\}$ and estimate D_* in Section 2.1 below. We will use the notation of $D_* = D_0 = D_0(A_0)$ in Section 2.1.
- Case II: If both the linear and nonparametric components are not compact, we then estimate both A_* and D_* in Section 2.2 below. Note that the notation of $D_* = D_0(A_*)$ will be used in Section 2.2.
- Case III: If model (3) is already compact, then $A_* = A_0$ and $D_* = D_q$. For this case, no model selection is needed.
- Case IV: If the nonparametric component of model (3) is already compact but the parametric component is not component, we take $D_* = D_q$ and then estimate A_*. As this is a special case of Case II with $D_* = D_q$ and the detailed discussion for this case is very similar but less difficult than that for Case II, we shall not discuss it in detail.

2.1. *Cross–validation criterion for nonparametric regressors*

Assume that the data set $\{(Y_t, U_t, X_t) : t \geq 1\}$ satisfies model (3). In this section, we assume that the linear component is already compact in the selection of nonparametric regressors.

Assume that the data set $\{(Y_t, U_t, X_{tD}) : t \geq 1\}$ satisfies

$$Y_t = U_t^\tau \beta(D) + \phi_D(X_{tD}) + \epsilon_{tD}, \tag{5}$$

where $\{\epsilon_{tD}\}$ is a sequence of errors, $\beta(D) = (\beta_1(D), \ldots, \beta_p(D))^\tau$ is a vector of unknown parameters, and $\phi_D(\cdot)$ is an unknown function over $I\!R^{d_D}$. Note that $\beta(D)$ is still a vector of p unknown parameters that may depend on D.

In order to ensure that model (5) is identifiable for each given $D \in \mathcal{D}$, one needs to define (see §1.2 of Härdle, Liang and Gao 2000)

$$\beta(D) = \{E\left(U_t - E[U_t|X_{tD}]\right)\left(U_t - E[U_t|X_{tD}]\right)^\tau\}^{-1} E\left(U_t - E[U_t|X_{tD}]\right)\left(Y_t - E[Y_t|X_{tD}]\right)$$

and

$$\phi_D(X_{tD}) = \phi_D(X_{tD}, \beta(D)) = E\left\{(Y_t - U_t^\tau \beta(D))|X_{tD}\right\} = \phi_1(X_{tD}) - \phi_2(X_{tD})^\tau \beta(D), \tag{6}$$

under Assumption 2.1(i) below, where $\phi_1(X_{tD}) = E[Y_t|X_{tD}]$ and $\phi_2(X_{tD}) = E[U_t|X_{tD}]$.

For any $D \in \mathcal{D}$, let $\psi_D(U_t, X_{tD}) = U_t^\tau \beta(D) + \phi_D(X_{tD})$ and $\Psi(U_t, X_t) = E[Y_t|U_t, X_t]$. Assumption 2.1 below imposes some existence and uniqueness conditions on model (5).

Assumption 2.1. (i) Assume that $\Delta_D = E\left\{U_t - E[U_t|X_{tD}]\right\}\left\{U_t - E[U_t|X_{tD}]\right\}^\tau$ is a positive definite matrix with order $d_D \times d_D$ for each given $D \in \mathcal{D}$.

(ii) Let $\mathcal{D}_1 = \{D \in \mathcal{D}$, such that $\psi_D = \Psi\}$ and $\mathcal{D}_0 = \{D_0 \in \mathcal{D}_1$, such that $|D_0| = \min_{D \in \mathcal{D}_1} |D|\}$. Assume that D_0 is the unique element of $\mathcal{D}_0$ and that $\phi_{D_0}(X_{tD_0})$ is an unknown nonparametric function.

It follows from (4)–(6) and Assumption 2.1 that for Case I we may define the true model as

$$Y_t = U_t^\tau \beta(D_0) + \phi_{D_0}(X_{tD_0}) + \epsilon_{tD_0}, \tag{7}$$

where $\epsilon_{tD_0} = Y_t - E[Y_t|U_t, X_t]$. Note that model (7) is a special case of (4) where $A_* = A_0$, $U_{tA_0} = U_t$, $\beta_* = \beta(D_0)$, $\phi_{D_0} = \phi_*$, $D_* = D_0$, $X_{tD_0} = X_{tD_*}$, and $\epsilon_{tD_0} = \epsilon_t^*$.

For the given D_0, we define the least squares estimator, $\tilde{\beta}(D_0, h)$, of $\beta(D_0)$ as the solution of (see §1.2 of Härdle, Liang and Gao 2000 for example)

$$\sum_{t=1}^{T} \left\{ Y_t - U_t^\tau \tilde{\beta}(D_0, h) - \hat{\phi}\left(X_{tD_0}, \tilde{\beta}(D_0, h)\right) \right\}^2 = \min!, \tag{8}$$

where

$$\hat{\phi}(X_{tD}, \beta) = \sum_{s=1}^{T} W_D(t,s)(Y_s - U_s^\tau \beta) \quad \text{with} \quad W_D(t,s) = \frac{K_D((X_{tD} - X_{sD})/h)}{\sum_{l=1}^{T} K_D((X_{tD} - X_{lD})/h)},$$

T is the number of observations, K_D is a multivariate kernel function, and h is a bandwidth parameter satisfying $h = h_T \to 0$ as $T \to \infty$.

It follows from (7) that

$$\tilde{\beta}(D, h) = (\tilde{\Sigma}(D, h))^+ \sum_{t=1}^{T} \tilde{U}_t(D, h)(Y_t - \hat{\phi}_1(X_{tD}, h)), \tag{9}$$

where $(\cdot)^+$ denotes the Moore–Penrose inverse,

$$\tilde{\Sigma}(D, h) = \sum_{t=1}^{T} \tilde{U}_t(D, h) \tilde{U}_t(D, h)^\tau, \ \tilde{U}_t(D, h) = U_t - \hat{\phi}_2(X_{tD}, h),$$

$$\hat{\phi}_1(X_{tD}, h) = \sum_{s=1}^{T} W_D(t,s) Y_s \text{ and } \hat{\phi}_2(X_{tD}, h) = \sum_{s=1}^{T} W_D(t,s) U_s.$$

In order to select both h and D_0, we introduce several leave–one–out estimates. For any $D \in \mathcal{D}$, equations (6)–(7) suggest the leave-one-out estimator

$$\hat{\phi}_t(X_{tD}, \beta) = \hat{\phi}_{1t}(X_{tD}, h) - \hat{\phi}_{2t}(X_{tD}, h)^\tau \beta,$$

where $\hat{\phi}_{1t}(X_{tD}, h) = \sum_{s=1, s\neq t}^{T} W_D^{(-t)}(t,s) Y_s$ and $\hat{\phi}_{2t}(X_{tD}, h) = \sum_{s=1, s\neq t}^{T} W_D^{(-t)}(t,s) U_s$, in which

$$W_D^{(-t)}(t,s) = \frac{K_D((X_{tD} - X_{sD})/h)}{\sum_{l=1, l\neq t}^{T} K_D((X_{tD} - X_{lD})/h)}.$$

Then, we define the leave–one–out least squares (LS) estimator $\hat{\beta}(D, h)$ of $\beta(D)$ as the solution of

$$\sum_{t=1}^{T} \left\{ Y_t - U_t^\tau \hat{\beta}(D, h) - \hat{\phi}_t(X_{tD}, \hat{\beta}(D, h)) \right\}^2.$$

For any given $D \in \mathcal{D}$, the leave–one–out LS estimator is

$$\hat{\beta}(D,h) = (\tilde{\Sigma}(D,h))^{+} \sum_{t=1}^{T} \tilde{U}_t(D,h)(Y_t - \hat{\phi}_{1t}(X_{tD},h)), \tag{10}$$

where $\tilde{U}_t(D,h) = U_t - \hat{\phi}_{2t}(X_{tD},h)$, $\tilde{\Sigma}(D,h) = \sum_{t=1}^{T} \tilde{U}_t(D,h)\tilde{U}_t(D,h)^{\tau}$. It is noted that the LS estimator $\tilde{\beta}(D_0,h)$ of (9) is asymptotically equivalent to the leave–one–out least squares (LS) estimator $\hat{\beta}(D_0,h)$ of (10). In defining the following leave–one–out cross-validation, we use the latter.

We now introduce a version of the leave–one–out cross-validation, abbreviated as CV1. For any $D \in \mathcal{D}$, we define

$$\mathrm{CV1}(D,h) = \frac{1}{T}\sum_{t=1}^{T}\left\{Y_t - U_t^{\tau}\hat{\beta}(D,h) - \hat{\phi}_t(X_{tD},\hat{\beta}(D,h))\right\}^2 w(X_t), \tag{11}$$

where $w(\cdot)$ is a weight function defined on $I\!R^q$. Note that the proposed CV1(D,h) function extends the original CV1 function proposed in Cheng and Tong (1992) for the case where the focus is on the estimation of dimensionality.

Let $\hat{D}_0$ and $\hat{h}$ denote the estimators of D_0 and h, respectively, which are obtained by minimising the $CV1(D,h)$ function over $D \in \mathcal{D}$ and $h \in H_{TD}$, and written as

$$(\hat{D}_0,\hat{h}) = \mathrm{argmin}_{\{D\in\mathcal{D},\ h\in H_{TD}\}}\mathrm{CV1}(D,h), \tag{12}$$

where

$$H_{TD} = \left[a_D T^{-\frac{1}{4+|D|}-c_D}, b_D T^{-\frac{1}{4+|D|}+c_D}\right],$$

in which the constants a_D, b_D and c_D satisfy $0 < a_D < b_D < \infty$ and $0 < c_D < \frac{1}{2(4+|D|)}$.

We now state the first theorem of this review and its proof is available from Appendix B of Gao and Tong (2005).

THEOREM 2.1. *Assume that Assumption 2.1 holds. In addition, let Assumptions A.1–A.4 of Gao and Tong (2005) hold with $A = A_0$. Then*

$$\lim_{T\to\infty} P(\hat{D}_0 = D_0) = 1 \ \text{ and } \ \frac{\hat{h}}{h_0} \to_p 1$$

as $T \to \infty$, where h_0 is the minimizer of the mean average squared error (MASE) given by

$$\mathrm{MASE}(D_0,h) = \frac{1}{T}\sum_{t=1}^{T} E\left\{U_t^{\tau}\tilde{\beta}(D_0,h) + \hat{\phi}\left(X_{tD_0},\tilde{\beta}(D_0,h)\right) - U_t^{\tau}\beta(D_0) - \phi_{D_0}(X_{tD_0})\right\}^2.$$

It can be shown that $h_0 = C_{D_0}T^{-\frac{1}{4+|D_0|}}$ and $C_{D_0} > 0$ is a constant independent of T. Due to this property, instead of defining h_0 as the minimizer of certain MASE we shall use this explicit form for h_0 throughout the rest of the paper. Theorem 2.1 shows that the true and unique subset D_0 can be identified asymptotically. Moreover, the criterion can also determine the bandwidth asymptotically.

In Section 2.1, we have considered Case I where the linear component is already compact and then propose the leave–one–out cross-validation for the selection of nonparametric regressors. In Section 2.2 below, we consider the selection of both parametric and nonparametric regressors. Since for the selection of parametric regressors the leave–one–out cross-validation is asymptotically inconsistent (see Shao 1993), we need to consider using the leave–T_v–out cross-validation for the selection of parametric regressors. Moreover, because the theory of the leave–T_v–out cross-validation is different to that of the leave–one–out cross-validation and much more complicated, we consider Case II separately.

2.2. *CV criterion for the selection of parametric regressors*

As can be seen in Section 2.1, the selected $\hat{D}_0$ and $\hat{h}$ depend on A_0. Thus we can rewrite $\hat{D}_0 = \hat{D}_0(A_0)$ and $\hat{h} = \hat{h}(A_0)$. Let $\mathcal{A}$ denote all nonempty subsets of A_0. For $A \in \mathcal{A}$, let β_A be a column vector consisting of $\{\beta_i : i \in A\}$. Denote U_{tA} with $A = A_0$ by U_t and β_A with $A = A_0$ by $\beta = (\beta_1, \ldots, \beta_p)^\tau$.

To extend Assumption 2.1 to the case where both the linear and nonparametric components are not compact, one needs to restate some notation.

For each $A \in \mathcal{A}$ and $D \in \mathcal{D}$, define $\psi_{A,D}(U_{tA}, X_{tD}) = U_{tA}^\tau \beta_A + \phi_D(X_{tD})$ and $\Psi(U_t, X_t) = E[Y_t|U_t, X_t]$. The following assumption imposes some existence and uniqueness conditions on the true versions of A and D.

Assumption 2.2. (i) Let $\Delta_{A,D} = E\{U_{tA} - E[U_{tA}|X_{tD}]\}\{U_{tA} - E[U_{tA}|X_{tD}]\}^\tau$ be a positive definite matrix with order $d_D \times d_D$ for each given $A \in \mathcal{A}$ and $D \in \mathcal{D}$.

(ii) For each given $A \in \mathcal{A}$, let $\mathcal{D}_{1A} = \{D \in \mathcal{D}$, such that $\psi_{A,D} = \Psi\}$ and $\mathcal{D}_{0A} = \{D_0(A) \in \mathcal{D}_{1A}$, such that $|D_0(A)| = \min_{D\in\mathcal{D}_{1A}} |D|\}$. Assume that $D_0(A)$ is the unique element of $\mathcal{D}_{0A}$ and that $\phi_{D_0(A)}(X_{tD_0(A)})$ is an unknown nonparametric function for each given $A \in \mathcal{A}$.

Following Assumption 2.2, for each $A \in \mathcal{A}$ we can define the corresponding $D_0(A)$. Theorem 2.1 then shows that

$$\lim_{T\to\infty} P\left(\hat{D}_0(A) = D_0(A)\right) = 1 \text{ and } \frac{\hat{h}(A)}{h_0(A)} \to_p 1$$

as $T \to \infty$, where $h_0(A) = C_{D_0(A)} T^{-\frac{1}{4+|D_0(A)|}}$.

For simplicity and convenience, we introduce the following notation:

$$\begin{aligned}
\hat{\psi}_1(t, A) &= \hat{\phi}_1\left(X_{t\hat{D}_0(A)}, \hat{h}(A)\right) = \sum_{s=1}^{T} W_{\hat{D}_0(A)}(t, s) Y_s, \\
\hat{\psi}_2(t, A) &= \hat{\phi}_2\left(X_{t\hat{D}_0(A)}, \hat{h}(A)\right) = \sum_{s=1}^{T} W_{\hat{D}_0(A)}(t, s) U_{sA}, \\
\eta_{tA} &= U_{tA} - E[U_{tA}|X_{tD_0(A)}], \; \delta_{tA} = E[U_{tA}|X_{tD_0(A)}] - \hat{\psi}_2(t, A), \\
V_{tA} &= \eta_{tA} + \delta_{tA} = U_{tA} - \hat{\psi}_2(t, A), \; V_A = (V_{1A}, \ldots, V_{TA})^\tau, \\
\hat{\psi}_1(t) &= \hat{\psi}_1(t, A_0), \quad \hat{\psi}_2(t) = \hat{\psi}_2(t, A_0), \\
\eta_t &= U_t - E[U_t|X_{tD_0}], \quad \delta_t = E[U_t|X_{tD_0}] - \hat{\psi}_2(t), \\
V_t &= \eta_t + \delta_t = U_t - \hat{\psi}_2(t), \quad V = (V_1, \ldots, V_T)^\tau, \\
Z_t &= Y_t - \hat{\psi}_1(t) \text{ and } Z = (Z_1, \ldots, Z_T)^\tau,
\end{aligned} \tag{13}$$

where $D_0 = D_0(A_0)$ is as defined in Assumption 2.1.

Because some of the components of β may be zero, the following model

$$Y_t = U_{tA}^\tau \beta_A + \phi_{D_0(A)}(X_{tD_0(A)}) + \epsilon_{tA} \text{ where } \epsilon_{tA} \text{ is an error process,} \tag{14}$$

might be more compact than model (7) given by $Y_t = U_t^\tau \beta(D_0) + \phi_{D_0}(X_{tD_0}) + \epsilon_{tD_0}$. Note that $\beta(D)$ signifies that $\beta(D)$ may depend on D while the notation of β_A means that β_A is a subset of β.

As mentioned earlier, for each $A \in \mathcal{A}$ it is natural to estimate each $D_0(A)$ by $\hat{D}_0(A)$. The definition of $\hat{\phi}(X_{tD}, \beta)$ of (8) then suggests estimating (see §1.2 of Härdle, Liang and Gao 2000 for example)

$$\phi_{D_0(A)}\left(X_{tD_0(A)}\right) = \phi_{D_0(A)}(X_{tD_0(A)}, \beta_A)$$

by

$$\hat{\phi}\left(X_{t\hat{D}_0(A)}, \beta_A\right) = \hat{\psi}_1(t, A) - \hat{\psi}_2(t, A)^\tau \beta_A.$$

Thus, using (13), model (14) can be rewritten as

$$Y_t - \hat{\psi}_1(t, A) = \beta_A^\tau \left(U_{tA} - \hat{\psi}_2(t, A)\right) + \phi_{D_0(A)}(X_{tD_0(A)}) - \hat{\phi}\left(X_{t\hat{D}_0(A)}, \beta_A\right) + \epsilon_{tA}$$

$$= V_{tA}^\tau \beta_A + \epsilon_{tA} + o_p(1)$$

using the fact that the rate of uniform convergence of $\hat{\phi}\left(X_{t\hat{D}_0(A)}, \beta_A\right)$ to $\phi_{D_0(A)}\left(X_{tD_0(A)}\right)$ is of order $o_p(1)$ (see Theorem 3.2.2 of Härdle, Liang and Gao 2000 for example).

This suggests using a linear model of the form

$$Y_t - \hat{\psi}_1(t, A) = V_{tA}^\tau \beta_A + \epsilon_{tA} \tag{15}$$

to approximate model (14) in the selection of A without changing the true version of A. Obviously, there are $2^p - 1$ possible models of the form (15), each of which corresponds to a subset A and is defined by $\mathcal{M}_A$. The dimension of $\mathcal{M}_A$ is defined to be d_A, the number of predictors in $\mathcal{M}_A$. If we know whether each component of β is zero or not, then the models $\mathcal{M}_A$ can be classified into two categories:

- Category I: At least one nonzero component of β is not in β_A.
- Category II: β_A contains all nonzero components of β.

Clearly, the models in Category I are incorrect models, and the models in Category II may be inefficient because of their unnecessarily large sizes. The optimum model, denoted by $\mathcal{M}_*$, is the model in Category II with the smallest dimension.

Let A_* correspond to $\mathcal{M}_*$. For Case II, we may define the true model as

$$Y_t = U_{tA_*}^\tau \beta_{A_*} + \phi_*(X_{tD_*}) + \epsilon_t^*,$$

where $D_* = D_0(A_*)$, $\phi_* = \phi_{D_*} = \phi_{D_0(A_*)}$, and ϵ_t^* is as defined in (4). Note that this is the true model we have assumed in (4) for Case II.

Thus, in order to determine the true model (4) for Case II, one needs to estimate A_*. The selection of A is carried out by using the data $\{(Z_t, V_t) : t = 1, 2, \ldots, T\}$ satisfying

$$Z_t = V_t^\tau \beta + \epsilon_t,$$

where $\{\epsilon_t\}$ is a sequence of errors. Under model $\mathcal{M}_A$, the least squares estimator of β_A is

$$\hat{\beta}_A = (V_A^\tau V_A)^+ V_A^\tau Z,$$

where Z and V_A are as defined in (13).

Using model $\mathcal{M}_A$ fitted based on the data $\{(Z_t, V_t) : t = 1, 2, \ldots, T\}$, the average squared prediction error is

$$L_T(A) = \frac{1}{T}\sum_{t=1}^{T}\left[Z_t - V_{tA}^\tau \hat{\beta}_A\right]^2 = \frac{1}{T}\left(Z - V_A\hat{\beta}_A\right)^\tau \left(Z - V_A\hat{\beta}_A\right)$$
$$= \frac{1}{T}\epsilon^\tau \epsilon + \frac{1}{T}\epsilon^\tau P_A \epsilon + \frac{1}{T}(V\beta)^\tau R_A(V\beta) + \frac{2}{T}\epsilon^\tau R_A(V\beta), \qquad (16)$$

where $\epsilon = (\epsilon_1, \ldots, \epsilon_T)^\tau$, $P_A = V_A\left(V_A^\tau V_A\right)^+ V_A^\tau$, $R_A = I_T - P_A$, and I_T is the identity matrix of order T.

It follows from (16) that the conditionally expected average squared error is

$$R_T(A, V) = E[L_T(A)|V] = \frac{1}{T}E[\epsilon^\tau \epsilon|V] + \frac{1}{T}E[\epsilon^\tau P_A \epsilon|V] + \frac{1}{T}(V\beta)^\tau R_A(V\beta)$$
$$+ \frac{2}{T}E\left[\epsilon^\tau R_A(V\beta)|V\right] = \sigma_\epsilon^2 + \frac{1}{T}d_A\sigma_\epsilon^2 + \Delta_{T,A}, \qquad (17)$$

with probability one, where $\sigma_\epsilon^2 = E[\epsilon^\tau \epsilon]$ and $\Delta_{T,A} = \frac{1}{T}(V\beta)^\tau R_A(V\beta)$.

When $\mathcal{M}_A$ is in Category I, we assume that

$$\liminf_{T\to\infty} \Delta_{T,A} > 0 \text{ in probability.} \qquad (18)$$

When $\mathcal{M}_A$ is in Category II, it follows from (16) and (17) that because $V\beta = V_A\beta_A$,

$$L_T(A) = \frac{1}{T}\epsilon^\tau \epsilon + \frac{1}{T}\epsilon^\tau P_A \epsilon + \frac{2}{T}\epsilon^\tau R_A(V\beta) \text{ and } R_T(A, V) = \frac{1}{T}(T - d_A)\sigma_\epsilon^2.$$

We now propose our cross-validation criterion for the selection of $A \in \mathcal{A}$. Suppose that we split the data set into two parts: $\{(Z_t, V_t) : t \in S\}$ and $\{(Z_t, V_t) : t \in S^c\}$, where S is a subset of $\{1, 2, \ldots, T\}$ containing T_v integers and S^c is its complement containing T_c integers, $T_v + T_c = T$. The model $\mathcal{M}_A$ is fitted using the construction data $\{(Z_t, V_t) : t \in S^c\}$ and the prediction error is assessed using the validation data $\{(Z_t, V_t) : t \in S\}$, treated as if they were future values. The average squared prediction error is

$$\mathrm{CV}(T_v) = \mathrm{CV}_{A,S}(T_v) = \frac{1}{T_v}\left|\left|Z_S - \hat{Z}_{A,S^c}\right|\right|^2$$
$$= \frac{1}{T_v}\left|\left|(I_{T_v} - Q_{A,S})^+(Z_S - V_{A,S}\hat{\beta}_A)\right|\right|^2,$$

where $||x|| = \sqrt{x^\tau x}$ for a vector x, Z_S is the column vector containing the components of Z indexed by $t \in S$, $V_{A,S}$ is the $T_v \times d_A$ matrix containing the rows of V_A indexed by $t \in S$, $\hat{Z}_{A,S^c}$ is the prediction of Z_S using the construction data and the least squares method under model $\mathcal{M}_A$, $Q_{A,S} = V_{A,S}(V_A^\tau V_A)^+ V_{A,S}^\tau$, and $\hat{\beta}_A$ is as defined before.

The $\mathrm{CV}_{A,S}(T_v)$ function is called the leave–T_v–out cross–validation, abbreviated as $\mathrm{CV}(T_v) = \mathrm{CV}T_v$. From the computational point of view, the simplest $\mathrm{CV}T_v$ is the one with $T_v \equiv 1$ and $S = \{t\}$; that is, the CV1. As the CV1 is asymptotically inconsistent, we adopt the following Monte Carlo $\mathrm{CV}T_v$ in the selection of A.

Randomly draw a collection $\mathcal{R}$ of b subsets of $\{1, 2, \ldots, T\}$ that have size T_v and select a model by minimizing

$$\mathrm{MCCV}(A, T_v) = \frac{1}{b}\sum_{S\in\mathcal{R}} \mathrm{CV}_{A,S}(T_v) = \frac{1}{bT_v}\sum_{S\in\mathcal{R}}\left|\left|Z_S - \hat{Z}_{A,S^c}\right|\right|^2. \qquad (19)$$

This method is called the Monte Carlo CVT_v, abbreviated as MCCVT_v, as (19) is obtained by randomly splitting the data b times and averaging the squared prediction errors over the splits. We now have the following theorem.

THEOREM 2.2. *Assume that Assumption 2.2 holds. Let Assumptions A.1–A.5 of Gao and Tong (2005) hold. Then we have the following conclusions:*

(i) *If $\mathcal{M}_A$ is in Category I, then there exists $R_T \geq 0$ such that*

$$\text{MCCV}(A, T_v) = \frac{1}{T_v b} \sum_{S \in \mathcal{R}} \epsilon_S^\tau \epsilon_S + \Lambda_{T,A} + o_p(1) + R_T,$$

where $\epsilon_S = V_S - Z_S \beta$ and $\Lambda_{T,A} = \frac{1}{T}(\eta\beta)^\tau (I_T - \eta_A(\eta_A^\tau \eta_A)^+ \eta_A^\tau)\eta\beta$.

(ii) *If $\mathcal{M}_A$ is in Category II, then*

$$\text{MCCV}(A, T_v) = \frac{1}{T_v b} \sum_{S \in \mathcal{R}} \epsilon_S^\tau \epsilon_S + \frac{d_A}{T_c}\sigma_\epsilon^2 + o_p\left(\frac{1}{T_c}\right).$$

(iii) *Consequently,*

$$\lim_{T \to \infty} P(\text{the selected model is } \mathcal{M}_*) = 1.$$

Let $\hat{A}$ correspond to the selected model. Then, Theorems 2.1 and 2.2 imply the following main theorem of this review.

THEOREM 2.3. *Assume that the conditions of Theorem 2.2 hold. Then*

$$\lim_{T \to \infty} P(\hat{A} = A_*,\ \hat{D}_0(\hat{A}) = D_*)) = 1 \text{ and } \frac{\hat{h}(\hat{A})}{h_0(A_*)} \to_p 1$$

as $T \to \infty$, where $h_0(A_*) = C_{D_*} T^{-\frac{1}{4+|D_*|}}$.

The proofs of Theorems 2.2 and 2.3 are available from Appendices B and C of Gao and Tong (2005). Meanwhile, examples of implementation of Theorem 2.3 are also given in Gao and Tong (2005).

3. Discussion

We have discussed the proposed the two–step semiparametric cross–validation selection method. In addition to this approach, Dong, Gao and Tong (2007) propose a semiparametric penalty function–based model selection criterion by incorporating some essential features of the CV1 selection method for the choice of both the parametric and the nonparametric regressors in model (3). The main objective of Dong, Gao and Tong (2007) is to propose a new selection criterion, establish the associated theory and demonstrate the key feature of easy implementation of the proposed semiparametric penalty function method by using two simulated examples. Chapter 4 of Gao (2007) provides some detailed discussion about the advantages and disadvantages of the semiparametric cross–validation method proposed in Gao and Tong (2004) and the semiparametric penalty function method proposed in Dong, Gao and Tong (2007).

References

1. Auestad, B., and Tjøstheim, D. (1990) Identification of nonlinear time series: first order characterization and order determination. *Biometrika*, **77**, 669–687.
2. Avramidis, P. (2005) Two–step cross–validation selection method for partially linear models. *Statistica Sinica*, **15**, 1033–1048.
3. Cheng, B. and Tong, H. (1992) On consistent nonparametric order determination and chaos (with discussion). *J. Roy. Statist. Soc. Ser. B*, **54**, 427–449 and 451–474.
4. Cheng, B. and Tong, H. (1993) Nonparametric function estimation in noisy chaos. *Developments in Time Series Analysis* (ed. T. Subba Rao), 183–206. Chapman and Hall, London.
5. Cheng, B. and Tong, H. (1994) Orthogonal projection, embedding dimension and sample size in chaotic time series from a statistical perspective. With a discussion by R. J. Bhansali, P. M. Robinson and A. Kleczkowski and replies by H. Tong. *Philos. Trans. Roy. Soc. London Ser. A* **348**, 325–341.
6. Dong, C., Gao, J. and Tong, H. (2007) Semiparametric model selection in partially linear time series. *Statistica Sinica*, **17**, 99–114.
7. Gao, J. (2007) *Nonlinear Time Series: Semiparametric and Nonparametric Methods.* Monographs on Statistics and Applied Probability Volume **108**. Chapman & Hall/CRC, London.
8. Gao, J., Anh, V., and Wolff, R. C. L. (2001) Semiparametric approximation methods in multivariate model selection. *Journal of Complexity*, **17**, 754–772.
9. Gao, J. and Tong, H. (2004) Semiparametric nonlinear time series model selection. *J. Roy. Statist. Soc. Ser. B*, **66**, 321–336.
10. Gao, J. and Tong, H. (2005) Nonparametric and semiparametric regression model selection. Unpublished paper available from `http://www.adelaide.edu.au/directory/jiti.gao`
11. Härdle, W., Liang, H. and Gao, J. (2000) *Partially Linear Models.* Springer Series in Contributions to Statistics. Physica–Verlag, New York.
12. Shao, J. (1993) Linear model selection by cross–validation. *J. Amer. Statist. Assoc.*, **422**, 486–494.
13. Tjøstheim, D. and Auestad, B. (1994a) Nonparametric identification of nonlinear time series: projections. *J. Amer. Statist. Assoc.*, **89**, 1398–1409.
14. Tjøstheim, D. and Auestad, B. (1994b) Nonparametric identification of nonlinear time series: selecting significant lags. *J. Amer. Statist. Assoc.*, **89**, 1410–1419.
15. Tschernig, R. and Yang, L. (2000) Nonparametric lag selection for time series. *J. Time Ser. Anal.*, **21**, 457–487.
16. Vieu, P. (1994) Choice of regressors in nonparametric estimation. *Computat. Statist. & Data Anal.*, **17**, 575–594.
17. Vieu, P. (1995) Order choice in nonlinear autoregressive models. *Statistics*, **26**, 307–328.
18. Vieu, P. (2002) Data–driven model choice in multivariate nonparametric regression. *Statistics*, **36**, 231–245.
19. Yang, L. and Tschernig, R. (2002) Non–and semiparametric identification of seasonal nonlinear autoregression models. *Econometric Theory*, **18**, 108–1448.
20. Yang, Y. (1999) Model selection for nonparametric regression. *Statistica Sinica*, **9**, 475–500.
21. Yao, Q. and Tong, H. (1994) On subset selection in nonparametric stochastic regression. *Statistica Sinica*, **4**, 51–70.

An Introduction to a Paper by Bing Cheng and Howell Tong: "On Consistent Nonparametric Order Determination and Chaos (with Discussion)"

DAG TJØSTHEIM

Department of Mathematics, University of Bergen
Johs. Brunsgt. 12, 5008, Bergen, Norway
E-mail: Dag.Tjostheim@math.uib.no

The starting point of the paper by Cheng and Tong is a chaotic dynamic system. However an additional noise term is included making this into a "noisy" chaotic system, or in non-chaotic terminology, simply a nonlinear time series model

$$Z_t = F(Z_{t-1}, \dots, Z_{t-d}) + \varepsilon_t. \tag{1}$$

This is model (2.1) of the paper. Both F and d are unknown, and the theoretical problem in the paper is mainly that of determining the embedding dimension d. However, the problem is not only *introduced* in a chaotic framework, its *solution* is also sought integrated in chaos theory discussing other dimension concepts, attractors and limit cycles, and there are several illuminating real data examples.

In this brief introduction I will try to put the main part of the paper; that is, the determination of d, into perspective and briefly state some further developments.

First note that for a given d, it is not obvious that every lag $t-1, \dots, t-d$ should be included in the function F. For example, if seasonality is involved with a seasonal lag of 12, say, typically only a few intermediate lags would be needed, so that the model may be stated as

$$Z_t = F(Z_{t-i_1}, \dots, Z_{t-i_d}) + \varepsilon_t \tag{2}$$

where $i_1, \dots, i_d$ and F are unknown. This is particularly important in the nonlinear nonparametric case because the curse of dimensionality puts strong limitations on the number of lags that should go into the model if one wants to estimate F reasonably well. Equation (2) makes it possible that i_d is quite large, whereas d is small. Even if d is moderate and the estimate of $F(z_1, \dots, z_d)$ is poor for an arbitrary point $\mathbf{z} = [z_1, \dots, z_d]$, it is conceivable, though, that a good estimate of $i_1, \dots, i_d$ can be obtained. This is because in the algorithm for determining $i_1, \dots, i_d$ one only needs to evaluate $\widehat{F}(\cdot)$ at the observation points $[Z_{t-i_1}, \dots, Z_{t-i_d}]$.

In the linear situation, determining the order and/or selecting relevant explanatory variables are managed by such devices as the Akaike's FPE (Akaike 1969, Akaike 1970) or AIC (Akaike 1973) criteria, the Mallows (1973) C_p, or the BIC criterion (Rissanen 1978 and Schwarz 1978). Much less has been done in a nonparametric and nonlinear environment such as that posed by (1). As far as I know, the order determination problem was first attacked in a fully nonparametric context in Auestad and Tjøstheim (1990). They sought to generalize the FPE-criterion to a nonlinear and nonparametric framework such as in equation (1).

The basic idea is the same as in Akaike's papers in that one introduces a new process $\{Y_t\}$ independent of $\{Z_t\}$ but having identical properties. Next, the mean square error (the FPE) is evaluated when using the nonparametric conditional expectation constructed from $\{Z_t\}$ as a nonlinear predictor of $\{Y_t\}$. The evaluation of this error is rather more complex than in the linear case. The details can be found in Tjøstheim and Auestad (1994b), where it is also indicated how the AIC can be evaluated using essentially identical arguments.

Cheng and Tong, in the present paper, use a cross validation approach instead of evaluating the FPE, and a very nice feature of the paper is a comparison of the two approaches showing that asymptotically they are essentially equivalent. I refer to the present paper, Cheng and Tong (1993), and Yao and Tong (1994) for an outline of the theory and practical examples of the cross validation procedure. If the lag structure of the conditional variance is to be determined, cross validation is a particularly interesting alternative. The paper by Gao, Anh and Wolff (2001) extends the nonparametric cross validation selection method to a semiparametric setting for the determination of a partially linear model.

It is indicated in Tjøstheim and Auestad (1994b) how a nonparametric analogy to the linear AIC could be set up, and how the algorithm could be applied to the conditional variance in addition to the conditional mean. The procedure works fairly well on the chosen examples, but there are some unsolved theoretical and practical problems, and the method must be used with care. For example, there is a clear tendency of overestimating the number of lags needed in the model.

The early work on nonlinear order dependence has been followed up, extended and deepened by the present paper, Cheng and Tong (1993), Granger and Lin (1994), Tjøstheim and Auestad (1994a), Vieu (1994, 1995, 2002), Yao and Tong (1994). Recent contributions include Tschernig and Yang (2000), Yang and Tschernig (2002) and Gao and Tong (2004), of which the last one is especially concerned with selecting lags and variables in additive and semiparametric models. An application to asset markets is given by Jansen et al. (2006). An order determination problem is also appearing in deciding the dimension in a projection pursuit problem with an interesting contribution by Xia et al. (2002).

The concept of consistency has played a certain role in developing the various versions of the linear parametric criteria. It is well-known that in the linear case neither FPE nor AIC are consistent: if the true model is AR(d); that is, autoregressive of order d, then the order estimate $\widehat{d}$ will not converge to d with probability one. Actually, there is a positive probability of over-estimating the order asymptotically. Alternative criteria such as the BIC and LIL (Hannan and Quinn 1979) impose stronger penalties on the number of variables and yield consistency. I refer to Koreisha and Yoshimoto (1991) for a review and to Teräsvirta and Mellin (1986) for a sequential testing point of view.

In the nonparametric case, both the FPE and the cross-validated criteria are consistent as pointed out in the present paper, in Cheng and Tong (1993), in Vieu (1994), and more recently in Tschernig and Yang (2000) for the FPE case. However, Tschernig and Yang go further. They show that the probability of over-estimation goes to zero much more slowly than the corresponding error of under-estimation, and they use this to suggest a corrected (still consistent) criterion where higher lags are penalized harder than shorter ones. In an extensive simulation study conducted by the authors this criterion works better than the other available versions, including those of Tjøstheim and Auestad (1994b).

The curse of dimensionality has already been mentioned as an obstacle in nonparametric order selection. If there are reasons to believe, for example based on additivity tests, that the data are well approximated by an additive model, one could try a selection procedure based on minimizing

$$\mathsf{E}\{Z_t - g_1(Z_{t-i_1}) - \cdots - g_d(Z_{t-i_d})\}^2 \tag{3}$$

where the functions g_i could be estimated by the marginal integration procedure (Tjøstheim and Auestad (1994a) or by back-fitting or smoothed backfitting (Mammen et al. 1999, and Nielsen and Sperlich 2004). Again, one could use an approach based on the FPE-criterion or the one based on cross validation. In this case the penalty factor would be of order $(Th)^{-1}$ as $T \to \infty$ as compared to $(Th^j)^{-1}$ for a model of order j in equation (2). Here T is the number of observations and h is the bandwidth. Both the additive and the semiparametric case have been analyzed by Gao and Tong (2004).

References

1. Akaike, H. (1969). Fitting autoregression for prediction. Annals of the Institute of Statistical Mathematics **22**, 243-247.
2. Akaike, H. (1970). Statistical predictor identification. Annals of the Institute of Statistical Mathematics **22**, 203-217.
3. Akaike, H. (1973). Information theory and an extension of the maximum likelihood principle, in Petrov and F. Caski (eds), Second International Symposium in Information Theory, Akademiai Kiado, 203-217.
4. Auestad, B. and Tjøstheim, D. (1990). Identification of nonlinear time series: first order characterization and order determination. Biometrika **58**, 525-534.
5. Cheng, B. and Tong, H. (1992). On consistent nonparametric order determination and chaos (with discussion), Journal of the Royal Statistical Society Series B **54**, 427-449 and 451–474.
6. Cheng, B. and Tong, H. (1993). On residual sums of squares in nonparametric autoregression. Stochastic Processes and Their Application, **48**, 157-174.
7. Gao, J., Anh, V. and Wolff, R.C.L. (2001). Semiparametric approximation methods in multivariate model selection. Journal of Complexity, **17**, 754-772.
8. Gao, J. and Tong, H. (2004). Semiparametric nonlinear time series model selection . Journal of the Royal Statistical Society Series B, **66**, 321-336.
9. Granger, C.W.J. and Lin, J.L. (1994). Using the mutual information coefficient to identify lags in nonlinear time series models. Journal of Time Series Analysis **15**, 371-384.
10. Hannan, E.J. and Quinn, B. (1979). The determination of the order of an autoregression. Journal of the Royal Statistical Society Series B **41**, 190-195.
11. Jansen, D., Li, Q., Wang, Z. and Yang, J. (2006). The impact of fiscal policy on asset markets. Preprint, Department of Economics, Texas A&M University.
12. Koreisha, S. and Yoshimoto, A. (1991). A comparison among identification procedures for autoregressive moving average models. International Statistical Review **51**, 37-57.
13. Mallows, C. (1973). Some comments on C_p. Technometrics **15**, 661-675.
14. Mammen, E., Linton, O. and Nielsen, J.P. (1999). The existence and asymptotic properties of a backfitting projection algorithm under weak conditions. Annals of Statistics 27, 1443-1490.
15. Nielsen, J.P. and Sperlich S. (2004). Smooth backfitting in practice. Journal of the Royal Statistical Society Series B **67**, 43-61.
16. Rissanen, J. (1978). Modeling by shortest data description. Automatica **14**, 185-207.
17. Schwarz, G. (1978). Estimating the dimension of a model. Annals of Statistics **6**, 461-464.
18. Teräsvirta, T. and Mellin, I. (1986). Model selection and model selection tests in regression models. Scandinavian Journal of Statistics **13**, 159-171.
19. Tjøstheim, D. and Auestad, B. (1994a). Nonparametric identification of nonlinear time series: Projections. Journal of the American Statistical Association **89**, 1398-1409.
20. Tjøstheim, D. and Auestad, B. (1994b). Nonparametric identification of nonlinear time series: selecting significant lags. Journal of the American Statistical Association **89**, 1410-1419.

21. Tschernig, R. and Yang, L. (2000). Nonparametric lag selection in time series. Journal of Time Series Analysis **21**, 457-487.
22. Vieu, P. (1994). Choice of regressors in nonparametric estimation. Computational Statistics and Data Analysis **17**, 575-594.
23. Vieu, P. (1995). Order choice in nonlinear autoregressive models. Statistics, **26**, 307-328.
24. Vieu, P. (2002). Data-driven model choice in multivariate nonparametric regression. Statistics, **36**, 231-245.
25. Xia, Y., Tong, H., Li, W.K. and Zhu, L-X. (2002). An adaptive estimation of dimension reduction space (with discussion). Journal of the Royal Statistical Society Series B **64**, 363-410.
26. Yang, L. and Tschernig, R. (2002). Non- and semiparametric identification of seasonal nonlinear autoregression models. Econometric Theory **18**, 1408-1448.
27. Yao, Q. and Tong, H. (1994). On subset selection in non-parametric stochastic regression. Statistica Sinica **4**, 51-70.

Adv. Appl. Prob. **17**, 666–678 (1985)
Printed in N. Ireland

ON THE USE OF THE DETERMINISTIC LYAPUNOV FUNCTION FOR THE ERGODICITY OF STOCHASTIC DIFFERENCE EQUATIONS

K. S. CHAN* AND H. TONG, * *The Chinese University of Hong Kong*

Abstract

We have shown that within the setting of a difference equation it is possible to link ergodicity with stability via the physical notion of energy in the form of a Lyapunov function.

STABILITY

1. Introduction

In this paper, we are exclusively interested in stochastic difference equations of the form

$$(1.1)\qquad X_{n+1}=T(X_n)+e_{n+1},\qquad n\geqq 0,\qquad T:\mathbb{R}^m\to\mathbb{R}^m.$$

X_n takes values in $\mathbb{R}^m$. Let $\mathscr{B}_m$ be the class of Borel sets of $\mathbb{R}^m$ and μ_m the Lebesgue measure. Then $(\mathbb{R}^m, \mathscr{B}_m, \mu_m)$ is the state space of (1.1). The random forcing terms, $\{e_{n+1}\}$, on the right-hand side of (1.1) are assumed to be of either one of the following forms:

(1.2a) i.i.d.; the marginal distribution is absolutely continuous and has a positive p.d.f. $f(\cdot)$ over $\mathbb{R}^m$;

(1.2b) $e_n=\begin{pmatrix} e'_n \\ 0 \\ \vdots \\ 0\end{pmatrix}$ with e'_n i.i.d., each having an absolutely continuous distribution and the p.d.f. $f(\cdot)$ is positive everywhere in $\mathbb{R}$.

Now, we assume (1.2a) holds. Let $A\in\mathscr{B}_m$ and $x\in\mathbb{R}_m$. Let $P(x, A)$ be the transition probability function. Then

$$(1.3a)\qquad P(x,A)=\int_{A-T(x)} f(t)\mu_m(dt).$$

Received 31 May 1984; revision received 1 October 1984.
* Postal address: Department of Statistics, University Science Centre, The Chinese University of Hong Kong, Shatin, N. T., Hong Kong.

Hence, $\{X_n\}$ with $P(x, A)$ defined in (1.3a) forms a Markov chain with the state space $(\mathbb{R}^m, \mathcal{B}_m, \mu_m)$.

Suppose (1.2b) holds. Let $A \in \mathcal{B}_m$ and $x \in \mathbb{R}^m$. Let

$$x = \begin{pmatrix} x_1 \\ x_2 \\ \vdots \\ x_m \end{pmatrix}.$$

Define

$$A_x = \left\{ y = \begin{pmatrix} y_1 \\ \vdots \\ y_m \end{pmatrix} \in \mathbb{R}^m : y_i = x_i,\ i \geqq 2,\ y \in A \right\}.$$

Let $\nu : \mathbb{R}^m \to \mathbb{R}$ be the projection map onto the first coordinate, i.e., $\nu(y) = y_1$. Then

$$P(x, A) = \int_{\nu(A_{Tx}) - (Tx)_1} f(t)\, dt, \tag{1.3b}$$

where $(Tx)_1$ is the first coordinate of Tx and we have written Tx for $T(x)$.

Again, $\{X_n\}$ with $P(x, A)$ defined in (1.3b) forms a Markov chain with state space $(\mathbb{R}^m, \mathcal{B}_m, \mu_m)$. (Note that in (1.3a) and (1.3b) we have abused the use of f, which represents two different entities.)

For the case where $T(X_n) = TX_n$, T being a companion matrix, it is well known that (1.1) defines an asymptotically *stationary* time series if the characteristic roots of T all lie inside the unit circle. It is equally well known that *exactly* the same condition ensures the *stability* of the solution of the deterministic equation associated with (1.1) in which e_{n+1} is replaced by the zero vector. *Is this situation merely a coincidence*?

At a deeper level, in the theory of a Markov chain over a general state space, it is known that *ergodicity* (in a suitable sense) of an irreducible chain may be established by identifying a 'centre' towards which there is a 'mean drift'—the so-called Foster condition. (See for example Tweedie (1975).) On the other hand, in the theory of a (non-linear) difference equation, *stability* (in a suitable sense) of the solution may be investigated by studying the behaviour of a generalized energy associated with the equation: roughly speaking, when the trajectory moves towards the asymptotic solution (cf. the centre), a dissipation of the generalised energy (cf. the mean drift) is essential for stability to be attained. (See, for example LaSalle (1976).) Again, the basic ideas seem rather strikingly similar. *Is this situation merely a coincidence again*?

In this paper, we aim to show that the Lyapunov function, which may be interpreted as a generalized energy, plays a significant role in studying not only the stability (a well-known fact) of a deterministic difference equation but also the ergodicity of a stochastic difference equation.

2. A simple criterion for non-null compact sets to be small

It is of interest to establish conditions under which (1.1) is geometric ergodic. We shall employ notations and terminology adopted in Tweedie (1983a). For the general theory, we refer to Tweedie (1975), (1976), (1983a), (1983b) and Nummelin and Tuominen (1982). Let $A, B, K \in \mathscr{B}_m$ with non-zero μ_m-measure. K always denotes a compact set. For non-null compact sets to be small, it is sufficient that T in (1.1) is continuous. However, we may relax this condition to some extent.

Consider (1.1) with e_n of the form (1.2a). Then $\{X_n\}$ is μ_m-irreducible and aperiodic. Suppose T is compact (i.e. T sends compact sets into relatively compact sets). Suppose $f(\cdot)$ is lower semi-continuous. Then, clearly,

$$\inf_{x\in K} P(x, A) > 0.$$

Hence, K is small.

We now consider the case where X_n satisfies (1.1) with e_n of the form (1.2b). Let $(\mathbb{R}^m, \mathscr{B}_m, \mu_m)$ be the state space. Then $P(x, A)$ is of the form (1.3b). We assume further that T is of a more restricted form, i.e.,

$$T(x) = \begin{pmatrix} h(x) \\ x_1 \\ \vdots \\ x_{m-1} \end{pmatrix}, \text{ where } x = \begin{pmatrix} x_1 \\ x_2 \\ \vdots \\ x_m \end{pmatrix} \in \mathbb{R}^m$$

and h is a measurable function from $\mathbb{R}^m$ to $\mathbb{R}$. Suppose $m = 2$ and $A = (a_1, b_1) \times (a_2, b_2)$, an open interval in $\mathbb{R}^2$. Then

$$\begin{aligned} P^2(x, A) &= \int P(y, A) P(x, dy) \\ &= \int_{a_2}^{b_2} f(y_2 - h(x_1, x_2)) P(y, A)\, dy_2 \end{aligned}$$

where

$$y = \begin{pmatrix} y_2 \\ x_1 \end{pmatrix}.$$

However,

$$P(y, A) = \int_{a_1}^{b_1} f(y_1 - h(y_2, x_1))\, dy_1.$$

Hence, by Fubini's theorem,

$$P^2(x, A) = \iint_A f(y_2 - h(x_1, x_2)) f(y_1 - h(y_2, x_1))\, dy_1\, dy_2. \tag{2.1}$$

Since for fixed $x \in \mathbb{R}^2$ both the right- and left-hand sides define a Borel measure on the Borel sets and they are equal over the rectangles, the equality holds for all $A \in \mathcal{B}_2$. Evidently, the same idea goes through in higher dimension and we have

$$p^m(x, A) = \int_A \prod_{j=1}^m f'_j(y) \mu_m(dy), \tag{2.2}$$

where $y = (y_1, y_2, \cdots, y_m)'$, $x = (x_1, x_2, \cdots, x_m)'$,

$$f'_m(y) = f(y_m - h(x_1, x_2, \cdots, x_m)),$$
$$f'_i(y) = f(y_i - h(y_{i+1}, \cdots, y_m, x_1, \cdots, x_i)), \qquad m > i \geqq 1.$$

Formula (2.2) is very useful. It is clear that $P^m(x, A) > 0$, $\forall x \in \mathbb{R}^m$. Therefore $\{X_n\}$ is μ_m-irreducible and aperiodic. Suppose h is compact and $f(\cdot)$ is lower semi-continuous. Then

$$\inf_{x \in K} P^m(x, A) > 0.$$

Hence, K is small.

3. From deterministic stability to ergodicity: a precursor

In this section, we apply the above framework to a particular case, the so-called SETAR model in non-linear time series analysis. (For a comprehensive introduction to this kind of model, see Tong (1983).)

$$h(x_1, x_2, \cdots, x_m) = c_i + \sum_{j=1}^m a_{ij} x_j \quad \text{if} \quad r_{i-1} \leqq x_d < r_i, \tag{3.1}$$

where $\{-\infty = r_0 < r_1 < \cdots < r_l = +\infty\}$ is an ordered partition of $\mathbb{R}$, $d \leqq m$, c_i and a_{ij} are constants. The function h of (3.1) is the autoregressive function for the full SETAR model. If we define $T: \mathbb{R}^m \to \mathbb{R}^m$ by

$$T(x) = \begin{pmatrix} h(x) \\ x_1 \\ \vdots \\ x_{m-1} \end{pmatrix} \quad \text{where} \quad x = \begin{pmatrix} x_1 \\ x_2 \\ \vdots \\ x_m \end{pmatrix} \in \mathbb{R}^m,$$

then this T together with e_n of the form (1.2b) constitutes the Markovian state-space equation for the full SETAR model. Specifically, we have the following setup:

$$X_n \text{ takes values in } \mathbb{R}^m.$$

$$X_{n+1} = T(X_n) + e_n, \tag{3.2a}$$

$$\xi_n = (1, 0, \cdots, 0)X_n, \tag{3.2b}$$

where

$$T(x) = \begin{pmatrix} h(x) \\ x_1 \\ \vdots \\ x_{m-1} \end{pmatrix}$$

with $h(\cdot)$ as defined in (3.1) and

$$e_n = \begin{pmatrix} e_n' \\ 0 \\ \vdots \\ 0 \end{pmatrix},$$

e_n''s being i.i.d. zero-mean random variables and each having an absolutely continuous distribution, the density of which is lower semi-continuous and positive everywhere in $\mathbb{R}$.

It is then clear that $h(\cdot)$ is compact. From the results of Sections 1 and 2, we see that the chain $\{X_n\}$ is μ_m-irreducible and aperiodic, and non-null compact sets are small sets. Henceforth, let $\|.\|$ denote the Euclidean norm.

Lemma 3.1. If $\max_i \sum_j |a_{ij}| < 1$ and e_n' possesses first absolute moment, then (3.2a) is geometrically ergodic.

Proof. Let

$$z = \begin{pmatrix} z_1 \\ \vdots \\ z_m \end{pmatrix} \in \mathbb{R}^m.$$

As $\max_i \sum_j |a_{ij}| < 1$, $\exists p_1 > p_2 > \cdots > p_m > 0$ such that $\max_i \sum_j |a_{ij}| (p_1/p_j) < \theta < 1$ for some θ. Moreover, θ may be chosen such that $\theta > (p_{i+1}/p_i)$. Define $g : \mathbb{R}^m \to \mathbb{R}$ by $g(z) = 1 + \max_i |z_i| \, p_i$. Then,

$$\int g(z)P(x, dz) \leqq C + \theta g(x).$$

Since $m \|z\| \leqq g(z) \leqq M \|z\| + 1$, for some $0 < m < M$, therefore for $r \geqq$

$2C/m(1-\theta)$ and some $B>0$,

(i) $\int g(z)P(x, dz)<B<+\infty$, $\|x\|\leqq r$

(ii) $\int g(z)P(x, dz)<\frac{1}{2}(1+\theta)g(x)$, $\|x\|>r$.

By Theorem 4 of Tweedie (1983a), $\{X_n\}$ is geometrically ergodic.

4. Some general results

Suppressing the random forcing term e_{n+1} in (1.1), we obtain

$$X_{n+1}=T(X_n) \qquad n\geqq 0, \tag{4.1}$$

which will be called the associated deterministic difference equation or the deterministic part of the stochastic difference equation (1.1). The study of (4.1) may be viewed as a stepping-stone to, or even the 'bone' of, the study of (1.1). To start with, if the range of T is bounded, it is clear that (1.1) is ergodic. Boundedness is one form of stability of the dynamics of (4.1). Since the stability theory of (4.1) is well known, we omit all details but refer the readers to, for example, Kalman and Betram (1960), Halanay (1963) and LaSalle (1976).

Remark 4.1. It is known that the existence of a continuous Lyapunov function $V(x)$ near the origin implies the uniform asymptotic stability of the origin. For a precise statement, see for example Kalman and Bertram (1960). The converse is also true. Moreover, if T is Lipschitz-continuous near the origin, then the Lyapunov function constructed is also Lipschitz-continuous. However, it is easily seen from the proof given in Halanay (1963) that for the existence of a continuous Lyapunov function near the origin, T only needs to be continuous.

Theorem 4.2. Let $\{X_n\}$ satisfies (1.1). Let T be continuous and homogeneous (i.e. $T(cx)=cT(x)$, $\forall c>0, x\in\mathbb{R}^m$). Let the origin, O, be a fixed point of T. In the case of e_n satisfying (1.2a), we assume that $\int \|t\| f(t)\mu_m(dt)<+\infty$. If e_n satisfies (1.2b), we assume that $\int |t|\, f(t)\, dt<+\infty$ and

$$T(x)=\begin{pmatrix} h(x) \\ x_1 \\ \vdots \\ x_{m-1} \end{pmatrix}.$$

Then the existence of a continuous Lyapunov function, V, in a neighbourhood of the origin implies the geometric ergodicity of (1.1).

Proof. Let $W\subseteq\mathbb{R}^m$. We denote the closure of W by $\bar{W}$ and its boundary by ∂W. Without loss of generality, let V be defined over the closure of the unit ball. Let $m_0 = \inf_{\|x\|=1} V(x)$. Let G be the maximal connected component of $\{x : V(x) < \frac{1}{2}m_0\}$ which contains the origin. Then $T(\bar{G}) \subset G$. Let

$$g(x) = \inf\{r \geqq 0, x \in rG\}, \qquad x \in \mathbb{R}^m,$$

where $rG = \{rx, x \in G\}$. Then $g(x)$ is well defined and it may be easily checked that g has the following properties:

(i) $g(cx) = cg(x), \forall c > 0$

(ii) $\exists 0 < m < M < +\infty$ such that $m\|x\| \leqq g(x) \leqq M\|x\|$

(iii) $(x/g(x)) \in \partial G$

(iv) $\exists \varepsilon > 0, 0 < \theta < 1$ such that $\forall x \in \partial G, y \in \mathbb{R}^m$

$$\|y - Tx\| < \varepsilon \Rightarrow y \in G \quad \text{and} \quad g(y) < \theta$$

$$(\text{i.e.} \quad \text{dist}\,(y, T(\partial G)) < \varepsilon \Rightarrow g(y) < \theta).$$

Now, for model (1.1) with e_n satisfying (1.2a) and $\int \|t\| f(t)\mu_m(dt) < +\infty$, it holds that

$$\int g(y)P(x, dy) = \int g(T(x) + t)f(t)\mu_m(dt)$$

$$= \int_{\|t/g(x)\| < \varepsilon} g(x)g\left(T\left(\frac{x}{g(x)}\right) + \frac{t}{g(x)}\right) f(t)\mu_m(dt)$$

$$+ \int_{\|t/g(x)\| \geqq \varepsilon} g(x)g\left(T\left(\frac{x}{g(x)}\right) + \frac{t}{g(x)}\right) f(t)\mu_m(dt).$$

The first term is less than $\theta g(x)$. The second term is $C + g(x) \cdot \xi(x)$ for some $C > 0$.

Here, $|\xi(x)| < B < +\infty, \forall x$ and $\xi(x) \to 0$ as $\|x\| \to +\infty$. Let $h(x) = g(x) + 1$. Let r be such that

$$|\xi(x)| < \tfrac{1}{4}(1 - \theta) \quad \text{for} \quad \|x\| > r.$$

Then for $r_0 = \max(r, 4(C+1)/(1-\theta)m), \exists B'$, such that

(i) $\int h(y)P(x, dy) < B' < +\infty, \|x\| \leqq r_0$

(ii) $\int h(y)P(x, dy) < \frac{1}{2}(1 + \theta)h(x)\ \|x\| > r_0$.

Hence, Theorem 4 in Tweedie (1983a) shows that $\{X_n\}$ is geometric ergodic. The case when e_n satisfies (1.2b) is similarly proved.

Corollary 4.3. Suppose $\{X_n\}$ satisfies (1.1) and the conditions in Theorem 4.2 hold. Define

$$\xi_n = (1, 0, \cdots, 0)'X_n.$$

Then ξ_n equipped with the marginal distribution of the first component of X_n is strictly stationary.

Corollary 4.4. Under the same conditions as in Corollary 4.3, if e_n possesses the kth absolute moment (i.e. $\int \|t\|^k f(t)\mu_m(dt) < +\infty$ if (1.2a) holds, or $\int |t|^k f(t)\, dt < +\infty$ if (1.2b) holds), then the stationary distribution of $\{\xi_n\}$ has finite kth absolute moment.

Proof. Use the test function $H(y) = (g(y))^k + 1$ and then it may be verified that the conditions in Theorem 3 of Tweedie (1983b) are satisfied.

On appealing to our earlier remark, we obtain the following result.

Theorem 4.5. Under the same conditions as in Theorem 4.2, the uniform asymptotic stability of the origin (and thus global uniform asymptotic stability) implies the existence of a continuous Lyapunov function near the origin and therefore the geometric ergodicity of $\{X_n\}$.

5. Some extensions and examples

Theorems 4.2 and 4.5 place some restrictions on T. The weakest assumption on T is that of homogeneity. We mention some easy extensions. Suppose T is merely compact but can be decomposed into two parts, namely,

$$T = T_h + T_d,$$

where T_h is homogeneous and continuous and T_d is of bounded range. We consider the 'component' of (1.1) given by

$$X_{n+1} = T_h(X_n) + e_n, \qquad n \geqq 0. \tag{5.1}$$

Then we can apply Theorems 4.2 and 4.5 to (5.1). It is clear that the conclusion then holds also for $\{X_n\}$ satisfying (1.1).

Since the stability of a deterministic dynamical system is a very well-researched area, the link which we have just established should prove useful in the study of ergodicity of the associated stochastic system. We now give some examples illustrating the use of Theorem 4.2.

We assume that $\{X_n\}$ satisfies (1.1) with e_n satisfying (1.2b). Moreover, let

$$x = \begin{pmatrix} x_1 \\ x_2 \\ \vdots \\ x_m \end{pmatrix} \in \mathbb{R}^m \quad \text{and} \quad T(x) = \begin{pmatrix} h(x_1, x_2, \cdots, x_m) \\ x_1 \\ \vdots \\ x_{m-1} \end{pmatrix}$$

where $h(\cdot)$ is from $\mathbb{R}^m$ to $\mathbb{R}$.

Example 1.

$$h(x) = \sum_{i=1}^{m} a_i x_i + l \tag{5.2a}$$

where the a_i's and l are constants.

Decompose $h(x)$ into the sum of $h_h(x)$ and $h_d(x)$, where

(5.2b) $$h_h(x)=\sum_i a_i x_i,$$

and

(5.2c) $$h_d(x)=l.$$

Clearly $h(x)$ is compact. Now 0 is uniformly asymptotically stable with respect to (5.2b) iff all the roots of the characteristic equation (i.e. $x^m-\sum a_i x^{m-i}=0$) have magnitudes less than 1. Thus, from Theorem 4.5, we see that $\{X_n\}$ satisfying (1.1) with the autoregressive function $h(x)$ defined by (5.2a) is geometrically ergodic if all the characteristic roots have magnitudes less than unity.

Example 2.

(5.3a) $$h(x)=\sum_{i=1}^{m}(a_i+b_i\exp(-\gamma x_i^2))x_i.$$

Similarly, we have

(5.3b) $$h_h(x)=\sum_{i=1}^{m}a_i x_i$$

(5.3c) $$h_d(x)=\sum_i b_i\exp(-\gamma x_i^2)x_i.$$

It follows exactly as in Example 1 that $\{X_n\}$ satisfying (1.1) with the autoregressive function $h(x)$ defined in (5.3a) is geometrically ergodic if all the characteristic roots of (5.3b) have magnitude less than unity. This model is a variant of the exponential autoregressive model introduced by Ozaki (1980). His original model replaces each x_i in the exponent of Equation (5.3a) by x_1. It seems that the ergodicity of the latter model remains an open problem unless $m=1$.

Example 3. Consider the first-order SETAR model. Here

(5.4a) $$h(x)=\begin{cases}c_1+\phi_1 x & \text{if } x>r_1,\\ c_2+\phi_2 x & \text{otherwise.}\end{cases}$$

Then

(5.4b) $$h_h(x)=\begin{cases}\phi_1 x & \text{if } x>0\\ \phi_2 x & \text{otherwise}\end{cases}$$

(5.4c) $$h_d(x)=h(x)-h_h(x).$$

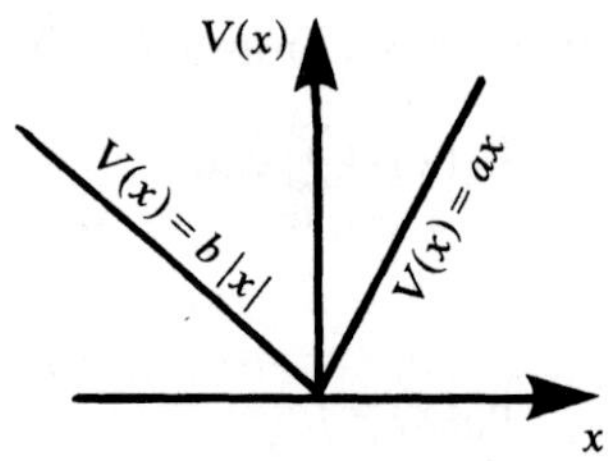

Figure 1. a and b are any real numbers such that $a>0, b>0, 1>\phi_1>-a/b, 1>\phi_2>-b/a$.

It is clear that a necessary and sufficient condition for the uniform asymptotic stability of the origin with respect to the deterministic difference equation corresponding to $h_h(x)$ of (5.4b) is

$$\phi_1<1, \qquad \phi_2<1, \qquad \phi_1\phi_2<1.$$

In fact, an appropriate Lyapunov function for (5.4b) is as shown in Figure 1.

Petruccelli and Woolford (1984) have proved that this condition is both necessary and sufficient for the ergodicity of the associated first-order threshold autoregressive model.

Example 4.

(5.5) $$T(x)=\begin{cases} Ax & \text{if } \ x\in\Omega, \\ Bx & \text{otherwise,}\end{cases}$$

where $\Omega\subseteq\mathbb{R}^m$, and is bounded, and $A=(a_{ij})$ and $B=(b_{ij})$ are matrices of constants. Some sufficient conditions for the uniform asymptotic stability of the origin with respect to (5.5) are given in LaSalle (1976). We give one easily checked condition. Let $|A|=(|a_{ij}|)$ and $|B|=(|b_{ij}|)$. If C is such that $c_{ij}\geqq \max(|a_{ij}|, |b_{ij}|)$ and the eigenvalues of C have magnitudes less than unity, then the origin is asymptotically stable. Thus, this condition is sufficient for the geometric ergodicity of (1.1) with T of the form (5.5) and e_n of the form (1.2a). For unbounded Ω, a Lyapunov function, V, may be constructed such that $V(x)=V(|x|)$ and $V(|x|)>V(|y|)$ if $|x|>|y|$, where $|x|=(|x_1|,\cdots,|x_m|)'$ (cf. LaSalle (1976), p. 17). From this V, geometric ergodicity follows on using standard arguments.

Another way to remove the restriction of the homogeneity of T in Theorems 4.2 and 4.5 is to set stronger conditions on the stability of the deterministic part. Suppose T is Lipschitz and the origin is exponential—asymptotically stable in the large. (For further details in this area, see for example, Yoshizawa (1966).)

Theorem 5.1. Suppose $\exists M>0$ such that $\forall x, y\in\mathbb{R}^m$

$$\|Tx-Ty\|\leqq M\|x-y\|.$$

Let $x(n; x_0)$ be the solution of (4.1) with x_0 as the starting point. Suppose $\exists c>0, K>0$ such that

$$\|x(n; x_0)\| \leqq Ke^{-cn} \|x_0\|, \quad \forall n.$$

Moreover, e_n possesses appropriate moments, i.e. $\int \|t\| f(t)\mu_m(dt) < +\infty$ or $\int |t| f(t)\, dt < +\infty$.

Then $\{X_n\}$ satisfying (1.1) with e_n of the form (1.2a) or (1.2b) is geometrically ergodic.

Proof. As in Yoshizawa (1966), we define

$$g(x) = \left(\sup_{\tau \geqq 0} \|x(\tau; x)\| e^{qc\tau}\right) + \gamma$$

where $0<q<1$ and $\gamma>1$.

Then $g(x)$ satisfies

(i) $\|x\| + \gamma \leqq g(x) \leqq K\|x\| + \gamma$.

(ii) $|g(x) - g(y)| \leqq L\|x-y\|, \forall x, y \in \mathbb{R}^m$,

(iii) $g(Tx) - g(x) \leqq -\alpha g(x) + \gamma(1-(1/e^{qc}))$, for some L, α, positive constants.

It is easily seen that

$$\|x\| + \gamma \leqq g(x) \leqq \gamma + \sup_{\tau \geqq 0} K . \exp(-(1-q)c\tau) \|x\| \leqq \gamma + K\|x\|.$$

Now, for (ii) and the determination of L, let β be such that $K = \exp((1-q)c\beta)$. If $\tau \geqq \beta$, then $K\exp(-(1-q)c\tau)\|x\| \leqq \|x\|$ and hence $g(x) = \gamma + \sup_{0\leqq\tau\leqq\beta} \|x(\tau; x)\| \exp(qc\tau)$. Therefore, if $x, y \in \mathbb{R}^m$

$$\begin{aligned} |g(x) - g(y)| &\leqq \sup_{0\leqq\tau\leqq\beta} \|x(\tau; x) - x(\tau; y)\| e^{qc\tau} \\ &\leqq e^{qc\beta} M^\beta \|x-y\|, \end{aligned}$$

where M is the Lipschitz constant for T.

For (iii),

$$\begin{aligned} g(Tx) &= \gamma + \sup_{\tau \geqq 0} \|x(\tau; Tx)\| e^{qc\tau} \\ &= \gamma + \sup_{\tau \geqq 0} \|x(\tau+1; x)\| e^{qc(\tau+1)} . \frac{1}{e^{qc}} \end{aligned}$$

$$\Rightarrow g(Tx) - g(x) \leqq -\alpha g(x) + \gamma\alpha, \quad \text{where} \quad \alpha = 1 - \frac{1}{e^{qc}}.$$

Then, as before, we conclude that $\{X_n\}$ is geometric ergodic.

We may note that e^{qc} is related to the (geometric) rate of convergence of $P^n(x, .)$ to the invariant measure.

6. Discussion

Results obtained so far encourage us to take the view that a systematic approach to prove (geometric) ergodicity of a stochastic difference equation via a Lyapunov function for its deterministic part is conceptually satisfying and practically useful. However, our results are quite modest. Deeper results should be possible. For example, we may consider replacing the random driving term e_{n+1} of (1.1) not merely by a zero vector but by a general deterministic 'control vector'. Even more generally, we may consider the wider class given by

$$X_{n+1} = T(X_n, e_{n+1}).$$

As far as stochastic differential equations are concerned, significant progress has already been made in this respect in the last two decades: see for example Arnold and Kliemann (1983). It is hoped that the significance of these results will be rendered more transparent to applied probabilists when they are 'translated' into the discrete-time case.

Acknowledgement

We are grateful to the referee for drawing to our attention the fact that significant results are available in the continuous-time literature and for numerous helpful comments and suggestions which have greatly improved the presentation of the paper.

References

Arnold, L. and Kliemann, W. (1983) Qualitative theory of stochastic systems. In *Probabilistic Analysis and Related Topics* **3.** ed. A. T. Barucha-Reid. Academic Press, New York.

Halanay, A. (1963) Quelques questions de la théorie de la stabilité pour les systèms aux differences finites. *Arch. Rat. Mech. Anal.* **12,** 150–154.

Kalman, R. E. and Bertram, J. E. (1960) Control system analysis and design via the "Second method" of Lyapunov II: Discrete-time systems. *Trans. A.S.M.E., J. Basic Engng.* D **82,** 394.

LaSalle, J. P. (1976) *The Stability of Dynamical Systems.* SIAM, Philadelphia, Pa.

Nummelin, E. and Tuominen, P. (1982) Geometric ergodicity of Harris recurrent Markov chains with application to renewal theory. *Stoch. Proc. Appl.* **12,** 187–202.

Ozaki, T. (1980) Non-linear time series models for non-linear random vibrations. *J. Appl. Prob.* **17,** 84–93.

Petruccelli, J. D. and Woolford, S. W. (1984) A threshold AR (1) model. *J. Appl. Prob.* **21,** 270–286.

Tong, H. (1983) *Threshold Models in Non-Linear Time Series Analysis.* Lecture Notes in Statistics **21,** Springer-Verlag, Heidelberg.

Tweedie, R. L. (1975) Sufficient conditions for ergodicity and recurrence of Markov chains on a general state space. *Stoch. Proc. Appl.* **3,** 385–403.

Tweedie, R. L. (1976) Criteria for classifying general Markov chains. *Adv. Appl. Prob.* **8,** 737–771.

Tweedie, R. L. (1983a) Criteria for rates of convergence of Markov chains, with application to queueing theory. In *Papers in Probability, Statistics and Analysis*, ed, J. F. C. Kingman and G. E. H. Reuter. Cambridge University Press, Cambridge.

Tweedie, R. L. (1983b) The existence of moments for stationary Markov chains. *J. Appl. Prob.* **20,** 191–196.

Yoshizawa, T. (1966) *Stability Theory by Liapunov's Second Method*. Publications of the Mathematical Society of Japan, No. 9.

Thoughts on the Connections Between Threshold Time Series Models and Dynamical Systems

DAREN B. H. CLINE

Department of Statistics
Texas A&M University
College Station TX 77843-3143, USA
E-mail: dcline@stat.tamu.edu

When introducing threshold time series models some 30 years ago, Howell Tong noted possible close connections with certain dynamical systems. This idea has generated much interest and study. One such connection compares stability of dynamical systems with that of nonlinear time series. I examine this relationship, noting when their conclusions agree and identifying parallels even when they diverge.

1. Introduction

It has now been three decades since Howell Tong first pointed to possible close connections between nonlinear time series and dynamical systems. At the time he had a few answers, some partial, but also many insightful and impelling questions. In particular, he both introduced the idea of threshold time series models (Tong (1977, 1978, 1983); Tong and Lim (1980)) and described their apparently intimate relationships with certain dynamical systems. (A personal review of this history is in Tong (2007).) With Lim, Tong showed that many features seen in threshold models may be explained by corresponding features in a related dynamical system. The presence of limit cycles was especially exciting as it could explain the psuedo-periodic nature of the sunspot and Canadian lynx data.

The questions Tong raised have since inspired much fruitful research. For example, there inevitably came a cottage industry of identifying which threshold models have stationary distributions (Chan and Tong (1985, 1986, 1994); Chan et al (1985); Chen and Tsay (1991); Brockwell et al (1992), and many others). The literature for this is immense and I shall not attempt to survey it (nor even pretend to know it all). Suffice to say that a number of techniques have been utilized: backward recursion, domination by known stable models, comparison to dynamical systems and application of a Foster-Lyapunov drift condition. The latter approach has been the most successful, although its greatest benefit often is realized only through very delicate construction of the so-called test function.

The drift condition approach resulted from the shrewd capture of ideas produced by Tweedie (1975, 1976), extending Foster (1953) and others, at about the same time as the introduction of the threshold model. (Tweedie subsequently returned the favor by highlighting nonlinear time series in his book with Meyn (1993).) In particular, Chan and Tong (1985), Chan et al (1985) and Chen and Tsay (1991) demonstrated the precision of the method by delineating the exact parameter spaces for simple threshold autoregression models. Chan (1989, 1990) provided an excellent introductory overview of this approach. (See also Jones (1976).)

Indeed, the very term "Foster-Lyapunov drift condition" results from the union of previously separate fields of mathematics: Markov chain theory (Foster) and dynamical systems

(Lyapunov). Chan and Tong's insight, simply stated, was not only that the Foster drift condition promoted by Tweedie was parallel to the Lyapunov function method of dynamical systems but also that they often had corresponding conclusions. As it has turned out, however, not all is as simple as it first seemed to be.

In this tribute to the success of the threshold model, I will reinforce the very real, but often subtle, connection of stability of a time series with that of discrete time dynamical systems and I also will illustrate the distinctions between the two. Again, I shall not recite all the extensive literature, except to point to the more seminal papers. My discussion partly expands on ideas that have been around for some time and also leads into some new thoughts.

The paper is organized as follows: section 2 explains what is meant by a dynamical system used as a point of comparison to a nonlinear time series. Section 3 describes parallel sets of drift conditions and section 4 looks into the meaning of a Lyapunov exponent for drift stability. Section 5 delves deeper for cases such as a threshold autoregression model, and specifically considers one model suggested for the Canadian lynx data. Finally, section 6 compares and distinguishes stability of bilinear and GARCH models with the above. A few short proofs are provided.

2. Noisy Dynamical Systems

As the objective is to compare stability of nonlinear time series with that of dynamical systems, I start with some standard terminology.

A (discrete time) *dynamical system* (cf. Martelli (1999), for example) is a deterministic sequence with initial value x_0 that satisfies

$$x_t = F(x_{t-1}) = F^t(x_0) \tag{1}$$

under the recursion $F^t(x) = F(F^{t-1}(x))$. An *attractor*, loosely speaking, is a set such that the sequence eventually is contained in any open covering of the set and its *basin of attraction* is the set of initial values x_0 giving rise to such a sequence. Often F is assumed to be continuous, or even continuously differentiable, but such an assumption is impossible if the system to be compared to a threshold time series.

A *nonlinear autoregressive time series* of order p is defined by

$$\xi_t = f(\xi_{t-1}, \ldots, \xi_{t-p}, \epsilon_t)$$

for an iid sequence ϵ_t. Often the model is expressed by

$$\xi_t = f(\xi_{t-1}, \ldots, \xi_{t-p}) + \sigma(\xi_{t-1}, \ldots, \xi_{t-p})\epsilon_t, \tag{2}$$

where $f(x)$ and $\sigma(x)$ are locally bounded and $\sigma(x)$ is locally bounded away from 0. The *(self-exciting) threshold autoregression model* is (2) when f is piecewise linear and σ is piecewise constant. The boundaries between the regions of linearity are called *thresholds*. However, for stability purposes it is more suitable to consider the state process $X_t = (\xi_t, \ldots, \xi_{t-p+1})$, which is a time homogeneous Markov chain. Even more generally, any homogeneous Markov chain can be expressed as a *stochastic recursion*

$$X_t = F(X_{t-1}, e_t) = F^t(X_0, e_1, \ldots, e_t), \tag{3}$$

where the recursion formula is

$$F^t(X_0, e_1, \ldots, e_t) = F(F^{t-1}(X_0, e_1, \ldots, e_{t-1}), e_t)$$

and the errors e_t are iid.

When X_t is the state vector for (2), the chain is expressed as

$$X_t = F(X_{t-1}) + \sigma(X_{t-1})e_t \tag{4}$$

with $F(X_t) = (f(X_t), \xi_{t-1}, \ldots, \xi_{t-p+1})$ and $e_t = (\epsilon_t, 0, \ldots, 0)$. In this case (1) is said to be the *skeleton* of (4) and X_t is said to *clothe* x_t.

For the sake of clarity I will talk about a deterministic dynamical *system* and a time series (stochastic recursion) *model.*

One important distinction is that, whereas the system may not be irreducible (i.e., it has multiple attractors with disjoint basins of attraction), the time series usually is irreducible (in the stochastic sense). For example, it has long been known that if ϵ_t has a density locally bounded away from 0 and σ is bounded and bounded away from 0 then the state process for a nonlinear autoregression of the form (2) is a ϕ-irreducible, aperiodic T-chain (Chan (1993); Cline and Pu (1999a)). In particular, this means that while the time series may mimic the dynamical system in one basin of attraction for a time, it is certain to move eventually to any others that exist.

3. Relevant Notions of Stability

Henceforth, $\mathbb{X} \subset \mathbb{R}^m$ is the state space for either x_t or X_t and $\|\cdot\|$ is a suitable norm on $\mathbb{X}$. There are several notions of stability for a dynamical system (cf. La Salle (1976); Tong (1990)), most of which are versions of "Lyapunov stability" and are concerned with sensitivity to initial conditions. However, I am concerned here with stability of time series in the sense that a stable process always returns to some bounded set no matter how large it may become in the meantime. In essence, I am interested in various notions of stability that can lead to "recurrence".

For this purpose, therefore, a dynamical system is said to have *regular stability* if there exists $M < \infty$ such that

$$\limsup_{t\to\infty} x_t = \limsup_{t\to\infty} F^t(x_0) \le M \quad \text{locally uniformly in } x_0. \tag{5}$$

In other words, the set $\{x : \|x\| \le M\}$ is a strong attractor for the system. Note that I am avoiding the continuity assumptions typical in the dynamical systems literature.

Verification of stability frequently relies on finding a *Lyapunov test function* $V(x)$ satisfying an appropriate drift condition. Here, I suggest the following *regular drift condition*: suppose there exist $M, K < \infty$, a locally bounded nonnegative function $V(x)$ with $V(x) \to \infty$ as $\|x\| \to \infty$ and a nonnegative function $g(c)$, continuous and strictly increasing off $[0, M]$, such that

$$V(F(x)) \le \begin{cases} K & \text{if } \|x\| \le M, \\ V(x) - g(\|x\|) & \text{if } \|x\| > M. \end{cases} \tag{6}$$

The "drift" $g(\|x\|)$ ensures that the sequence cannot stay large indefinitely.

Theorem 3.1. *Assume $F(x)$ is locally bounded. Then the dynamical system is regularly stable iff a regular drift condition holds.*

Proof. Let $\delta > 0$. If the system is regularly stable then (6) holds with $V(x) = \sum_{t=0}^{\infty}(\|F^t(x)\| - M - \delta)_+$, $g(c) = (c - M - \delta)_+$ and M replaced with $M + \delta$. Note that

(5) implies that this expression for $V(x)$ has finitely many positive terms for each x and is locally bounded.

Conversely, if (6) holds, let

$$L_1 = \inf_{||x||\le M}(V(x) - g(||x||)), \quad L_2 = \max(\max_{c\le M} g(c), K - L_1) \quad \text{and}$$
$$M_1 = \max\{c : g(c) \le L_2\}.$$

Define $g_1(c) = (g(c) - L_2)_+$ and it follows that $V(F(x)) \le V(x) - g_1(||x||)$ for all x. Also, g_1 is continuous, strictly increasing off $[0, M_1]$ and vanishes on $[0, M_1]$. Iterating the revised drift condition results in

$$0 \le V(F^n(x)) \le V(x) - \sum_{t=0}^{n-1} g_1(||F^t(x)||) \quad \text{for all } n \ge 1.$$

It follows that $g_1(||F^t(x)||) \to 0$ as $t \to \infty$. This convergence is locally uniform in x since V and g_1 are locally bounded. Thus, due to the nature of g_1, $\limsup_{t\to\infty} F^t(x) \le M_1$ locally uniformly in x, verifying (5). □

Regular stability is not the only boundedness concept for dynamical systems; the weaker "Lagrange stability" is more commonly mentioned. But the form of its drift condition does resonate with a similar condition for Markov chains (see below), hence my mention of it here. I do not know whether Theorem 3.1 is a familiar result to those who study discrete time dynamical systems, though Halanay and Răsvan (2000) have a remarkably similar result for what they call "uniform" Lyapunov stability.

Often of greater interest is *exponential stability* of a dynamical system, namely that for some $\rho < 1$, $M, K < \infty$ and $n \ge 1$,

$$||F^n(x)|| \le \begin{cases} K & \text{if } ||x|| \le M, \\ \rho^n ||x|| & \text{if } ||x|| > M. \end{cases} \tag{7}$$

Clearly, an exponentially stable system also is regularly stable. A corresponding *exponential drift condition* can be stated: there exists $\rho < 1$, $M, K < \infty$ and a Lyapunov test function $V(x)$ such that

$$V(F(x)) \le \begin{cases} K & \text{if } ||x|| \le M, \\ \rho V(x) & \text{if } ||x|| > M. \end{cases}$$

Theorem 3.2. *Assume* $\frac{||F(x)||}{1+||x||}$ *is bounded. Then the dynamical system is exponentially stable iff an exponential drift condition holds with* $d_0||x||^r \le V(x) \le 1 + d_1||x||^r$ *for some positive* r, d_0, d_1.

Proof. Necessity of the drift condition is shown with Lyapunov function $V(x) = \sum_{t=0}^{n-1} \rho^{n-t-1}||F^t(x)||$. Sufficiency is verified by iterating the drift condition sufficiently many times and using the requirement $d_0||x||^r \le V(x) \le 1 + d_1||x||^r$. □

A strong form of stability is, of course, *ergodicity* (cf. Sinai (2000, chapter 1)). I will not characterize ergodicity of a dynamical system, except to say that when I refer to it henceforth I mean at least that there exists a unique probability measure μ such that, for any bounded measurable h,

$$\lim_{n\to\infty} \frac{1}{n} \sum_{t=0}^{n-1} h(x_t) = \int_{\mathbb{X}} h(x)\,\mu(dx) \quad \text{for almost all initial values } x_0. \tag{8}$$

The measure μ is called *invariant* analogously to the invariant measure of a Markov process, and in fact $\int_{\mathbb{X}} h(x)\,\mu(dx) = \int_{\mathbb{X}} h(F(x))\,\mu(dx)$. When the system is ergodic with a limit cycle, μ is uniform on the points of the cycle. If the system has multiple disjoint basins of attraction then it may be ergodic within each (that is, ergodic when restricted to each basin).

Now I will turn to notions of stability for a Markov chain (stochastic recursion), and specifically to those that echo the definitions above. Again, stability only requires returning to some bounded set. Thus a chain could be stable in this sense yet transient if, for example, it shrinks toward a set of lower dimension. Ordinarily, this is prevented by irreducibility but I will not be concerned with that issue here. For the results below, define stopping times

$$\tau_M = \inf\{t \ge 1 : ||X_t|| \le M\} \quad \text{and} \quad \sigma_M = \inf\{t \ge 0 : ||X_t|| \le M\}.$$

A stability condition analogous to the regular stability described above is *g-regularity*: there exists nonnegative, locally bounded $g(x)$ and $M, K < \infty$ such that

$$E\Big(\sum_{t=0}^{\tau_M - 1} g(X_t) \mid X_0 = x\Big) < \begin{cases} K & \text{if } ||x|| \le M, \\ \infty & \text{if } ||x|| > M. \end{cases} \tag{9}$$

One distinction is that this condition does not imply that an open covering of the set $C = \{x : ||x|| \le M\}$ is absorbing. However, if X_t is ϕ-irreducible, C is petite (cf. Meyn and Tweedie (1993)) and $g(x) \ge \delta > 0$ for all $||x|| > M$ then (9) implies $E(\tau_M | X_0 = x) < \infty$ for all x and thus X_t is positive recurrent. Moreover, if the process also is aperiodic then there is a stationary distribution π, such that

$$\lim_{n\to\infty} \sup_{|h| \le \max(g,1)} |E(h(X_n) \mid X_0 = x) - \int_{\mathbb{X}} h(x)\,\pi(dx)| = 0 \quad \text{for all } x \in \mathbb{X}. \tag{10}$$

In particular $E(g(X_t)) = \int_{\mathbb{X}} g(x)\,\pi(dx)$ is finite for the stationary distribution.

Again, there is a corresponding *g-regular drift condition.* For some $M, K < \infty$ and nonnegative, locally bounded $V(x), g(x)$,

$$E(V(X_1) \mid X_0 = x) \le \begin{cases} K & \text{if } ||x|| \le M, \\ V(x) - g(x) & \text{if } ||x|| > M. \end{cases} \tag{11}$$

Compare (11) with (6).

Theorem 3.3. *X_t is g-regular iff the g-regular drift condition holds.*

Proof. (See also Meyn and Tweedie (1993).) If X_t is g-regular then the drift condition holds with $V(x) = E(\sum_{t=0}^{\sigma_M - 1} g(X_t) \mid X_0 = x)$.

Conversely, there is no loss in assuming that

$$E(V(X_1) \mid X_0 = x) \le V(x) - g(x) + K1_{||x|| \le M}.$$

Then, since $||X_t|| > M$ for $0 < t < \tau_M$,

$$E\Big(\sum_{t=0}^{\tau_M - 1} g(X_t) \mid X_0 = x\Big) \le V(x) - E(V(X_{\tau_M}) \mid X_0 = x)$$

$$+ KE(\sum_{t=0}^{\tau_M-1} 1_{||X_t||\le M} \mid X_0 = x)$$
$$\le V(x) + K1_{||x||\le M}.$$

Hence X_t is g-regular. □

Finally, *exponential stability* for a Markov chain X_t: there exist $n \ge 1$, $\rho < 1$, $r > 0$ and $M, K < \infty$ such that

$$E(||X_n||^r \mid X_0 = x) \le \begin{cases} K & \text{if } ||x|| \le M, \\ \rho^n ||x||^r & \text{if } ||x|| > M. \end{cases} \tag{12}$$

And an *exponential drift condition* for X_t: there exist $\rho < 1$, $M, K < \infty$ and nonnegative, locally bounded function $V(x)$ such that

$$E(V(X_1) \mid X_0 = x) \le \begin{cases} K & \text{if } ||x|| \le M, \\ \rho V(x) & \text{if } ||x|| > M. \end{cases} \tag{13}$$

Not surprisingly, these are equivalent with a further constraint on V. They also imply g-regularity with $g(x) = (1-\rho)V(x)$. Exponential stability can be used to show geometric rates of convergence in (10) (Meyn and Tweedie (1993)) but it is a not necessary condition for that.

Theorem 3.4. *Assume* $\sup_{x\in\mathbb{X}} E((\frac{1+||X_1||}{1+||x||})^\delta \mid X_0 = x) < \infty$ *for some* $\delta > 0$. *Then* X_t *is exponentially stable iff the exponential drift condition holds with* $d_0||x||^r \le V(x) \le 1+d_1||x||^r$ *for some positive* r, d_0, d_1.

Proof. In this case, if (12) holds then the test function has the unenviable form $V(x) = (\prod_{t=0}^{n-1} E(||X_t||^r \mid X_0 = x))^{1/n}$, which is locally bounded. The proof of (13) uses an extended form of Hölder's inequality. If $||x|| > M$ then

$$E(V(X_1) \mid X_0 = x) = E((\prod_{t=0}^{n-1} E(||X_{t+1}||^r \mid X_1))^{1/n} \mid X_0 = x)$$
$$\le (\prod_{t=0}^{n-1} E(||X_{t+1}||^r \mid X_0 = x))^{1/n}$$
$$= (E(||X_n||^r/||x||^r \mid X_0 = x))^{1/n} V(x) \le \rho V(x).$$

Despite the immense inefficiency of Hölder's inequality, this argument cannot be improved upon. As in the proof of Theorem 3.2, the converse follows by iteration. See Meyn and Tweedie (1993), Cline and Pu (1999a) and Cline (2007) for related results. □

Given the similarity of these notions of stability and their apparent agreement for some threshold autoregression systems/models, Tong asked (in various works), if it is coincidental that stability of the stochastic recursion corresponds to stability of its dynamical skeleton. Obviously it is not, but the connection unfortunately is not at all clear cut. For starters, the (optimal) test functions mentioned in the discussion above are not so similar. Secondly, the intuitive notion that errors are negligible when $||X_0||$ is large presumes some sort of continuity – and the piecewise continuity of a threshold model may not be enough. For example, to apply the test function for exponential stability of a dynamical system directly in order to show exponential stability of its clothed stochastic recursion, there must at least exist a *continuous homogeneous* function $G(x)$ such that $|F(x) - G(x)| = o(||x||)$ (cf. Chan and Tong (1985)). The homogeneity is usually not a problem but the continuity of G rules out most threshold models. It even rules out most smooth transition (STAR) models as well. When very large in value, a STAR process is not very different from some threshold process and therefore the method for verifying stability of STAR models invariably is the same as that for related threshold models (Chan and Tong (1986); Chen and Tsay (1993); An and Huang (1996)). I will return to Tong's question in section 5.

4. The Lyapunov Exponent of Drift

Another point of comparison between dynamical systems and time series models is the concept of a (top or first) Lyapunov exponent. But there is divergence here from the standard definition because stability as I have described it is not the same as the local "Lyapunov stability" of a dynamical system with a continuous F. Instead, I will examine the notion of an *exponent of drift* in this section.

But first let us take a quick look at the usual definition of a Lyapunov exponent for the dynamical system (1). This value is intended to measure the sensitivity of the system to initial conditions:

$$\lambda(x) = \lim_{n\to\infty} \frac{1}{n} \log(||D_{F^n}(x)||), \tag{14}$$

where D_{F^n} is the derivative (Jacobian) matrix for F^n and $|| \cdot ||$ is the matrix (operator) norm induced by the norm on $\mathbb{X}$. (Sometimes the spectral radius is used as the matrix norm.) That the limit in (14) exists is a consequence of the multiplicative ergodic theorem for "random" matrices (Furstenberg and Kesten (1960); Osledec (1968)) which in turn are corollaries to Kingman's (1973) subadditive ergodic theorem. For a thorough discussion, see Sinai (2000, chapter 1).

Under sufficient regularity, if the system is ergodic with invariant measure μ then the exponent does not depend on x or on the choice of norm. Obviously, F must be continuously differentiable as well. When the system has multiple attractors so that it is not irreducible (in parallel to the stochastic sense), each basin of attraction has its own exponent.

If the system is scalar ($m = 1$) and ergodic, the Lyapunov exponent is in fact

$$\lambda = \int_{\mathbb{X}} \log(|F'(x)|)\, \mu(dx). \tag{15}$$

This may be seen by applying the chain rule and the ergodic property (8). Basically,

$$\begin{aligned} \lim_{n\to\infty} \frac{1}{n} \log(|D_{F^n}(x_0)|) &= \lim_{n\to\infty} \frac{1}{n} \log(\prod_{t=0}^{n-1} |F'(x_t)|) = \lim_{n\to\infty} \frac{1}{n} \sum_{t=0}^{n-1} \log(|F'(x_t)|) \\ &= \int_{\mathbb{X}} \log(|F'(x)|)\, \mu(dx). \end{aligned} \tag{16}$$

One must be careful about extending (15) to the multiple dimension setting. Let $\rho(A)$ be the spectral radius of a matrix A. Then

$$\lambda \leq \int_{\mathbb{X}} \log(\rho(D_F(x)))\, \mu(dx) \leq \int_{\mathbb{X}} \log(||D_F(x)||)\, \mu(dx),$$

with strict inequality except in very special cases. An interesting question is what representation should replace (15), if any. For dynamical systems that converge to a limit point or to a limit cycle, $x_0, x_1, \ldots, x_N = x_0$, the invariant measure is uniform on the points of the cycle and $\lambda = \frac{1}{N}\rho(D_{F^N}(x_0))$ which is not in integral form.

There has been a great deal of interest both in defining an analogous value for the model (3) and in estimating it. Tong's (1995) very fine review and the discussion that followed it presented a number of viewpoints. Chan and Tong (2001) gave an in-depth discussion of statistical approaches for determining a measure of sensitivity to the initial state. I will not recount all the work that has been done because I wish to take the concept of an exponent of stability in a very different direction.

Suppose $F(x)$ is locally bounded and define

$$\gamma = \liminf_{n\to\infty} \limsup_{||x||\to\infty} \frac{1}{n} \log(||F^n(x)||/||x||). \tag{17}$$

Observe that γ is the infimum of $\log \rho$ such that (7) holds for some n, M. Clearly, the system is exponentially stable iff $\gamma < 0$. Therefore, γ is referred to as the *Lyapunov exponent of drift*. Under some regularity, γ may be expressed with limits,

$$\gamma = \lim_{n\to\infty} \lim_{||x||\to\infty} \frac{1}{n} \log(||F^n(x)||/||x||).$$

The definition in (17) does not require F to be continuous, let alone differentiable.

In the case of a linear system with $F(x) = Ax$, where A is an irreducible matrix, $\gamma = \log \rho(A)$. But notice also that definition (14) exists trivially with $\lambda = \log \rho(A)$ independent of x if A is irreducible, even if the system is not stable. Thus, for irreducible linear systems, $\gamma = \lambda$.

Extending the notion in (17) to the stochastic model (3), define

$$\gamma = \liminf_{n\to\infty} \limsup_{||x||\to\infty} \frac{1}{n} E(\log(||X_n||/||X_0||) \mid X_0 = x). \tag{18}$$

Actual limits frequently exist. Again, no differentiability at x is assumed and, in particular, this exponent differs from the initial value sensitivity such as that discussed in Chan and Tong (2001). Note that (12) implies $\gamma < 0$ holds (since $\gamma \le \log \rho$). In fact, if $E(|\log(||X_1||)| \mid X_0 = x)$ is locally bounded then exponential stability is equivalent to $\gamma < 0$. (A proof is implicit in Cline and Pu (1999a); see also Cline (2007).) On the other hand, irreducibility and $\gamma > 0$ implies $||X_n|| \to \infty$ almost surely. Thus, as it is for dynamical systems, γ is a critical value for stability and again it will be called the *Lyapunov exponent of drift*.

It is easy to see, moreover, that $\gamma = \log \rho(A)$ for the linear model, $X_t = AX_{t-1} + e_t$, with irreducible A, and thus the linear model and its skeleton, $x_t = Ax_{t-1}$, have the same exponent of drift. In particular, if X_t is the state vector of an irreducible AR(p) time series then e^γ is the modulus of the largest root of the characteristic equation. As mentioned above, $\gamma = \lambda$ for the skeleton.

For nonlinear models, however, γ differs from λ for the associated skeleton. Nevertheless, there are some important parallels between the two. For example, just as the chain rule in (16) made it possible to express λ as a longterm average, the use of telescoping ratios converts (17) and (18) to longterm averages. This will be exploited in the sections to come.

The major question of interest is whether γ has the same value for both the dynamical system and its clothed nonlinear time series. The answer is known to be in the affirmative for certain cases (next section) and in the negative for some specialized cases but, as far as I know, the question remains open more generally. An additional question is whether there is a representation for γ, analogous to (15), that is valid for both the time series and its skeleton. Interestingly, a representation may exist in the stochastic setting even when the time series does not have a dynamical system skeleton (section 6).

5. Limit Cycles and Chaos

Stable dynamical systems have limit points, limit cycles and/or chaotic behavior. Hence, a threshold time series model that is a dynamical system cloaked with noise might be expected to have similar behavior. They do but, as mentioned earlier, if the system has multiple attractors then an irreducible time series will mimic one limit cycle or chaos of the system and eventually will move to another. All this presumes the error variance $\sigma(x)$ is relatively small.

For example, the celebrated Canadian lynx time series exhibits a cyclic behavior with a period of about 9 years (cf. Tong (1990), among others). Numerous models have been suggested to account for this; among them is a threshold AR(2) model fit by Tong to the log-transformed values. The model is

$$\xi_t = \begin{cases} 0.62 + 1.25\,\xi_{t-1} - 0.43\,\xi_{t-2} + 0.195\,\epsilon_t & \text{if } \xi_{t-2} \leq 3.25, \\ 2.25 + 1.52\,\xi_{t-1} - 1.24\,\xi_{t-2} + 0.25\,\epsilon_t & \text{if } \xi_{t-2} > 3.25. \end{cases} \tag{19}$$

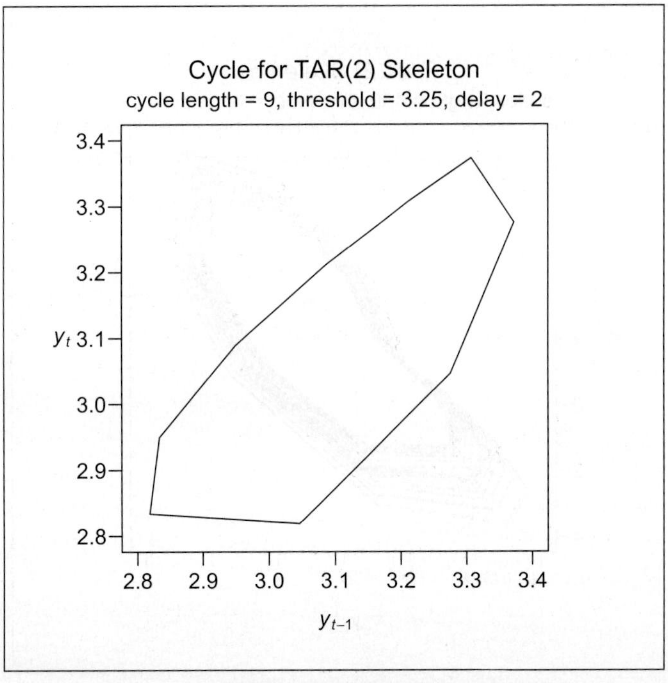

Figure 1. Limit cycle for the threshold AR(2) skeleton of the log-lynx data fit by Tong (1990).

The state vector is $X_t = (\xi_t, \xi_{t-1})$. The skeleton for this model is $x_t = (y_t, y_{t-1})$ where

$$y_t = \begin{cases} 0.62 + 1.25\, y_{t-1} - 0.43\, y_{t-2} & \text{if } y_{t-2} \leq 3.25, \\ 2.25 + 1.52\, y_{t-1} - 1.24\, y_{t-2} & \text{if } y_{t-2} > 3.25. \end{cases} \tag{20}$$

The skeleton does indeed have a cycle with period equal to 9 (Figure 1). (Cycles are found numerically: by iterating $x_t = F(x_{t-1})$ until repetitions occur, using several starting values to find all the attractors. Convergence usually is quite fast.)

Although the fact that the skeleton has a limit cycle strongly suggests that the time series model is stable, one cannot ascertain this simply by studying the limit cycle. Indeed, if the errors are expected to be a negligible constraint on stability then so are the intercepts (and, equally, the threshold). But the limiting behavior of the skeleton is quite sensitive to these values. See Figure 2 in comparison to Figure 1. When the threshold value in (20) is changed to 0, the skeleton appears to be chaotic but returns to roughly the same region every 10th or 11th iteration.

Instead, stability of the system (and hopefully of the time series) is determined by its behavior when it is very large, as measured by the drift exponent γ. One way to visualize this is to separate out the relative change in magnitude from the polar direction, keeping in mind that the intercepts and threshold are negligible for this exercise. To this end, assume that $F^*(x)$ is a homogeneous function such that $||F^*(x) - F(x)|| = o(||x||)$, as $||x|| \to \infty$, and define the *homogeneous system* $x_t^* = F^*(x_{t-1}^*)$. Continuity of F^* is not required. The polar direction and the relative change of magnitude for x_t^* are

$$\begin{aligned} \theta_t &= x_t^*/||x_t^*|| = F^*(\theta_{t-1})/||F^*(\theta_{t-1})|| \quad \text{and} \\ w_t &= ||x_t^*||/||x_{t-1}^*|| = ||F^*(\theta_{t-1})||, \end{aligned} \tag{21}$$

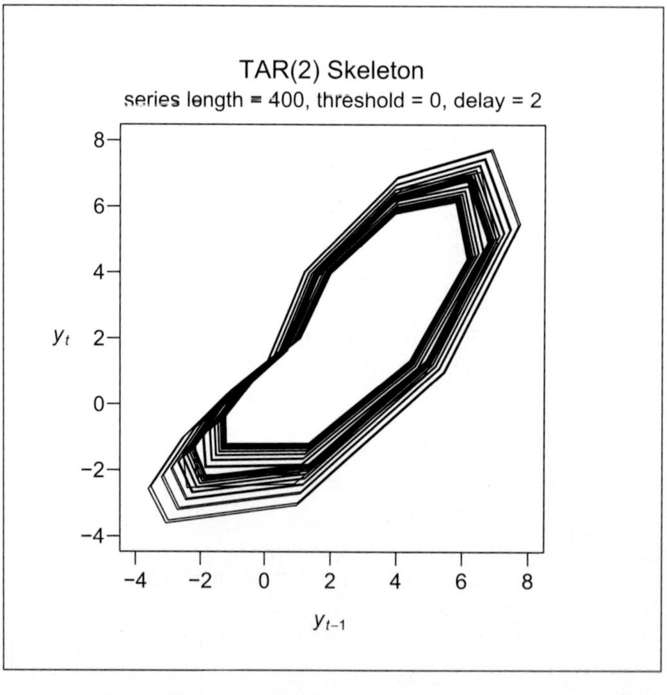

Figure 2. Apparently chaotic track for the log-lynx skeleton when the threshold is set to 0.

respectively. Thus, θ_t is a dynamical system (called the *collapsed system*) on the unit m-sphere Θ. Since it is confined to a compact set, it necessarily is regularly stable and typically it is ergodic within each of its basins of attraction. This leads to the following characterization of γ.

Theorem 5.1. *Suppose θ_t is ergodic with invariant probability measure μ. Then the Lyapunov exponent of drift for x_t is*

$$\gamma = \lim_{n\to\infty} \frac{1}{n} \log(\prod_{t=1}^{n} w_t) = \int_{\Theta} \log(||F^*(\theta)||)\, \mu(d\theta). \tag{22}$$

In particular, if $\theta_0, \theta_1, \ldots, \theta_N = \theta_0$ defines a limit cycle for θ_t then

$$\gamma = \frac{1}{N} \sum_{t=0}^{N-1} \log(||F^*(\theta_t)||). \tag{23}$$

More generally, if θ_t has l attractors with respective invariant measures $\mu_1, \ldots, \mu_l$ then

$$\gamma = \max_{1\leq i\leq l} \int_{\Theta} \log(||F^*(\theta)||)\, \mu_i(d\theta). \tag{24}$$

Compare (22) with (15).

Proof. It is easy to verify that definition (17) and the condition $||F^*(x) - F(x)|| = o(||x||)$ guarantee that x_t and x_t^* have the same Lyapunov exponent. Then (22) simply follows from (17), (21) and the ergodic theorem. Since the invariant measure for a limit cycle is uniform, (23) is a more explicit representation in that case.

When θ_t has multiple attractors, (17) implies the exponent for each basin of attraction must be computed and then γ takes the maximum value. (In the linear case, this corresponds to A being a reducible matrix.) □

To apply this result to the log-lynx system (20), let $\theta_t = (\theta_{t,1}, \theta_{t,2})$ and

$$\begin{aligned} y_t^* &= \begin{cases} 1.25\,\theta_{t-1,1} - 0.43\,\theta_{t-1,2} & \text{if } \theta_{t-1,2} \leq 0, \\ 1.52\,\theta_{t-1,1} - 1.24\,\theta_{t-1,2} & \text{if } \theta_{t-1,2} > 0, \end{cases} \\ \theta_t &= \frac{(y_t^*, \theta_{t-1,1})}{||(y_t^*, \theta_{t-1,1})||}. \end{aligned} \tag{25}$$

The threshold for the collapsed system is defined by $\theta_{t,2} = 0$. Actually, the result should be applied to $y_t' \overset{\text{def}}{=} y_t - 3.25$ so that the condition $||F^*(x) - F(x)|| = o(||x||)$ is clearly met. More generally (and this is especially relevant for time series), all that really matters is that θ_t ultimately stays away from the threshold. In the case θ_t has a limit cycle, this simply means that no points of the cycle are on the threshold.

Corollary 5.1. *Let $C \subset B$ be closed and open subsets, respectively, of Θ such that the condition $||F^*(x) - F(x)|| = o(||x||)$ holds for $x/||x|| \notin C$, $||x|| \to \infty$, and $\mu(B) = 0$ (or $\mu_1(B) = \cdots = \mu_l(B) = 0$). Let $x_t^* = F^*(x_{t-1}^*)$ and define (θ_t, w_t) by (21). Then the conclusions of Theorem 5.1 hold.*

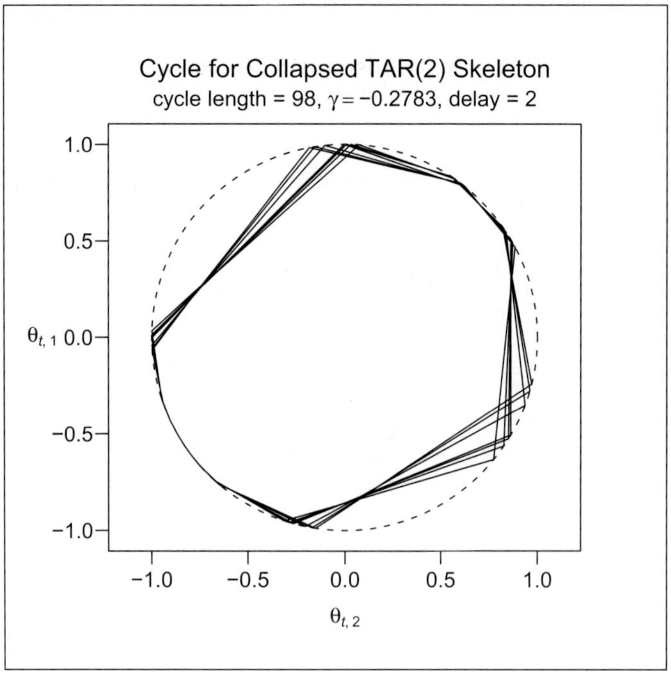

Figure 3. Limit cycle with period 98 for the collapsed log-lynx threshold AR(2) model.

Returning to the example (19), (20) and (25), one may easily find that θ_t has a limit cycle with period 98 (by numerical computation) and that no points are on the threshold, although one is quite close. This limit cycle is shown in Figure 3. The computed value for γ is -0.2783, substantiating the prior belief that the system is exponentially stable. One also can visualize how much the homogeneous system x_t^* expands or contracts at each point in the cycle by computing w_t and picturing it in the plot. See Figure 4.

Now, the ultimate question is whether Theorem 5.1 or Corollary 5.1 also describe the Lyapunov exponent of drift for the process X_t, as defined in (4). This actually remains open to date, as far as I know, but I suggest that the following at least is true without much more in the way of regularity assumptions.

Theorem 5.2 (Conjectured). *Suppose X_t is defined by (4) with skeleton (1), the dynamical systems x_t and x_t^* satisfy all the conditions of Corollary 5.1, F and σ are piecewise continuous and σ is bounded and locally bounded away from 0. Then the Lyapunov exponent of drift for the process X_t is given in (24).*

Threshold autoregression models are a special case for which there are results (Tjøstheim (1990); Cline and Pu (1999b); Boucher and Cline (2007)), epitomized as Corollary 5.2 below. Assume ξ_t is a threshold-like autoregression such that its state vector X_t is given by the following.

$$X_t = \sum_{j=1}^{k} (a_{j0} + A_j X_{t-1} + \sigma_j e_t)\, 1_{X_{t-1} \in C_j} + g(X_{t-1}) + h(X_{t-1}) e_t \tag{26}$$

where $C_1, \ldots, C_k$ are disjoint cones with nonempty interiors in $\mathbb{R}^p$ and $A_1, \ldots, A_k$ are the companion matrices for k linear models. Also, assume $g(x) = o(||x||)$ and $h(x) = o(||x||)$, as

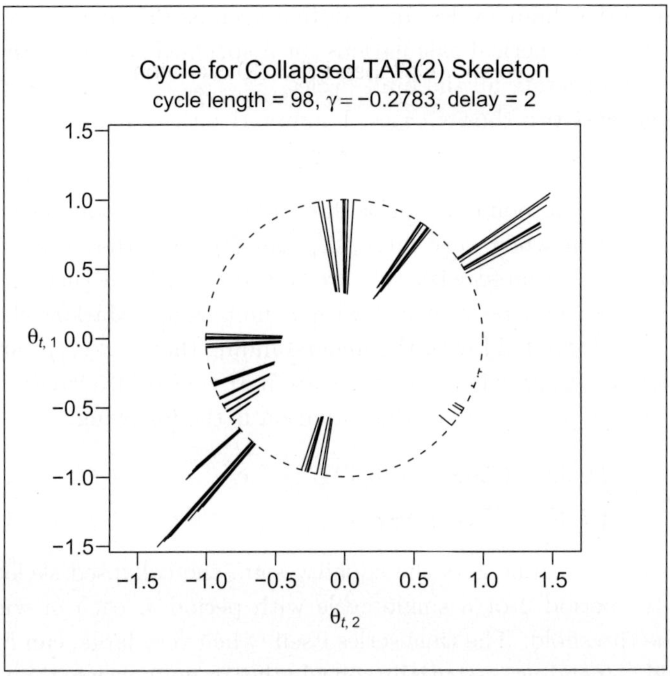

Figure 4. Points of the limit cycle for the collapsed log-lynx threshold AR(2) model showing their respective effects. The endpoints of the line segments are θ_t and $w_{t+1}\theta_t$.

$||x|| \to \infty$. (This includes many STAR models.) The collapsed dynamical skeleton clearly is

$$\theta_t = x_t^*/||x_t^*|| \quad \text{with} \quad x_t^* = \sum_{j=1}^{k} A_j \theta_{t-1} 1_{\theta_{t-1} \in C_j}. \tag{27}$$

Suppose $\theta_{i,0}, \ldots, \theta_{i,N_i-1}$ is a limit cycle with corresponding companion matrices $A_{i,1}, \ldots, A_{i,N_i}$. If $x_0^* = \theta_{i,0}$ (so that the collapsed system follows the cycle exactly) then

$$A_{i,N_i} \cdots A_{i,1}, \theta_{i,0} = x_N^* = ||x_N^*||\, \theta_{i,0}$$

Therefore $A_{i,N_i} \cdots A_{i,1}$, is proportional to an identity matrix, implying

$$\rho(A_{i,N_i} \cdots A_{i,1}) = ||x_N^*|| = \prod_{t=1}^{N_i} w_t = \prod_{t=0}^{N_i-1} ||F^*(\theta_t)||,$$

and this leads to the following result.

Corollary 5.2. *Suppose the attractors for the collapsed skeleton (27) consist only of limit points and limit cycles, and each individual limit point or cycle point is in the interior of some C_j. Then the Lyapunov exponent of drift for X_t, as defined in (26), is*

$$\gamma = \max_{1 \le i \le l} \frac{1}{N_i} \log(\rho(A_{i,N_i} \cdots A_{i,1})),$$

where the maximum is taken over all the limit points/cycles.

Identifying the actual limit cycles for a system such as (27) can be very tricky. Aside from careful but ad hoc numerical calculations for a specified set of parameters, I know of no algorithm that can generate all the limit cycles.

There remain at least two thorny cases. I summarize them here as they have yet to be fully dealt with.

(i) If a limit point or cycle point lies on a threshold (e.g., the boundary between two of the cones) then, even when very large, X_t can fall on either side of the boundary. If the probabilities associated with this are (essentially) fixed then the limit cycles of the collapsed system can be replaced with a finite state Markov chain on Θ. The results above should still hold with the understanding that $\mu_1, \dots, \mu_l$ are the stationary distributions for this chain when the states are partitioned into irreducible subsets. An example of a time series leading to this situation is the following.

$$\xi_t = \begin{cases} 0.62 - 1.56\,\xi_{t-1} - 0.42\,\xi_{t-2} + \epsilon_t & \text{if } \xi_{t-1} + \xi_{t-2} \leq 1.2, \\ 2.25 - .75\,\xi_{t-1} + \epsilon_t & \text{if } \xi_{t-1} + \xi_{t-2} > 1.2. \end{cases}$$

Depending on which regime gets the equality sign, the collapsed skeleton has either a single cycle with period 2 or a single cycle with period 4, each of which has a point precisely on the threshold. The time series itself, when very large, can fall on either side of the threshold, depending (essentially) on whether or not ϵ_t is less than the difference of the threshold and the intercept. This means the finite state Markov chain can randomly switch (with the appropriate probabilities) from one cycle to the other whenever it hits the cycle point on the threshold.

(ii) If at least one attractor for the collapsed system is strange (i.e., the system is chaotic) then its invariant measure very likely will be positive for any open set containing the thresholds. This certainly ruins the arguments above, but exactly what should replace them is still unclear. An apparent example, close to the log-lynx model above, is

$$\xi_t = \begin{cases} 0.62 + 1.25\,\xi_{t-1} - 0.52\,\xi_{t-2} + 0.195\,\epsilon_t & \text{if } \xi_{t-2} \leq 3.25, \\ 2.25 + 1.55\,\xi_{t-1} - 1.25\,\xi_{t-2} + 0.25\,\epsilon_t & \text{if } \xi_{t-2} > 3.25. \end{cases}$$

6. Bilinear and GARCH Time Series

The nature of the stability problem changes immensely for bilinear and GARCH time series. In particular, the "error terms" are no longer negligible. Thus, there is no dynamical system skeleton that may be analyzed. Nevertheless, an analogous representation exists for the exponent of drift for the Markov state process. Actually, these are special cases of models of quite longstanding interest.

Specifically, consider the random coefficient (RC) model

$$X_t = A(e_t)X_{t-1} + B(e_t). \tag{28}$$

Here $(A(e_t), B(e_t))$ is a stationary sequence of random matrices and random vectors, respectively. It is well known that, under appropriate irreducibility, (28) is ergodic iff

$$\lim_{n\to\infty} \frac{1}{n} E(\log(||\prod_{t=0}^{n-1} A(e_t)||)) < 0,$$

where $||A||$ denotes the matrix norm induced by $||x||$. This is another instance of the use of methods like the multiplicative ergodic theorem (Furstenberg and Kesten (1960)) or Kingman's (1973) subadditive ergodic theorem. With little further effort, one may demonstrate also that (18) exists as a limit and

$$\gamma = \lim_{n\to\infty} \frac{1}{n} E(\log(|| \prod_{t=0}^{n-1} A(e_t)||)) \tag{29}$$

so that ergodicity again depends on the drift exponent. Note that γ is (usually much) less than $E(\log(\rho(A(e_t))))$. Additionally,

$$\lim_{n\to\infty} \frac{1}{n} \log(|| \prod_{t=0}^{n-1} A(e_t)||) \overset{\text{a.s.}}{=} \lim_{n\to\infty} \frac{1}{n} E(\log(|| \prod_{t=0}^{n-1} A(e_t)||)), \tag{30}$$

which suggests a means for estimating γ. Typically, one even can provide an almost sure representation of X_t in terms of past $(A(e_t), B(e_t))$.

The result has been noted and applied to various time series models, including bilinear time series (Pham (1993)) and GARCH time series (Bougerol and Picard (1992a,b)). It can also be used to show stability for Markov regime switching autoregression models and shock-driven threshold autoregression models. In these two cases, the A_i's are chosen independently of the current value of the series, unlike the self-exciting threshold model (26). The so-called double AR model can be expressed as a RC model (Ling (2007)) *if the errors have normal distribution.*

Regrettably, (29) and (30) are cumbersome and inefficient for actual calculations involving bilinear and GARCH models. Furthermore, they cannot be applied to a self-exciting threshold GARCH model or to a model with both AR and GARCH components because such models cannot be embedded into RC models. Both of these faults can be overcome by expressing γ once again as a longterm average (Cline and Pu (2004); Cline (2007)). Strangely enough, given the lack of a skeleton, the solution is to mimic the idea of (22).

First express the state process as

$$X_t = B(X_{t-1}/||X_{t-1}||, e_t)||X_{t-1}|| + C(X_{t-1}, e_t), \tag{31}$$

where $||B(x/||x||, u)||$ and $||C(x, u)||$ are bounded by $K(1 + ||u||)$, $K < \infty$. Noting that the first term is homogeneous in X_{t-1}, define the *homogeneous state process,*

$$X_t^* = B(X_{t-1}^*/||X_{t-1}^*||, e_t)||X_{t-1}^*||,$$

and its related *collapsed process* and change in magnitude,

$$\begin{aligned} \theta_t &= X_t^*/||X_t^*|| = B(\theta_{t-1}, e_t)/||B(\theta_{t-1}, e_t)|| \quad \text{and} \\ W_t &= ||X_t^*||/||X_{t-1}^*|| = ||B(\theta_{t-1}, e_t)||. \end{aligned} \tag{32}$$

Assume also that $B(\cdot, \cdot)$ is piecewise continuous in the first component and θ_t strongly prefers (in some sense) to stay within the regions of continuity. This includes ordinary GARCH, threshold GARCH and the most popular bilinear models, as well as RC models. With some additional regularity (unstated, but see Cline and Pu (2004) and Cline (2007)), we have the following representation of γ which is very much like (22).

Theorem 6.1. *Suppose X_t and θ_t are given by (31) and (32), respectively. Under appropriate irreducibility and other regularity assumptions, θ_t is a uniformly ergodic Markov* *chain with some stationary distribution* μ. *Moreover, the Lyapunov exponent of drift for the original process* X_t *is*

$$\gamma = E(\log(W_t)) = \int_\Theta E(\log(||B(\theta, e_1)||))\,\mu(d\theta). \tag{33}$$

Theorem 6.1 also suggests a simple method of evaluating γ for a given model, namely by simulating (θ_t, W_t) and directly estimating $E(\log(W_t))$.

References

1. An, H.Z. and Huang, F.C. (1996). The geometrical ergodicity of nonlinear autoregressive models, *Stat. Sinica* 6, 943–956.
2. Boucher, T.R. and Cline, D.B.H. (2007). Stability of cyclic threshold and threshold-like autoregressive time series models, *Stat. Sinica* 17, 43–62.
3. Bougerol, P. and Picard, N.(1992a). Stationarity of GARCH processes and some nonnegative time series, *J. Econom.* 52, 115–127.
4. Bougerol, P. and Picard, N. (1992b). Strict stationarity of generalized autoregressive processes, *Ann. Probab.* 20, 1714–1730.
5. Brockwell, P.J., Liu, J. and Tweedie, R.L. (1992). On the existence of stationary threshold autoregressive moving-average processes. *J. Time Series Anal.* 13, 95–107.
6. Chan, K.-S. (1989). A note on the geometric ergodicity of a Markov chain, *Adv. Appl. Probab.* 21, 702–704.
7. Chan, K.-S. (1990). Deterministic stability, stochastic stability, and ergodicity, Appendix 1 in *Non-linear Time Series Analysis: A Dynamical System Approach*, by H. Tong, Oxford University Press (London).
8. Chan, K.-S. (1993). A review of some limit theorems of Markov chains and their applications, *Dimensions, Estimation and Models*, ed. by H. Tong, World Scientific (Singapore), 108–135.
9. Chan, K.-S., Petruccelli, J.D., Tong, H. and Woolford, S.W. (1985). A multiple threshold AR(1) model, *J. Appl. Probab.* 22, 267–279.
10. Chan, K.-S. and Tong, H. (1985). On the use of the deterministic Lyapunov function for the ergodicity of stochastic difference equations, *Adv. Appl. Probab.* 17, 666–678.
11. Chan, K.-S. and Tong, H. (1986). On estimating thresholds in autoregressive models, *J. Time Series Anal.* 7, 179–190.
12. Chan, K.-S. and Tong, H. (1994). A note on noisy chaos, *J. Royal Stat. Soc.* 56, 301–311.
13. Chan, K.S. and Tong, H. (2001). *Chaos: a statistical perspective*, Springer-Verlag.
14. Chen, R. and Tsay, R.S. (1991). On the ergodicity of TAR(1) process, *Ann. Appl. Probab.* 1, 613–634.
15. Chen, R. and Tsay, R.S. (1993). Functional-coefficient autoregressive models, *J. Amer. Stat. Assoc.* 88, 298–308.
16. Cline, D.B.H. (2007). Stability of nonlinear stochastic recursions with application to nonlinear AR-GARCH models, *Adv. Appl. Probab.* 39, 462–491.
17. Cline, D.B.H. and Pu, H.H. (1999a). Geometric ergodicity of nonlinear time series, *Stat. Sinica* 9, 1103–1118.
18. Cline, D.B.H. and Pu, H.H. (1999b). Stability of nonlinear AR(1) time series with delay, *Stoch. Proc. Appl.* 82, 307–333.
19. Cline, D.B.H. and Pu, H.H. (2004). Stability and the Lyapounov exponent of threshold AR-ARCH models, *Ann. Appl. Probab.* 14, 1920–1949.
20. Foster, F.G. (1953). On the stochastic matrices associated with certain queueing processes. *Ann. Math. Stat.* 24, 355–360.
21. Furstenberg, H. and Kesten, H. (1960). Products of random matrices, *Ann. Math. Stat.* 31, 457–469.
22. Halanay, A. and Răsvan, V. (2000). *Stability and Stable Oscillations in Discrete Time Systems*, Gordon and Breach Science Publishers (Amsterdam).

23. Jones, D.A. (1976). Nonlinear autoregressive processes. Ph.D thesis, University of London.
24. Kingman, J.F.C. (1973). Subadditive ergodic theory, *Ann. Probab.* 1, 883–899.
25. La Salle, J.P. (1976). *The Stability of Dynamical Systems*, CMBS 25, Society for Industrial and Applied Mathematics (Philadelphia).
26. Ling, S. (2007). A double AR model: structure and estimation, *Stat. Sinica* 17, 161–175.
27. Martelli, M. (1999). *Introduction to Discrete Dynaimcal Systems and Chaos*, John Wiley & Sons.
28. Meyn, S.P. and Tweedie, R.L. (1993a). *Markov Chains and Stochastic Stability*, Springer-Verlag (London).
29. Osledec, V.I. (1968). A multiplicative ergodic theorem: Liapunov characteristic numbers for dynamical systems, *Trans. Moscow Math. Soc.* 19, 197–231.
30. Pham, D.T. (1993). Bilinear times series models, *Dimensions, Estimation and Models*, ed. by H. Tong, World Scientific Publishing (Singapore), 191–223.
31. Sinai, Y.G., ed. (2000). *Dynamical Systems, Ergodic Theory and Applications*, Springer-Verlag.
32. Tjøstheim, D. (1990). Non-linear time series and Markov chains, *Adv. Appl. Probab.* 22, 587–611.
33. Tong, H. (1977). Discussion of a paper by A.J. Lawrance and N.T. Kottegoda, *J. Roy. Stat. Soc. (series A)* 140, 34–35.
34. Tong, H. (1978). On a threshold model. In *Pattern Recognition and Signal Processing* (ed. C.H. Chan), Sijthoff and Noordhoff (Amsterdam).
35. Tong, H. (1983). *Threshold Models in Nonlinear Time Series Analysis*, Notes in Statistics, No. 21, Springer Verlag (Heidelberg).
36. Tong, H. (1990). *Non-linear Time Series Analysis: A Dynamical System Approach*, Oxford University Press (London).
37. Tong, H. (1995). A personal overview of nonlinear time series from a chaos perspective (with discussion). *Scan. J. Statist.* 22, 399–445.
38. Tong, H. (2007). Birth of the threshold time series model, *Stat. Sinica* 17, 8–14.
39. Tong, H. and Lim, K.S. (1980). Threshold autregression, limit cycles and cyclical data (with discussion), *J. Roy. Stat. Soc. (series B)* 42, 245–292.
40. Tweedie, R.L. (1975). Sufficient conditions for ergodicity and recurrence of Markov chains on a general state space. *Stoch. Proc. Appl.* 3, 385–403.
41. Tweedie, R.L. (1976). Criteria for classifying general Markov chains. *Adv. Appl. Probab.* 24, 542–574.

 Vol. 22: 399–445, 1995

A Personal Overview of Non-linear Time Series Analysis from a Chaos Perspective*

HOWELL TONG

University of Kent

ABSTRACT. A personal overview of non-linear time series from a chaos perspective is given in an informal but, it is hoped, informative style. Recent developments which, in a radically new way, formulate the notion of initial-value sensitivity with special reference to stochastic dynamical systems are surveyed. Its practical importance in prediction is highlighted and its statistical estimation included by appealing to the modern technique of locally linear non-parametric regression. The related notions of an embedding dimension and correlation dimension are also surveyed from the statistical stand-point. It is shown that deterministic dynamical systems theory, including chaos, has much to offer to the subject. In return, some current results in the subject are summarized, which suggest that some of the standard practice in the former may have to be revised when dealing with real noisy data. Several open problems are identified.

Key words: attractors, chaos, correlation dimension, dynamical systems, embedding dimension, initial-value sensitivity, Kullback–Leibler information, Lyapunov exponent, noise amplification, skeleton, threshold principle

1. Introduction

The new field of deterministic chaos has been hailed as a revolution of thoughts and attracting ever increasing attention outside statistics. It has aroused the attention of many scientists and technologists from diverse disciplines including mathematics (both pure and applied), physics, computation, engineering, biology, neurology, economics and many others and has become a truly multi-disciplinary area of research. It has even captured the imagination of the general public. The concept of chaos has found applications in a vast number of areas, ranging from meteorology to climatology, from cryptography to optimization, from animal population dynamics to epidemiology, from turbulence to flames, from electrocardiography to electroencephalography, from structural engineering to vibrations, and many others (see e.g. Hao, 1990; Drazin & King, 1992; Grenfell *et al.*, 1994; Titterington, 1994). As may be expected with any new field, some of the applications (e.g. structural engineering and turbulence) are genuinely established as important whilst others (e.g. economics, electroencephalography and epidemiology) are still at a tentative stage.

As far as the statisticans are concerned, the subject of deterministic chaos tends to provoke different reactions. Some of them find it totally alien and even suspicious (see e.g. Ozaki, 1990; Thompson & Tapia, 1990, esp. p. 251). They might have formed the impression that the theory attempts to explain almost all random phenomena by purely deterministic systems and tend to take their leave at this point because their training has convinced them of the limitations of determinism in analysing real data. However, critiques are always necessary for the healthy growth of a new subject; they enable us to prune away the inessential and misleading branches.

* The contents of this paper were presented as a Special Invited Lecture at the 15th Nordic Conference on Mathematical Statistics, Lund, Sweden, August 1994, and at the DMV (German Mathematical Society) Seminar programme in July 1994.

There are these other statisticians, perhaps forming the majority, who are vaguely aware of the fact that sensitivity to initial conditions in a deterministic dynamical system can lead to randomness. For example, they may not strongly argue against modelling the tossing of a coin by a purely mechanical system. At the same time they have ample experience of the sensitive dependence of the outcome (heads or tails) on the initial strength of the toss. If it were not for this sensitive dependence, statisticians would surely not use coin-tossing as their standard randomization device. Despite this intimate connection, they seem to be reluctant to invest much time or energy to follow the voluminous publications in the physical science literature on what might be crudely described as "deterministic randomness", perhaps because of the following reasons.

(i) The unfamiliar and sometimes even forbidding language of dynamical systems theory. (Readable and succinct accounts are, however, available in e.g. Ott, 1993; Ruelle, 1989.)
(ii) The unclear prospects to them of any direct relevance to statistics.

Finally, there is an increasing number of statisticians who recognize the vast potential to be gained through a proper cross-fertilization between deterministic chaos and statistics (see e.g. Cox & Smith, 1953; Bartlett, 1990; Tong, 1990, 1992 and others to be referred to shortly). At the same time, they have not lowered their vigilance when faced with any claim that low dimensional deterministic chaos has been detected in real time series data be they from astrophysics, biology, economics, meteorology or other fields. Nevertheless, despite a fairly slow start, sufficient progress has been made over the past decade or so, which lends support to the following statements (Tong, 1992).

(i) Many of the ideas surrounding chaos have direct and sometimes quite profound contributions to statistics.
(ii) The statisticians have an important role to play in clarifying and deepening the understanding of the notion of chaos in a stochastic environment.
(iii) The statisticians have much to offer in real data analysis with a view to extracting chaotic signals in noisy data. Indeed, the dynamicists have reinvented some of the tools long known to the statisticians. Tong (1992) has listed some significant examples and argued that better communications between the two groups will be beneficial to both.

The collections of papers edited by Tong & Smith (1992) and Grenfell *et al.* (1994), the review articles by Chatterjee & Yilmaz (1992), Berliner (1992), Isham (1993), Jensen (1993) and Cutler (1993) and the books by Tong (1990) and Chan & Tong (1995) provide some relevant references in what we may call the statistical analysis of non-linear time series from a chaos perspective. In this paper, the former are referred to simply as non-linear time series analysis. It should be pointed out that in the dynamical systems literature, the same term has been used even when the statistical content is minimal.

On the one hand, it is generally accepted that deterministic dynamical systems can generate chaos, that is highly erratic behaviour reminiscent of realizations of a random process. On the other hand, statistics is the study of chance. Now, since both chance and chaos are expressions of randomness, it is not surprising that they should have much in common: the consequence of sensitivity to initial conditions. Indeed, Poincaré (1905) fully recognized this. Unfortunately, in the study of deterministic dynamical systems, environmental and dynamic noise tends to be suppressed or, at most, plays a secondary role, whilst in the study of statistics, the deterministic dynamic kernel of the random generating mechanism tends to give way to the more macroscopic characterizations such as the mean functions, the covariance functions, the spectral functions and so on. Although Laplace has often been described as the protagonist of determinism, he recognized the dual role of probability. He said, "The curve described by a simple molecule of air or any gas is regulated in a manner as certain as the planetary orbits; the only difference between them lies in our ignorance. *Probability relates partly to our ignorance, partly to our knowledge*" (our italics) (Laplace, 1814).

It is the thesis of this paper that a stochastic dynamical system, in the form of a non-linear time series model, provides a natural environment for a proper intercourse between chaos and statistics, thereby bringing about greater realism to dynamical systems. The style will be informal but references will be given whenever necessary. The aim is not for comprehensiveness (hence the title of the paper) but instead to give a flavour of those aspects of non-linear time series which are closely related to dynamical systems theory. It is hoped that the forthcoming book by Chan & Tong (1995) might fill in many of the missing details.

2. Discrete-time dynamical systems

2.1. Attractors

We shall restrict our discussion in this paper to discrete-time dynamical systems partly because we shall be concerned with the statistical analysis of digitized data. Another reason is the fact that the statistical analysis of continuous-time stochastic dynamical systems is not as well developed to date.

Let us start with the deterministic case. First, we note that it is almost impossible to give a precise definition of deterministic chaos which at the same time encapsulates all that the term implies in the diverse literature of chaos. Deterministic chaos is a phenomenon in a non-linear dynamical system. It does not exist in linear systems. It can be generated either in a continuous-time system or a discrete-time system. For the former, the state vector has to be of dimension no less than three if it is described by a non-linear differential equation. No such condition is necessary for the latter. We now introduce informally a minimal nomenclature of deterministic dynamical systems in discrete time by following the informal guide of Tong & Smith (1991). (For a rigorous account, see e.g. Ott, 1993.) Let X_t denote a state vector in R^d. A discrete-time dynamical system may be described by a difference equation:

$$X_t = F(X_{t-1}), \tag{1}$$

with $X_0 \in R^d$ and for $t \geqslant 1$. Here F is a vector-valued function. We shall also call F a *map*. It is well known that for linear F, then generically speaking, as $t \to \infty$, either $|X_t| \to \infty$ (the unstable case) or $|X_t| \to$ a constant, say c, such that $c = F(c)$ (the stable case). (We omit the non-generic case typified by $X_t = X_{t-1}$.) The above holds for all initial values $X_0 \in R^d$. Note that $X_t = F^{(t)}(X_0)$, $F^{(t)}$ being the t-fold composition of F. For non-linear F, besides the above possibilities, there are at least three more.

(i) As $t \to \infty$, $X_t \to \Gamma_1 = \{c_1, \ldots, c_p\}$ such that $F(c_1) = c_2, \ldots, F(c_{p-1}) = c_p$, $F(c_p) = c_1$. Γ_1 is called a *limit cycle of period p*, assuming that p is the smallest such integer >0. A limit cycle of period 1 is also called a *limit point*.

(ii) As $t \to \infty$, the limiting behaviour is such that X_t is the sum, or some other smooth function, of a finite number of periodic functions with non-commensurate periods. We refer to this case as the *quasi-periodic* case and denote it by Γ_2.

(iii) As $t \to \infty$, $X_t \to \Gamma_3$, where Γ_3 is a "non-degenerate" closed subspace of R^d. (In a sense, we may think of cases (i) and (ii) as degenerate but not atypical.) Deterministic *chaos* is associated with this case. It is also related to a *strange attractor* although a strict definition of a strange attractor may be rather esoteric. Note that the quasi-periodic process is an exmaple of a process which is aperiodic but not chaotic.

For non-linear F we have to be more specific about the initial condition X_0. Depending on X_0, the 'ultimate' state may be different. Thus, for $i = 1, 2, 3$, let B_i denote the totality of initial states X_0 which iterate to Γ_i. We usually assume that B_i is non-trivial (say it has a positive Lebesgue measure). We call Γ_i an *attractor* with *basin of attraction* B_i. Loosely, an attractor may be classified as a limit point, a limit cycle, a quasi-periodic attractor or a chaotic attractor (or simply chaos). We generally assume that Γ_i cannot be further decomposed into disjoint attractors. As geometric objects, Γ_is may be assigned a dimension in such a way that a point has dimension 0, a line segment has dimension 1 and more exotic objects may have a fractional dimension (the so-called fractals). We shall not delve into the various definitions of dimension (see e.g. Cutler, 1993).

2.2. Lyapunov exponents

The fundamental nature of chaos is surprisingly simple. It comes about if (i) the dynamical system is globally bounded and (ii) there is sensitivity to the initial value X_0 when iterating with F, i.e. there is local instability. These are necessary but not sufficient conditions for the generation of chaos. Loosely speaking, if a globally bounded system is locally unstable, then there is the possibility that no matter how close two initial values are to each other, they will lead to drastically different consequences (usually called *realizations*, *orbits* or *trajectories*). In this case, the ultimate state cannot be a limit point, a limit cycle or a quasi-periodic state. Instead, it is chaotic and each realization is then almost indistinguishable from that of a stochastic process. An example, made almost immortal by May (1976), is the so-called logistic map inspired by animal population dynamics:

$$X_t = \alpha X_{t-1}(1 - X_{t-1}), \tag{2}$$

where, $t = 1, 2, \ldots, X_0 \in [0, 1]$ and $\alpha \in (0, 4]$. Perhaps the most well-known member of this family is when $\alpha = 4$. Here, we can actually solve the equation explicitly (a rare event) to yield, for $t = 1, 2, \ldots,$

$$X_t = \sin^2 (2^t \omega_0 \pi), \tag{3}$$

where ω_0 is determined by X_0. The multiplier 2 in (3) acts like a shift-register to the right and before long all significant figures of ω_0 will be lost. Note that this model has an invariant measure defined by a Beta (0.5, 0.5) distribution on (0, 1).

The standard notion in deterministic dynamical systems theory which quantifies initial-value sensitivity is the *Lyapunov exponent*. For simplicity of notation, let us consider the case $d = 1$. (The discussion can be generalized to cover $d > 1$.) Let X_0 and X'_0 denote two nearby initial points in R. Then, after n iterates,

$$X'_n - X_n = F^{(n)}(X'_0) - F^{(n)}(X_0) \tag{4}$$

$$\approx \dot{F}^{(n)}(X_0)(X'_0 - X_0), \tag{5}$$

where $F^{(n)}$ denotes the n-fold composition of F and the over-dot denotes the differential operator. By the chain rule,

$$\dot{F}^{(n)}(X_0) = \dot{F}(X_0)\dot{F}(X_1) \ldots \dot{F}(X_{n-1}). \tag{6}$$

If the right-hand factors are of comparable size, then $\dot{F}^{(n)}(X_0)$ increases (or decreases) exponentially with n. Therefore, considering the average rate of change, we have

$$|X'_n - X_n| \approx \exp(n\lambda(X_0))|X'_0 - X_0| \tag{7}$$

where

$$\lambda(X_0) = \lim_{n \to \infty} \ln |\dot{F}^{(n)}(X_0)|^{1/n}, \tag{8}$$

assuming that the limit exists. We call $\lambda(X_0)$ the *local Lyapunov exponent at* X_0. If $\lambda(X_0)$ is independent of X_0, then we may under general conditions have the *global Lyapunov exponent* $\lambda = E \ln |\dot{F}(X)|$, where the expectation is with respect to an appropriate invariant measure induced by F. For example, for the logistic map with parameter $= 4$, $\lambda = \ln 2 > 0$. In general, λ is invariant under a one-to-one differentiable co-ordinate transformation. Further, we note that the exponential separation of $X'_n - X_n$ in terms of $\exp(n\lambda(X_0))$ is physically meaningful only for small to at most medium n, in view of the infinitesimal nature of the arguments. Moreover, global boundedness will rule out any excessive separation.

2.3. *Stochastic difference equations*

Now, we have seen that the deterministic dynamical systems theory studies the movement in a noise-free environment from one state to the next, as time evolves. However, as we have consistently argued (e.g. Tong, 1990; Cheng & Tong, 1992), in reality observations rarely evolve according to system (1) simply because stochastic noise is ubiquitous, which may arise as a result of one or more of the following sources: (i) our model is invariably inexact and toy models such as the logistic map, the Hénon map and so on are fine to provide us with insights into various non-linear structures, but we do not think that they should be taken too seriously as far as the modelling of real data is concerned; (ii) there are always unexpected external random disturbances; (iii) measurements are often inexact. It is thus more realistic to replace the above states by random variables and the dynamics by a Markovian model such as

$$X_t = F(X_{t-1}, e_t), \tag{9}$$

where $t \in Z_+$, $F\colon R^{2d} \to R^d$, $\{e_t\}$ is a sequence of independent and identically distributed d-dimensional random vectors and e_t is independent of X_s, $0 \leqslant s < t$. We shall call $\{e_t\}$ the *dynamic noise*. (It is also called the system noise or intrinsic noise.) Following Tong (1990), we shall refer to (1) as the *skeleton* of model (9), by an abuse of notation (i.e. $F(X) = F(X, 0)$.) We sometimes refer to the situation as "clothing the skeleton (1) to produce the stochastic model (9)". Needless to say, there are situations for which model (9) is not realistic. For convenience of analysis later on, we shall further assume that the dynamic noise is additive so that equation (9) reduces to the *model with additive noise*

$$X_t = F(X_{t-1}) + e_t, \tag{10}$$

where $t \in Z_+$ and we have abused the notation F. If

$$X_t = (Y_t, Y_{t-1}, \ldots, Y_{t-d+1})^{\mathrm{T}}, \tag{11}$$

$$F(X_t) = (f_d(X_t), Y_t, \ldots, Y_{t-d+2})^{\mathrm{T}} \tag{12}$$

and

$$e_t = (\varepsilon_t, 0, \ldots, 0)^{\mathrm{T}}, \tag{13}$$

then (10) implies

$$Y_t = f_d(X_{t-1}) + \varepsilon_t, \tag{14}$$

for $t \in Z_+$. Obviously, model (14) is the well-known *non-linear autoregressive model of order d*. In fact, we shall relax the independence assumption on $\{\varepsilon_t\}$ but will only require $E[\varepsilon_t \mid Y_0, \ldots, Y_{t-1}] = 0$ for $t \geqslant 1$. Conversely, (14) can be written in the form of (9) by "vectorizing" the Y_ts. We shall consistently use X_t^d to denote the stacked vector $(Y_t, Y_{t-1}, \ldots, Y_{t-d+1})^T$. If the dimension d is unambiguous, we may drop the superscript and write simply X_t.

In the noise-free case, the (integral) dimension d is often associated with the notion of embedding in topology. The basic idea is quite simple: to view a one-dimensional object, say a loop unambiguously (this is the keyword) we only need to live in R^d, $d \leqslant 3$. By unambiguity we mean roughly that there is a one-to-one map from the object (attractor) to R^d which preserves differential information. (Going beyond 3 will enable us to view the same object equally unambiguously but 3 will be sufficient to guarantee unambiguity.) For example, if the loop twists into the figure 8 but is non-self-intersecting, then we need to go all the way to 3. In other cases, a lower dimension may often suffice. More generally, to view an attractor say Γ (as a geometric object) unambiguously, we need to live in a d-dimensional space where $d \leqslant 1 + 2 \dim (\Gamma)$. Again in specific cases, we can often get away with a smaller d than $1 + 2 \dim (\Gamma)$. In short, $1 + 2 \dim (\Gamma)$ is the smallest dimension which will guarantee unambiguous viewing for all attractors of dimension $\dim \Gamma$, however "weird". This is the basic content of the celebrated Takens' theorem (Takens, 1981), which extends the classic embedding theorem of Whitney in topology to dynamical systems. As commented by Takens (private communications), "Such a result was in the air!" In this connection, we may mention Mañé (1981) and Tong & Lim (1980). In the dynamical systems literature, the ambient space in which we do the viewing is called the embedding space and its dimension d is called the embedding dimension. For our purpose, we shall reserve the term the embedding dimension to refer to the smallest dimension which guarantees unambiguous viewing. Just as going beyond the embedding dimension will not yield any additional information about the geometric structure of the attractor Γ, going beyond the order of a non-linear autoregressive model will add nothing to the probabilistic structure of the stochastic process. Thus, the two concepts are linked at least at this level. Cheng & Tong (1994) and Takens (1994) give further discussions. We shall return to the statistical estimation of d from time series observations in section 4.2.

Another interesting connection between a non-linear autoregressive model and its skeleton has been discussed by Chan & Tong (1985, 1994), who have proved that, under appropriate conditions, a skeleton which admits an attractor can be "clothed" (namely by additive dynamic noise) to yield an ergodic stochastic process. One of the conditions requires that the attractor should be "sufficiently attractive", an idea which has connection with the concept of hyperbolicity (which rules out maps F with $\dot{F}(X)$ having complex eigenvalues on the unit circle) and the shadowing property (which roughly says that given a noisy trajectory from the initial condition X_0, it is possible to find a slightly different initial condition X_0', such that the true (i.e. noise-free) trajectory from X_0' shadows the noisy trajectory from X_0) in dynamical systems theory as well as the idea of exponential stability in stability theory (e.g. Tong, 1990). Another condition stipulates that, depending on the character of the attractor, the dynamic noise might have to be state-dependent and have compact support, the permissible size of which typically depends on the geometry of the attractor including its domain of attraction. This suggests that when drawing inference from the skeleton of a fitted non-linear autoregressive model about the existence of attractors we should pay attention to their "attractiveness" by, for instance, estimating their basins of attraction (e.g. Chan & Tong, 1994) and the size of the noise. Further, it is known that an attractor may admit infinitely many invariant measures and yet the physically relevant measure should be close to the invariant measure of the system with "small" dynamic noise. The idea seems to go back to Kolmogorov (see ch. 8 of Ruelle, 1989). Takens (1994) is also relevant in the context of a connection between a non-linear autoregressive model and its skeleton. In particular, he has mentioned that for a hyperbolic skeleton the impacts from a small dynamic noise and from a small measurement noise are indistinguishable.

3. Initial-value sensitivity in stochastic dynamical systems

3.1. Identical noise realizations

It is perhaps quite natural to start with the Lyapunov exponent defined for a deterministic dynamical system, namely $\lambda = E \ln |\dot{F}(X)|$, where for notational convenience, the system is temporarily assumed to be one-dimensional here. The discussion extends to the general case with an obvious change of notation. It is also assumed to be ergodic. Instead of taking the expectation with respect to the invariant measure of the deterministic dynamical system, a number of workers, e.g. Crutchfield *et al.* (1982), Kifer (1986), Herzel *et al.* (1987), Gerrard in association with Tong (in an unpublished handout at the SERC Edinburgh International Workshop on Non-linear Time Series in 1989) and Nychka *et al.* (1992), have suggested replacing the invariant measure by that of the stochastic dynamical system (assumed to exist). Herzel *et al.* (1987) seems to be the first to suggest that the modified Lyapunov exponent measures the separation of the trajectories originated from two nearby initial values when disturbed by the same noise realization. Dechert & Gencay (1990, 1993) and Nychka *et al.* (1992) have discussed the estimation of λ for "noisy" data, using the neural network model based method. Specifically, they have used a functional form of f_d that is motivated by the so-called single-hidden-layer-feed-forward neural network:

$$f_d(z_1, \ldots, z_d) = \beta_0 + \sum_{j=1}^{q} \beta_j \psi\left(\sum_{i=1}^{d} w_{ij} z_i + \omega_{0j} \right),$$

where the β_j and w_{ij} are real parameters to be estimated by e.g. least squares, and ψ is typically a sigmoidal-type function, e.g. $\psi(x) = 1/(1+\exp(-x))$. Substituting these estimates in f_d gives an estimate of it, which we denote $\hat{f}_d$. The Lyapunov exponents are then obtained from the derivative of $\hat{f}_d$. Consistency of the estimates has been claimed by Nychka *et al.* (1992), who have additionally proposed the use of the thin plate spline method extensively discussed by Wahba (1990). They illustrated both methods with the marten fur annual records (on a logarithmic scale) of the Hudson Bay Company over the period 1820–1900, with the conclusion of a negative λ. Using a different approach, Cheng & Tong (1992) have lent some support to the above conclusion. Given the level of noise and the short data length, a positive λ would have been less plausible. However, in view of the increasing interest in "detecting chaos" in ecological data, we would caution against attaching too much scientific value to this kind of analysis unless either there is a substantive model to back it up or we have a substantially greater amount of clean data than is ordinarily available.

From the point of view of interpretation, we must ask the question, is it realistic to assume that the same random shocks/noise sequence will be applied as excitations even if we start a dynamical system with different initial values? If the random shocks are "very small", then the above interpretational problem is probably not so serious because it is then plausible that the two invariant measures are "close" in some sense. However, how small is small? Quite possibly the answer depends on the geometry of the attractors of the skeleton. To date, this question seems to remain unaddressed. Moreover, Jensen (1993, p. 245) has pointed out that the identical-noise device produces different answers if we transform the data (e.g. reading the data on a logarithmic scale as in the marten example); in other words the Lyapunov exponent so defined is not invariant under one-to-one differentiable co-ordinate transformations, in contrast to the noise-free case as discussed in section 2.2. However, it seems that Jensen's objection disappears if we restrict the one-to-one differentiable transformation to simple linear location-scale ones: $x \to ax + b$, a and b being real constants.

3.2. The noisy case

One of the more recent methods dealing with the noisy case is the local Lyapunov exponent (LLE) due to Wolff (1992), who considers $d = 1$ and suggests an estimate of the form

$$\lambda_{i,m} = \frac{1}{m}\frac{1}{n_i}\sum_{j \in S_i} \ln |(X_{i+m} - X_{j+m})/(X_i - X_j)|, \tag{15}$$

where $S_i = \{j: 0 < |X_i - X_j| \leqslant \delta\}$, $n_i = \#(S_i)$, $m \in \mathbf{Z}_+$ and $\delta > 0$. The ingenious idea is to estimate the Lyapunov exponent locally at X_i for lag m and a pre-specified δ representing the perturbation. Wolff (1992) has applied the technique to simulated data and studied the statistical properties of $\lambda_{i,m}$ when the data are from specified models. The models are limited to the case where the embedding dimension is 1. However, the following open issues seem to remain. (i) How do we extend the technique to embedding dimensions higher than 1? In principle, we may explore the use of the Euclidean norm of the state vectors. (ii) Investigate the general sampling properties of $\lambda_{i,m}$ or its simplified version (effected by replacing $X_i - X_j$ by δ), under minimal assumptions such as stationarity, mixing in some sense (e.g. absolute regularity) and finite absolute moments of appropriate order. Clearly the theory of U-statistics for dependent data is relevant. (cf. Aaronson *et al.* (1993) and Denker & Keller (1986).) (iii) Is the LLE invariant under a one-to-one co-ordinate transformation? We suggest that it is not.

3.2.1. The conditional distribution approach

Yao and Tong (1994a, b) have adopted a radically different approach. They consider the sensitivity of the conditional distribution, or one of its characteristics (e.g. the conditional mean), with respect to initial values. Let us consider the conditional distribution approach first. Suppose that the states remain bounded. First, we introduce a kind of "distance" over the conditional distributions with different initial values. This replaces the ordinary (i.e. Euclidean) distance over the states in a deterministic system. One natural "distance" is the (negative) mutual information related to the Kullback–Leibler information. Let $g_m(\cdot \mid x)$ denote the conditional density of X_m given $X_0 = x$. We suppose that $g_m(\cdot \mid x)$ is sufficiently smooth in x. For two neighbouring initial values x, $x + \delta \in R^d$, after time $m \geqslant 1$, the divergence of the conditional distribution of X_m is defined as

$$K_m(x; \delta) = \int \{g_m(z \mid x + \delta) - g_m(z \mid x)\} \ln \{g_m(z \mid x + \delta)/g_m(z \mid x)\}\, dz. \tag{16}$$

Note that stationarity is not required for this equation. Now, for small δ, we can expand the right-hand side using Taylor's series about x. This gives the approximation

$$K_m(x; \delta) = \delta^{\mathrm{T}} I_m(x)\delta + o(\|\delta\|^2), \tag{17}$$

where

$$I_m(x) = \int \dot{g}_m(z \mid x)\dot{g}_m^{\mathrm{T}}(z \mid x)/g_m(z \mid x)\, dz, \tag{18}$$

and $\dot{g}_m(z \mid x)$ denotes $dg_m(z \mid x)/dx$ and $\dot{g}_m^{\mathrm{T}}(z \mid x)$ its transpose. If we treat the initial value x as a parameter vector of the distribution, $I_m(x)$ is the Fisher's information matrix, which represents the information on the initial value $X_0 = x$ contained in X_m. Roughly speaking, the more information X_m brings, the more sensitively the distribution depends on the initial condition. The converse is also true. Suppose that there is a differentiable and one-to-one co-ordinate transformation ϕ from X_t to ξ_t for each t. Then standard statistical theory (see e.g. th. 2.4.1 of Kullback, 1967) gives that

$$K_m(x; \delta) = K_m^*(\phi(x); \phi(x+\delta) - \phi(x)), \tag{19}$$

where $K_m^*(\cdot\,;\cdot)$ denotes the K_m measure in the transformed co-ordinate system. Thus, just like the Lyapunov exponents, the sensitivity measure K_m is invariant under one-to-one differentiable co-ordinate transformations.

3.2.2. *An example*

Let us consider a simple example which allows exact calculations. In (14), let $d = 1$, $F\colon x \to \alpha x$ and ε_t have a Gaussian distribution with zero mean and variance σ^2. Clearly,

$$X_m = \varepsilon_m + \alpha X_{m-1} = \varepsilon_m + \alpha\varepsilon_{m-1} + \cdots + \alpha^{m-1}\varepsilon_1 + \alpha^m X_0. \tag{20}$$

Therefore, given $X_0 = x$, we have $X_m \sim \mathcal{N}(\alpha^m x, \sigma_m^2)$, where

$$\sigma_m^2 = \left(\frac{1-\alpha^{2m}}{1-\alpha^2}\right)\sigma^2, \tag{21}$$

for $\alpha \neq 1$, and

$$\sigma_m^2 = m\sigma^2, \tag{22}$$

for $\alpha = 1$. Then simple calculations yield that

$$K_m(x; \delta) = \delta^2\alpha^{2m}/\sigma_m^2. \tag{23}$$

Equation (23) shows a sensitivity measure which differs from the classical or neo-classical Lyapunov exponent λ in that it incorporates directly the effect of the dynamic noise in the form of a diffusion term σ_m^2 in order to adjust the impact of the disturbance δ on the drift term. Let us measure δ in units of σ (i.e. set $\Delta = \delta/\sigma$); then

$$K_m(x; \Delta) = \Delta^2\sigma^{2m}(1-\alpha^2)/(1-\alpha^{2m}). \tag{24}$$

Note that it is quite natural to use Δ because all measurements can only have limited accuracy, i.e. background noise of one kind or another is forever present. Note also that (23) and (24) are independent of x for this simple example.

To investigate the asymptotic behaviour of K_m, we consider three cases separately as follows.

(i) $|\alpha| < 1$: In this case, $K_m(x; \Delta) \to 0$ as $m \to \infty$. This mimics the behaviour of the globally stable skeleton, i.e. the case with the dynamic noise switched off. Thus, even after clothing the skeleton remains initial-value insensitive.

(ii) $|\alpha| > 1$: In this case, $K_m(x; \Delta) \to \Delta^2(\alpha^2 - 1)$ as $m \to \infty$. It is interesting to note that the limit is positive but finite. Thus, the stochastic model is sensitive to initial value; the sensitivity is clearly induced by the instability of the skeleton. (Recall that stationarity is not required in the definition of $K_m(x; \Delta)$.)

(iii) $|\alpha| = 1$: This has points of contact with the well-known unit-root model in econometrics. In this case, $K_m(x; \Delta) = \Delta^2/m \to 0$ as $m \to \infty$.

The exact result of (23) will not hold if we generalize $F: x \to \alpha x$ to $F: x \to \alpha(x)$ where $\alpha(\cdot)$ is more general than a linear function. It remains an interesting open conjecture that if δ and σ are sufficiently small then (23) will hold approximately but with α replaced by $d\alpha(x)/dx$.

For large σ^2, we envisage that the stochastic noise will generally dominate the whole system thereby submerging the impact of the skeleton. This is certainly true with (23). However, in general how large is large? This is likely to depend on m, F and the noise distribution.

3.2.3. *The conditional mean approach*

To consider the sensitivity of the conditional mean to initial condition, let $F_m(x) = E[X_m \mid X_0 = x]$, $x \in R^d$ and $m \geqslant 1$. For $\delta \in R^d$,

$$F_m(x+\delta) - F_m(x) = \dot{F}_m(x)\delta + o(\|\delta\|), \tag{25}$$

where $\dot{F}_m(x)$ denotes $dF_m(x)/dx^{\mathrm{T}}$, a $d \times d$ matrix. For a model with additive noise, $F_1(x) = F(x)$ and we have from (10) that

$$\begin{aligned} F_m(x) &= E\{F(X_{m-1}) \mid X_0 = x\} \\ &= E\{F(F(X_{m-2}) + e_{m-1}) \mid X_0 = x\} \\ &= E\{F((\cdots(F(x)+e_1)+\cdots)+e_{m-1}) \mid X_0 = x\}. \end{aligned} \tag{26}$$

By the chain rule, matrix differentiation of the right-hand side of (26) gives

$$\dot{F}_m(x) = E\left\{\prod_{k=1}^{m} \dot{F}(X_{k-1}) \mid X_0 = x\right\}, \tag{27}$$

where we have assumed that the differentiation under the integral sign is justified. We may interpret (25) and (27) as stochastic generalizations of their deterministic counterparts, namely (5) and (6) respectively. Similarly, if all the factors on the right-hand side of (27) are of comparable size, then roughly speaking an initial discrepancy of size $\|\delta\|$ will grow (or decay) exponentially with m. Pursuing this argument further, we may arrive at a notion (if it exists) which includes the classical Lyapunov exponent $\lambda(x)$ as a special case. For explicitness, let $d = 1$, although this is not essential. Define

$$\kappa(x) = \lim_{m\to\infty} \left\{\frac{1}{m} \ln |\dot{F}_m(x)|\right\} \tag{28}$$

$$= \lim_{m\to\infty} \left\{\frac{1}{m} \ln \left| E\left[\prod_{i=0}^{m-1} \dot{F}(X_i) \mid X_0 = x\right]\right|\right\}, \tag{29}$$

if the limits exist. Clearly, if $\kappa(x)$ exists, we have

$$|F_m(x+\delta) - F_m(x)| \approx \exp(m\kappa(x))|\delta|, \tag{30}$$

which is the stochastic extension of (7). Clearly, when $e_t \equiv 0$, $\kappa(x)$ reduces to $\lambda(x)$. To date, the existence of $\kappa(x)$ remains an open issue.

So far we have derived some indices which describe the initial-value sensitivities in a stochastic environment. However, this is still some distance from a rigorous definition of chaos for a stochastic dynamical system and we are not aware of any universally accepted and rigorous definition.

3.3. Noise amplification and prediction

Yao & Tong (1994a) have shown that a small dynamic noise could be amplified rapidly through the dynamics if the system is sensitive to initial values. To simplify our discussion, let us restrict ourselves for the moment to a one-dimensional system. Here, $\{\varepsilon_t, t \geqslant 1\}$ is a noise process with

$$E(\varepsilon_t \mid X_k, k < t) = 0, \quad \sigma^2 \equiv \text{Var}(\varepsilon_t) = \text{Var}(\varepsilon_t \mid X_k, k < t). \tag{31}$$

It follows that $E(\varepsilon_t \varepsilon_s \mid X_k, k < t) = 0$ for all $t > s$. We also assume that for all $t \geqslant 1$, $|\varepsilon_t| < \zeta$ (almost surely), where $\zeta > 0$ is a small constant. By Taylor's expansion, it is easy to see that for $m \geqslant 1$,

$$\begin{aligned} X_m &= F(X_{m-1}) + \varepsilon_m \\ &= F(F(X_{m-2}) + \varepsilon_{m-1}) + \varepsilon_m \\ &= F^{(2)}(X_{m-2}) + \varepsilon_m + \dot{F}(F(X_{m-2}))\varepsilon_{m-1} + O(\zeta^2) \\ &= \cdots \\ &= F^{(m)}(x) + \varepsilon_m + \dot{F}(F^{(m-1)}(x))\varepsilon_{m-1} + \cdots + \left\{\prod_{k=1}^{m-1} \dot{F}(F^{(k)}(x))\right\}\varepsilon_1 + O(\zeta^2). \end{aligned} \tag{32}$$

Let $\sigma_m^2(x) = \text{var}(X_m \mid X_0 = x)$, which monitors the performance of the least-square predictor, $F_m(x)$. Then $\sigma_1^2(x) \equiv \sigma^2$ and for $m > 1$,

$$\sigma_m^2(x) = \mu_m(x)\sigma^2 + O(\zeta^3), \tag{33}$$

where

$$\mu_m(x) = 1 + \sum_{j=1}^{m-1} \left\{\prod_{k=j}^{m-1} \dot{F}(F^{(k)}(x))\right\}^2. \tag{34}$$

Some remarks are now in order.

(i) The fact that $\sigma_m^2(x)$ is dependent on x shows that how well we can predict depends on where we are. Herein lie windows of opportunity for substantial reduction in prediction errors if the present state is in the right place of the state space. Linear predictors have failed completely to grasp this truth, known so well to the man in the street!

(ii) If $|\dot{F}(x)| > 1$ for a large range of values of x, $\mu_m(x)$ can be very substantial for moderate (and perhaps even small) m. The consequent and rapid increase of $\sigma_m^2(x)$ with m is a manifestation of noise amplification. In such cases, only very short range prediction can be entertained. Thus, how far ahead we can predict reliably also depends on the current position in the state space.

(iii) Noise amplification is for ever present in a stochastic system except for the most trivial cases (e.g. $\{X_t\}$ is a white noise process) because $\mu_m(x) > 1$ for almost all x and for all m.

(iv) Equation (34) implies that

$$\mu_{m+1}(x) = 1 + \mu_m(x)\{\dot{F}(F^{(k)}(x))\}^2. \tag{35}$$

Thus, $\mu_{m+1}(x) < \mu_m(x)$ if $\{\dot{F}(F^{(m)}(x))\}^2 < 1 - 1/\mu_m(x)$. By (33), it is possible that for such x and m, $\sigma^2_{m+1}(x) < \sigma^2_m(x)$. Thus, standing at certain parts of the state space, the error of the $(m+1)$-step-ahead least-square prediction could be smaller than that of the m-step-ahead least-square prediction with non-trivial probability. If $F(\cdot)$ is linear, $\dot{F}(\cdot)$ is a constant and the remainder on the right-hand side of (33) is zero. By assumption (31), the noise is homogeneous, which then implies that for linear $F(\cdot)$, $\sigma^2_m(x)$ does not depend on x and is monotonically increasing in m. Until quite recently, these properties of state-independence and monotonicity of σ^2_m, which are enjoyed by the linear case, have been taken for granted as being universal for all cases. Such a belief is clearly unfounded and is the result of being a slave to linearity for too long.

(v) Almost for the first time, we are now beginning to be in a position to address the following really important practical issues in prediction: (a) Given the present position of the state, can we trust our prediction enough to make a sensible decision (e.g. whether to invest or not)? (b) Given the present position of the state, what are the lead times for "reliable" prediction? Note that these issues are typically state-dependent. Smith (1994a) has addressed similar issues for the noise-free and low-noise situations.

3.4. A decomposition theorem

Consider the stochastic dynamical system

$$Y_t = f(Y_{t-1}, Y_{t-2}, \ldots, Y_{t-d}) + \varepsilon_t \tag{36}$$

$$= f(X_{t-1}) + \varepsilon_t, \tag{37}$$

where $f: R^d \to R$ is smooth, and $\{\varepsilon_t, t \geqslant 1\}$ satisfies (31). As usual, X_t denotes the vector $(Y_t, Y_{t-1}, \ldots, T_{t-d+1})^{\mathrm{T}}$. Let $\hat{f}_m$ be any mean-square consistent estimator of $f_m = E[Y_m \mid X_0]$ based on the observations $(X_d, X_{d+1}, \ldots, X_N)$. If we think that our current position is at $x \in R^d$, then our natural m-step-ahead prediction would be $\hat{f}_m(x)$. How well this prediction performs in the mean-square sense is, of course, measured by $\sigma^2_m(x)$. Suppose that the true current position is at $x + \delta \in R^d$ instead, $\|\delta\|$ being small. What effect would this have on the prediction performance? This is a relevant question in practice because we rarely know where we are exactly. Yao & Tong (1994a) have proved that, under general conditions, the following decomposition holds:

$$\lim_{N \to \infty} E\{[Y_{N+m} - \hat{f}_m(x)]^2 \mid X_N = x + \delta\} = \sigma^2_m(x+\delta) + \{\delta^{\mathrm{T}} \dot{f}_m(x)\}^2 + R_m, \quad \text{(almost surely)}, \tag{38}$$

where $\dot{f}_m(x) = df_m(x)/dx^{\mathrm{T}}$, and $R_m = o(\|\delta\|^2)$ as $\|\delta\| \to 0$.

A few remarks are now in order.

(i) In the presence of inexact information concerning the current position x, the mean-squared error of prediction is inflated by the factor $\{\delta^{\mathrm{T}} \dot{f}_m(x)\}^2$, which is clearly related to the sensitivity of the underlying skeleton via $\dot{f}_m(x)$. Specifically $\dot{f}_m(x)$ is equal to the transpose of the first row vector of the matrix $\dot{F}_m(x)$ in (25). In this sense, dynamical system considerations have benefited statistics.

(ii) In return, statistics has highlighted the significance of the dynamic noise, without which the term $\sigma^2_m(\cdot)$ would be absent. Note that the first and second terms on the right-hand side of (38) are of the orders σ^2 and $\|\delta\|^2$ respectively. If f is linear (and stable), then $\dot{f}_m(x)$ is a constant vector with norm less than one and the term $\{\delta^{\mathrm{T}}\dot{f}_m(x)\}^2$ can therefore be ignored if $\|\delta\| \ll \sigma$. However, for a system with large $\dot{f}_m$ over some parts of the state space, the term $\{\delta^{\mathrm{T}}\dot{f}_m(x)\}$ can be quite substantial.

4. Statistical estimation

We now turn our attention to the estimation, from the observations $\{Y_t: 1 \leqslant t \leqslant N\}$, of the functions $f_m(\cdot)$, $\sigma_m^2(\cdot)$ and $K_m(x; \delta)$ and the dynamical system invariants such as the order d and others. We shall concentrate on the dynamic noise case only. Berliner (1991, 1992) and Lele (1994) have considered the measurement noise case.

4.1. Locally linear non-parametric regression

We can approach the estimation of the above functions and invariants in at least two ways: the parametric and the non-parametric. In the former approach, we may use the plug-in method to obtain estimates from the fitted model. Tong (1990) has given a fairly comprehensive account of parametric non-linear time series models and modelling. Tjøstheim (1994) has argued quite elegantly that the non-parametric approach has an important role to play in non-linear time series analysis and the recent results of Casdagli (1992), Cheng & Tong (1992, 1994), Sugihara & May (1990), Yao & Tong (1994a, b) and others suggest that this continues to be the case from the chaos perspective. As an illustration, for the estimation of d a non-parametric approach would be more appropriate because a parametric model might introduce some unquantifiable bias. Against the non-parametric approach, we must mention the well-known curse of dimensionality and the fact that they tend to produce smooth-looking effects out of "nothing".

To begin with, a well-known estimate of $f_m(\cdot)$ is the Nadaraya–Watson kernel estimate:

$$\tilde{f}_m(x) = T_0(x)S_0(x)^{-1}, \tag{39}$$

where

$$S_0(x) = \frac{1}{N-m}\sum_{t=1}^{N-m} P_d\left(\frac{x-X_t}{h}\right), \tag{40}$$

$$T_0(x) = \frac{1}{N-m}\sum_{t=1}^{N-m} Y_{t+m}P_d\left(\frac{x-X_t}{h}\right). \tag{41}$$

Here and elsewhere, $P_d(\cdot)$ denotes a smoothing kernel in the form of a well-behaved probability density function on R^d and $h = h(d; N)$ is a bandwidth satisfying the standard conditions. Clearly, $\tilde{f}_m(\cdot)$ is the minimizer of the following weighted sum of squares over the space of well-behaved functions:

$$\sum_{t=1}^{N-m} \{Y_{t+m} - f_m(x)\}^2 P_d\left(\frac{x-X_t}{h}\right). \tag{42}$$

The method of the locally linear non-parametric regression (see e.g. Fan, 1992) was initially motivated to reduce the bias of the estimate $\tilde{f}_m(\cdot)$ by modifying the above weighted sum to

$$\sum_{t=1}^{N-m} \{Y_{t+m} - f_m(x) - \dot{f}_m^T(x)(X_t - x)\}^2 P_d\left(\frac{x-X_t}{h}\right), \tag{43}$$

leading to

$$\hat{f}_m'(x) = \{T_0(x) - S_1^T(x)S_2^{-1}(x)T_1(x)\}\{S_0(x) - S_1^T(x)S_2^{-1}(x)S_1(x)\}^{-1}, \tag{44}$$

$$\hat{f}_m'(x) = \{S_1(x)T_0(x)S_0^{-1}(x) - T_1(x)\}\{S_2(x) - S_1(x)S_1^T(x)S_0^{-1}(x)\}^{-1}, \tag{45}$$

$$S_1(x) = \frac{1}{N-m}\sum_{t=1}^{N-m}(x - X_t)P_d\left(\frac{x - X_t}{h}\right), \tag{46}$$

$$S_2(x) = \frac{1}{N-m}\sum_{t=1}^{N-m}(x - X_t)P_d\left(\frac{x - X_t}{h}\right)(x - X_t)^T, \tag{47}$$

$$T_1(x) = \frac{1}{N-m}\sum_{t=1}^{N-m}(x - X_t)Y_{t+m}P_d\left(\frac{x - X_t}{h}\right). \tag{48}$$

(For technical reasons, we may sometimes replace $S_0(x)$ by $S_0(x) + h^2$ on the right-hand side of (44), which has no material effect for large N.) We thus gain a useful estimate of $\hat{f}_m'(\cdot)$ as a spin-off. Our experience suggests that a numerical differentiation of $\tilde{f}_m(\cdot)$ does not usually lead to a useful estimate of $\dot{f}(\cdot)$. On the other hand, it is possible to consider locally quadratic alternatives and so on. It would be an interesting exercise to investigate if computation-time/improvement considerations might show that local linearity is a reasonable compromise. We may similarly obtain an estimate for $E[Y_m^2 \mid X_0 = x]$ and from which an estimate for $\sigma_m^2(\cdot)$. Yao & Tong (1994a, b) have shown that these estimates are consistent under general conditions and have also illustrated their finite sample behaviour with simulations and real data.

Fan *et al.* (1993) have shown that, by choosing an appropriate weighted sum of squares, the locally linear non-parametric regression methodology can be used to obtain consistent estimates of $K_m(x; \delta)$. The key lies in observing that, for small h,

$$E\left[\frac{1}{h}P_1\left(\frac{Y - y}{h}\right)\middle| X = x\right] \approx g(y \mid x). \tag{49}$$

They have also studied the central limit properties.

Clearly, like almost all non-parametric function fitting, of which the above estimation problem is one, we cannot avoid the curse of dimensionality. Thus the above estimates are practically useful only for small d (see also Bosq & Guégan, 1994; Cheng and Tong, 1993).

4.2. *Order determination*

Clearly, when $m = 1$, $\tilde{f}_1(\cdot)$ is an estimate of the autoregressive function denoted by $f_d(\cdot)$ in (14), where the suffix d denotes the dimension of X_t. For consistency, let us denote $\tilde{f}_1(\cdot)$ by $\tilde{f}_{1,d}(\cdot)$ or more simply (by abusing the notation) $\tilde{f}_d(\cdot)$ henceforth. Using $\tilde{f}_d(\cdot)$, we may obtain an "estimate" of the innovation sequence ε_t, say $\hat{\varepsilon}_{t,d}$, more traditionally called the fitted residuals. Obviously, a normalized sum of squares of the fitted residuals, say $RSS(d)$, monitors the goodness of fit of the model: $Y_t = \tilde{f}_d(X_{t-1}) + \hat{\varepsilon}_t$ to the data. By penalizing the $RSS(d)$ in a manner similar to Akaike's final prediction error, Auestad & Tjøstheim (1990) obtained a criterion, which they also called the FPE-criterion for the determination of the order. Around the same time, Tong (1990) proposed a criterion based on the cross-validation approach: first delete X_{t-1} in computing T_0 and S_0 for $\tilde{f}_d$ to get $\tilde{f}_{d,\setminus t}$ say, then replace $\hat{\varepsilon}_{t,d}$ by the modified residual $\check{\varepsilon}_{t,d} = Y_t - \tilde{f}_{d,\setminus t}(X_{t-1})$ in $RSS(d)$ to get $CV(d)$, say. Cheng & Tong (1992) have proved that, under general conditions, minimizing $CV(d)$ over a suitable set of positive integers leads to a consistent estimate of an optimal order for bounded time series. By establishing a connection between the CV-criterion and the above FPE-criterion, they have also proved the consistency of the estimate obtained by the latter method. For "over-sampled" data, we may have to delete k observations, where $k > 1$ (see e.g. Cheng & Tong, 1993). However, the optimal choice of k seems an open question.

Yao & Tong (1994c) have extended the CV-method to subset stochastic regressor selection and Tjøstheim (1994) and Tjøstheim & Auestad (1994) have summarized the various extensions of the FPE-method to similar problems. Of particular note is the fact that we can now adequately handle the estimation of the delay parameter (assumed to be an integer), say τ, in a non-linear open-loop system typified by $V_t = g(U_{t-\tau}) +$ noise, where g is non-linear with an unknown functional form. Linear methods such as cross-spectral analysis or cross-correlation analysis would be powerless even for the simple case where g is quadratic, U_t has a symmetric distribution and the noise is white and has a symmetric distribution! (see Yao & Tong, 1994c).

As mentioned above, $\tilde{f}_d(\cdot)$ suffers from the curse of dimensionality. On top of this is the well-known fact that neither the FPE-criterion (as given by Akaike) nor its cross-validatory equivalent (in an asymptotic sense) gives a consistent estimate of d for bounded time series (i.e. Y_t is almost surely finite), when we know that the model is a linear autoregression. Yet, we now get consistency despite the fact that (i) we still use cross-validation and (ii) we do not even know the functional form of the autoregression! This surprising result (there is one more to come) is firstly due to the benefit of kernel smoothing. Recall that a similar kernel smoothing produces a consistent estimate of the spectral density from the periodogram, which is an inconsistent estimate. Secondly, when estimating d, Cheng & Tong (1994) have shown that "faithfulness" of $\tilde{f}_d(\cdot)$ to $f_d(\cdot)$ is only of secondary importance. What is really important is the simple geometric fact typified by the cylinder depicted by $(E[Y_t \mid Y_{t-1}, Y_{t-2}], Y_{t-1}, Y_{t-2})$ if $\{Y_t\}$ is a first order non-linear autoregressive process. The determination of d thus becomes an exercise of cylinder hunting, which turns out to be quite manageable, so much so that we even get consistency. Needless to say, all the usual cautionary remarks (e.g. Tong, 1990) regarding the use of model selection/order determination tools apply here.

Although consistency is theoretically comforting, we must enquire its relevance in specific cases. In the example of an open-loop system, a consistent estimate of the time-delay may enable us to design a more efficient controller in the context of control engineering. Using the wrong delay could lead to instability of the whole control system! In the context of chaos, a consistent estimate of the order may lead to a better assessment of the variability of the estimate of the attractor dimension (to be described shortly). On the other hand, our experiences so far suggest that the precise value of d might not be critical in assessing the initial-value sensitivity in stochastic dynamical systems. However, we have no theoretical result to support this statement.

Naturally, we would expect to pay a price somewhere. The price lies in a greater sample size requirement than the case when we know that the model is for instance linear. How much greater? Before answering this question, we note that it is well known that an exponential sample size is required to produce a usable estimate of the correlation dimension, say θ, of the attractor, which is considered a very important invariant to estimate by the dynamicists. A conservative estimate has actually put it as high as 42^{θ} although some moderation of the base number seems possible without sacrificing the accuracy too much. (By definition,

$$\theta = \lim_{r \to 0} \frac{\ln \Pr(\|X - Y\| \leqslant r)}{\ln r},$$

where the max norm is used and X and Y are independently and identically distributed with the ergodic probability measure of the dynamical system. The limit is assumed to exist.) Note that in the physical literature, sample sizes of the order of 10^6 and beyond are not uncommon. Against this background, the second surprise is that the sample size requirement for the CV or FPE determination of d is ordinarily only quadratic in d. Specifically, Cheng & Tong (1994) have given the empirical formula

$$\text{order} \leqslant \frac{\sqrt{N} \times (\text{“failure rate”})^2}{\text{normalized dynamic noise variance}}. \tag{50}$$

(The "failure rate" is set by the user; for example, he might be prepared to tolerate 1 "wrong" order estimate in 20.) This may be compared with their suggestion which replaces $\sqrt{N}$ by N in the above empirical formula if the model is known to be a linear autoregression.

4.3. *Attractor dimension*

If we give a dynamicist (or a time series analyst) time series data say $\{Y_1, \ldots, Y_N\}$, which are either real or artificially generated by a map unknown to him, almost the first thing he would do is to obtain $\{X_1^{(d)}, \ldots, X_N^{(d)}\}$ by the usual stacking (ignoring the edge effect for convenience); he calls this a delay-coordinate construction for which Takens' embedding theorem provides the theoretical background. More often than not he would follow this by estimating the correlation dimension, θ, with a view to saying something about the existence or otherwise of an exotic attractor of a low (perhaps fractional) dimension. Probably the single most frequently used statistic in the physical literature related to chaos is the so-called Grassberger–Procaccia sample correlation integral, which may be defined as

$$C_{N,d}(r) = \frac{2\,\#\{(i,j)\colon 1 \leqslant i < j \leqslant N,\ \|X_i^{(d)} - X_j^{(d)}\| \leqslant r\}}{N(N-1)}, \tag{51}$$

where the max norm is used. The essence of the Grassberger–Procaccia method is to plot $\ln C_{N,d}(r)$ against $\ln r$ and obtain an estimate of θ by using the slope of a suitably fitted straight line on the log–log plot over a suitably chosen range of the abscissae. This exercise is repeated over a range of trial values of d until some sort of stability is achieved. The decision process involved is actually quite delicate. Cutler (1993) has given an accessible and fairly comprehensive account of the estimation of fractal dimensions including the correlation dimension. We have already mentioned the exponential sample size requirement in the last sub-section. In practical terms, the calculations of R. L. Smith (1992) suggest that to achieve a failure rate of 10%, the sample size requirement for the determination of θ would be roughly 5×10^3 for $\theta = 5$, 3×10^6 for $\theta = 10$ and 2.5×10^9 for $\theta = 15$. (Note that the figure of $10^{1.60}$ obtained by L. A. Smith (1988) was based on different assumptions from the ones used by R. L. Smith (1992).) In contrast, the simulation studies of Cheng & Tong (1994) suggest that to achieve the same failure rate, the sample size requirement for the determination of d would be roughly 500 for $d = 4$ and $< 1.5 \times 10^3$ for $4 < d \leqslant 13$. We should, of course, emphasize the fact that the above numerical values were obtained under different assumptions. Nevertheless, they lend some support to the statement that to achieve a reasonably low failure rate in the determination, we would need substantially more observations for θ than for d. We have argued elsewhere (Cheng & Tong, 1992) that logically speaking we should determine the dimension in which the attractor lives before we determine the dimension of the attractor. Sugihara & May (1990) seem to share the same view. More importantly, the comparatively modest sample size requirement for the estimation of d should tilt the balance in favour of estimating d first on statistical grounds.

The popularity of correlation dimension estimation in the dynamical systems literature is perhaps understandable in view of the fact that the Fortran algorithm of Grassberger (1990) and others have made the sample correlation integral "computationally readily accessible". On the theoretical front, convergence results are now available for the sample correlation integral constructed from an observed time series which is a projection from an ergodic dynamical system or a stochastic process (see e.g. Pesin, 1993; Serinko, 1995; Aaronson *et al.*, 1993). Perhaps more importantly from an estimation or practical viewpoint, Denker & Keller (1986) have established standard errors as well as asymptotic normality under suitable mixing conditions. More recently, Cutler (1994) has studied the whole question more systematically for stationary time series.

Despite the tremendous amount of interests and numerous results, especially in the dynamical systems literature, we believe that the practical importance of the correlation dimension for real data is rather limited. It is our view that other dynamical system invariants, e.g. the embedding dimension, the information dimension (viewing a chaotic dynamical system as an information creation system) and some others, deserve more attention in the dynamical systems literature, relative to the correlation dimension. In fact, it is known that for some noise-free maps the correlation dimension equals the embedding dimension, e.g. for the logistic map with coefficient equal to 4, they are both equal to one. It is also known by a result due to Wolff (1990), Hansen (1992) and more generally Cutler (1994) that in a stochastic process

$$\frac{\ln \Pr(\|X - Y\| \leqslant r)}{\ln r}$$

may increase without bound as $r \to 0$. (Cutler, in private communications, has informed us that it is generally believed among the dynamicists that this is the rule rather than the exception.) Thus, when it comes to the analysis of real (i.e. dynamically noisy) data, it remains to be seen whether we need estimate the correlation dimension at all and, even if we do, as routinely as the dynamicists have been doing.

4.4. Map reconstruction

4.4.1. Substantive modelling and black-box modelling

Given observations $\{Y_1, \ldots, Y_N\}$, we are often interested in estimating the underlying dynamics. The dynamicists describe this activity as map reconstruction. There are essentially two separate approaches: the substantive approach and the black-box approach. In the substantive approach, we know from subject matter considerations the functional form of the underlying map. The dynamics can sometimes take the form of differential equations, partial differential equations, functional equations or others. In this approach, with the functional form given by substantive considerations, a primary use of statistics is the fitting of an appropriate member from the (usually parameterized) family of functions sharing this same functional form. The problem is reduced to parameter estimation. In the physical and biological literature, the unknown parameters are sometims determined (rather than estimated in the conventional statistical sense) by reference to external information coupled with the trial-and-error device; the fitted model is then used to generate artificial realizations. Visual comparison of these with the real data is then used to check if there are any marked discrepancies. The emphasis here is on qualitative match rather than quantitative goodness of fit. Sometimes the scales are ignored during the matching exercise by means of suitable standardization.

Quite often, subject matter considerations may lead to a rather complicated set of equations (e.g. the delayed simultaneous partial differential equations used to model the magnetohydrodynamics of sunspot creation) which do not readily yield a usable functional form for the dynamics. At other times, we simply have incomplete knowledge of the subject matter (e.g. in most areas of the so-called "soft" sciences such as economics, finance and others). In these circumstances, the black-box approach provides a pragmatic alternative to the substantive approach. We emphasize that only a substantive model, if available, can represent a comprehensive understanding of the dynamics. Having said this, we note that a black-box model does have its use. For example, it can provide a check on the substantive model (if available) by comparing its predictions with those given by the latter. The following analogy might be suggestive. If we view the more analytical approach of modern Western medicine as an example of a substantive approach, then the intuitive approach of traditional Chinese medicine (with a two thousand-year-old history) is an example of a black-box approach. The fact that these two types of medicine have been in peaceful co-existence with each other since the introduction of the former into China, Japan, Korea and Southeast Asia, tends to suggest the advantage of having both approaches.

We shall briefly survey the black-box approach and divide the models there into roughly three classes: (i) global, (ii) local and (iii) semi-local. Of course, the division is not always clear-cut. The focus here is on the estimation of the conditional mean function, i.e. $f_d(\cdot)$ of (14) and the problem is essentially one of extracting information from high dimensional data, bearing in mind the curse of dimensionality. How successful each method is in handling the curse seems to require more research. The area has expanded extremely rapidly in recent years and we shall only mention some of the major developments. We shall omit neural networks partly because, within the context of non-linear time series, they tend to address mainly prediction problems. Other omissions are obviously unavoidable.

4.4.2. Global function approximations

Global function approximations certainly include most of the parametric models discussed in Tong (1990). The recent results of Chan & Tong (1994) might revive interests in the class of polynomial autoregressive models driven by white noise with finite support since such models can be stationary. A relevant example is Cox (1977), who suggested a quadratic autoregressive model for the Canadian lynx data, which was further analysed in Chan & Tong (1994). We note that several standard maps are polynomials in the state variable.

4.4.3. Local function approximations

In recent years, the physical literature sees substantial developments in local function approximations. These are commonly based on the idea of 'divide and rule', i.e. decomposing the state space into regimes and then fitting a simple (sub) model to each regime. Following Tong (1987, 1990), we may refer to this idea by the name the threshold principle. Historically, the new paradigm of local function approximation in the context of non-linear time series analysis and the modelling of chaotic data was probably inaugurated by the threshold autoregressive model introduced by Tong in a series of papers, starting from 1977 (see Tong, 1990, for an historical account). His basic idea was the introduction of an indicator time series $\{J_t\}$, which effects the state space decomposition. To emphasize the role of J_t, he has used the symbolism $\{X_t; J_t\}$. He has stressed the flexibility of $\{J_t\}$ in order to accommodate as broadly as possible the almost unlimited number of ways of decomposing the state space. At one extreme, $J_t = 1$, all t, in which case we have a global model. If, in particular, the submodel is a linear model, then we have a (globally) linear model. At the other extreme, $\{J_t\}$ may be a time series unaffected by the outcomes of the time series $\{X_t\}$. For example, $\{J_t\}$ may be a "hidden Markov chain" (see e.g. Tong, 1983, 1990). This then gives what is sometimes called the Markov switching model in the econometric literature (see e.g. Hamilton, 1989) as a special case.

Recently, there have been exciting new implementations of the threshold principle. Farmer & Sidorowich (1987), Mees (1989), Sugihara & May (1990) and others have used the nearest neighbours to divide the state space. Lewis & Stevens (1991) and Lewis & Ray (1993) have adapted Friedman's multivariate adaptive regression splines (MARS) to effect a recursive partition of the state space to produce autoregressive functions which are continuous (but not differentiable) in the form of piecewise polynomials. Both methods can be computationally quite intensive, however.

4.4.4. Semi-local function approximations

A local function approximation enjoys the advantage of adhering to the local shape of an arbitrary surface associated with high-dimensional dynamics. The disadvantage of the approach is that if taken to the extreme it can lead to a grossly over-parameterized model lacking in a compact description. A global function approximation gives a compact description at the risk of larger badness-of-fit. Therefore in applying the threshold principle much attention should be paid to the number of regimes. In a sense, a threshold model with moderate number of regimes (relative to the sample size) could be regarded as a model sitting somewhere between a highly localized function approach and a global one.

Casdagli (1989) has borrowed the idea of radial basis function from numerical analysis and developed it into what is sometimes described as a semi-local function approximation method. It is currently quite popular with the dynamicists. He has suggested that the method is a global interpolation technique with good localization properties. The essential ingredient is to express

$$\hat{f}_d(x) = \sum_{j=1}^{N_c} \lambda_j \phi(\|x - x_j^c\|), \tag{52}$$

where $\phi(r)$ is a radial basis function of the form such as $\exp(-r^2/c)$, r, r^3, $\sqrt{r^2+c}$ and others. Here, $\{x_j^c : j = 1, \ldots, N_c, x_j^c \in R^d\}$ is a set of centres, whose choice is at the disposal of the user and often holds the key to the success of the method. The simplest form of the centres is to choose each centre from a learning set which has been put aside in a manner similar to cross-validation. L. A. Smith (1992) and Judd & Mees (1994) have given more sophisticated forms of the centres. Note that there is a link between the radial basis method and the Nadaraya–Watson estimate. If the Gaussian kernel is used, then the latter estimate is equivalent to the former with $\phi(r) = \exp(-r^2/2)$.

5. Some recent applications

The four references at the start of this paper and others quoted in later parts of this paper contain many examples of practical applications of chaos. However, most of these are from the deterministic standpoint. There is clearly a need for much more substantial input from the statistical community. Below we shall select only a few recent examples, all except one of which have substantial statistical content, to illustrate the potential.

It seems that epidemiological data, such as measles data from New York (monthly data from 1928 to 1963) and other cities have provided fertile material for continuing discussions in both the statistical and the dynamical systems literature. For a while, the focus was on the question of the existence of "chaos" in the measles epidemic. Given the absence of a generally accepted definition of chaos in either a deterministic system or a stochastic system, the issue tended to be concerned with the existence of a high-dimensional attractor vs a low-dimensional attractor, with the latter taken to indicate a predominantly deterministic system. More recent interests tend to focus on the impact of immunization on the measles dynamics and the possible spatial interaction of measles dynamics over neighbouring sites. For some references, see e.g. Bartlett (1990), Sugihara & May (1990), Grenfell (1992), Isham (1993), several of the papers edited by Tong & Smith (1992) and by Grenfell *et al.* (1994), and the references therein. The analysis of the famous New York measles data is made particularly difficult by the fact that the data suggest quite a high (about 7) dimensional (ambient) state space and we have only about 432 data points. As Grenfell (1992) has rightly pointed out, we should not spare our efforts in the analysis of well-documented measles records available in the developed countries because they might in due course lead to an understanding of the measles dynamics to such an extent that it could begin to have an impact on the health care in developing countries.

Population biologists are quite fascinated by animal population cycles and seem to show increasing interests in using some of the recent non-linear time series techniques beyond those described in e.g. Royama (1992). As an interesting historical aside (see e.g. Tong, 1990) the population biologists' Moran diagram/reproductive curve pre-dated the dynamicists' delay map but post-dated Yule's plot for the sunspot numbers, which the latter mistakenly attributed to Wölfer with the result that even now many subsequent papers have used the wrong name too! We have made some passing comments on the detection of initial-value sensitivities in animal population data in section 3.1. The classic Canadian lynx data have been repeatedly (some might say over) analysed perhaps due to its "unusually long record" in this context. More recently, Yao & Tong (1994a) have shown that for these data some parts of the state space are more predictable than others. Unfortunately, many animal population data sets tend to be rather short, often fewer than 100 points.

Recently, the new area of non-linear signal processing and communications is fast emerging with great promise. So far the area seems to be barely visited by the statisticians. Given the long history of interaction between signal processing and statistics, we suggest that new opportunities should not be missed. In the volume edited by Titterington (1994), the paper by Broomhead and the paper by Mars, Chen & Thorneley have given the general flavour. For example, we may use a chaotic signal to mask our intended signal before its transmission, the chaotic mechanism being known to both the transmitter and the receiver only. Imagine the potential of this method of cryptography! It was Christian Huygens who first discovered the phenomenon of synchronization by observing that two clocks placed on a piece of soft wood became in step whilst they had previously been slightly out of step when hung on a wall. Engineers are experimenting with synchronization using chaotic signals.

6. Conclusion

As we have said elsewhere, there is no doubt that chaos takes us to the Kingdom of the True, the Good and the Beautiful. He who goes on a pilgrimage to the Kingdom will be greatly rewarded.

Acknowledgements

This paper was partially supported by a grant from the EPSRC of the United Kingdom under their Complex Stochastic Systems initiative. I thank Sir David Cox and Professors Kung-sik Chan, Colleen Cutler and Floris Takens for comments.

References

Aaronson, J., Burton, R., Dehling, H., Gilat, D., Hill, T. & Weiss, B. (1993). Law of large numbers for U-statistics of dependent observations. Preprint.

Auestad, B. & Tjøstheim, D. (1990). Identification of nonlinear time series: first order characterization and order determination. *Biometrika* **77,** 669–688.

Bartlett, M. S. (1990). Chance or chaos (with discussion). *J. Roy. Statist. Soc.* **A153,** 321–347.

Berliner, L. M. (1991). Likelihood and Bayesian prediction for chaotic systems. *J. Amer. Statist. Assoc.* **86,** 938–952.

Berliner, L. M. (1992). Statistics, probability and chaos. *Statist. Sci.* **7,** 69–122.

Bosq, D. & Guégan, D. (1994). Nonparametric estimation of the chaotic function and the invariant measure of a dynamical system. Technical Report INSEE-ENSAE, 92241 Malakoff Cedex, France.

Casdagli, M. (1989). Nonlinear prediction of chaotic time series. *Physica D* **35,** 335–356.

Casdagli, M. (1992). Chaos and deterministic versus stochastic nonlinear modelling. *J. Roy. Statist. Soc. Ser. B* **54,** 303–328.

Chan, K. S. & Tong, H. (1985). On the use of the deterministic Lyapunov function for the ergodicity of stochastic difference equations. *Adv. Appl. Probab.* **17,** 666–678.

Chan, K. S. & Tong, H. (1994). A note on noisy chaos. *J. Roy. Statist. Soc. Ser. B* **56,** 301–311.

Chan, K. S. & Tong, H. (1995). *Chaos: a statistical approach.* Springer-Verlag, Heidelberg, in press.

Chatterjee, S. & Yilmaz, M. R. (1992). Chaos, fractals and statistics. *Statist. Sci.* **7,** 49–121.

Cheng, B. & Tong, H. (1992). Consistent nonparametric order determination and chaos—with discussion. *J. Roy. Statist. Soc. Ser. B* **54,** 427–449, 451–474.

Cheng, B. & Tong, H. (1993). Nonparametric function estimation. In *Developments in time series analysis* (ed. T. SubbaRao), 183–206. Chapman & Hall, London.

Cheng, B. & Tong, H. (1994). Orthogonal projection, embedding dimension and sample size in chaotic time series from a statistical perspective. *Phil. Trans. Roy. Soc. (Lond.)* **A348,** 325–341.

Cox, D. R. (1977). Discussion of papers by Campbell and Walker, Tong and Morris. *J. Roy. Statist. Soc. Ser. A* **140,** 453–454.

Cox, D. R. & Smith, W. L. (1953). The superposition of several strictly periodic sequences of events. *Biometrika* **40,** 1–11.

Crutchfield, J. P., Farmer, J. D. & Huberman, B. A. (1982). Fluctuations and simple chaotic dynamics. *Phys. Rev.* **92,** 45–82.

Cutler, C. D. (1993). A review of the theory and estimation of fractal dimension. In *Dimension estimation and models* (ed. H. Tong), 1–107. World Scientific, Singapore.

Cutler, C. D. (1994). A theory of correlation dimension for stationary time series. *Phil. Trans. Roy. Soc. (Lond.)* **A348,** 343–355.

Dechert, W. D. & Gencay, R. (1990). Estimating Lyapunov exponents with multilayer feedforward network learning. Technical Report, Department of Economics, University of Houston, TX.

Dechert, W. D. & Gencay, R. (1993). Lyapunov exponents as a nonparametric diagnostic for stability analysis. In *Nonlinear dynamics, chaos and econometrics* (eds M. Hashem Pesaran & S. M. Potter), 33–52. Wiley, New York.

Denker, M. & Keller, G. (1986). Rigorous statistical procedure for data from dynamical systems. *J. Statist. Phys.* **44,** 67–93.

Drazin, P. G. & King, G. P. (1992). *Interpretation of time series from nonlinear systems.* North-Holland, Amsterdam.

Fan, J. (1992). Design-adaptive nonparametric regression. *J. Amer. Statist. Assoc.* **87,** 998–1004.

Fan, J., Yao, Q. & Tong, H. (1993). Estimating measures of sensitivity to initial values in nonlinear stochastic systems with chaos. Technical Report, Institute of Mathematics and Statistics, University of Kent, UK, December 1993.

Farmer, J. D. & Sidorowich, J. J. (1987). Predicting chaotic time series. *Phys. Rev. Lett.* **59,** 845.

Grassberger, P. (1990). An optimized box-assisted algorithm for fractal dimensions. *Phys. Lett.* **50,** 345–349.

Grenfell, B. T. (1992). Chance and chaos in measles dynamics. *J. Roy. Statist. Soc. Ser. B* **54,** 383–398.

Grenfell, B. T., May, R. M. & Tong, H. (1994). *Phil. Trans. Roy. Soc. (Lond.)* **A348** Discussion Meeting on Chaos and Forecasting (2–3 March 1994) organized and edited by B. T. Grenfell, R. M. May & H. Tong, pp. 325–530.

Hamilton, J. D. (1989). A new approach to economic analysis of non-stationary time series and business cycles. *Econometrika* **57,** 357–384.

Hansen, M. B. (1992). The behaviour of the correlation integral in the nonlinear time series case. Preprint.

Hao, B. L. (1990). *Chaos II.* World Scientific, Singapore.

Herzel, H-., Ebeling, W. & Schumeister, Th. (1987). *Z. Naturforsch.* **42,** 136.

Isham, V. (1993). Statistical aspects of chaos: a review. In *Networks and chaos—statistical and probabilistic aspects* (eds O. E. Barndorff-Nielsen, J. L. Jensen & W. S. Kendall), 251–300. Chapman & Hall, London.

Jensen, J. L. (1993). Chaotic dynamical systems with a view towards statistics: a review. In *Networks and chaos—statistical and probabilistic aspects* (eds O. E. Barndorff-Nielsen, J. L. Jensen & W. S. Kendall), 201–250. Chapman & Hall, London.

Judd, K. & Mees, A. (1994). On selecting models for nonlinear time series. Technical Report, Department of Mathematics, University of Western Australia.

Kifer, Y. (1986). *Ergodic theory of random transformations.* Birkhäuser, Basel.

Kullback, S. (1967). *Information theory and statistics.* Dover, New York.

Laplace, P. S. (1814). *Oeuvres complètes.* In *Encyclopedia of statistical sciences* (eds S. Kotz & N. L. Johnson), vol. 4, p. 472. Wiley, New York.

Lele, S. (1994). Estimating functions in chaotic systems. *J. Amer. Statist. Assoc.* **89,** 512–516.

Lewis, P. A. W. & Ray, B. K. (1993). Nonlinear modeling of multivariate and categorical time series using multivariate adaptive regression splines. In *Dimension estimation and models* (ed. H. Tong). World Scientific, Singapore.

Lewis, P. A. W. & Stevens, J. G. (1991). Nonlinear modeling of time series using multivariate adaptive regression splines (MARS). *J. Amer. Statist. Assoc.* **86,** 864–877.

Mañé, R. (1981). On the dimension of the compact invariant sets of certain nonlinear maps. In Dynamical systems and turbulence (eds D. A. Rand & L.-S. Young), *Lecture Notes in Mathematics* **898,** 230–242. Springer Verlag, New York.

May, R. M. (1976). Simple mathematical models with very complicated dynamics. *Nature* **261,** 459–467.

Mees, A. I. (1989). Modelling complex systems. Technical Report, Mathematics Department, University of Western Australia.

Nychka, D., Ellner, S., Gallant, A. R. & McCaffrey, D. (1992). Finding chaos in noisy systems. *J. Roy. Statist. Soc. Ser. B* **54,** 399–426.

Ott, E. (1993). *Chaos in dynamical systems.* Cambridge University Press, Cambridge.

Ozaki, T. (1990). Contribution to the discussion of M. S. Bartlett's paper. *J. Roy. Statist. Soc. Ser. A* **153,** 337–338.

Pesin, Y. B. (1993). On rigorous mathematical definitions of correlation dimension and generalized spectrum for dimensions. *J. Statist. Phys.* **71,** 529–547.

Poincaré, H. (1905). *Science and hypothesis.* Reprinted in student edn by Dover, New York, in 1952.

Royama, T. (1992). *Analytical population dynamics.* Chapman & Hall, London.

Ruelle, D. (1989). *Chaotic evolution and strange attractors.* Cambridge University Press, Cambridge.

Serinko, R. J. (1995). A consistent approach to least squares estimation of correlation dimension in weak Bernoulli dynamical systems. *Ann. Appl. Probab.* (to appear).

Smith, L. A. (1988). Intrinsic limits on dimension calculations. *Phys. Lett.* **A133,** 283–288.

Smith, L. A. (1992). Identification and prediction of low dimensional dynamics. *Physica* **D58,** 50–76.

Smith, L. A. (1994). Nonlinear prediction and local optimal forecasting from time series. *Phil. Trans. Roy. Soc. (Lond.)* **A348,** 371–378.

Smith, R. L. (1992). Estimating dimension in noisy chaotic time series. *J. Roy. Statist. Soc. Ser. B* **54,** 329–352.

Sugihara, G. & May, R. M. (1990). Nonlinear forecasting as a way of distinguishing chaos from measurement errors in time series. *Nature* **344,** 734–741.

Takens, F. (1981). Detecting strange attractors in turbulence. In Dynamical systems and turbulence (eds D. A. Rand & L.-S. Young), *Lecture Notes in Mathematics* **898,** 366–381. Springer Verlag, New York.

Takens, F. (1994). Analysis of non-linear time series with noise. Technical Report, Department of Mathematics, Groningen University, 20 March 1994.

Thompson, J. R. & Tapia, R. A. (1990). *Nonparametric function estimation, modeling and simulation.* SIAM, Philadelphia.

Titterington, D. M. (1994). *Proceedings: complex stochastic systems and engineering.* Wiley, Chichester.

Tjøstheim, D. (1994) Non-linear time series: a selective review. *Scand. J. Statist.* **21,** 97–130.

Tjøstheim, D. & Auestad, B. (1994). Nonparametric identification of nonlinear time series: selecting significant lags. *J. Amer. Statis. Assoc.* **89,** 1410–1419.

Tong, H. (1983). *Threshold models in non-linear time series analysis*. Springer Verlag, Heidelberg.
Tong, H. (1987). Nonlinear time series models of regularly sampled data: a review. In *Proceedings of the First World Congress of the Bernoulli Society* (eds Y. V. Prohorov & V. V. Sazonov) **2,** 355–367. VNU Science Press, Amsterdam.
Tong, H. (1990). *Non-linear time series: a dynamical system approach*. Oxford University Press, Oxford.
Tong, H. (1992). Some comments on a bridge between nonlinear dynamicists and statisticians. *Physica D* **58,** 299–303.
Tong, H. & Lim, K. S. (1980). Threshold autoregression, limit cycles and cyclical data (with discussion). *J. Roy. Statist. Soc. Ser B* **42,** 245–292.
Tong, H. & Smith, R. L. (1991). An informal guide to chaos. Unpublished handout given at the Chaos Day meeting of the Royal Statistical Society on 16 October 1991.
Tong, H. & Smith, R. L. (1992). Royal Statistical Society Meeting on Chaos, *J. Roy. Statist. Soc. Ser B* **54,** 301–474.
Wahba, G. (1990). *Spline models for observational data*. SIAM, Philadelphia.
Wolff, R. C. L. (1990). A note on the behaviour of the correlation integral in the presence of a time series. *Biometrika* **77,** 689–697.
Wolff, R. C. L. (1992). Local Lyapunov exponents: looking closely at chaos. *J. Roy. Statist. Soc. Ser B* **54,** 353–372.
Yao, Q. & Tong, H. (1994a). Quantifying the influence of initial values on non-linear prediction. *J. Roy. Statist. Soc. Ser B* **56,** 701–725.
Yao, Q. & Tong, H. (1994b). On prediction and chaos in stochastic systems. *Phil. Trans. Roy. Soc. (Lond.)* **A348,** 357–369.
Yao, Q. & Tong, H. (1994c). On subset selection of stochastic regression model. *Statist. Sinica* **4,** 51–70.

Received October 1994, in final form March 1995

H. Tong, Institute of Mathematics and Statistics, University of Kent, Canterbury, Kent CT2 7NF, United Kingdom.

DISCUSSION AND COMMENTS

K. S. CHAN

University of Iowa

My congratulations to Professor Tong for writing such a thought-provoking and timely overview on non-linear time series from a chaos perspective. My comments will be confined to an issue related to quantifying the sensitivity to initial conditions for stochastic dynamical systems. For simplicity, I only consider the one-dimensional case, although some of the discussions below can be readily extended to the higher dimensional case. As pointed out by Professor Tong in his overview, there are, at least, two approaches to the aforementioned task. In one approach, the dynamical system starts with different initial conditions, but with identical dynamic noise sequence exciting the system. In other words, keeping the dynamic noise sequence fixed, we analyse the effect of the initial conditions on the divergence of the trajectories. Suppose that the dynamics is driven by (10): $X_t = F(X_{t-1}) + e_t$. Using this approach, the Lyapunov exponent is defined by

$$\lambda(X_0) = \lim_{N\to\infty} \frac{1}{N} \sum_{t=1}^{N} \log |\dot{F}(X_{t-1})|.$$

If the underlying stochastic process is ergodic and $\max(0, \log|\dot{F}(X)|)$ is integrable w.r.t. the invariant measure, then $\lambda(X_0)$ exists, and equals a constant λ independent of the initial condition X_0 and the dynamic noises. Professor Tong mentioned an objection raised by Jensen (1993) to this approach: that λ so defined is not invariant under one-to-one co-ordinate transformations.

However, if we transform X_t to $Y_t = g(X_t)$, the transition equation becomes

$$Y_t = g(F(g^{-1}(Y_{t-1})) + e_t)$$

which is generally different from (10), but is of a different kind like (9), i.e. the dynamic noise is not additive. It then seems that a more appropriate definition for quantifying the sensitivity to initial conditions (under the scenario of identical noise sequence) is

$$\tilde{\lambda}(Y_0) = \lim \frac{1}{N} \sum_1^N \log \left| \frac{\partial G}{\partial Y}(Y_{t-1}, e_t) \right| = \lim \frac{1}{N} \log \left| \frac{\partial Y_N}{\partial Y_0} \right|.$$

It seems that this definition is more in accord with the definition of the Lyapunov exponent given by Kifer (1986) for the case of random maps. Again, under ergodicity, $\tilde{\lambda}(Y_0)$ exists, and equals a constant $\tilde{\lambda}$ independent of the initial condition and the dynamic noises. If the dynamic noise is additive, the two definitions of the Lyapunov exponent are formally equivalent. It can be checked that the Lyapunov exponent so defined is invariant under one-to-one co-ordinate transformations. An outline of the proof follows. Let $Z_t = h(Y_t)$ and $Y_t = G(Y_{t-1}, e_t)$. Then $Z_t = h(G(h^{-1}(Z_{t-1}), e_t)) = \tilde{h}(Z_{t-1}, e_t)$. From the chain rule, we have

$$\frac{\partial Z_t}{\partial Z_{t-1}} = \dot{h}(Y_t) \frac{\partial G}{\partial Y}(Y_{t-1}, e_t) / \dot{h}(Y_{t-1}),$$

and hence

$$\frac{\partial Z_N}{\partial Z_0} = \dot{h}(Y_N) \frac{\partial Y_N}{\partial Y_0} \bigg/ \dot{h}(Y_0).$$

Therefore,

$$\lim \frac{1}{N} \log \left| \frac{\partial Z_N}{\partial Z_0} \right| = \lim \frac{1}{N} \log \left| \frac{\partial Y_N}{\partial Y_0} \right|$$

in probability and hence a.s. by ergodicity. This completes the proof of the invariance of the new definition of the Lyapunov exponent.

The more intriguing question is to understand the pros and cons of this approach and the conditional distribution approach pioneered by Professor Tong and Dr Yao. Hopefully, Professor Tong's excellent review will prompt further research needed to clarify the situation.

D. R. COX
Nuffield College, Oxford

Professor Tong's impressively wide-ranging paper links two fascinating fields, non-linear time series and deterministic dynamical systems. I want to sound one note of caution and to ask two questions.

It has been suggested that the idea of chaos has application in epidemic modelling, in rainfall modelling for hydrology and in economic time series, especially financial series. These certainly seem fields where non-linear aspects are important, but is there really clear evidence that chaos in the technical sense has a lot to offer? After all these are hardly isolated systems likely to be encapsulated in relatively simple deterministic equations.

Are there implications of chaos for the study of point processes?

Is it possible that chaos theory throws light on that mysterious issue, the role of probability in the foundations of quantum mechanics?

COLLEEN D. CUTLER
University of Waterloo

I would like to congratulate Professor Tong on his informative and inspiring review. I will make two comments. Professor Tong has proposed a general stochastic model comprised of a deterministic skeleton and a random error sequence. We may associate and estimate parameters (e.g. Lyapunov exponents, embedding dimension, attractor dimension) with either the skeleton of the system or with the stochastic system itself (although in the latter case some of these parameters might be open to both definition and interpretation). It seems clear to me in cases of either measurement noise or of hyperbolic exponentially stable systems with dynamic noise (as described by Tong at the end of section 2) that parameters determined by the skeleton are particularly relevant to analysis of the system. However, in some cases of dynamic noise, the perturbed stochastic system may exhibit very different asymptotics from that of the skeleton. In these cases parameters associated with the stochastic system itself would seem more important (although understanding the skeleton for purposes such as short-term prediction or measuring sensitivity to initial conditions would still be useful). We need to find appropriate ways of describing and quantifying the stochastic system itself.

My second comment concerns the uses of the various possible definitions of dimension and finally the distinction between deterministic and stochastic systems. Professor Tong has raised the question of the necessity of routinely estimating correlation dimension. There is no doubt that this quantity has been over-used in practice and sometimes assigned an importance out of proportion to reality. Correlation dimension seems to be estimated for two distinct purposes. Very often, the experimentalist really seems to be interested in the embedding dimension (in particular, whether or not it is finite, and then whether or not it is small) and finds the Grassberger–Procaccia algorithm a convenient way of getting at this information. We now know that for various reasons, some of a statistical nature (to do with convergence of estimators) and some of an analytic nature (to do with the actual mathematical meaning of correlation dimension and structure of time series), this algorithm does not always lead to a correct bound on the embedding dimension (even if we follow the "rule" $1 + 2$ (correlation dimension)). There are now numerous methods available for getting at the embedding dimension directly (one suggested by Cheng & Tong (1994), using a particular definition of embedding dimension, and others reviewed in Abarbanel *et al.* (1993)) and certainly these should be used if embedding dimension is the quantity of interest. However, correlation dimension is sometimes estimated along with several other dimensions (such as information dimension and, more generally, the entire spectrum of Rényi dimensions) in order to obtain information about the structure of the natural measure over the attractor; the way these various dimension quantities differ among themselves provides clues to the non-uniformity of the distribution. This is connected to ideas from multifractal analysis, and an introductory review can be found in Ott *et al.* (1994).

Finally, I am not certain that one can always satisfactorily estimate embedding dimension before proceeding to other dimensions. The discrete-spectrum Gaussian process

$$X(t) = \sum_{k=1}^{\infty} k^{-\alpha/2}(A_k \cos(kt) + B_k \sin(kt)) \tag{1}$$

discussed in Cutler (1994) provides a curious example. While globally infinite-dimensional in every sense of the word, this process disintegrates into uncountably many ergodic components, each of which almost surely (with respect to the stationary measure of the process) lives on a subset of finite fractal dimension (an attractor?) in the space of continuous functions. This dimension can be determined theoretically as well as computed numerically from a realization of the time series; it is $2/(\alpha - 1)$. However, I am not so certain of how to assign embedding dimension here. If we take the approach of Cheng & Tong (1994) we might argue that the embedding dimension should still be regarded as infinity, since we continue to gain information about predicting the future of a realization when conditioning on more and more past observations (for "most" choices of the time lag). However, I suspect that traditional definitions of embedding, restricted to the ergodic component itself, might yield a finite embedding dimension. This illustrates some of the difficulties, or at least confusion, that could arise in attempting to make inferences about prediction, embedding dimension, and determinism in a process based on a single realization. I don't believe that we can always count on ergodicity.

References

Abarbanel, H. D. I., Brown, R., Sidorowich, J. J. & Tsimring, L. S. (1993). The analysis of observed chaotic data in physical systems. *Rev. Mod. Phys.* **65,** 1331–1392.

Cutler, C. D. (1994). A theory of correlation dimension for stationary time series. *Phil. Trans. Roy. Soc.* **A348,** 343–355.

Ott, E., Sauer, T. & Yorke, J. A. (eds) (1994). *Coping with chaos*. Wiley, New York.

D. GUÉGAN

CREST, France

H. Tong's paper is very interesting and it gives a good general view of the past years' work on the approach on dynamical stochastic systems. Indeed as Tong does it, as from section 3, it seems to me more just to speak of an overview of the statistical theory in the present case—given the statistical described results—rather than in the context of chaotic systems. Thus, if a discrete time dynamical system described by a difference equation such as

$$X_t = \varphi(X_{t-1}) \tag{1}$$

where $X_0 \in \mathbb{R}^d$, φ is some non-linear function, represents a deterministic chaos, when φ satisfies specific conditions, the whole of the approach described by Tong in his paper does not concern itself with model (1), but a system as the one introduced in (10) in his paper. So, it seems there currently is an ambiguity concerning the works on deterministic chaotic models.

Indeed, if as Tong pointed out there is no universal definition of a deterministic chaos, it can, however, be characterized relatively well provided some precautions are taken. Thus we shall say that a system as (1) is a chaotic deterministic system if it is sensitive to initial conditions and/or if there is a strange attractor, and if there is an invariant ergodic probabilistic measure μ defined on $\mathbb{R}^d$, $\mathscr{B}(\mathbb{R}^d)$, where $\mathscr{B}(\mathbb{R}^d)$ is the Borel σ-field of $\mathbb{R}^d$, such that if: (1) $\mu(\varphi^{-1}(B)) = \mu(B)$, for any $\mathscr{B}$ in $\mathscr{B}(\mathbb{R}^d)$; (2) $\mu^{-1}(B) = B$, where $B \in \mathscr{B}(\mathbb{R}^d)$, then $\mu(B) = 0$ or 1.

In most of the work developed by Tong and the papers he quoted, nothing is concerned with model (1), as everywhere methods relative to stochastic systems where noise appears additive, are developed. The ambiguity of this approach lies in the fact that the authors often mentioned "noisy chaos" and apply mathematical concepts developed for deterministic dynamical systems to these stochastic processes. The academic papers that stem from their theory is indeed useful and interesting, but the use of the word "chaos" in "noisy chaos" is ambiguous in so far as it is known that the very presence of a noise in (1)—however small the noise may be—makes process (1) lose any chaotic characteristics.

If the development in statistical research in chaotic systems has had such an impact, it is because it took place at a time when theoreticians and practicians realized that the use of non-linearities allowed a better knowledge of the mechanism of real data. The study of stochastic systems as

$$X_t = \varphi(X_{t-1}) + \varepsilon_t \tag{2}$$

where φ is some non-linear function, ε_t a sequence of i.i.d random variables, is well known now from a non-parametric point of view. The developments presented in Tong's paper often take up these works with modifications concerning hypotheses on φ in (2) or on the noise $\{\varepsilon_t\}$, so as to obtain results on the various estimates.

One of the problems that seems fascinating to us concerning the study of a chaotic deterministic system as (1) is to find out whether it is possible to point out chaos from time series and also to get results permitting one to decide between stochastic and deterministic systems. To get these results we can consider the analytical approach and try to calculate the Lyapunov exponents to measure their positiveness, so as to detect the sensitivity of system (1) to initial conditions. We can also consider the dimension approach and try to show, for (1), the existence of a strange attractor whose dimension is fractal. We are going to focus on the former approach.

We are first going to assume that these observations are realizations of random variables and that therefore, $\{X_t\}$, $t \in \mathbb{Z}$, is being observed, following a system as (1). We assume that there exists, for these observations, an invariant measure μ which has a density f with respect to the Lebesgue measure. So a first indispensable step consists in estimating the invariant measure. Such an estimate using the kernel method has been built by Bosq & Guégan (1994a) under ergodicity hypotheses. The assumptions that we have considered allow us to adapt the classical techniques of the computation of expectations and variances from the stochastic case to the deterministic system (1). Then we obtain the rate of convergence for this estimate.

Let us now proceed with the identification of a deterministic chaotic system from time series. For this we need to have an estimate of the chaotic function φ in (1). We have developed three methods.

First, we construct an estimate of φ in the following manner. As the functional relationship (1) implies that the joint distribution of (X_t, X_{t-1}) is singular with respect to the Lebesgue measure, thus a suitable density estimate for this joint distribution will explode in the neighbourhood of the graph of φ and will vanish elsewhere. We use this property to construct an estimate $\hat{\varphi}_n^{(1)}$ of φ. Using hypotheses concerning the ergodicity of (1) we obtain the rate of convergence for $\hat{\varphi}_n^{(1)}$, see Bosq & Guégan (1994a).

Second, we have obtained an estimate that is simpler and easier to handle than the one described above. For this, we no longer assume that our observations are random variables characterized by a law, but we very simply take them as discrete "physical" observations without any stochasticity notion. Then a naïve estimate $\hat{\varphi}_n^{(2)}$ for φ can be obtained by the nearest neighbour or by an interpolation between the two nearest neighbours. Delecroix *et al.* (1994) have established the convergence of these estimates.

Third, it is also possible to build another estimate for φ defined in (1) based on the regressogram approach. In that case the proofs are completely different from what is currently being done. For here no stochasticity hypothesis on random variables is used, see Delecroix *et al.* (1995).

To build convergent estimates for Lyapunov exponents the two latter approaches are better than the former thanks to their easy implementation. We prove the convergence of the Lyapunov exponents in the two latter cases and on simulations we show how fast they converge in spite of the problem of the choice of bandwidth h_n in the regressogram approach, see Delecroix *et al.* (1994, 1995).

All the approaches developed earlier can only be used for real data if an efficient estimate for the embedding dimension is available. This is essential to calculate the Lyapunov exponents in dimension k. So far the technique that has been most used is the one of Grassberger & Procaccia as Tong reminds us. However, it is well known that this technique is not robust. In the case where the observations $\{X_t\}$ are random variables, and follow a system as (1), using the same method as the one developed for $\hat{\varphi}_n^{(1)}$, Bosq & Guégan (1994b) construct a consistent estimate for d, see Guégan & Léorat (1995) for applications.

Of course, because of this deterministic chaos approach we develop, many problems are left unsolved. The very existence of noise in observational data that are supposedly generated by chaos is a difficult problem. For the moment no proper step taking that noise into account has been defined. So what is important to remember, actually, is that the statistical approach to chaos so far has provided us with a better understanding of the non-linearity of the data to be studied.

References

Bosq, D. & Guégan, D. (1994a). Nonparametric estimation of the chaotic function and the invariant measure of a dynamical system. *Statist. Probab. Lett.* To appear.

Bosq, D. & Guégan, D. (1994b). Estimation of the embedding dimension of a dynamical system. Working Paper, CREST, No. 9451.

Delecroix, M., Guégan, D. & Léorat, G. (1994). Detecting deterministic chaos from observational data. Working Paper, CREST, No. 9450.

Delecroix, M., Guégan, D. & Léorat, G. (1995). Using regressogram for estimation in chaotic system. In preparation.

Guégan, D. & Léorat, G. (1995). Consistent estimation to detect chaos in financial data. Working paper.

JENS LEDET JENSEN

University of Aarhus

Howell Tong has given a review that seems quite close to my own opinions on this subject. My main objection is the three lines of conclusion that I somehow feel do not quite capture the tone of the paper.

In a dynamical system X_n is a function of X_0, i.e. $X_n = F^{(n)}(X_0)$, where F is some mapping on the state space. In the mathematical description of chaotical dynamical systems there are two main objects: the Lyapunov exponent describing the average exponential rate of divergence of trajectories, and the strange attractor describing the long run position of the system. The Lyapunov exponent is an average over the strange attractor and a positive value points to the impossibility of long-term predictions, but gives no information on the possibility of short-term predictions. This is often described as sensitivity to the initial conditions. The strange attractor is a very complicated object, typically with a fractal structure.

When there is noise in the system, to be specific think of $X_{n+1} = G(X_n, \varepsilon_{n+1})$ with the ε_is an i.i.d. sequence, we no longer have X_n as a function of X_0 and the above quantities have no strict meaning. Basically, we have infinite sensitivity to initial conditions in the sense that X_{i+m} will not be close to X_{j+m} even though X_i and X_j are arbitrarily close, and the attractor is no longer strange, the system occupies an open subset of the state space. There is, however, one natural way of defining a Lyapunov exponent, namely by considering X_n a random function of X_0. For a particular fixed realization $\{\varepsilon_1, \varepsilon_2, \ldots,\}$ of the noise sequence we have that X_n is a function of X_0 and we can speak of the average exponential rate of divergence of trajectories. Precisely because it is the average rate the particular sequence $\{\varepsilon_i\}$ does not influence the result. I refer to Kiefer (1988) for an account of results of this type. Again, this gives us information on the long-term behaviour, whereas for prediction purposes we would be interested in a measure like

$$\tau_m(x_0, \delta) = E|G(G(\cdots(G(x_0+\delta, \varepsilon_1), \varepsilon_2)\cdots), \varepsilon_m) - G(G(\cdots(G(x_0, \varepsilon_1), \varepsilon_2)\cdots), \varepsilon_m)| \tag{1}$$

with δ small and m a small integer. When we have no model for $G(\cdot, \cdot)$ there seems to be no direct way of estimating $\tau_m(x_0, \delta)$. The suggestion of Wolff mentioned in section 3.2 seems not so helpful since X_{i+m} as a function of X_i is not defined through the same ε_k-sequence as X_{j+m} considered as a function of X_j. Actually, $\lambda_{i,m}$ in (15) in Tong's paper will presumably tend to infinity for $n \to \infty$ and for δ small. Removing the numerical sign in (1) and letting $\delta \to 0$ we get the asymptotic formula

$$\delta E\{\dot{G}(X_0, \varepsilon_1)\dot{G}(X_1, \varepsilon_2)\cdots\dot{G}(X_{m-1}, \varepsilon_m)\},$$

where $\dot{G}$ is the derivative of $G(x, \varepsilon)$ w.r.t. x. Note that this equals (27) in the paper in the case of additive noise so that in a sense sections 3.1 and 3.2.3 consider the same quantity. I should add though that I find the latter linearization argument dangerous and prefer to evaluate (1) by simulation. Let me add here that I have doubts as to the value of the measure $\kappa(x)$ in (28) of Tong's paper. If we have an ergodic Markov chain then we often have that the convergence to the stationary distribution is exponentially fast in which case $\kappa(x)$ will be negative. Thus in this case $\kappa(x)$ gives us no useful information on the rate of separation for small m. This can also be illustrated through example 3.2.2(i). Here α^m gives useful information for fixed m while the limit $\lim_m K_m(x, \Delta) = 0$ is uninformative.

The conclusion that I feel is implicit in Tong's paper is that for the kind of data a statistician is usually confronted with, the mathematical theory of *chaotic dynamical systems* can only be an inspiration, and that the proper heading for the statistician is *non-linear time series analysis*. The inspiration, or facts, we have got from chaos theory is the importance of non-linearity. Typically there is so much noise in the data, or we have only modelled the most important aspects, so that the non-linear dynamic cannot be separated from the influence of the noise. To be specific let us briefly discuss the nature of the measles data mentioned in the paper. This is a spatial contact process where the number M_{n+1} of new measles cases have a Poissonian nature with the intensity dependent on M_n and on the number at risk R_n. A simple model could look somewhat like:

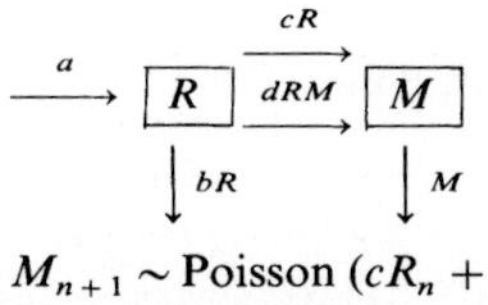

(2)

$$M_{n+1} \sim \text{Poisson}\,(cR_n + dR_nM_n), \quad R_{n+1} = R_n + a - bR_n - M_{n+1},$$

where a, b, c and d are constants and the arrows indicate rates of transfer. With a Poisson model for M_n the variance equals the mean, i.e. the variance is quite large, and it does not seem sensible to think of this as a dynamical system with a small noise component.

How can we analyse a non-linear time series. Quite often it is unrealistic to think that we observe the state variable X_n in the system $X_{n+1} = G(X_n, \varepsilon_{n+1})$. Rather we will have an observable $Y_n = h(X_n)$. As an example $X_n = (R_n, M_n)$ in the measles case and $Y_n = M_n$. For a dynamical system we handle this problem by using the embedded variable $Y_n^d = (Y_n, Y_{n-1}, \ldots, Y_{n-d+1})$ as described in section 2.3. In the time series case we hope that for some d we can describe Y_n as a non-linear autoregressive process of order d. I wonder if Howell Tong knows of any results saying that the model $X_{n+1} = G(X_n, \varepsilon_{n+1})$ leads to an autoregressive model for Y_n? If not it seems that d has no clear physical meaning, but rather becomes a product of the class of models being fitted. It seems to me that in many applications one should only think of d as a tool in the process of making good predictions. Speaking about the dimension let me make an even more provoking statement: with a finite number of data points, n say, it is impossible to estimate d without making model assumptions. The simple reason is that with d of the order $\log_{10} n$ (which is usually not very big!) the n points Y_j^d will be isolated points in $\mathbb{R}^d$ and we can find a function f_d such that we have a perfect fit $Y_{j+1} = f_d(Y_j^d), j \leq n$. The usual argument is now that we use a test set and take d as small as possible but such that we get the best possible predictions from f_d. However, this only reflects the smoothness of the function class used for estimating f_d. As long as we have made no assumptions we cannot know if we can use the isolated points $Y_j^d, j \geq n$, to say what the value of f_d may be at points in between. Of course I do not mean to imply that we should not estimate d. However, it seems sensible to me to either introduce some model assumption or at least to make some explicit restrictions on f_d, or to use some other arguments to determine a useful value of d. As an example consider the measles data again corresponding to observing M_n in (2). For the model in (2) the "state space" is of dimension 2 and it seems reasonable then to consider $d = 2$. The intuitive idea is that (M_n, M_{n-1}) allows us to estimate R_{n-1} as $M_n/(dM_{n-1} + c)$ and next to estimate R_n as $R_{n-1} - bR_{n-1} + a - M_n$. With (M_n, R_n) at hand we can then predict M_{n+1}. If instead of the model (2) we imagine that we have two independent populations developing according to (2) and that we only observe the total number of measles cases we would presumably need to take $d = 4$. In the New York measles data we have 432 data points, and if we work non-parametrically we are trying to estimate a function from $\mathbb{R}^4$ to $\mathbb{R}$ based on 432 points! In my view this is not advisable. (Tong mentions the value $d = 7$ for the New York measles data which to me seems far beyond what the data can possible tell. As a side remark it is not clear to me if one has estimated d as though $V(M_n) = \sigma^2$ or as $V(M_n) = EM_n$?) If we have only a small number of data points and we imagine that d is large ($d > 2$) I feel that one has to use parameterized models.

I have a feeling that some of the points raised above appear repeatedly in Tong's paper in the phrase "the curse of dimensionality". Perhaps Tong will comment on whether this is indeed the case.

Reference

Kiefer Y. (1988). *Random perturbations of dynamical systems.* Birkhäuser, Boston.

SØREN JOHANSEN

University of Copenhagen

I have learned a lot by reading the paper by Professor Tong and by listening to the lectures. It is useful to have a survey like this which gives the non-expert an opportunity to see what is going on in another field and ask some questions that come to mind when the topic is seen from the outside.

The first question is concerned with the model

$$\Delta X_t = \beta f(X_{t-1}) + \varepsilon_t, \quad t = 1, \ldots, T$$

which describes the changes in the level to a non-linear function of the level. This is clearly a stochastic dynamical system in the sense of (9). The Gaussian maximum likelihood estimator is

$$\hat{\beta} = \sum_{t=1}^{T} \Delta X_t f(X_{t-1}) \Big/ \sum_{t=1}^{T} f(X_{t-1})^2.$$

In this context it is natural to ask questions like: does $\hat{\beta}$ converge to β? Is $\sqrt{T}(\hat{\beta} - \beta)$ asymptotically normal or does it have another limit distribution?

These are the usual questions that concern inference, and it would be useful to know if the theory of chaos has a role to play here. Does the Lyapunov exponent of the skeleton have any relations to the inferential questions? Does the structure of the attractor set determine properties of the limit distribution? etc.

Note that the usual boundedness condition on dynamical systems is not quite consistent with the tradition for Gaussian errors in regression models. Specifically the above regression model has no meaning if f is the logistic map and the errors are Gaussian.

The paper contains results on the non-parametric estimation of the response function, but it would be nice to have the parametric results as well.

Another question that puzzles me in this area is the relation between chaos and stochasticity. If we want to study the above regression model and the properties of the estimator $\hat{\beta}$, we can do so by simulation. The usual random number generators have the form

$$\varepsilon_{t+1} = g(\varepsilon_t),$$

hence we replace the stochastic dynamical system by the deterministic dynamical system

$$\begin{pmatrix} X_t \\ \varepsilon_{t+1} \end{pmatrix} = \begin{pmatrix} f(X_{t-1}, \varepsilon_t) \\ g(\varepsilon_t) \end{pmatrix}.$$

Does this mean that there are no stochastic dynamical systems, only deterministic ones? If the answer is that there is in fact something called a stochastic dynamical system, then the consequence seems to be that we cannot investigate it by simulation.

What is the relation between the invariant measure for the univariate stochastic system and the bivariate deterministic system?

Is a stochastic system just a mathematical limit of a bivariate system with two Lyapunov coefficients, where one of them tends to infinity?

A regression model that it would be extremely useful to have inferential methods for is

$$\Delta X_t = \alpha f(\beta' X_{t-1}) + \varepsilon_t.$$

This is a non-linear error correction model where estimation could be performed by projection pursuit methods, but what about inference, which is complicated by the non-stationarity of $\alpha'_\perp X_t \alpha'_\perp X_0 + \alpha'_\perp \Sigma_{s=1}^{t} \varepsilon_s$.

The above questions are mainly related to topics not treated in the paper, but the paper has a large number of results and ideas. I find in particular the results of lag length determination impressive. I am surprised that such general systems can be handled. I find that in linear models it is often a useful idea to rethink the information set, in the sense that if we observe long lags or other strange behaviour, it is often due to an omitted variable. If this is included in the analysis the lag length reduces with a simpler analysis as a result.

My final comment is concerned with the statistical formulation of initial value sensitivity, as the sensitivity of the conditional density of X_{t+h} given $X_t = x$, to changes in x. This to me appears the correct formulation, and the relation to the Fisher information when x is considered a parameter is a nice result.

In the linear framework one investigates the prediction $\hat{X}_{t+h} = E(X_{t+h} \mid X_t)$ by the so-called impulse response curves, which are just

$$\partial \hat{X}_{i,\,t+h} / \partial X_{j,\,t}, \quad h = 1, 2, \ldots$$

which show how a change in component j of X_t influences the prediction of $X_{i,\,t+h}$. Such plots would be interesting for non-linear systems.

We were told in the lecture that if we do not remember anything else, at least we should remember that

"Conditional variance is important".

The reason for this is that in non-linear models the prediction error depends on where we happen to be in the observation space. Thus there are "windows of opportunity".

In my opinion linear models do not represent reality, but they often give a good description of some basic behaviour of the time series. The moment we leave linearity we have many choices of which model to choose.

I have yet to find an example in statistics where many people agree that the one and only non-linear formulation has been found. I feel that "the window of opportunity", where the prediction variance is small could be very sensitive to the choice of non-linear model, and that it would require even more careful inferential methods for non-linear models.

A. J. LAWRANCE
University of Birmingham

Howell Tong begins his paper by being concerned that some time series people have been suspicious of chaos and even found it alien—perhaps from leaving our comfortable planet of linearity, or should I say from leaving their flat earth. Personally, I first escaped from such earthly attraction in the 1970s with Peter Lewis when we constructed non-Gaussian time series models and in so doing were launched into the fringes of non-linear space. This is the universe into which Howell Tong has travelled further than many, and the paper traces the path of his rocket. But has it yet reached another planet, and is there life there? I think the answer has to be that our chaotic time series lander is hovering over the surface, knows theoretically how to land, but has not quite engaged with the alien friends below. I had better leave this analogy for a moment.

I admire the papers' concern with statistical matters of chaos, which draws the subject ever closer into non-linear stochastic models. I am uneasy with the notion of initial sensitivity when stochastic effects are already present to take its place. Yet the sensitivity of the conditional expectation, because of its relevance to prediction, is surely important and this does depend on the deterministic and probably chaotic aspects of the structural part of the model. The realization of possible windows of predictive opportunity is surely profound. But here, and in other parts of non-linear time series, I feel a frustration at the lack of simple convincing and necessarily over-simple black-box models: these could help us to understand and demonstrate the advantages and effects of chaos and non-linearity relative to our well-known linear friends. There is mention in the paper of global, local and semi-local such models, but is there an equivalent in simplicity to the first order AR(1) model of linear time series? It might not be of the additive noise type, rather having state-dependent noise. There is still relatively little emphasis on estimating the deterministic component in the additive model, but anyway I wonder whether this is a natural model form on the non-linear planet.

Howell Tong's paper is wide-ranging, stimulating and enjoyable. To close my own discussion, I would like to emphasize his passing references to the contributions which chaotic ideas are making to random number generation, communication systems and cryptography—we should look more to Japan. In these areas it is likely that idealized simple models will naturally arise and find much use.

BLAKE LEBARON

University of Wisconsin

Howell Tong has his own special style and approach to chaos and non-linear times which this paper provides a nice survey of. Tong bridges the gap between the time series and dynamical systems literature showing that the two can complement each other in the analysis of many real world problems. This approach is extremely useful for fields such as economics, where we cannot escape the impact of some stochastic elements in our models. In my comments I would like to highlight some of Tong's most important points which might get missed by the casual reader.

In section 3, Tong stresses the importance of sensitive dependence for chaotic system analysis. The main tool that is used to analyse the phenomenon is the Lyapunov exponent. While Lyapunov exponents are a well-defined concept for deterministic dynamical systems, a satisfying equivalent for stochastic systems has not been found. There are several contenders, but none provide the comfortable universal acceptance of the deterministic Lyapunov exponent in the context of systems influenced by outside random shocks. There are some very tricky issues involved here, and Tong addresses several of these. Some systems whose deterministic skeleton is non-chaotic will exhibit sensitive dependence-like properties in the presence of noise. Measures that try to take this into account need to address several important theoretical questions about sensitive dependence.

Do we mean that two trajectories with nearby starting values will end up arbitrarily far apart when subject to the same stochastic shocks, or do we mean that some statistical features of stochastic trajectories, either conditional means or densities, will spread apart? My opinion is that in most fields the latter of these two definitions makes the most sense.

Section 3.3 stresses the important feature of state-dependent variance. Although not necessarily related to chaos, this feature is one of the most important reminders about non-linear systems that people can take away from the non-linear dynamics literature. I have found it to be an important feature of foreign exchange rate (LeBaron, 1992). Researchers looking at usual unconditional measures of predictive accuracy can be greatly fooled when non-linearities are involved. As Tong correctly mentions, this may be a fact that people use in their daily lives, but is ignored by almost all time series analysis.

Tong mentions another non-linear dynamics tool, dimension analysis, in several parts of the paper. He generally expresses a very cautious tone in all cases. I strongly agree with him on this, and would stress that in economics it has proved to be almost useless, and probably somewhat dangerous in terms of misdiagnoses of evidence for chaos. Estimating correlation dimensions is an interesting thing to do for many beginning chaos scientists, but all should be cautioned about the dangers of blindly feeding data into a dimension analysis procedure. I think Tong's cautions on this subject should probably be stronger.

The controversies Tong suggests between black-box and substantive approaches exist in all fields, economics included. It is at times difficult to make strong economic conclusions without the use of substantive or theoretical models connected to empirical results. This is countered by the fact that any substantive model is a drastic simplification of reality, and cannot fully represent the richness contained in the dynamics of any given data sets. I think both black-box and substantive approaches remain useful in all fields. However, an interesting question remains about the limits of what we can understand using a strictly black-box approach. For example, if we are interested in the question of how much variability of a system (stock market, population, etc.) is coming from internal structure, vs exogenous shocks, we will probably have to take a stand on some substantive issues. Most importantly, our units of measurement will matter, since this may affect our conclusions in the stochastic realm. In the completely deterministic world we are free of this problem since most measures are invariant to most transformations. Neither of these two approaches should be dropped from our menu of techniques, but the limitations of each should always be remembered.

The usefulness of chaos to many fields will depend on our ability to understand how to bring random shocks into connection with non-linear dynamical models. Few sciences have the luxury of being able to ignore exogenous shocks. I think many researchers in these fields will find Tong's survey, and some of his other papers, a useful guide on how chaos affects how we should think about time series.

Reference

LeBaron, B. (1992). Forecast improvements using a volatility index. *J. Appl. Econometrics* **7**, S137–S150.

T. OZAKI

I find Professor Tong's paper very intriguing. My comments to the paper are as follows.

I am pleased to know that a stochastic model such as (10) or its continuous time version, i.e. stochastic dynamical system, can now be accepted as a chaos model. I had thought that, in the view of dynamicists and statisticians,

$$x_t = F(x_{t-1}) \quad \text{and} \quad x_t = F(x_{t-1}) + e_t$$

are very different. If dynamicists did support the stochastic model, either in discrete time or in continuous time, time series analysts would have welcomed them at a much earlier stage. It is well known that since the late 1970s non-linear time series analysts have been interested in the identification problem of those non-linear time series models and non-linear stochastic dynamical systems (see Ozaki, 1985). While dynamicists indulge in minor inferential problems such as estimation of the Lyapunov exponent, estimation of the fractal dimension, estimation of the embedding dimension and noise reduction of trajectory by smoothing, a direct parametric method for the identification of non-linear stochastic dynamical system models has been introduced already using non-linear filtering techniques and has been used in many applications (Ozaki, 1994).

Professor Tong seems to see a fruitful future for statistics in following dynamicists' work on chaos. However, if the dynamicists are serious about the business of solving real problems with real data, it may be dynamicists who find their fruitful future in following the work of statisticians.

References

Ozaki, T. (1985). Nonlinear time series models and dynamical systems. In *Handbook of statistics*, **5** (eds E. J. Hannan *et al.*), 25–83. North-Holland, Amsterdam.

Ozaki, T. (1994). The local linearization filter with application to nonlinear system identifications. In *Proceedings of the First US/Japan Conference on the Frontiers of Statistical Modeling: an informational approach* (ed. H. Bozdogan), 217–240. Kluwer Academic, Amsterdam.

DOUGLAS W. NYCHKA, STEPHEN ELLNER, and BARBARA A. BAILEY

North Carolina State University

A. R. GALLANT

University of North Carolina

Chaotic systems have two seemingly contradictory properties: sensitive dependence on initial conditions (SDIC), and mixing. SDIC says that where you are "now" is very important; mixing says that where you are "now" has no effect in the long run. Some of the new statistics introduced by Professor Tong are related to SDIC, while others relate to mixing. We believe that it is important for statisticians to remember that SDIC and mixing are related but not equivalent. SDIC is regarded in the dynamics literature as a defining property of chaos, because mixing by itself is not unique to chaotic systems.

For the system (9) we can view X_t as a function of X_0 and the random shocks $\{e_1, e_2, \ldots, e_t\}$. SDIC is the property that if X_0 is varied a little, with all else fixed (including the e_j), X_t varies a lot for large t. Alternatively, we can take (9) at face value as a stochastic process and ask how the distribution of X_t depends on X_0. Mixing means that if we vary X_0 (or its distribution), the distribution of X_t is not much changed for large t. Paradoxically, SDIC tends to promote rapid mixing, because the set of trajectories starting near any point X_0 quickly spreads out over a large region. Because the same is true for any other initial point, the trajectories that pass through a given region at time t could have started at many different locations. The system quickly "forgets" where it started, thus leading to the mixing property.

The sensitivity measures presented in sections 3.2.1–3.2.3 relate to mixing: how the conditional measure $P(X_t \in A \mid X_0 = x)$ depends on x. Given that these are presented as sensitivity measures, one might be tempted to interpret large values of $K_m(x, \delta)$ or of $|F_m(x + \delta) - F_m(x)|$ as indicating chaos, and rapid convergence of $K_m(x, \delta)$ to zero as indicating a non-chaotic system. However, as illustrated in Fig. 1, the mixing property does not distinguish between mixing due to SDIC, and mixing due to random shocks. Because SDIC amplifies the impact of the random shocks, in this example system the conditional mean is less dependent on the initial value x in the chaotic regime. Sections 3.3 and 3.4 give important results for prediction, but for higher-noise cases it may be preferable to compute the measures along actual trajectories of the system rather than trajectories of the "skeleton".

As a complement to the suite of measures proposed by Professor Tong, we have found it useful to consider the measures of SDIC based on products of map Jacobians at the observed data (as in (6)). The (theoretical) infinite product gives the 'neoclassical' Lyapunov exponent; finite products can be used to calculate local Lyapunov exponents, generalizing Wolff's (1992) approach to higher dimensions and to systems with noise (Bailey, 1995). These local exponents give important information about the heterogeneity of the system, such as identifying regimes where the system exhibits SDIC even if overall the system is not chaotic. The dynamics of measles epidemics (Ellner *et al.*, 1995), and many animal populations (Ellner & Turchin, 1995) appear to have this behaviour. The definition in (6), which only involves derivatives of F, is a special case of the general definition (given by Kifer, 1986). Equation (6) is only appropriate for i.i.d. additive noise (10). The general definition, which applies to systems with possibly non-additive noise (such as (9)), does have the property of invariance under non-linear co-ordinate transformations. Definition (6) does not, because the noise will generally not be additive in both scales. Jensen's (1993) results produce a seeming contradiction because he assumed additive noise on both the original and transformed scales.

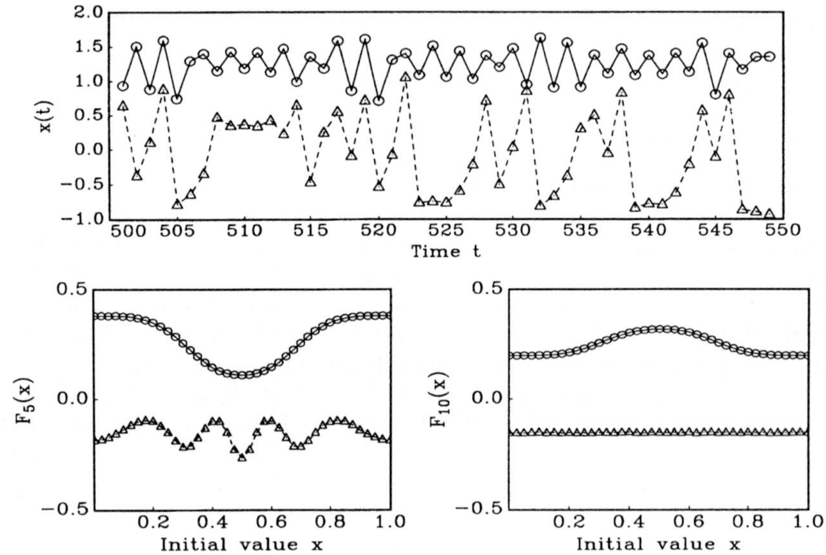

Fig. 1. Dynamics and conditional mean for the system $x_{t+1} = \alpha \cos(\pi x_t) + \sigma e_t$, with $\alpha = 0.5$ (○, solid line) and $\alpha = 0.89$ (△, dashed line); the errors e_t are i.i.d. normal (0, 1) and $\sigma = 0.1$ for both values of α. Upper panel shows a typical segment of the time series, and lower panels show the conditional mean function $F_m(x) = E(X_m \mid X_0 = x)$ for $m = 5$ (lower left panel) and $m = 10$ (lower right panel), estimated by Monte Carlo simulation with $n = 250{,}000$ replicates. The skeleton system for $\alpha = 0.5$ has a stable two-point cycle that is still evident in the dynamics with noise and accounts for the form of the conditional mean $F_m(x)$. The skeleton system for $\alpha = 0.89$ is apparently chaotic in numerical simulations, and the noisy system mixes rapidly so that the conditional mean is essentially independent of the initial point after $m = 10$ time steps.

In closing we would like to highlight some experiences in transferring these ideas to analysing data. While neural net models may seem exotic, we feel that they provide the best chance for avoiding the curse of dimensionality (Barron, 1991a, b) that afflicts kernel and local regression estimators. We have also found it useful to blend substantive models for the system based on scientific principles with "black box" empirical techniques. For example, in fitting a model to time series of disease case reports, it is very helpful to use a bivariate model that also tracks the number of susceptible individuals for the disease. Compared with purely "black box" models, these semi-mechanistic models have higher out-of-sample forecasting accuracy on historical data on measles, and have much lower variability (over different cities) in measures that characterize the dynamics, because they incorporate information in addition to that present in the time series. Finally, given the discussion above, it is always imperative to check residuals to confirm additive noise structure.

References

Bailey, B. A. (1995). Local Lyapunov exponents: predictability depends on where you are. To appear in: *Proceedings of the Ninth International Symposium in Economic Theory and Econometrics* (eds W. Barnett, A. Kirman & M. Salmon). Cambridge University Press, Cambridge.

Barron, A. R. (1991a). Complexity regularization with application to artificial neural networks. In: *Nonparametric function estimation and related topics* (ed. G. Roussas), 561–576. Kluwer Academic, Amsterdam.

Barron, A. R. (1991b). Approximation and estimation bounds for artificial neural networks. In: *Proceedings of the Fourth Annual Workshop on Computational Learning Theory* (eds M. K. Varmuth & L. Valiant), 243–249. Morgan Kaufmann, San Mateo, CA.
Ellner, S. & Turchin, P. (1995). Chaos in a noisy world: new methods and evidence from time series analysis. *Amer. Naturalist* **145,** 343–375.
Ellner, S., Gallant, A. R. & Theiler, J. (1995). Detecting nonlinearity and chaos in epidemic dynamics. In: *Epidemic models: their structure and relation to data* (ed. D. Mollison), 229–247. Cambridge University Press, Cambridge.

L. R. SMITH
Mathematical Institute, University of Oxford

It is a great pleasure to contribute to the discussion of Professor Tong's paper. As a physicist, I am struck by the symmetry between disciplines: many physicists are, at best, suspicious of statistical approaches to the analysis of physical dynamical systems. Reflecting Professor Tong's observation of statisticians, some physicists have formed the impression that statisticians attempt to explain almost purely deterministic phenomena by random systems and take their leave at this point, as the models contain "no physics". Both the present paper, and Professor Tong's previous work, have contributed to the partial lifting of this confusion in both camps. In this brief comment, I would like to note some phenomena which arise even when a perfect model of the system is available, but the initial conditions are not known precisely, a situation where probability relates solely to our ignorance.

Predictive models discussed in section 4.4.3 effectively perform interpolation in state space, and great care has been taken to account for "noise" on the learning data, usually by invoking a least-squares approach. Presumably, the initial condition upon which each prediction is based is also uncertain, but there has been a tendency to evaluate the prediction of deterministic systems by comparing a single forecast with a single verification trajectory. An alternative approach is to examine the trajectories of an ensemble of initial conditions (each consistent with a given observation). This approach provides a better evaluation of the model, more information about the quality of the prediction, and reveals how linear intuitions can lead us astray when interpeting non-linear dynamics. In fact, I believe each of the five remarks of section 3.3 is illustrated in Fig. 2.

This figure reflects the evolution of a perfect ensemble of initial conditions evolving under a perfect model of the set of three ODEs known as the Lorenz equations (Lorenz, 1963). The initial distribution is a collection of 1024 points which are observationally identical: at 8-bit resolution, these points all lie in the same cube in phase space. Time increases vertically, and at each point in time, the distribution of the values of the variable x for the ensemble is shown. Thus the figure represents the evolution of the probability density function under a perfect model: the only uncertainty in this analysis is due to the quantization of the initial condition (see Smith, 1994 for more details).

Initially the distribution spreads out as one might expect, remaining mono-modal; but at $t \approx 0.50$ the distribution sharpens! It is easier to predict 0.50 into the future than 0.25; this window of opportunity is unexpected in linear systems, but was shown to occur in non-linear systems by Tong & Moeanaddin (1988). For the Lorenz system, its existence can be established analytically (see Ziehmann-Schlumbohm *et al.*, 1995). Meteorologists call this phenomenon "return of skill" and its existence in weather forecasting is hotly contested.

Similarly, as the distribution quickly becomes multi-modal, the standard deviation becomes useless as a measure of the quality of the forecast. At $t \approx 3.50$, for example, the standard deviation is significantly greater than its asymptotic value; ensemble forecasts

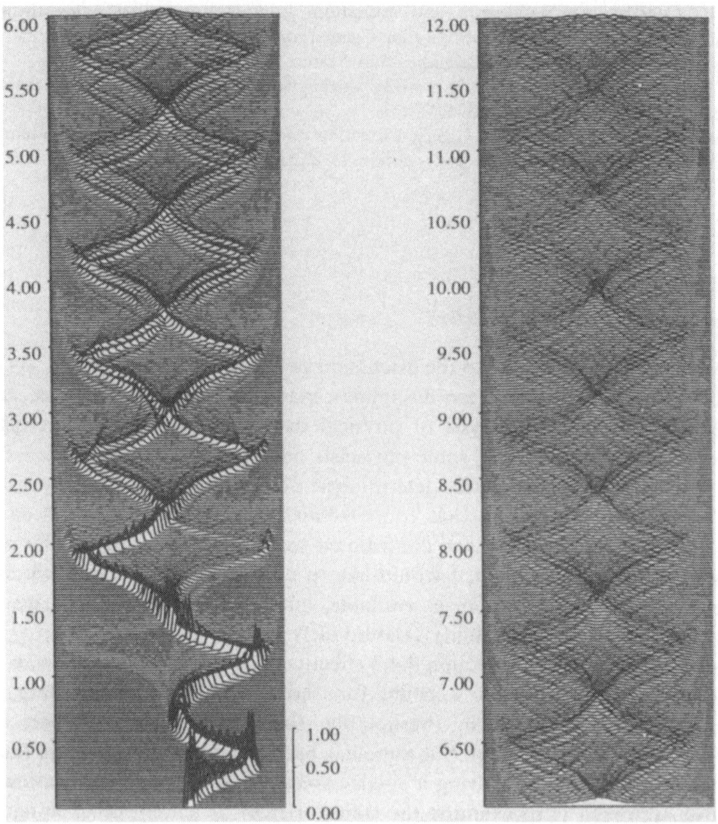

Fig. 2. The time evolution of the probability density function (PDF) corresponding to an 8-bit observation on the Lorenz attractor. Time, t, increases from the bottom to the top, the two columns show $0 < t < 6$, and $6 < t < 12$. At each time step, the probability density for the variable x is shown, the scale for the PDF (i.e. from 0 to 1) is located just to the right of the first column.

indicate that an extreme value of x will be observed, although extreme positive and extreme negative values are almost equally likely. This illustrates the difficulty we may get into by evaluating non-linear prediction models in a least-squares sense: as the distribution is roughly symmetric, least-squares solutions will tend to predict central values, where there is no chance of actually observing the system. In this case, evaluating a time series of forecast/verification pairs in the least-squares sense provides a poor measure of the quality of the model.

As time increases, the (commercial) value of the forecast becomes harder to assess. Eventually, the ensemble will become indistinguishable from randomly chosen observations of x (i.e. the projection of the invariant measure); at this point the forecast is useless. But the transition to this asymptotic distribution is not monotonic, and its duration can be long compared to the typical dynamic time scales of the system. It also varies tremendously with the location of the initial condition in state space.

I repeat that this uncertainty is due only to quantization of the initial observation: we have a perfect model, employ a perfect ensemble of initial conditions, each of which is consistent with both the long-term dynamics and the observational "noise". This simple deterministic case provides only a hint of phenomena available in truly stochastic, non-linear systems. Professor Tong's paper provides an overview of the statistical questions which arise in both perfect and imperfect cases, as with the analysis of physical systems; these offer a wide variety of additional interesting, open problems.

References

Lorenz, E. N. (1963). Deterministic nonperiodic flow. *J. Atmos. Sci.* **20,** 130–141.

Smith, L. A. (1994). Visualising predictability with chaotic ensembles. In *Advanced signal processing: algorithms, architectures and implementations* (ed. F. T. Luk), 2296, 293–304. SPIE, Bellingham, WA.

Tong, H. & Moeanaddin, R. (1988). On multi-step non-linear least squares prediction. *The Statistician* **37,** 101–110.

Ziehmann-Schlumbohm, C. Fraedrich, K. & Smith, L. A. (1995) Ein internes Vorhersagbarkeitsexperiment im Lorenz-Modell. *Meterologische Zeitschrift N.F.* **4,** 16–21.

R. L. SMITH
Cambridge University

Howell Tong has written an excellent survey of non-linear time series analysis from the point of view of dynamical systems. In the six years since Tong himself organized a workshop in Edinburgh on this theme, there have been many developments well represented by the collections of papers edited by Tong & Smith (1992), Drazin & King (1992) and Grenfell *et al.* (1994); see also Isham (1993) and Jensen (1993). Nevertheless, the subject still seems to be regarded with some suspicion by both statisticians and dynamicists. I think there are a number of reasons for this, not the least of them being that, in a subject where so much of the work necessarily has a strong computational component, the distinction between computation and theory can become very blurred. For example, many of the "well-known results" for the properties of systems such as the Hénon map are still known only by computer simulation and not by rigorous mathematical proof. Tong's survey draws attention to numerous topics on which new theory—not merely new computational expertise—is being developed, and also makes clear several areas where new theory is needed.

From the point of view of "what statistics can offer to the development of dynamical theory" my own view remains: a very great deal! Most of the proven "successes" of dynamical systems are in areas such as astrophysics and fluid dynamics, where the equations may be hard to solve but, at least, everyone agrees what the equations are. More subtle applications to fields such as economics, epidemiology and physiology require (a) gaining information from real data about the dynamics of the system, and (b) expanding the theory of dynamical systems to include random disturbances—since in these fields, unlike fluid mechanics, it is scarcely believable that any tractable set of equations could *exactly* describe the dynamics; we have to allow for some essentially random component. However, research on these questions is fundamentally statistical in nature and so points towards a clear need for the kind of theory Howell Tong has described here.

Turning to more technical issues raised by the paper, Tong makes a big distinction, as he has in some previous papers, between estimating the embedding dimension and estimating the fractal dimension of the attractor. While agreeing with the distinction, I am not convinced that it is a matter of "either ... or". The two dimension concepts reflect quite different aspects of the system and I do not feel that it is a question of which is more important, or more easily estimated.

Perhaps the following artificial example will illustrate some of the distinctions. Suppose we generate random variables (X_1, X_2, X_3) according to some continuous distribution on $\mathbb{R}^3$, and then successively

$$X_4 = f(X_1, X_2, X_3),$$

$$X_5 = f(X_2, X_3, X_4),$$

$$\ldots$$

$$X_{12} = f(X_9, X_{10}, X_{11}),$$

where f is some smooth non-linear function, to create a 12-vector $\mathbf{X} = (X_1, \ldots, X_{12})$. Now repeat the procedure may times to obtain an i.i.d. sample in $\mathbb{R}^{12}$. (In reality, it is more likely that we would be working with a single long time series, and that we would have some noise component as well, but I am setting it up this way for simplicity.) Now suppose we estimate the "dimension" of this set of vectors using either the embedding dimension or the fractal dimension techniques. Either should, if applied correctly to a large enough sample, give the answer 3. Note, however, that the correlation dimension is invariant under either of the following transformations:

(i) replace X_j by $X_{\sigma(j)}$ where σ is some permutation of the integers 1–12,
(ii) replace $\mathbf{X}$ by $\mathbf{Y} = g(\mathbf{X})$ where g is a continuous (maybe even linear) function from $\mathbb{R}^{12}$ to itself.

It is clear that these transformations would destroy any direct attempt to estimate an embedding dimension by the methods used by Tong and his co-authors. So we would expect the fractal dimension to be the harder of the two to estimate: it reflects more general properties of the data.

Moreover, the discussion so far is concerned solely with "static" aspects of dimension. When we consider the dynamics as well, the distinction becomes even stronger. When we say that, for the Hénon map, the embedding dimension is 2, we are referring solely to a finite-dimensional property of the system, in the sense that any succession of three or more consecutive observations can be expressed as a deterministic function of the first two. On the other hand, the (computer-based) statement that the correlation dimension is about 1.2 is a statement about the long-run behaviour of the system and so something completely different.

None of this is intended to discredit the excellent work that Cheng & Tong have done on the determination of an embedding dimension. It is certainly an important practical problem, and one that has, until their work, lacked any rigorous statistical solution. However, I do think that there are reasons why fractal dimensions have attracted the bulk of the attention in the past and will continue to do so in the future. More work is needed on the scientific interpretation of fractal dimensions and I have followed Colleen Cutler's (1993, 1994) work with particular interest for this reason.

Perhaps before leaving the subject of dimension estimation, I might be allowed to make some points of clarification! The sample size requirements quoted in R. L. Smith (1992) are for a very specific model—independent Gaussian data—and the figures would be completely different even for correlated Gaussian data, let alone something like the Hénon attractor. L. A. Smith's (1988) computations were again for a specific set-up and were, I believe, intended to illustrate some of the difficulties of dimension estimation in general rather than to provide a specific prescription. That is certainly the way I think of these results. On the statistical properties of dimension estimates, these all stem from the fact that the correlation integral (51) is an example of a U-statistic, and Denker & Keller (1986) proved a central limit theorem for this by extending the well-known asymptotic normality of U-statistics to the case of dependent random variables. However, this is for fixed r and $N \to \infty$. For estimating the correlation dimension, it is natural to look for limit theorems as $r \to 0$ at the same time as $N \to \infty$. This involves us in Poisson approximations for U-statistics. The first paper to give a theorem of this form was, I believe, Silverman & Brown (1978), though Barbour & Eagleson (1984) simplified and strengthened their result by making use of the Stein–Chen method; cf. Barbour *et al.* (1992). In R. L. Smith (1992) I sketched an outline of this and subsequently Wolff (1994) made more explicit the application of Stein–Chen in this context. At the moment, I sense that the interplay between all these different approaches is not generally appreciated.

Finally I would like to turn to section 3. Despite all the work on fractal dimensions, it is clear that they are not the only dynamical quantity that it is of interest to measure, and quantities that try to measure the predictability of a system are very important. For that reason, I welcome attempts such as the work on Kullback–Leibler information, which aim to explore these issues more thoroughly. On the subject of local Lyapunov exponents, Wolff's one-dimensional definition (section 3.2) may be reinterpreted in a setting which uses non-parametric regression methods to reconstruct the dynamics, and then estimates the LLE from the derivatives of the estimated map. From this point of view, the method generalizes naturally to higher dimensions. Recently Lu (1994) has obtained theoretical asymptotic expressions for the bias and variance of LLE estimators, when the method of function reconstruction is local polynomial regression. Lu's results exploit and generalize recent results of J. Fan, M. Wand and D. Ruppert on this method of function reconstruction. So far they have not been tried out in any practical examples but they do suggest one way to obtain some rigorous statistical results for the estimation of these important exponents. Much of the attention on fractal dimensions up until now has, I suspect, been due to the existence of a simple algorithm (Grassberger–Procaccia) for estimation. However, ease of estimation is not the same thing as scientific importance, and as we learn more about how to estimate other dynamical quantities, I would expect them to be studied more extensively than fractal dimensions. Tong's paper provides much food for thought on all of these matters.

References

Barbour, A. D. & Eagleson, G. K. (1984). Poisson convergence for dissociated statistics. *J. Roy. Statist. Soc. Ser. B* **46**, 397–402.

Barbour, A. D., Holst, L. & Janson, S. (1992). *Poisson approximation*. Oxford University Press, Oxford.

Lu, Z.-Q. (1994). Estimating Lyapunov exponents in chaotic time series with locally weighted polynomial fit. PhD thesis, Department of Statistics, University of North Carolina at Chapel Hill.

Silverman, B. W. & Brown, T. C. (1978). Short distances, flat triangles and Poisson limits. *J. Appl. Probab.* **15,** 815–825.

Wolff, R. C. L. (1994). Independence in time series: another look at the BDS test. *Phil. Trans. Roy. Soc. (Lond.)* **A348,** 383–395.

RODNEY C. L. WOLFF

Queensland University of Technology

We foreigners are never brought so rudely face to face with non-linearity as when we visit London! The man in the street (cf. section 3.3(i)) will abandon all regard for linearity as he traipses along a gentle deceptive curve. He can thank Professor Tong's view of chaos and non-linearity for restoring his orientation!

To remark on the local Lyapunov exponent (LLE) (section 3.2 and Wolff, 1992), this prototype statistic indeed invites more sophistication. I would suggest that replacing the S_is by the natural kernel estimator analogue would render a statistic more amenable to rigorous statistical analysis, in conjunction with the theory of U-statistics as Professor Tong suggests. Invariance of the LLE under one-to-one co-ordinate transformations may well not hold; however, this need not impede their use. Actual numerical values of the LLE are not informative—as, I claim, is the case with Yao & Tong's (1994a, b) conditional distribution approach—but relative magnitudes are. Change of co-ordinates may deform the scale non-linearly, but orderings of the complete set of LLEs ought to remain fixed. For $d = 1$, the sum of the sample LLEs should converge to the global Lyapunov exponent as $\Sigma n_i \to \infty$ and $\delta \to 0$ at an appropriate rate. The beauty of Yao & Tong's method is that issues of the convergence of δ are much less messy. Moreover, the cross-validatory application of LLEs—perhaps their safest and most effective use—should remain just as useful after a one-to-one co-ordinate transformation, even if the LLEs themselves are not invariant. It would be interesting to see if Yao & Tong's statistics perform similarly in such an application.

Professor Tong identifies disciplines in which the concept of chaos has enjoyed well-founded applications (section 1). Such areas, however, will invite further statistical study. The existence of parametric models suggests usefulness of likelihood methods. These have been explored partially by Berliner (1991) and Geweke (1990), the latter's bizarre findings having their roots in the possibly unnatural assumptions about the noise process. Even if precise parametric forms are not available, local likelihood (Hastie & Tibshirani, 1987) can enable piecewise linear models to be fitted ... and here is a link with threshold (piecewise linear) autoregressive time series models, in accordance with Professor Tong's perspective. The interpretation of a hypothesis test in the presence of deterministic chaos and the applicability of standard sampling distributions will require close and careful attention. Hall & Wolff (1995) have already considered a special case of the latter problem, but a personal overview of these issues from Professor Tong would be warmly welcomed!

References

Geweke, J. (1990). Inference and forecasting for deterministic non-linear time series observed with measurement errors. In *Non-linear dynamics and evolutionary economics* (eds R. Day & P. Chen).

Hall, P. & Wolff, R. C. L. (1995). On the strength of dependence of a time series generated by a chaotic map. *J. Time Ser. Anal.*, in press.

Hastie, T. & Tibshirani, R. (1987). Local likelihood estimation. *J. Amer. Statist. Assoc.* **82,** 559–567.

Wolff, R. C. L. (1992). Local Lyapunov exponents: looking closely at chaos (with Discussion). *J. Roy. Statist. Soc. Ser. B* **54,** 353–371.

REJOINDER TO THE DISCUSSION

HOWELL TONG

I would like to thank the organizers of the conference for the opportunity of presenting the series of lectures, the local organizers for their charming hospitality and to the discussants for their kind words and thought-provoking comments and questions. The convergence of opinions among many of the discussants and myself is at such an unexpectedly rapid rate that I do not detect any significant sensitivity to initial conditions. Perhaps the 1989 Edinburgh SERC/RSS International Workshop, to which Professor Richard Smith has referred, prepared the ground well because I can count no fewer than seven amongst the discussants here who were participants at the Workshop. (Need I say that our Editor was also present?) As one would expect, the convergence rate is not uniform and I shall presently turn to these local variations as well as the many challenging questions raised by my learned colleagues. Naturally, I think I would have to limit my scope.

1. The deterministic and stochastic divide?

Professor Jensen understands correctly that for me chaos theory is a means to an end. I doubt if I would pass the 'cricket test' even if I were to emigrate to the Land of Chaos. My ideal destination is the Kingdom of the True, the Good and the Beautiful where all the secrets of Randomness are revealed. I share completely the scepticism of Sir David regarding the likely impact of chaos for complex (i.e. non-isolated) systems. It is precisely because simplification of one type or another is almost always inevitable when we try to analyse a truly complex system that I prefer using models such as (9) and (10) to (1) in the analysis of real time series data. Professor LeBaron has reinforced this point with special reference to economic systems. Professor Guégan seems to appreciate the necessity for the former but prefers to embrace fundamentalism, considering the former ambiguous but without elaboration. At the same time, I am pleased to note the apparent support of Professor Ozaki for (10). I had previously expected that he would see the advantage of integrating the skeleton (1) with the stochastic model (10) because he used to support doing so when the skeleton admits a periodic attractor. Professor Cutler is quite right in stressing the fact that whilst in hyperbolic exponentially stable systems with dynamic noise the skeleton is particularly revealing as shown by Chan & Tong (1994), in some other cases the clothed stochastic model may well exhibit very different asymptotics from that of the skeleton. In this connection, Takens (1994) is very relevant, which illustrates the fact that our dynamical systems friends have a lot of wisdom to offer to us statisticians. Naturally, we statisticians have a lot of wisdom to offer to them too in return. Professors Lawrance and Richard Smith and Dr Wolff have expounded the latter most eloquently. Personally I am always enormously thankful for continuing to benefit from two great cultures.

Of course, when it comes to the analysis of a particular real time series, I agree with Professor Guégan that it would be interesting to have some idea as to whether it is the deterministic randomness (the YANG?) or the stochastic randomness (the YIN?) that plays the more dominant role. In this connection, Yao & Tong (1995) have recently shown that by minimizing

$$\frac{1}{n}\sum_{t=1}^{n}\{\hat{f}_1(X_t)-f_1(X_t)\}^2 W(X_t),$$

where $\hat{f}_1(x)$ is given by (44) and $W(\cdot)$ is a weight function (taking $d=1$ for simplicity), they obtain the optimal bandwidth $\tilde{h}_n$

$$\tilde{h}_n=\frac{\alpha}{n^{1/5}}\left\{\frac{\int \sigma_1^2(x)W(x)\,dx}{\int \{\ddot{f}_1(x)\}^2 P_1(x)W(x)\,dx}\right\}^{1/5},$$

where α is a constant depending only on the kernel P_1. Now, clearly $\tilde{h}_n \approx 0$ if the system is ODS, i.e. non-linear and 'operationally deterministic' (i.e. $\sigma_1^2(x)\approx 0$, for all x). On the other hand, $\infty > \tilde{h}_n > 0$ if the system is SS, i.e. non-linear and stochastic. For a linear stochastic system, $\tilde{h}_n$ is infinite. By establishing that $(\hat{h}_n-\tilde{h}_n)/\tilde{h}_n \to 0$ in probability for a data-driven $\hat{h}_n$ (e.g. by the CV method), we have used $\hat{h}_n$ as a pivotal statistic in a bootstrap setup to discriminate between an ODS and an SS with very encouraging results. For example, experimenting with the tent map in both forward time and backward time, our pivotal statistic readily identifies the former as ODS and the latter as SS.

2. The Lyapunov exponent

Professor Nychka *et al.* regard initial-value sensitivity and mixing as two seemingly contradictory properties of chaos. I beg to differ. If they note that initial-value sensitivity says that "where you are now is very important *for the short term*" then their apparent contradiction will disappear. It is this short-term effect which leads to the long-term deterministic randomness and mixing if the latter holds. Thus, the two notions complement (rather than contradict) each other. Once this point is appreciated, the position taken by Yao & Tong (1994a esp. sect. 2, 1994b) should be clear to them. Specifically, we focus on the effect on short-to-medium-term prediction due to the initial-value sensitivity of the conditional distributions or conditional means. As far as the identical-noise-realizations approach is concerned, I agree with Professor Chan and Professor Nychka *et al.* that Professor Jensen's objection disappears if the additivity restriction is dropped. The $\tilde{\lambda}(Y_0)$ described by Professor Chan seems to be related to the later part of the discussion by Professor Nychka *et al.* It is clear to many that the classical definition of the Lyapunov exponent is not intended to be useful for the purpose of prediction, not even short-term prediction. In this connection, Dr Wolff's attempt, (15), is laudable. He has now further given some 'sophistry'. I accept his point that generally it is the equivariance of $\lambda_{i,m}$ that matters. I am impressed with his anticipation of Professor Jensen's query concerning the possibility of $\lambda_{i,m}$ going to ∞; I tend to agree with Dr Wolff that provided one is careful in controlling the rate of $\Sigma n_i \rightarrow \infty$ and $\delta \rightarrow 0$, his $\lambda_{i,m}$ will not explode. His point about recasting part of the definition of his $\lambda_{i,m}$ in a kernel estimator form would bring it closer to the conditional mean approach and is clearly related to the work of Dr Lu mentioned by Professor Richard Smith.

Professor Jensen has given a variation of the same theme as the conditional mean approach. I agree with him that $\kappa(x)$ in (28) is unlikely to enjoy any intrinsic value as indeed we have never claimed that it does (see sect. 2.4 of Yao & Tong, 1994a). Nevertheless, it remains an intellectual curiosity as to its value, which he seems to have conjectured negative under geometric ergodicity. I would welcome the opportunity to study his proof! I agree with Professor Chan that more work needs to be done in order to further assess the various approaches. At present, I am inclined to share Professor LeBaron's belief.

Professor Johansen has raised a very challenging question (his second question). Interestingly, I have also previously raised a similar question! (see the chapter entitled "An overview on chaos" in Titterington (1994)). I cannot give a complete answer. However, I shall try to indicate some current and related developments. To simplify discussion, let us assume that his $f(X_{t-1}, \varepsilon_t)$ is additive, i.e. it takes the form of $f(X_{t-1}) + \varepsilon_t$. In this case, the two-dimensional deterministic dynamical system has the Lyapunov spectrum (assumed to exist) consisting of λ_f, λ_g, where λ_f is the Lyapunov exponent of the system $X_t = f(X_{t-1})$ and λ_g (typically positive) is that of the system $\varepsilon_t = g(\varepsilon_{t-1})$. Our present investigation suggests that how well the two-dimensional deterministic dynamical system simulates the stochastic counterpart possibly depends on λ_g, the entropy and correlation dimension associated with the system $\varepsilon_t = g(\varepsilon_{t-1})$ as well as the strength of dependence of the data generated from the system. For example, suppose $f(x) = \beta x$. Stockis & Tong (1995) have found that how well the standard sampling theory applies to the nominal maximum likelihood estimator $\hat{\beta}$ tends to depend on the above-mentioned factors for the mapping $\varepsilon \mapsto g(\varepsilon)$. Specifically, they have used the $\varepsilon_t s$ generated by the logistic map $\varepsilon \mapsto \theta\varepsilon(1-\varepsilon)$. (All the values of θ are chosen to lie in the "chaotic regime".) They have found that the sampling properties of $\hat{\beta}$ depend on the value of θ used and can depart substantially and unsmoothly from the standard sampling theory. In particular,

$$\hat{\beta} - \beta \approx \sum \varepsilon_t X_{t-1} \Big/ \sum X_{t-1}^2 = \sum \varepsilon_t \{\varepsilon_{t-1} + \beta\varepsilon_{t-2} + \beta^2 \varepsilon_{t-3} + \cdots\} \Big/ \sum X_{t-1}^2.$$

Thus, the autocorrelation function of $\{\varepsilon_t\}$ play a vital role here. Using Hall & Wolff (1995), as my best compliment, we expect and notice negative bias when $\theta = 3.98$, 3.825 and 3.58; when $\theta = 4$ the standard sampling theory applies quite well. Turning to the more philosophical sub-questions raised by Professor Johansen, I am inclined to believe that (i) there is a stochastic dynamical system and (ii) we can create a man-made dynamical system to mimic the former at least to the extent that central-limit-theorem based statistical inference could be justified provided we tune the parameters of the latter with sufficient care. However, I do not think that a stochastic dynamical system is necessarily obtained if $\lambda_g \to \infty$ because the (Takens') embedding dimension of the above deterministic dynamical system remains equal to 2, which by Takens' theorem (or its extension) limits the (box-counting) dimension of the attractor to a small finite number; this is typically not the case with a stochastic dynamical system.

3. Dimensions

Professor Jensen has raised the question as to whether observing $\{X_t\}$ of (9) via $Y_t = h(X_t)$ would lead to an autoregressive model for $\{Y_t\}$. The answer is in the affirmative if h is one-to-one. I am unclear about his "provoking statement" because I would have thought that (39)–(48) have spelt out the restrictions on the class of functions for f_d. I am not unaware of the pitfalls of free lunches. He also seems to take me to task on the choice of $d = 7$ for the New York measles data, but Cheng and I (1992, esp. p. 441 and p. 443) did stress that this CV choice was only adopted tentatively and with several caveats. It is well known that in any non-parametric (or rather infinite-dimensional) function estimation, the curse of dimensionality can cause us serious problems. What Cheng and I have found is that the problem is fortunately less serious for non-linear autoregressive order determination because of the cylinder effect. Unlike correlation dimension estimation and some others, here we are not concerned with the "finer structure" of the data, a point also touched upon by Professor Richard Smith, and hence the problem becomes easier.

Professor Cutler has discussed, in the present context, the problem of inference of a non-ergodic process on the basis of one realization. As she has given that for her model (53) the fractal dimension corresponding to each ergodic component is $2/(\alpha - 1)$, therefore by Takens' theorem (with the usual caveats and/or its recent extension), there corresponds a finite embedding dimension. Consequently, on the basis of one realization, the CV method will also lead to a finite estimated embedding dimension.

Turning to the fractal dimension of the attractor, Professor Cutler shares my view that it has been over-used in practice and sometimes assigned an importance out of proportion to reality. What I have done in this and previous papers is an attempt to restore some proper balance. I believe that the voice has now been heard. There are really no fundamental differences between the views of Professor Richard Smith and mine. His 'artificial' example highlights the fact that correlation dimension is essentially a spatial concept whilst embedding dimension is essentially a temporal concept. It is therefore not surprising that if one tampers with the temporal sequence one loses all temporal information. Similarly, if I intercept Professor Richards Smith's data and add some dynamic noise to $X_4, \ldots, X_{12}$, with different variances for the different 12-vector **X**s, before passing them back to him, I doubt if he could get a good estimate for the correlation dimension any more. However, I can for the embedding dimension.

4. Modelling and prediction

There is clearly considerable consensus of opinions that non-linearity reveals features unknown to the linear diehards. As one of the early birds to non-linearity, Professor Lawrance seems to bemoan the difficulties in converting our linear diehards. Curiously, the woman in the street needs no conversion! To her (especially if the location happens to be in Hong Kong), all I need to say is that, other things being equal, she is right now in and around the state at which a window of opportunity is open to her. She will then happily take care of the rest of the RISK business. Can any linear diehard ever do that? The point is that the lady is free and has never had her feet bound! Dr Lenny Smith has demonstrated very elegantly the greater realism of non-linearity in this respect. Equally, Professor Johansen should take comfort in knowing that the decomposition theorem of section 3.4 assures us that the window of opportunity is open to any mean-square consistent estimator $\hat{f}_m$. Naturally, it is up to the individual to exploit the opportunity as best (s)he can; each may have his/her favourite models and I certainly have mine. In any non-linear (i.e. free) society, one can be spoiled for choice and I would not apologize for disappointing Professors Johansen and Lawrance that usually the choice is not unique. Why should it be anyway and would that be desirable?

5. Point processes, quantum mechanics and non-linear-unit roots

Of course, I knew that I could not escape some tough questioning from Sir David! Before answering his first question, let me recall his own point process paper with W. L. Smith (1953), which was characteristically ahead of its time. First, let me translate the famous *Gleichverteilungssatz* of H. Weyl (apparently also independently discovered by P. Bohl and W. Sierpinski about the same time), which forms the basis of their work: "For the deterministic dynamical system: $x_n = x_{n-1} + \theta$, $x_0 = 0$, $(n = 1, 2, \ldots)$, where we observe $y_n = x_n \bmod 1$, the time series $\{y_n : n = 1, 2, \ldots\}$ is uniformly distributed on $[0, 1)$ if and only if θ is irrational". This is a special case of Arnold's circle map (see e.g. Ott, 1993). Now, consider a logistic map $x_n = x_{n-1}(1 - x_{n-1})/2$ over the field of algebraic numbers and observe $y_n = x_n \bmod 8$. In this case, starting with $x_0 = 14$, the observed time series is $\{6, 5, 6, 1, 0, 0, 0, 0, 0, 0, 0, 0, 4, 6, 1, 4, \ldots\}$. Note that $|x_4 - 0|_2 = 2^{-10}$, where we have used the p-adic metric $|\ |_p$. (By definition, for p prime, $|u - v|_p = p^{-r} \Leftrightarrow u \equiv v \bmod p^r$ and $u \not\equiv v \bmod p^{r+1}$.) Note also the long sequence of zeros in the y-series, which begins to capture, albeit not yet completely, features of point processes. I understand from my number theorist colleague, Dr C. F. Woodcock, it is possible to study p-adic dynamical systems, which can lead to arithmetic chaos. This seems to me a truly fascinating subject, which may well have potential implications for the study of point processes. In a sense, Sir David might have already answered his own question for me more than forty years ago!

Sir David's second question has to do with quantum mechanics. Curiously, Professor Takens has also raised a similar question in private communications with me. I am not aware of any definitive result but here are some of my unprofound thoughts. It seems to me that between (1) and (9), which model is appropriate might depend on the level of resolution, with the former corresponding to infinite resolution. If we hypothesize a finite upper bound for resolution, then we might obtain a kind of uncertainty principle. Moreover, following Kifer (1986) we can think of (9) as consisting of an ensemble of deterministic dynamical systems, one for each realization of the dynamic noise. This interpretation reminds me of the concept of multiverse in quantum mechanics. At a technical level, there are puzzling features. For example, since the Schrödinger equation is really linear I wonder if and how 'non-linear quantum dynamics' has been formulated. It is also unclear to me if and how initial-value sensitivity has been developed in quantum chaos either. Wouldn't it be simply magnificent if the conditional distribution/mean approach has something to offer in this respect?

Professor Johansen's other question concers a very important area, so far scarcely trodden. The inferential problems are technically quite daunting. For example, even for the special case of a piecewise linear f, Pham *et al.* (1991) have only managed to obtain strong consistency for the least squares estimates of the "slope parameters", leaving open the problems of the convergence rate and the limit distribution. More recently, prompted by my collaborator, the biologist Professor M. Carwley, Professor K. S. Chan and I (1995) have considered the related problem of testing for non-linear unit roots. As an illustration, let us consider the famous cyclical Soay sheep annual data (1955–93) from the island of Hirta, for which I have detected some non-linearity in their population dynamics. In order to test the biologically important hypothesis of "density independence" over the depressed population regime, we are led to test for a unit root in this regime. The following threshold model on the log scale has been fitted: $X_t = 0.63 + 0.82X_{t-1} + \varepsilon_t$ if $X_{t-1} \leqslant 3.05$, and $= 3.03 + \varepsilon_t$ otherwise. Here, $\varepsilon_t \sim$ i.i.d. (0, 0.01). The standard error of the estimated coefficient of X_{t-1} is 0.22. Chan and I (1995) have shown that by conditioning on the sample size in depressed regime, we can apply the Dickey–Fuller test to this regime to test H_0: the coefficient of X_{t-1} is 1. The test is not significant at the 10% level of significance. Perhaps I might be permitted to end on a potentially controversial note: despite the huge resources invested by the economists/econometricians in the so-called unit root problem, they might have missed the main target because of their pre-occupation with linearity. I understand that some of them are beginning to ask the question: is there a unit root problem?

References

Chan, K. S. & Tong, H. (1995). On tests for nonlinear unit roots. (In preparation.)

Pham, D. T., Chan, K. S. & Tong, H. (1991). Strong consistency of the least squares estimator for a non-ergodic threshold autoregressive model. *Statist. Sinica* **1**, 361–369.

Stockis, J.-P. & Tong, H. (1995). A note on pseudo-random numbers and statistical inference. (In preparation.)

Yao, Q. & Tong, H. (1995). On bandwidths for dependent data. (In preparation.)

Crossing the Bridge Backwards: Some Comments on Early Interdisciplinary Efforts

COLLEEN D. CUTLER
Department of Statistics and Actuarial Science, University of Waterloo
Waterloo, Ontario, N2L 3G1, Canada
E-mail: cdcutler@uwaterloo.ca

When we see a natural style, we are quite surprised and delighted, for we expected to see an author and we find a man.

– Blaise Pascal

1. Introduction

It is a real pleasure to be asked to contribute to this volume in honour of Professor Howell Tong and his work. As we all know, Howell brings not only incredible talent and creativity to his work, but generosity, enthusiasm, and an unmistakable uniqueness of style. I was reminded of this once again while reading "A personal overview of non-linear time series analysis from a chaos perspective" (Tong, 1995) as well as Howell's recent and delightful article "Birth of the threshold time series model" (Tong, 2007). The dubious notion of double-blind refereeing would most certainly fail miserably in Howell's case, even if all names and dates were changed to protect the guilty. (It seems Pascal can add accurate long range forecasting to his list of accomplishments.)

Howell's dedication to producing important results within the field of statistics is matched only by his determination to bring together disciplines such as statistics, the physical sciences, and econometrics to pursue common goals, especially the modelling and forecasting of nonlinear and chaotic time series. His enthusiasm for the subject is unparalleled. I recall visiting him once at the University of Kent in Canterbury in 1994. I had been completely unable to sleep on the plane and was jet lagged and exhausted beyond belief. Howell was excited to talk about chaos, fractals, and nonlinearity. He took me out for supper and wasted no time getting into his favourite topic while my head, unfortunately, did a decidedly linear dive into my dinner plate.

Professor Kung-Sik Chan has asked me to talk a bit about the three papers "On the use of the deterministic Lyapunov function for the ergodicity of stochastic difference equations" (Chan and Tong, 1985, Adv. Appl. Probab.), "Some comments on a bridge between nonlinear dynamicists and statisticians" (Tong, 1992, Physica D), and "A personal overview of non-linear time series analysis from a chaos perspective" (Tong, 1995, Scand. J. Statistics). The first is a technical paper applying ideas from dynamical systems to problems in stochastic difference equations. The last two are review-style articles designed to cross-fertilize (to build bridges) between disciplines.

2. "On the Use of the Deterministic ..." (Adv. Appl. Prob. 1985)

In this paper Professors Chan and Tong use the Lyapunov function to connect ergodicity of a stochastic difference equation of the form

$$X_{n+1} = T(X_n) + \epsilon_{n+1}$$

with the stability of the solution to the associated deterministic equation

$$X_{n+1} = T(X_n).$$

This paper is intriguing for several reasons in addition to the result itself. It must represent one of the earliest efforts of statisticians to understand the relationship between a nonlinear deterministic equation and an associated nonlinear stochastic time series. The reference list is striking by the conspicuous absence of citations to the statistical and time series literature. The only exceptions are Tong (1983) and Ozaki (1980). Indeed, we see a subject in its infancy. It is also interesting to note that Howell had not yet introduced the term "skeleton" to describe the deterministic mapping $T(X)$. Rather, the authors use the more cumbersome terms "associated deterministic difference equation" and "deterministic part of the stochastic difference equation". However, the suggestion is made that "bone" might be an appropriate descriptive label. (Apparently Howell ultimately decided that there was no point in settling for just a bone when you could have the whole skeleton.)

This paper has given rise to over 60 citations thus far, and it is interesting to note that the first citation did not appear until 1990, five years after the publication of the paper. Even given customary one or two-year publication lags, this suggests that the paper "slept" for a time while the statistical community caught up with these new ideas. A representative mix of citations is given by Saikkon (2007), Ling et al. (2007), Cline and Pu (2004), Chen and Chen (2000), Cline and Pu (1999), Lu (1998), An and Huang (1996), Bhattacharya and Lee (1995), and that famous first citing paper Tjøstheim (1990).

3. "Some Comments On A Bridge ..." (Physica D, 1992)

This article, published in a prestigious physics journal dedicated to nonlinear phenomena, was primarily an inspired attempt to persuade dynamicists that statisticians and statistical methods had much to offer in the way of potential solutions to problems in dynamics. This initiative was no small undertaking because the physics community had developed a thriving statistical sub-culture of its own and was accustomed to solving its own statistical problems; see, for example, Kennel and Isabelle (1992) and Theiler et al. (1992). (Howell alluded to as much when in the acknowledgments of the paper he thanked the participants of a NATO Advanced Workshop for "being so tolerant towards the odd statistician in their company".) Howell's paper discussed four main areas where statisticians might have something to offer dynamicists. One was the Principle of Parsimony and the dangers of overfitting. The second was a description of the relationship of Principal Components Analysis and the Karhunen-Loéve expansion to the Singular Value Decomposition used in dynamics, including a discussion of potential further uses and statistical difficulties. Third was a discussion of threshold models as the basis for local function approximations and how techniques such as Multivariate Adaptive Regression Splines (MARS) could be utilized. Howell saw nonparametric time series modelling as particularly fertile grounds for interdisciplinary work. Finally there was discussion of the important fact that a purely deterministic system (and its associated parameters, for example, Lyapunov exponents) need not behave in the same way as the related stochastic system (and its parameters) resulting from adding either measurement or system noise to the deterministic system. The problem of interpretation of these different parameters was pointed out. There does not seem to have been much reaction within the physics community to this perhaps rather audacious paper (probably for the reasons already mentioned). However, a smattering of diverse papers did arise which made use of Howell's comments, for example LePape et al. (1997), Sayar et al. (1997), and Mendes and Billings (1998).

4. "A Personal Overview of ..." (Scand. J. Stat, 1995)

As a result of a Special Invited Lecture at the 15th Nordic Conference on Mathematical Statistics in Lund, Sweden (1994), Professor Tong prepared an overview discussion paper on nonlinear time series and chaos for the Scandinavian Journal of Statistics. This paper was a kind of mirror image, albeit a much more ambitious one, of the earlier one in Physica D. Its main goal was to persuade statisticians of the value of including concepts, techniques, and problems from chaotic dynamics in their time series and forecasting research. In particular, Howell hoped that the important chaos notion of "sensitivity to initial conditions" would resonate with statisticians. He also emphasized his belief that employing ideas from chaos to nonlinear stochastic systems (and vice versa) would lead to profitable advances in each. In his continual effort to bridge these two fields he enjoyed so much, Howell stated "it is the thesis of this paper that a stochastic dynamical system, in the form of a non-linear time series model, provides a natural environment for a proper intercourse between chaos and statistics, thereby bringing about greater realism to dynamical systems." However, even beyond its persuasive goals, the paper was also an opportunity to highlight for the statistical community the many "chaotic" successes already enjoyed by Howell, his co-workers, and a small number of other nonlinearly like-minded statisticians. The number and credentials of the discussants on the paper were impressive; they consisted of K.S. Chan, D.R. Cox, myself, D. Guégan, J.L. Jensen, S. Johansen, A.J. Lawrance, B. LeBaron, T. Ozaki, D.W. Nychka, S. Ellner, B.A. Bailey, A.R. Gallant, R.L. Smith, R.C.L. Wolff, and, last but not least, a lone but brave and statistically-minded physicist, Lenny Smith.

I can only mention some of the topics covered in a paper which can truly be said to be packed with ideas and results. The paper features an excellent introduction to nonlinear dynamical systems, attractors, chaos, and sensitivity to initial conditions, a presentation easily accessible to even the novice. Stochastic dynamical systems are introduced and the question of initial-value sensitivity raised. Various methods of quantifying this are discussed, including the local Lyapunov exponents of Wolff (1992) and the conditional distribution approach of Yao and Tong (1994 a,b). Problems of prediction are considered, as are techniques for order determination (embedding dimension) and correlation dimension. Map reconstruction and associated local function approximations, including the threshold methods of Tong (1990), are covered in detail.

The paper has generated over 40 citations in the statistics, econometrics, and statistical computing literature. It would be impossible to list them all here, but the breadth, diversity, and quality of these papers can be seen from the following representative sample: Fan et al.(1996), Lin and Pourahmadi, M. (1998), Clements and Smith (1999), Cai et al.(2000), Tsai and Chan (2000), Golia and Sandri (2001), Bask and de Luna (2002), Lai and Chen

(2003), Huang and Shen (2004), and McMillan (2007).

References

1. An, H.Z. and Chen, S.G. (1997). A note on the ergodicity of non-linear autoregressive models. *Stat. Probab. Letters* **34**, 365-372.
2. Bask M., de Luna X. (2002). Characterizing the degree of stability of non-linear dynamic models. *Studies in Nonlinear Dynamics and Econometrics.* **6**.
3. Bhattacharya, R. and Lee, C.H. (1995). On Geometric Ergodicity of Nonlinear Autoregressive Models. *Stat. Probab. Letters* **22**, 311-315.
4. Cai, Z.W., Fan, J.Q., and Yao, Q.W. (2000). Functional-coefficient regression models for non-linear time series. *J. Amer. Statist. Assoc,* **95**, 941 – 956.
5. Chan, K.S. and Tong, H. (1985). On the use of the deterministic Lyapunov function for the ergodicity of stochastic difference equations. *Adv. Appl. Probab.,* **17,** 666-678.
6. Chen M. and Chen G.M. (2000). Geometric ergodicity of nonlinear autoregressive models with changing conditional variances. *Canad. J. Statist.* **28**, 605-613.
7. Clements, M.P. and Smith, J. (1999). A Monte Carlo study of the forecasting performance of empirical SETAR models. *J. Appl. Econometrics,* **14,** 123–141.
8. Cline, D.B.H. and Pu, H.M.H. (1999). Geometric ergodicity of nonlinear time series. *Statistica Sinica* **9**, 1103-1118.
9. Cline, D.B.H. and Pu, H.M.H. (2004). Stability and the Lyapounov exponent of threshold AR-ARCH models. *Ann. Appl. Probab.* **14**, 1920-1949.
10. Fan, J.Q., Yao, Q.W., and Tong, H. (1996). Estimation of conditional densities and sensitivity measures in nonlinear dynamical systems *Biometrika,* **83,** 189-206.
11. Golia S. and Sandri M. (2001). A resampling algorithm for chaotic time series. *Statistics and Computing,* **11,** 241-255.
12. Huang, J.H.Z., Shen, H.P. (2004). Functional coefficient regression models for non-linear time series: A polynomial spline approach. *Scand. J. Statist.* **31,** 515-534.
13. Kennel, M.B. and Isabelle, S. (1992). Method to determine possible chaos from colored noise and to determine embedding parameters. *Phys. Rev. A,* **46**, 3111–3118.
14. Lai D.J., Chen G.R., (2003). Distribution of the estimated Lyapunov exponents from noisy chaotic time series. *J. Time Ser. Anal.* **24,** 705-720.
15. Le Pape, G., Giacomini, H., Swynghedauw, B. and Mansier, P. (1997). A statistical analysis of sequences of cardiac interbeat intervals does not support the chaos hypothesis. *J. Theor. Biol.* **184,** 123–131.
16. Lin, T.C. and Pourahmadi, M. (1998). Nonparametric and non-linear models and data mining in time series: A case-study on the Canadian lynx data *J. Royal Statist. Soc. Series C,* **47,** 187-201.
17. Ling, S.Q., Tong, H, and Li, D. (2007). Ergodicity and invertibility of threshold moving-average models. *Bernoulli* **13**, 161-168.
18. Lu, Z.D. (1998). On the geometric ergodicity of a non-linear autoregressive model with an autoregressive conditional heteroscedastic term. *Statistica Sinica* **8**, 1205-1217.
19. McMillan, D.G. (2007). Non-linear forecasting of stock returns: Does volume help? *Int. J. Forecasting,* **23**, 115-126.
20. Mendes, E.M.A.M. and Billings, S.A. (1998). On overparametrization of nonlinear discrete systems. *Int. J. Bifurcation and Chaos,* **8**, 535–556.
   Ozaki, T. (1980). Non-linear time series models for non-linear random vibrations. *J. Appl. Probab.* **17**, 84–93.
21. Saikkonen, P. (2007). Stability of mixtures of vector autoregressions with autoregressive conditional heteroskedasticity. *Statistica Sinica.* **17**, 221-239.
22. Sayar, M., Demirel, M.C., and Atilgan, R. (1997). Dynamics of disordered structures: effect of non-linearity on the localization. *J. Sound and Vibration,* **205**, 372–379.
23. Theiler, J., Eubank, S., Longtin, A., Galdrikian, B., and Farmer, J.D. (1992). Testing for linearity in time series: the method of surrogate data. *Physica D,* **58,** 299–303.

24. Tjøstheim, D. (1990). Nonlinear time series and Markov chains. *Adv. Appl. Probab.* **22**, 587-611.
25. Tong, H. (1983). *Threshold Models in Non-Linear Time Series Analysis.* Lecture Notes in Statistics **21**, Springer-Verlag, Heidelberg.
26. Tong, H. (1990). *Non-linear Time Series: A Dynamical Systems Approach.* Oxford University Press, Oxford.
27. Tong, H. (1992). Some comments on a bridge between nonlinear dynamicists and statisticians, *Physica D,* **58**, 299–300.
28. Tong, H. (1995). A personal overview of non-linear time series analysis from a chaos perspective. *Scand. J. Statist.,* **22,** 399–421.
29. Tong, H. (2007). Birth of the threshold time series model. *Statistica Sinica,* **17**, 8–14.
30. Tsai, H.H. and Chan, K.S. (2000). Testing for nonlinearity with partially observed time series. *Biometrika,* **87,** 805-821.
31. Wolff, R.C.L. (1992). Local Lyapunov exponents: looking closely at chaos. *J. Royal Statist. Soc. Ser. B* **54**, 353–372.
32. Yao, Q. and Tong, H. (1994a). Quantifying the influence of initial values on non-linear prediction. *J. Royal Statist. Soc. Ser. B* **56**, 701–725.
33. Yao, Q. and Tong, H. (1994b). On prediction and chaos in stochastic systems. *Philos. Trans. Royal Soc. London* **A348**, 357–369.

Reflections from Re-Reading Howell Tong's 1995 Paper "A Personal Overview of Non-Linear Time Series Analysis from a Chaos Perspective"

TONY LAWRANCE

Department of Statistics, University of Warwick, Coventry
Warwickshire CV4 7AL, United Kingdom
E-mail: A.J.Lawrance@warwick.ac.uk

This paper picks up some points in Howell Tong's work over the past 38 years which have common cause with the author's own work over the same period. Among these are an interest in time series reversibility, chaotic time series and applications of time series. In particular, chaotic communications engineering research, only then developed enough for a brief final mention in Tong's 1995 overview paper, is illustrated by a subsequent contribution to chaotic time series modelling and by some of the author's research on the performance of chaotic communication systems.

1. Introduction

I am happy to contribute to this volume celebrating Howell Tong's 65th birthday; we go back a long way. Howell has said he was a research student in the audience of my first research seminar in February 1970, given at the then University of Manchester Institute of Science and Technology. Since then I have been in and out of time series but have kept an admiring technical interest in his work and a continuing personal friendship. A couple of times our interests have become tangential, although never quite chordal with a joint publication. Our closest common area has been in the reversibility or otherwise of time series. I recall he gave a seminar in Birmingham some time in 1991 concerning his threshold models when I asked a question about time-reversibility. I had been working at the time on the reversal link between autoregressive processes and congruential random number generators, work which would later appear in Lawrance (1992). After a few days my question had resulted in a short note, Tong & Cheng (1992), reversing other map processes, a topic I continued in Lawrance & Spencer (1998). More recently, Howell's interest turned to multivariate reversibility, Chan *et al* (2006). He has also pointed out some connections of my own work with Peter Lewis, Lawrance & Lewis (1985), to his early threshold models. In this paper I have chosen to follow a few threads of his 1995 discussion paper, Tong (1995), in the *Scandinavian Journal of Statistics*, giving his views on time series from a personal chaos perspective. Quite coincidentally, at about that time I was getting interested in chaotic time series and their role in communications engineering. On re-reading my dusty incomplete photocopy of his discussion paper, I was quite surprised to see my own comments and a final fleeting reference to chaos communications. Thus, I am motivated to make a contribution to this volume concerning statistical dependency in chaos and bit error rate analysis in chaotic communications. I rather regret not having re-read Tong (1995) until now. Together with the discussants' contributions, it is something of a milestone in defining the area.

2. Contrasting Views on Chaotic Time Series Modelling

The traditional view of chaotic time series is from discrete time dynamical systems in which an important concept is the so-called chaotic map, a function subject to the conditions set out in Section 2 of Tong (1995). Particular well-known examples are the logistic map, the Bernoulli-shift map and the tent map, and a newer map is presented in Section 4. The sensitivity to initial conditions of series generated by such maps, as evident from their positive Lyapunov exponents, is a strong signifier of their chaotic nature. One used to be worried about the precise mathematical definition of chaos, but a more pragmatic view now seems to prevail, with which I agree. I think I can say that one of Howell's major contributions was to bring the use of such maps into the noisy statistical world; the maps were then seen as the deterministic part replacing the linear part of an autoregressive time series model. This caused all sorts of trouble! How to get at the chaotic part? At the same time statistical time series was breaking out of its linear shackles, in both structural and distributional directions, the latter having been an early concern of mine, Lawrance & Lewis (1980). A theme of several discussants in Tong (1995) is that Howell's approach is more one of nonlinear time series than chaos, and thus by implication that the chaotic aspect is somewhat of a secondary issue. But Tong's approach has been to develop a conditional distribution approach and develop its sensitivity to initial conditions, surely a virtuous aim. The approach then continues to statistical non-parametric estimation of the chaotic or non-linear skeleton, as he calls it, of the model. Thus, I agree with discussants Guegan and Jensen that the approach blurs a sharp distinction between chaotic and non-linear time series, but I do not see that as a necessarily bad thing. In fact, from my own communication engineering perspective which uses chaotic maps operationally, there are circumstances in which noise is added to a chaotic map process, but the saving grace here is that the chaotic generator is part of the engineering design and so does not need estimating. My own choice of emphasis is to take a statistical approach to chaotic map sequences without any noise, but as if one does not know their method of generation, and I will briefly develop this theme in the next paragraph. Further, it is interesting that the final paragraph of Tong (1995) points to the possibilities of using chaos in communications and cryptosystems, both of which have been realized in the last decade; for instance, in cryptosystems note Kocarev's work, Bergamo *et al* (2005). The final Section 4 will give a flavour of my own contributions in the communications area.

3. Statistical Aspects of Chaotic Sequences

Suppose that $\tau(x), c \le x \le d$ is a typical chaotic map, as illustrated in Figure 1, and assumed without further ado. Next, in order to realize statistical aspects of chaotic sequences, consider a chaotic sequence of random variables $X_0, X_1, \ldots$, by which I mean a sequence of random variables such that X_0 and $\tau(X_0)$ have the same or *invariant distribution* and subsequent variables satisfy the equation

$$X_{i+1} = \tau(X_i), i = 0,1,\ldots \tag{1}$$

Without knowing the method of generation, $X_0, X_1, \ldots$ is a stationary and invariantly distributed sequence of dependent random variables. I like to think of it as the antithesis of an IID sequence; the dependence is so extreme that each variable is functionally related to the previous one, and so all are functions of X_0. This is the random variable model of a numerical chaotic sequence and I shall briefly exhibit its statistical features of interest, with the general dependency results coming via efforts in chaos communications research.

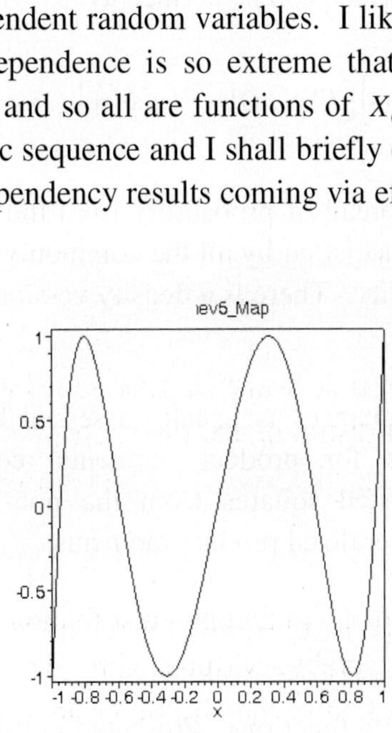

Figure1. A typical chaotic map.

What is well-known is that the invariant distribution of the sequence must satisfy the condition mentioned previously for the distribution of X_0. This leads to the so-called Perron-Frobenius equation, a probability balancing act, which can be stated in pdf $f(x)$ or distribution function $F(x)$ forms. For explicitness of presentation, both need the pre-image functions $g_i^\tau(x)$, $i = 1,2,\ldots,k$ which are the assumed k inverse functions of the map satisfying $\tau\{g_i^\tau(x)\} = x, i = 1,2,\ldots,k$. Then the Perron-Frobenius equation can be explicitly stated as

$$F(x) = \sum_{i=1}^{k} sign\{g_i^{\tau\prime}(c+)\}\left[F\{g_i^\tau(x)\} - F\{g_i^\tau(c)\}\right] \tag{2}$$

or

$$f(x) = \sum_{i=1}^{k} f\{g_i^{\tau}(x)\} \,|\, g_i^{\tau\prime}(x) \,| \tag{3}$$

where $g_i^{\tau\prime}(x)$ is the derivative function of $g_i^{\tau}(x)$.

Although the invariant distribution is an essential statistical property, the general dependency structure of the chaotic random variable sequence is more interesting, although less well-known outside of chaos communication research. A property identified by Khoda (1977) for chaotic maps which makes the dependency structure tractable is *equi-distributivity*, which can be defined as

$$k^{-1}F(x) = sign\left\{g_i^{\tau\prime}(c+)\right\}\left[F\{g_i^{\tau}(x)\} - F\{g_i^{\tau}(c)\}\right], \quad i = 1,2,\ldots,k\,. \tag{4}$$

It is a condition for the equi-spread of probability from the invariant distribution under the branches of the map and is satisfied by all the commonly used chaotic maps, so costs very little in terms of applicability. There is a density version of the condition as well, as introduced by Kohda.

Dependency in time series sequences is usually assessed by autocorrelations and the general dependency result is for product moments, conveniently also including autocorrelations of mean-adjusted squares from the non-linear world. The theory addresses calculation of the generalized product moments

$$E\{a(X_t)b(X_{t+s})\}, \; s = 1,2,\ldots. \tag{5}$$

where $a(.), b(.)$ can be any simple functions. Presented in this way, further details of the derivation may be gleaned from Lawrance & Balakrishna (2001), but the basic form of the final result is

$$E\{a(X_t)b(X_{t+1})\} = E_X\left\{b(X)\frac{1}{k}\sum_{i=1}^{k} a\{g_i^{\tau}(X)\}\right\}. \tag{6}$$

Notice two points, first, that if the pre-image sum on the right-hand side is constant, then there is no correlation on the left-hand side, and secondly, that although the result is for lag one, it can be generalized to lag s because $b(X_{t+s})$ can be cast as $b\{\tau^{(s-1)}(X_{t+1})\}$, just a more complicated function of X_{t+1}. Whilst (6) is explicit, there is a more useful iterative form

$$E\{a(X_t)b(X_{t+s})\} = E\left\{\left[\frac{1}{k}\sum_{i=1}^{k} a\{g_i^{\tau}(X_t)\}\right]b(X_{t+s-1})\right\}, \; s = 1,2,\ldots \tag{7}$$

This has been used to calculate autocorrelations of the $\{X_t\}$ and of $\{(X_t-\mu)^2\}$, the mean μ adjusted squares, for well-known and for not so well-known maps. For instance, for Bernoulli-shift maps these autocorrelation functions are $(1/2)^s$ and $(1/4)^s$, respectively. For the logistic map, both these autocorrelation functions are zero. When used in practical generation of sequences, the chaotic aspect will still be visible from scatter plots of adjacent values, reflecting that joint distributions in chaotic sequences are degenerate.

The general result (7) is also useful when the functions $a(x), b(x)$ are discretizing functions, such as binary functions useful in communications theory, Khoda (1977), and which destroy the chaotic characteristics of the sequence. Discretizing is also relevant when considering congruential random number generators as being derived from chaotic shift maps, Lawrance (1992). Extensions of the approach to bivariate maps has been presented in Hilliam & Lawrance (2004).

Regretfully, it is hard to imagine that results here could be available when noise is added to the right side of (1), unless Howell knows better….

4. Communications and Chaos

In this final section, I want to pick up Howell's remark at the end of Tong (1995) about the potential of applying chaos to communications. At that time the idea that it was possible to synchronize chaotic sequences was emerging to some excitement and its potential in communications was aroused by a now well-known paper of Pecora and Carrol (1990). This did create a considerable stir but gradually the practical difficulty of synchronizing chaotic electronic circuits became apparent and communication systems which did not need perfect synchronization were investigated, including *chaos shift-keying*, the one I will briefly soon mention The picture may now be changing back because of the possibility of synchronizing lasers leading to potential chaotic laser-based communication systems, as demonstrated for instance by Uchida and his group, Ozaki *et al* (2006). Indeed, some of my own current work is concerned with analyzing very large sequences from pairs of synchronized lasers. The analysis of performance in chaos communication systems is interesting statistically because it mixes chaos and noise and because the decoding of signals is essentially an operation of statistical estimation. Moreover, the performance of such systems is measured by bit error rate, essentially the probability of a decoding error. I hope to demonstrate these aspects in the following paragraphs. From a communications point of view, chaos-based systems offer security and spread spectrum advantages from chaotic sequences replacing conventional sinosoidal waves.

The broad aim of the communication modelling I am about to outline is to assess design and performance characteristics rather than represent the mass of technical electronic issues which are needed to practically implement the systems. Engineers in the area refer to the modelling as *discrete time baseband equivalent modelling*, but to statisticians it is modelling in terms of random variables or stochastic processes. A communication system to be practical must serve many users and receivers, which I shall refer later to as *multi-user*, but to simplify and condense presentation, only a single-user and receiver system will be initially outlined. This is the so-called *chaos shift-keying system* and is usually concerned with transmission and receipt of binary bit messages; it is illustrated by a block diagram in Figure 2, as apparently obligatory in engineering publications. Fundamental points are that a chaotic segment of spreading length N is used to transmit each bit b, and that the theory focuses on a single bit, not on a sequence of bits. The

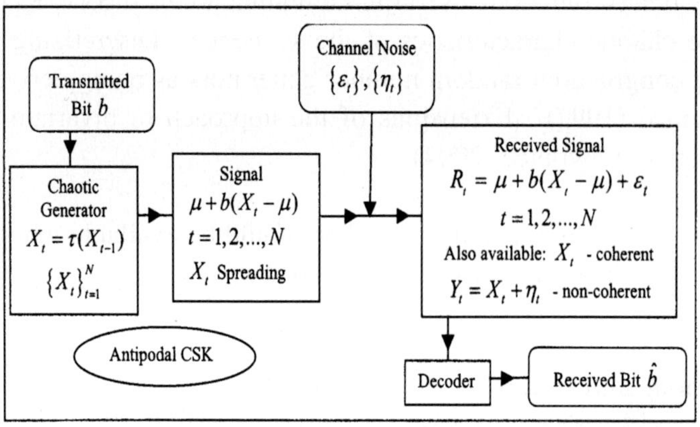

Figure 2. Block diagram of a chaos shift-keying communication

sequence is transmitted unchanged when, say, a $b = +1$ is transmitted and is multiplied by -1 if a $b = -1$ bit is transmitted, a so-called *antipodal* effect. At the receiver, the transmitted sequence is received with Gaussian white channel noise ε of variance σ^2 added to each member of the sequence; there is not enough information to decode the bit value; and some further knowledge of the spreading sequence is required. In the so-called *coherent* version, it is assumed that the original chaotic sequence can be *exactly* reproduced at the receiver by some method. Originally, and in systems where the chaotic sequences were produced by electronic means, it was perceived that this could be done by chaotic synchronization, but with map-generated segments synchronized generators were assumed, or, the segments were first transmitted without the binary bit modifications, and also attracted channel noise. The latter type of system is termed *non-coherent* and is more challenging to analyze. With the additional information, decoding of the bit value is the statistical problem of estimating b, regarded as a parameter, although not always seen this way in the engineering literature. In the coherent case, the maximum likelihood estimator is based on the sign of the covariance between the two transmitted segments, as intuitively sensible, and called the *correlation decoder*, before this statistician came on the scene. Hopefully, the previous description will be enough to make sense of the block diagram in Figure 2, where μ denotes the mean of the invariant distribution and σ_X^2 will be its variance. Referring to Figure 2, the bit error rate derivation is outlined as follows, much as in Lawrance & Ohama (2003). The received bit sequence $R = (R_1, R_2, ..., R_N)$ is of the form

$$R_i = \mu + b(X_i - \mu) + \varepsilon_i, i = 1, 2, \ldots, N, \quad b = \pm 1 \tag{8}$$

and the correlation decoder is

$$C(X, R) \equiv \sum_{i=1}^{N} (X_i - \mu)(R_i - \mu), \tag{9}$$

with positive values indicating a $+1$ bit and negative values a -1 bit. It can be seen as the maximum likelihood estimate of a correlation coefficient which can only take the values ± 1, not a very usual statistical situation ! The customary communications performance measure is the *bit error rate* (BER) which for a $+1$ bit error is given by

$$BER_{+1} = P\{C(X, R) < 0 \mid b = +1\} = P\left(\sum_{i=1}^{N} (X_i - \mu)^2 + \sum_{i=1}^{N} \varepsilon_i (X_i - \mu) < 0 \right). \tag{10}$$

This is usefully the same as that for a -1 bit error in this chaos shift-keying case, and thus is the overall bit error rate. The BER (10) can be evaluated by noting that its inner term, as a linear function of the random variables $(\varepsilon_1, \varepsilon_2, \ldots, \varepsilon_N)$, is itself a Gaussian random variable conditional on $(X_1, X_2, \ldots, X_N)$. Then, with $\Phi(\cdot)$ denoting the distribution function of a standardized Gaussian variable, the probability of bit error (PBE), which is conditional on $(X_1, X_2, \ldots, X_N)$, can be deduced as

$$PBE(X_1, X_2, \ldots, X_N) = \Phi\left\{ -\sqrt{\sum_{i=1}^{N} (X_i - \mu)^2} \Big/ \sigma \right\}. \tag{11}$$

Unconditionally, the exact bit error rate is thus

$$BER(N) = E\left\{ \Phi\left(-\sqrt{\sum_{i=1}^{N} (X_i - \mu)^2} \Big/ \sigma \right) \right\} \tag{12}$$

and with the chaotic assumption becomes

$$BER(N) = \int_{x=c}^{d} \Phi\left(-\sqrt{\sum_{i=1}^{N} \left(\tau^{(i-1)}(x) - \mu \right)^2} \Big/ \sigma \right) f_X(x) dx, \tag{13}$$

a univariate integral. It is only in the chaotic case that the result is conveniently available as a univariate integral, and an integral which can be calculated for moderate values of N. More intuition can be gained from (12) by defining a *signal-noise ratio* as $SNR = N\sigma_X^2 / \sigma^2$, whereupon

$$BER(N) = E_X\left\{\Phi\left(-\sqrt{SNR}\sqrt{\frac{1}{N}\sum_{i=1}^{N}(X_i-\mu)^2\Big/\sigma_X^2}\right)\right\} \tag{14}$$

which for large N becomes $\Phi(-\sqrt{SNR})$. From applying Jensen's inequality, this is also seen to be the lower bound of (14), although previously in the engineering literature such results were at first optimistically regarded as exact, Lau & Tse (2003). Graphical illustrations of the result (14) are given in Figure 3 for three methods of spreading; it can be seen that Gaussian spreading is by far the worst and logistic is the best, although not optimal. Subsequent work has been directed at chaotic spreading which can approach

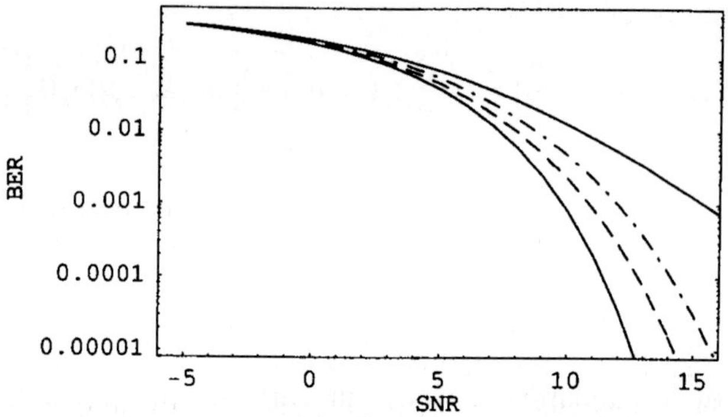

Figure 3. Bit error rate $BER(N)$ plotted against signal to noise ratio SNR with spreading factor $N = 5$ for the lower bound (lower solid line), logistic map spreading (dashed line), shift map spreading (dot-dashed line) and independent Gaussian spreading (upper solid line).

the lower bound. It turns out that the crucial condition is that of minimum lag 1 quadratic autocorrelation negativity of the chaotic sequence, and that a very satisfactory choice is the so-called *circular map*, displayed in Figure 4, introduced in Yao (2004), and developed optimally in Lawrance & Papamarkou (2006). This somewhat strange

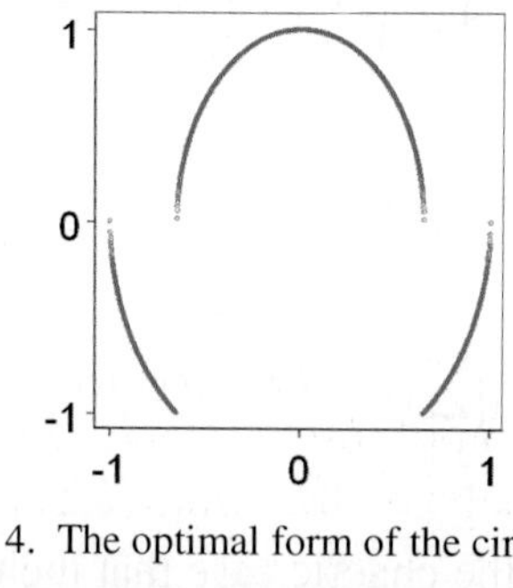

Figure 4. The optimal form of the circular map

creature gives autoun-correlated sequences which have lag 1 quadratic autocorelation of -0.722, not fully towards the relevant Frechet lower bound of -0.968, but giving close *BER* proximity to the lower bound curve in Figure 3.

In the corresponding *non-coherent* system, that is in the '*chaos plus noise*' case of Howell's terminology, the bit error rate becomes more interesting from a statistical point of view; the maximum likelihood decoder is not fully available, but the correlation decoder can still be used. Its bit error rate is given by the expression

$$BER(N) = E\left[P\left\{F_{N,N}\left(\frac{2}{\sigma^2}\sum_{i=1}^{N}(X_i-\mu)^2\right)<1\right\}\right] \tag{15}$$

where $F_{N,N}(.)$ is the distribution function of a statistical favourite, the non-central chi-squared distribution with (N,N) degrees of freedom. Whereas in the coherent case increasing the spreading brings the bit error rate closer and closer to the lower bound, in this non-coherent case there is an optimum amount of spreading N minimising (15). Intuitively, this is due to the balancing of gain from increased spreading against loss from the use of more inaccurately known spreading values.

This previous discussion has concerned singe-user systems. To be realistic but more complicated, multi-user systems have to be considered, as in Tam *et al* (2007). In such systems it is envisaged that the signals from different users travel additively through the same channel but are received with channel noise particular to the designated receiver. The signals of the other users then act as interference, equivalent to a second type of noise. Approaches to exact results analyzing these systems have been developed in Yao & Lawrance (2004), Lawrance & Yao (2007) and are continuing. Individual decoding in such coherent systems can be approached by maximum likelihood estimation and produces decoders which are enhancements of correlation decoders and generalizations of rake decoders, these latter being available in other communications contexts. For instance, under particular circumstances, the lower bound to the bit error rate generalizing the $\Phi(-\sqrt{SNR})$ result after (14) becomes

$$\Phi\left(-1\Big/\sqrt{\frac{1}{SNR}+\frac{1}{SIR}}\right), \tag{16}$$

where SIR is the *signal to interference ratio*, defined as $N/(L-1)$, and L is the number of users.

This has been just a personal flavour of results developed in chaos communications in the period since the appearance of Tong (1995). More recent communications developments are continuing with laser-generated chaos and the synchronization which is experimentally possible between lasers in different locations; these should lead to practical systems, useful in specialized applications, but not replacing the massive conventional public networks. Maybe when Howell writes his next discussion paper..., following on from his 1990 *reflections* paper, Tong (2002), I will have something more to add in this direction, and perhaps more statistical as well.

References

1. Bergamo, P., D'Arco, P., De Santis, A. and Kocarev, L. (2005). Security of public-key cryptosystens based on Chebyshev polynomials. *IEEE Transactions on Circuits and Systems-I: Regular Papers*, 52, 1382-1393.

2. Chan, K.S., Ho, L. and Tong, H. (2006). A note on time-reversibility of multivariate linear processes. *Biometrika*, 93, 221-227.

3. Hilliam, R. and Lawrance, A.J. (2004) The dynamics and statistics of bivariate chaotic maps in communications modelling. *International Journal of Bifurcation and Chaos,* 14, 4, 1177-1194.

4. Khoda, T. and Tsuneda, A. (1997). Statistics of chaotic binary sequences. *IEEE Transactions on Information Theory,* 43, 104-112.

5. Lau, F.C.M. and Tse, C.K. (2003). *Chaos-based digital communication systems.* Springer-Verlag, Heidelberg.

6. Lawrance, A.J. (1992). Uniformly distributed first order autoregressive time series models and multiplicative congruential random number generators. *Journal of Applied Probability*, 29, 896-903.

7. Lawrance, A. J. and Balakrishna, N. (2001). Statistical aspects of chaotic maps with negative dependency in a communications setting. *Journal of the Royal Statistical Society*, *Series B,* 63, 843-853.

8. Lawrance, A.J. and Lewis, P.A.W. (1980). An exponential autoregressive-moving average process EARMA (p,q). *J. R. Statist.* B, 42, 150-161.

9. Lawrance, A.J. and Lewis, P.A.W. (1985). Modelling and residual analysis of nonlinear autoregressive time series in exponential variables (with discussion). *Journal of the Royal Statistical Society*, *Series B,* 47, 165-202.

10. Lawrance, A. J. and Ohama, G. (2003). Exact calculation of bit error rates in communication systems with chaotic modulation. *IEEE Transactions on Circuits and Systems –I: Fundamental Theory and Applications*, 50, 1391-1400.

11. Lawrance, A.J. and Papamarkou, T. (2006). Optimal spreading sequences for chaos-based communications systems. *Proceedings of Nolta2007,* 208-211, 16-19 September, Vancouver, Canada.

12. Lawrance, A.J. and Spencer, N. (1998). Statistical aspects of curved chaotic map models and stochastic reversals. *Scandinavian Journal of Statistics,* 25, 371-382.

13. Lawrance, A.J. and Yao, J. (2007). Optimal demodulation in multi-user chaos shift keying communication. *Submitted for publication.*

14. Ozaki, M., Mihara, T., Someya, H., Uchida, A. and Yoshimori, S., (2006). *Proceedings of Nolta2006, 443-446,* 11-14 September, Bologna, Italy,

15. Pecora, L.M. and Carroll, T.L. (1990). Synchronization of chaotic systems. *Physical Review Letters*, 64, 821-824.

16. Tam, W. M., Lau, F.C.M. and Tse, C. K. (2007). *Digital communications with chaos*. Elsevier, Amsterdam.

17. Tong, H. (1995). A personal overview of nonlinear time series from a chaos perspective (with discussion). *Scandinavian Journal of Statistics*, 22, 399-445.

18. Tong, H. (2002). Nonlinear time series analysis since 1990: some personal reflections. *Acta Mathematicae Apllicatae Sinica, English Series*, 18, 177-184.

19. Tong, H. and B. Cheng, (1992). A note on one-dimensional chaotic maps under time reversal. *Advances in Applied Probability*, 24, 219-220.

20. Yao, J. and Lawrance, A.J. (2004). Bit error rate calculation for multi-user coherent chaos-shift-keying communications systems. *Transactions IEICE Fundamentals (Japan), E87-A, 2280-2291.*

21. Yao, J. and Lawrance, A.J. (2006). Performance analysis and optimization of multi-user differential chaos-shift-keying communication systems. *IEEE Transactions on Circuits and Systems –I: Regular Papers* 53, 9, 2075-2091.

Chaos Perspective of Nonlinear Time Series: A Selective Review

QIWEI YAO
Department of Statistics, London School of Economics
Houghton Street, London WC2A 2AC, UK
E-mail: q.yao@lse.ac.uk

This is a selective review on two of Howell Tong's papers [3,16] on the interplay between nonlinear (stochastic) time series and deterministic chaos.

1. Introduction

Howell Tong is an acknowledged leader in nonlinear time series analysis. He played a pioneering role in drawing strengths and inspirations from modern dynamical systems to time series analysis, which represented a new thinking in the late 1970 and early 1980. He has conducted path-breaking research in the dynamical system approach to nonlinear time series analysis since then. The major results are systematically presented in the monographs [14,4]. At the technical level, Howell constructs a wide class of nonlinear time series models by way of piecewise linearization. His threshold autoregressive model has wide applications, which won him the well-deserved Royal Statistical Society Guy Medal in Silver in 2007.

In this short review, we highlight some important contributions in two papers by Howell, namely, a personal overview [16] on nonlinear time series analysis from a chaos perspective, and a more theoretical paper[3] on the link between the ergodicity of stochastic difference equations and the physical notion of energy in the form of a Lyapunov function. Section 2 focuses on the important issue of the initial-value sensitivity addressed in the first paper. Section 3 deals with Chan and Tong[3].

2. Nonlinear Time Series and Chaos

Howell took the view: "a stochastic dynamical system, in the form of a non-linear time series model, provides a natural environment for a proper intercourse between chaos and statistics, thereby bringing about great realism to dynamical systems" (p. 401 of Tong[16]). Clearly his mind was on (stochastic) nonlinear time series. To understand his viewpoint, let us briefly remind ourselves the essence of (deterministic) chaos.

2.1. *Deterministic chaos*

A discrete-time deterministic dynamical system may be described by a difference equation

$$X_t = f(X_{t-1}), \quad t = 1, 2, \cdots, \tag{1}$$

where X_t is a state variable, and f is a real valued function. For simplicity, we assume X_t is a scalar. Suppose that the system starts at the initial value X_0 at time 0. At time t, it should land at $X_t = f(X_{t-1}) = f\{f(X_{t-2})\} = \cdots = f^{(t)}(X_0)$, where $f^{(t)}$ denotes the t-th fold composition of f. This looks extremely simple! However the very essence of chaos says that even if we know precisely both the initial value X_0 and the map f, we may have difficulties in identifying the position of X_t for even moderately large t *if the system (1) is chaotic*, due to the so-called *sensitivity to initial values*. This may only happen for some nonlinear f. There exists no universally accepted mathematical definition for chaos. However the fundamental nature of chaos is the sensitivity to initial values, i.e. two trajectories with nearby initial values may diverge from each other 'exponentially' fast. Furthermore this exponentially fast divergence is confined to a bounded set and the trajectory crawls around within the bounded set.

For the deterministic system (1), an effective way to measure the sensitivity to initial values is to use the Lyapunov exponent defined as

$$\lambda(X_0) = \lim_{t\to\infty} \log\{|\dot{f}^{(t)}(X_0)|^{1/t}\} = \lim_{t\to\infty} \frac{1}{t}\sum_{i=1}^{t} \log|\dot{f}(X_{i-1})| \tag{2}$$

provided that the limit on the RHS of the above expression exists. Here $\dot{f}$ denotes the derivative of f. Then it holds for any two nearby initial values X_0, X_0' and sufficiently large t that

$$|X_t' - X_t| \equiv |f^{(t)}(X_0') - f^{(t)}(X_0)| \approx \exp\{t\lambda(X_0)\}|X_0' - X_0|.$$

Thus when $\lambda(X_0) > 0$, the two trajectories diverge exponentially. To have a positive Lyapunov exponent is a necessary condition for the presence of chaos, which implies *the local instability* in the sense that a small shift in the initial value (such as rounding errors in computation) may lead to substantial departure from its original orbit. However another important characteristic of chaos is *the global stability*. This means, in the simplest case with a single non-trivial attractor, that for an infinitely long series $\{X_t\}$ generated from the chaotic system (1), the marginal distribution of X_t follows a probability measure called *invariance measure* determined entirely by the function f. In fact many chaotic systems exhibit a certain ergodicity in the sense that an average in time equals an average in space (according to the invariant measure). Consequently the Lyapunov exponent defined in (2) is equal to

$$\lambda \equiv \lambda(X_0) = E\{\log|\dot{f}(X_t)|\}, \tag{3}$$

which is a constant independent of the initial value X_0. In the above expression, the expectation is taken under the invariant measure.

2.2. *Sensitivity to initial values*

To appreciate the relevance of chaos to analyzing data subject to random errors, let us clothe the deterministic system (1) with dynamical additive noise:

$$X_t = f(X_{t-1}) + \varepsilon_t. \tag{4}$$

This is a nonlinear AR(1) model. Is the sensitivity to initial values still an issue here? If so, how does it interact with stochastic noise ε_t? Note that a fundamental difference in time series forecast is to forecast the near future instead of letting $t \to \infty$ as in (2). Nevertheless the chaotic behaviour of f still has a profound impact on the time evolution of the stochastic process $\{X_t\}$ even in the short term. Furthermore the impact is largely dictated by the derivative of f, although the Lyapunov exponent defined in (2) and (3) no longer has a direct bearing.

Let us consider a prediction problem. Suppose we are at time T with the observation $X_T = x$ while the true position is at $X_T = x + \delta$, where δ reflects a small error in the observation. We aim to predict the future X_{T+m} for $m \geq 1$. To understand the impact of the nonlinearity of f on the prediction, we assume that f is given. Then the least squares predictor is

$$f_m(x) = E(X_{T+m}|X_T = x).$$

It is easy to see that the mean squared prediction error (MSPE) can be decomposed as follows:

$$\begin{aligned} E[\{X_{T+m} - f_m(x)\}^2|X_T = x + \delta] &= \sigma_m(x+\delta)^2 + \{f_m(x) - f_m(x+\delta)\}^2 \\ &= \sigma_m(x+\delta)^2 + \{\dot{f}_m(x)\delta\}^2 + o(\delta^2), \end{aligned} \tag{5}$$

where

$$\sigma_m(x)^2 = \mathrm{Var}(X_{T+m}|X_T = x). \tag{6}$$

Note that

$$\begin{aligned} f_m(x) &= E\{f(X_{T+m-1}|X_T = x\} = E[f\{f(X_{T+m-2}) + \varepsilon_{T+m-1}\}|X_T = x] \\ &= E\{f((\cdots(f(x) + \varepsilon_{T+1})\cdots) + \varepsilon_{T+m-1})||X_T = x\}. \end{aligned}$$

By the chain rule,

$$\dot{f}_m(x) = E\{\prod_{k=1}^{m} \dot{f}(X_{T+k-1})|X_T = x\}. \tag{7}$$

If we assume all the noise are bounded by a small constant $\zeta > 0$, i.e. $|\varepsilon_t| \leq \zeta$ a.s., and $\mathrm{Var}(\varepsilon_t) = \sigma^2$ is a constant, argument similar to the above leads to

$$\sigma_m(x)^2 = \mu_m(x)\sigma^2 + O(\zeta^3), \tag{8}$$

and

$$\mu_m(x) = 1 + \sum_{j=1}^{m-1}\{\prod_{k=j}^{m-1} \dot{f}\{f^{(k)}(x)\}\}^2. \tag{9}$$

Combining (5) – (9) together, we see that the MSPE consists of two major terms: conditional variance $\sigma_m(x+\delta)^2$ and the error due to the shift in the initial value $\{\dot{f}_m(x)\delta\}^2$. The conditional variance is resulted from the accumulation of stochastic noise ε_{T+j} for $j = 1, \cdots, m$. Those errors are unevenly amplified according to a product function of $\dot{f}(\cdot)$ along the orbit $f^{(k)}(x)$. This reflects common sense: we are able to predict the future at sometime (or somewhere) better than other times! The second term is entirely due to the discrepancy between $f_m(x+\delta)$ and $f_m(x)$. For moderately small m and δ, this term is often negligible in practice. However for the map with large derivatives, $\dot{f}_m(x)$ may be adversely large; see (7). Equation (5) has been termed as a decomposition theorem[17]. In practice, we may estimate the predictor $f_m(\cdot)$ using local linear regression, and estimate the derivative $\dot{f}_m(\cdot)$ using local quadratic regression[9]. If we knew f_m, the conditional variance $\sigma_m(\cdot)^2$ could be estimated by a local linear regression based on the fact that $\sigma_m(x)^2 = E[\{(Y_{T+m} - f_m(x)\}^2|X_T = x]$. In fact, $\sigma_m(x)^2$ may be estimated asymptotically as if $f_m(\cdot)$ were given[8].

Since $f_m(\cdot)$ is a conditional expectation, $\dot{f}_m(\cdot)$ may be viewed as a measure for the sensitivity to initial value for conditional expectation. One natural question is to consider such a sensitivity for conditional distribution of X_{T+m} given X_T. Some initial attempts have been made[18,10], along with the nonparametric estimation for conditional density functions. Note that the estimation for conditional density functions is now a vibrant research area in econometrics and quantitative finance.

2.3. *Where to go from here?*

As stated above, Chaos exists within the contrast of the local instability and the global stability. The same might be said for a stochastic statistical model. As we have witnessed in the previous section, the thinking along the line of deterministic chaos does help us to understand, and therefore, to appreciate nonlinearity better. This is undoubtedly helpful for designing and performing statistical inference. However, how much more does chaos have to offer for the statistical inference for real data? Are the conventional statistical methods and techniques really useful in handling deterministic chaos? Those questions were largely open when Tong[16] was published more than ten years ago. In spite of some fruitful developments[6,7,12], there has been few fundamental breakthroughs in bridging the two areas together. The fundamental difficulty lies in the fact that we are dealing with two completely different animals. For example, likelihood based inference is almost irrelevant in deterministic systems (as conditional distributions are degenerate). The central limit theorem is only relevant for some special deterministic chaos[5,13]. In spite of the existence of the synergy between the two camps in describing and handling nonlinearity[15], it seems to me what chaos may offer in statistical inference for data subject to random noise is limited.

3. Ergodicity of Stochastic Difference Equations

Chan and Tong [3] deals with the ergodicity of stochastic difference equations using the deterministic Lyapunov functions. The paper was written for readers with the knowledge in Markov chains, ergodicity theory and also deterministic dynamic systems. More detailed account on the topic, including some further developments, may be found in Chan[1,2]. While those results may look more probabilistic, their impact on nonlinear time series analysis is immediate. See below.

Stationarity plays a fundamental role in the statistical inference for time series. While it is relatively easy to check stationarity in linear time series models, it is often a challenge to verify stationarity for nonlinear processes. It remains open to prove (or disprove) that some simple nonlinear autoregressive models admit stationary solutions. The common practice is to represent a time series as a (usually vector-valued) Markov chain and to establish that the Markov chain is ergodic. Stationarity follows from the fact that an ergodic Markov chain is stationary.

Let us consider a general form of nonlinear AR model

$$X_t = f(X_{t-1}, \cdots, X_{t-p}) + \varepsilon_t, \tag{10}$$

where $\{\varepsilon_t\}$ is a sequence of i.i.d. random variables, and ε_t is independent of $\{X_{t-k}, k \geq 1\}$. To embed it into a Markov model, put

$$\mathbf{X}_t = (X_t, \cdots, X_{t-p+1})^\tau, \qquad \boldsymbol{\varepsilon}_t = (\varepsilon_t, 0, \cdots, 0)^\tau,$$

and for $\mathbf{x} = (x_1, \cdots, x_p)^\tau \in \mathcal{R}^p$,

$$\mathbf{f}(\mathbf{x}) = \big(f(\mathbf{x}), x_1, \cdots, x_{p-1}\big)^\tau.$$

By (10), $\{\mathbf{X}_t\}$ is a Markov Chain defined as

$$\mathbf{X}_t = \mathbf{f}(\mathbf{X}_{t-1}) + \boldsymbol{\varepsilon}_t. \tag{11}$$

If we let

$$f(\mathbf{x}) = c_i + \sum_{j=1}^{p} a_{ij} x_j \qquad \text{for } r_{i-1} \le x_d < r_i, \tag{12}$$

where d is a positive integer smaller than p, $-\infty = r_0 < r_1 < \cdots < r_l = \infty$, c_i and a_{ij} are constant. Then (10) is a TAR model with l regimes. Chan and Tong[3] shows that if

$$\max_i \sum_{j=1}^{p} |a_{ij}| < 1, \tag{13}$$

and the probability density function of ε_t is positive on $\mathcal{R}$ with $E|\varepsilon_t| < \infty$, model (11) defines a geometrically ergodic Markov chain. Consequently we may conclude that the TAR model (10) and (12) admits a strictly stationary solution; see, for example, Theorem 2.2 of Fan and Yao[9].

The primary goal of Chan and Tong[3] is to show that the Lyapunov function plays a significant role in studying not only the stability of a deterministic difference equation but also the ergodicity of a stochastic difference equation. In fact they have derived the following result for general model (11). For a precise definition of the Lyapunov functions and the related results for the stability of deterministic systems, we refer to Kalman and Bertram[11].

Theorem. (Chan and Tong 1985)
Let $\mathbf{f}(\cdot)$ be continuous and homogeneous (i.e. $\mathbf{f}(c\mathbf{x}) = c\mathbf{f}(\mathbf{x})$ for any $c > 0$ and $\mathbf{x} \in \mathcal{R}^p$). Suppose that the probability density function of ε_t exists and is positive on $\mathcal{R}$, and $E|\varepsilon_t| < \infty$. Then the existence of a continuous Lyapunov function for the deterministic system $\mathbf{X}_t = \mathbf{f}(\mathbf{X}_{t-1})$ in a neighbourhood of the origin implies the geometric ergodicity of (10).

The theorem above places some restriction on the function $\mathbf{f}(\cdot)$, therefore also on $f(\cdot)$. For example, TAR model (12) is neither continuous nor homogeneous. Chan and Tong[3] provided some convenient ways to extend the ergodicity result above to more general models. For example, if $\mathbf{f}(\cdot)$ may be decomposed as follows:

$$\mathbf{f}(\mathbf{x}) = \mathbf{f}_h(\mathbf{x}) + \mathbf{f}_d(\mathbf{x}),$$

where $\mathbf{f}_h$ is continuous and homogeneous while $\mathbf{f}_d$ is of bounded range. We may then consider the 'component' model

$$\mathbf{X}_t = \mathbf{f}_h(\mathbf{X}_{t-1}) + \boldsymbol{\varepsilon}_t \tag{14}$$

for which the ergodicity can then be established by the theorem above. It is clear that the same conclusion also holds for $\{\mathbf{X}_n\}$ defined by (11). Along this line, several concrete examples were investigated in Chan and Tong[3]. For example, they show that the simple TAR model

$$X_t = \begin{cases} \alpha + X_{t-1} + \varepsilon_t, & X_{t-1} \le 0, \\ \beta + X_{t-1} + \varepsilon_t, & X_{t-1} > 0 \end{cases}$$

is ergodic if (and only if) $\alpha < 0 < \beta$. Note for this model, condition (13) does not hold.

It is a standard approach to establish stationarity for a nonlinear time series via the ergodicity of the associated Markov chain. A survey at an introductory level may be found in section 2.1.4 of Fan and Yao[9].

Acknowledgment

The author thanks Professor K. S. Chan for helpful comments and suggestions.

References

1. Chan, K.S. (1993a). Consistency and limiting distribution of a least squares estimator of a threshold autoregressive model. *Ann. Statist.* **21**, 520–533.
2. Chan, K.S. (1993b). A review of some limit theorems of Markov chains and their applications. In *Dimension Estimation and Models* (H. Tong, ed.). World Scientific, Singapore, pp. 108–135.
3. Chan, K.S. and Tong, H. (1985). On the use of the deterministic Lyapunov function for the ergodicity of stochastic difference equation. *Adv. Appl. Prob.* **17**, 666-678.
4. Chan, K.S. and Tong, H. (2001). *Chaos: A Statistical Perspective.* Springer, New York.
5. Chernov, N.I. (1995). Limit theorems and Markov approximations for chaotic dynamical systems. *Probab. Theory Relat. Fields*, **101**, 321-362.
6. Diks, C. (2003). Detecting serial dependence in tail events: a test dual to the BDS test. *Economics Letters*, **79**, 319-324.
7. Diks, C. (2004). The correlation dimension of returns with stochastic volatility. *Quantitative Finance*, **4**, 45-54.
8. Fan, J. and Yao, Q. (1998). Efficient estimation of conditional variance functions in stochastic regression. *Biometrika*, **85**, 645-660.
9. Fan, J. and Yao, Q. (2003). *Nonlinear Time Series: Nonparametric and Parametric Methods.* Springer, New York.
10. Fan, J, Yao, Q. and Tong, H. (1996). Estimation of conditional densities and sensitivity measures in nonlinear dynamical systems. *Biometrika*, **83**, 189-206.
11. Kalman, R.E. and Bertram, J.E. (1960). Control system analysis and design via the "Second method" of Lyapunov II: Discrete-time systems. *Trans. A.S.M.E., J Basic Engng. D*, **82**, 394.
12. Lawrance, A.J. and Balakrishna, N. (2001). Statistical aspects of chaotic maps with negative dependence in a communications setting. *J. Roy. Statist. Soc.* **B**, **63**, 843-853.
13. Stockis, J.-P. and Tong, H. (1998). On the statistical inference of a machine-generated autoregressive AR(1) model. *J. Roy. Statist. Soc.* **60**, 781-796.
14. Tong, H. (1990). *Non-linear Time Series: A Dynamical Systems Approach*, Oxford University Press, Oxford.
15. Tong, H. (1992). Some comments on a bridge between nonlinear dynamicists and statisticians. *Physica* **D**, **58**, 299-303.
16. Tong, H. (1995). A personal overview of non-linear time series analysis from a chaos perspective (with discussions). *Scand. J. Statist.* **22**, 399-445.
17. Yao, Q. and Tong, H. (1994). Quantifying the inference of initial values on nonlinear prediction. *J. Roy. Statist. Soc.* **B**, **56**, 701-725.
18. Yao, Q. and Tong, H. (1994). On prediction and chaos in stochastic systems. *Phil. Trans. Roy. Soc. (London)* **A**, **348**, 357-369.

Edinburgh Workshop on Nonlinear Time Series, 1989

(Left to right ignoring row number: Wai-keung Li, Ruey Tsay, Colin Sparrow, Russell Gerrard, John Lane, Murray Rosenblatt, Gudmundur Gudmundsson, John Petruccelli, Tze-Liang Lai, Tony Lawrance, Peter Robinson, Dominic Guegan, T. K. Brown, Pham Dinh Tuan, Timo Terasvirta, Rodney Wolff, Clive Granger, Peter Fisk, David Cox, Martin Casdagli, Jonathan Tawn, Tohru Ozaki, Granville Tunnicliffe Wilson, Howell Tong, Dag Tjostheim, Ed McKenzie, Peter Lewis, Richard Smith, Neville Davies, David Jones, Kung-Sik Chan, Zhao-Guo Chen)

Oxford in the rain, 1997

International Conference on Financial Statistics, Hong Kong, 1999

Akaike enjoying my after-dinner speech at a conference honouring him, Yokohama, 2003

Receiving the 2007 Guy Medal in Silver from President Tim Holt of the Royal Statistical Society

Chinese New Year dinner with friends at home in Canterbury, 2007

On the Analysis of Bivariate Non-stationary Processes

By M. B. PRIESTLEY and H. TONG

University of Manchester Institute of Science and Technology

[Read before the ROYAL STATISTICAL SOCIETY at a meeting organized by the RESEARCH SECTION on Wednesday, December 6th, 1972, Professor J. DURBIN in the Chair]

SUMMARY

In this paper, we propose a general definition of the *evolutionary* (time-dependent) *cross-spectrum* between two non-stationary processes and describe its physical interpretation. We also study the estimation of the evolutionary cross-spectrum at each time instant t from a *single* realization of a bivariate process. Further, we propose a definition (and a method of estimation) for the *coherency* (*spectrum*) between the two components of the bivariate process and show that the notion of *residual variance bound* first introduced in the analysis of bivariate stationary processes can be extended to that of non-stationary processes. As an application of the evolutionary cross-spectral analysis of bivariate non-stationary stochastic processes, we consider the estimation of the transfer function of a linear open-loop time-dependent system. Numerical illustrations of the estimation of a time-dependent transfer function are included.

Keywords: NON STATIONARY PROCESSES; OSCILLATORY PROCESSES; SEMI STATIONARY PROCESSES; EVOLUTIONARY SPECTRAL ANALYSIS; BIVARIATE PROCESSES; EVOLUTIONARY CROSS SPECTRA; COHERENCY; TIME DEPENDENT TRANSFER FUNCTIONS.

1. INTRODUCTION

IN a previous paper (Priestley, 1965a) we developed an approach to the spectral analysis of univariate non-stationary processes by introducing the notion of *evolutionary spectra*, that is, spectral functions which are time-dependent and admit a physical interpretation as *local* energy distributions. This approach to the study of non-stationary processes has facilitated the extension of the classical Wiener–Kolmogorov theory of prediction and filtering of stationary processes to the non-stationary case (Abdrabbo and Priestley, 1967, 1969), and Revfeim (1969) and Subba Rao (1970) have made use of evolutionary spectral theory in fitting non-stationary stochastic models with time-dependent parameters. Priestley (1969, 1971a) has introduced evolutionary spectral analysis into the study of stochastic control systems which are infected by non-stationary disturbances. On a more practical level, Hammond (1968) has used the evolutionary spectral approach in the analysis of jet engine noise. It is true to say that most of the above works are the results of the application of the evolutionary spectral theory of a *univariate* non-stationary process, with the exception of Abdrabbo and Priestley (1969), where a preliminary extension of the univariate theory to the bivariate case has been made. In this paper, we give a general definition of the *evolutionary cross-spectrum* and describe its physical interpretation. We consider also the estimation of the evolutionary cross-spectrum at each instant of time from a single relization of the bivariate process. Further, we propose a definition (and an estimate) of the *coherency* (*spectrum*) between the two components of the bivariate process and show that the notion of *residual variance bound* introduced in the analysis of bivariate stationary processes (Priestley, 1971b) can be extended to that of non-stationary processes. As an application of the evolutionary cross-spectral analysis of bivariate non-stationary stochastic processes, we describe a method of estimating the transfer function of a linear open-loop time-dependent system. Numerical illustrations of the estimation of the transfer function are included.

2. Definition of the Spectrum of a Non-stationary Process

It may, perhaps, be appropriate at this stage to discuss briefly the basic ideas underlying our approach to the definition of spectra for non-stationary processes. This is an intriguing problem, and various approaches have been discussed in the literature. (For a comprehensive review, see Loynes, 1968.) Most of these have attempted to characterize the spectral properties of a non-stationary process in terms of a "time-dependent spectrum", i.e. in terms of a spectral function which involves both "time" and "frequency" variables. (The main exception to this is the approach via Loève's "harmonizable representation" which leads instead to a "spectrum" which is a function of two "frequency" variables—see Priestley, 1965.) The notion of a time-dependent spectrum is quite a natural one, since non-stationarity implies, of course, that the probabilistic structure of the process itself changes with time.

The first attempt to define a time-dependent spectrum is due to Page (1952) who introduced the term *instantaneous power spectra*. For a process, $\{X(t)\}$, Page first introduces the quantity

$$g_T(\omega) = E\left|\int_0^T X(t)\exp(-i\omega t)\,dt\right|^2 \tag{2.1}$$

(effectively, the expected value of the periodogram on the interval $(0, T)$), and then defines the instantaneous power spectrum, $\rho_t(\omega)$, by writing

$$g_T(\omega) = \int_0^T \rho_t(\omega)\,dt, \tag{2.2}$$

so that

$$\rho_t(\omega) = (d/dt)\{g_t(\omega)\}. \tag{2.3}$$

However, some of the more recent studies have tried to follow the form of the Wiener–Khintchine theorem for stationary processes by defining a time-dependent spectrum as the Fourier transform of a *local* autocovariance function. Thus, for example, Mark (1970) defines an *instantaneous spectral density function*, $\psi(t, \omega)$, by

$$\psi(t,\omega) = \int_{-\infty}^{\infty} R^*(t,\tau)\exp(-i\omega\tau)\,d\tau, \tag{2.4}$$

where

$$R^*(t,\tau) = E[X(t-\tau/2)\,X(t+\tau/2)]. \tag{2.5}$$

In effect, $\psi(t, \omega)$ is the Fourier transform of $R^*(t, \tau)$, regarded as a function of τ with t fixed. However, having introduced the idea of a time-dependent spectrum, $f(t, \omega)$, say (however defined), we want to be able to interpret it physically in the same way as we interpret the spectrum of a stationary process—with the important proviso that

$f(t_0, \omega)$ will characterize the behaviour of the process only in the neighbourhood of the time instant t_0. Both the approaches described above, whilst interesting from the mathematical point of view, lack the required interpretation. The instantaneous power spectrum, $\{\rho_t(\omega)\,dt\}$ represents roughly the difference between the power distribution of the process over the interval $(0, t)$ and the interval $(0, t+dt)$, whereas the quantity we require is the power distribution of the process *within* the interval $(t, t+dt)$. On the other hand, the instantaneous spectral density function, $\psi(t, \omega)$, does not have any interpretation as a *power distribution* since, in particular, it may take negative values for certain processes (Mark, 1970).

In attempting to define a time-dependent spectrum which possesses a physical interpretation as a local distribution of power over frequency, the basic question we must answer is: what do we mean by "frequency"? This may seem a deceptively simple question, but its study is crucial. For, suppose we have constructed some function, $f(t, \omega)$, say, which is such that, for each t,

$$\operatorname{var}\{X(t)\} = \int_{-\infty}^{\infty} f(t, \omega)\, d\omega. \tag{2.6}$$

We certainly cannot conclude, on the basis of equation (2.6) alone, that $f(t, \omega)$ represents a decomposition of power over *frequency*, for in the above integral ω is merely a dummy variable and there is no reason why it should be in any way related to the physical concept of "frequency". This point is reinforced by reference to the "spectrum" $\psi(t, \omega)$ which does, in fact, satisfy an equation of the form (2.6) but, as has been noted, $\psi(t, \omega)$ does not possess the required physical interpretation.

Let us now consider the case of stationary processes. The reason why we can interpret the spectrum of a stationary process as a power/frequency distribution lies essentially in the fact that, if $\{X(t)\}$ is stationary, then the process itself has a spectral representation of the form

$$X(t) = \int_{-\infty}^{\infty} \exp(it\omega)\, dZ(\omega). \tag{2.7}$$

Heuristically, equation (2.7) means that a stationary process can be represented as a sum of sine and cosine waves with varying frequencies and (random) amplitudes and phases. We can then identify that *component* in $X(t)$ which has frequency ω, and meaningfully discuss the contribution of this *component* to the total power of the process. In the absence of such a representation we cannot immediately talk about "power distributions over frequency"—unless, of course, we first define a more general concept of "frequency" which agrees with our physical understanding. It is precisely this type of reasoning which forms the basis of our "evolutionary spectrum" approach, as explained in Section 3.

For a more detailed discussion of the various points raised in this section see Priestley (1968, 1971c).

3. UNIVARIATE NON-STATIONARY PROCESSES

We consider the class of (complex-valued) stochastic processes $\{X(t): t \in T\}$, which are trend free (i.e. $E[X(t)] = 0$, all t) and whose autocovariance functions $R(s, t) = E[X(s)\,X^*(t)]$ are not necessarily invariant under a shift in the parameter space T. We suppose that for each process there exists a family, $\mathscr{G}$, of functions $\{\phi_t(\omega)\}$ defined on the real line, and indexed by the suffix $t \in T$, and a measure μ on the real line, such that for each t, $\phi_t(\omega)$ is μ-square integrable, and for each s, $t \in T$ the autocovariance function $R(s,t)$ admits a representation of the form

$$R(s,t) = \int_\Lambda \phi_s(\omega)\,\phi_t^*(\omega)\,d\mu(\omega), \tag{3.1}$$

where $\Lambda = (-\infty,\infty)$ or $(-\pi,\pi)$ according to whether the process is continuous parameter (i.e. T is the set of real numbers) or discrete parameter (i.e. T is the set of integers). It is well known that corresponding to representation (3.1) for $R(s,t)$, the process $\{X(t)\colon t \in T\}$ admits a representation of the form

$$X(t) = \int_\Lambda \phi_t(\omega)\,dZ(\omega), \tag{3.2}$$

where $\{Z(\omega)\}$ is a process with orthogonal increments such that

$$E[dZ(\omega)\,dZ^*(\omega')] = \delta_{\omega,\omega'}\,d\mu(\omega). \tag{3.3}$$

The class $\mathscr{G}$ is usually too wide if we want to preserve the physical notion of frequency. Thus, we consider a sub-class, $\mathscr{F} \subset \mathscr{G}$, of functions each of which, for fixed ω (considered as a function of t), possesses a (generalized) Fourier transform whose modulus has an absolute maximum at $\theta(\omega)$, say. If θ is a single-valued function of ω, $\phi_t(\omega)$ may then be written in the form

$$\phi_t(\omega) = A_t(\omega)\exp(i\omega t), \tag{3.4}$$

where, for each fixed ω,

$$A_t(\omega) = \int_{-\infty}^{\infty} \exp(it\theta)\,dH_\omega(\theta), \tag{3.5}$$

with $dH_\omega(\theta)|$ having an absolute maximum at $\theta = 0$. Without loss of generality we shall standardize the functions $A_t(\omega)$ so that, for all ω, $A_0(\omega) = 1$, implying that $dH_\omega(\theta)$ is "normalized" in the sense that

$$\int_{-\infty}^{\infty} dH_\omega(\theta) = 1. \tag{3.6}$$

We may now interpret the variable ω as a "generalized frequency". It now follows that, with respect to the class $\mathscr{F}$, the process $\{X(t)\}$ admits an *evolutionary spectral representation* of the form

$$X(t) = \int_\Lambda A_t(\omega)\exp(i\omega t)\,dZ(\omega), \tag{3.7}$$

and correspondingly

$$R(s,t) = \int_\Lambda A_s(\omega)\,A_t^*(\omega)\exp\{i\omega(s-t)\}\,d\mu(\omega). \tag{3.8}$$

$\{X(t)\}$ is then termed an *oscillatory process* and the family $\mathscr{F}$ is likewise termed a *family of oscillatory functions*, $\phi_t(\omega)$ being termed an *oscillatory function.*

Within the class of oscillatory processes, which obviously includes, as a sub-class, the class of all second-order stationary processes, we can still describe a distribution of power over (generalized) frequency, although the distribution is now, in fact, local in character. Specifically, we define the *evolutionary power spectrum at time t with respect to the family* $\mathscr{F}$, $dF_t(\omega)$, by

$$dF_t(\omega) = |A_t(\omega)|^2\,d\mu(\omega). \tag{3.9}$$

From equation (3.8), we can easily deduce, for each t,

$$\operatorname{var}\{X(t)\} = R(t,t) = \int_{\Lambda} dF_t(\omega), \tag{3.10}$$

for all possible choices of $\mathscr{F}$. (We note that the choice of $\mathscr{F}$ is generally *not* unique. See Priestley (1965a) and Tong (1972) for more detailed discussions.) For a physical process $\{X(t)\}$, $\operatorname{var}\{X(t)\}$ may be interpreted as the total power dissipated by the process at time t and equation (3.10) then has the important physical interpretation as a decomposition of the local power over (generalized) frequency. Another important aspect of the evolutionary power spectrum lies in the fact that an estimate obtained essentially from a wave-analyser on a *single* realization of the process is really an estimate of the (weighted) average value of the spectrum in the neighbourhood of the time-instant t.

As we have pointed out previously, there usually ex‧‧ ‧ multitude of evolutionary spectral representations for a given oscillatory process $\{X(t)\}$ corresponding to all possible choices of the family $\mathscr{F}$. Thus, we would naturally want to seek a measure of preference over the class of possible families. Motivated by this, we consider the following sub-class of the class of oscillatory processes.

Definition 3.1. For each family $\mathscr{F}$, we define the function $B_{\mathscr{F}}(\omega)$ by

$$B_{\mathscr{F}}(\omega) = \int_{-\infty}^{\infty} |\theta|\,|dH_{\omega}(\theta)| \tag{3.11}$$

which is a measure of the *width* of $|dH_{\omega}(\theta)|$.

Definition 3.2. A family $\mathscr{F}$ of oscillatory functions is termed *semistationary* if the function $B_{\mathscr{F}}(\omega)$ is bounded for all ω, and the constant $B_{\mathscr{F}}$, defined by

$$B_{\mathscr{F}} = \left[\sup_{\omega}\{B_{\mathscr{F}}(\omega)\}\right]^{-1}, \tag{3.12}$$

is termed the *characteristic width* of the family $\mathscr{F}$.

Definition 3.3. A *semi-stationary process* $\{X(t)\}$ is defined as one for which there exists a semi-stationary family $\mathscr{F}$ in terms of which $\{X(t)\}$ admits a representation of the form (3.7).

Let $\mathscr{C}$ denote the class of semi-stationary families $\mathscr{F}$, in terms of each of which $\{X(t)\}$ admits a spectral representation.

Definition 3.4. We define the *characteristic width of the process* $\{X(t)\}$, B_X, by

$$B_X = \sup_{\mathscr{F}\in\mathscr{C}}\{B_{\mathscr{F}}\}. \tag{3.13}$$

Roughly speaking, $2\pi B_X$ may be interpreted as the maximum time interval over which the process may be treated as "approximately stationary".

4. DEFINITION OF EVOLUTIONARY CROSS-SPECTRA

Consider a bivariate process $\{(X(t), Y(t)) : t\in T\}$, in which each component is an oscillatory process. Then with an obvious notation, we can write

$$X(t) = \int_{\Lambda} A_{t,x}(\omega)\exp(i\omega t)\,dZ_x(\omega) \tag{4.1}$$

and

$$Y(t) = \int_{\Lambda} A_{t,y}(\omega) \exp(i\omega t)\, dZ_y(\omega), \tag{4.2}$$

where

$$\left.\begin{aligned} &E[dZ_x(\omega)\, dZ_x^*(\omega')] = E[dZ_y(\omega)\, dZ_y^*(\omega')] = E[dZ_x(\omega)\, dZ_y^*(\omega')] = 0 \\ &\qquad\qquad\qquad \text{for } \omega \neq \omega', \\ &E|dZ_x(\omega)|^2 = d\mu_{xx}(\omega), \quad E|dZ_y(\omega)|^2 = d\mu_{yy}(\omega) \\ \text{and} \\ &E[dZ_x(\omega)\, dZ_y^*(\omega)] = d\mu_{xy}(\omega). \end{aligned}\right\} \tag{4.3}$$

Definition 4.1. Let $\{\mathscr{F}_x, \mathscr{F}_y\}$ denote a vector family of oscillatory functions $\{\phi_{t,x}(\omega) \equiv A_{t,x}(\omega) \exp(i\omega t),\ \phi_{t,y}(\omega) \equiv A_{t,y}(\omega) \exp(i\omega t)\}$, and let $\{(X(t), Y(t))\}$ be a bivariate oscillatory process with components admitting representations of forms (4.1) and (4.2) with respect to the families $\mathscr{F}_x$ and $\mathscr{F}_y$ respectively. We define the *evolutionary power cross-spectrum at time t with respect to the families* $\mathscr{F}_x$ *and* $\mathscr{F}_y$, $dF_{t,xy}(\omega)$, by

$$dF_{t,xy}(\omega) = A_{t,x}(\omega)\, A_{t,y}^*(\omega)\, d\mu_{xy}(\omega). \tag{4.4}$$

We note that when we may choose $\mathscr{F}_x \equiv \mathscr{F}_y$, equation (4.4) takes on the special form

$$dF_{t,xy}(\omega) = |A_t(\omega)|^2\, d\mu_{xy}(\omega), \tag{4.5}$$

where $A_t(\omega) \equiv A_{t,x}(\omega) \equiv A_{t,y}(\omega)$. Abdrabbo and Priestley (1969) used (4.5) as their definition of evolutionary cross-spectrum for this special case. Further, if $\{X(t), Y(t)\}$ is a bivariate stationary process so that $\mathscr{F}_x$ and $\mathscr{F}_y$ may be chosen to be the family of complex exponentials, viz. $\mathscr{F}_x \equiv \mathscr{F}_y \equiv \{\exp(i\omega t)\}$, $dF_{t,xy}(\omega)$ reduces to the classical definition of the cross-spectrum. In addition, if $X(t) = Y(t)$ (in mean square), for all t, equation (4.4) reduces to equation (3.9). Since, for each t, we may write

$$dF_{t,xy}(\omega) = E[A_{t,x}(\omega)\, dZ_x(\omega)\, A_{t,y}^*(\omega)\, dZ_y^*(\omega)] \tag{4.6}$$

it follows that $dF_{t,xy}(\omega)$ possesses a physical interpretation similar to that of the cross-spectrum of a bivariate stationary process, namely, that it represents the average value of the product of the *amplitudes* of the corresponding frequency components in the two processes. Note, however, that in the non-stationary case these *amplitudes* are time dependent so that, correspondingly, the cross-spectrum is also time dependent.

Clearly $dF_{t,xy}(\omega)$ is complex valued, and in virtue of the Cauchy–Schwarz inequality we have immediately that

$$|dF_{t,xy}(\omega)|^2 \leqslant dF_{t,xx}(\omega)\, dF_{t,yy}(\omega), \quad \forall t \text{ and } \omega. \tag{4.7}$$

If the measure $\mu_{xy}(\omega)$ is absolutely continuous with respect to Lebesgue measure, we may write, for each t,

$$dF_{t,xy}(\omega) = f_{t,xy}(\omega)\, d\omega \tag{4.8}$$

(for almost all ω), where $f_{t,xy}(\omega)$ may be termed the *evolutionary cross-spectral density function.* We may now write

$$f_{t,xy}(\omega) = c_{t,xy}(\omega) - iq_{t,xy}(\omega), \tag{4.9}$$

and term the real-valued functions $c_{t,xy}(\omega)$ and $q_{t,xy}(\omega)$ the *evolutionary co-spectrum* and the *evolutionary quadrature spectrum* respectively. If the measures $\mu_{xx}(\omega)$ and

$\mu_{yy}(\omega)$ are absolutely continuous, we may similarly define the *evolutionary auto-spectral density functions*, $f_{t,xx}(\omega)$ and $f_{t,yy}(\omega)$ (Priestley, 1965a). The function, $W_{yx}(\omega)$, defined by

$$W_{xy}(\omega) = |f_{t,yx}(\omega)|/\{f_{t,xx}(\omega)\, f_{t,yy}(\omega)\}^{\frac{1}{2}} \tag{4.10}$$

is termed the *coherency* (*spectrum*) between $\{X(t)\}$ and $\{Y(t)\}$. Equivalently, using (3.9) and (4.5), (4.10) may be written

$$W_{yx}(\omega) = |E[dZ_y(\omega)\, dZ_x^*(\omega)]|/\{E|dZ_x(\omega)|^2\, E|dZ_y(\omega)|^2\}^{\frac{1}{2}}. \tag{4.10a}$$

We may thus interpret $W_{yx}(\omega)$ as the modulus of the correlation coefficient between $dZ_x(\omega)$, $dZ_y(\omega)$, or more generally, as a measure of the *linear* relationship between the corresponding components at frequency ω in the processes $\{Y(t): t \in T\}$ and $\{X(t): t \in T\}$—see Section 6. As we can see from (4.10a) this measure is *independent* of the time parameter t. On specializing to stationary processes, our definition of coherency coincides with the classical definition of coherency and thus we may view the former as a generalized version of the latter. We note that, in virtue of (4.7), $0 \leqslant W_{yx}(\omega) \leqslant 1$ and $W_{xy}(\omega) = W_{yx}(\omega)$, for all ω.

We shall sometimes refer to the following 2×2 complex-valued matrix as the *evolutionary spectral density matrix*:

$$\mathbf{f}_t(\omega) = \begin{bmatrix} f_{t,xx}(\omega) & f_{t,xy}(\omega) \\ f_{t,yx}(\omega) & f_{t,yy}(\omega) \end{bmatrix}. \tag{4.11}$$

Note that $\mathbf{f}_t(\omega)$ is clearly positive definite and hermitian.

5. Estimation of Evolutionary Cross-spectra

In this section, we assume that the processes are continuous parameter. For other parameter spaces, the following discussions can quite easily be adapted.

Suppose we are given *one* vector sample record of the bivariate semi-stationary process, $\{X(t), Y(t)\}$, say for $t \in [0, T_0]$ where T_0 denotes the length of the sample record. One method of estimating the evolutionary cross-spectrum $dF_{t,xy}(\omega)$ (assumed absolutely continuous) is based on an extension of the method described by Priestley (1965a, 1966) for dealing with the univariate case. We may briefly describe the method as follows.

Let $\{g(u)\}$ be a filter satisfying the usual conditions with

$$\Gamma(\omega) = \int_{-\infty}^{\infty} g(u) \exp(-iu\omega)\, du \tag{5.1}$$

denoting its *frequency response function* and

$$B_g = \int_{-\infty}^{\infty} |u|\, |g(u)|\, du \quad (<\infty), \tag{5.2}$$

characterizing its "width" (see Priestley, 1965a).

We may choose $\{g(u)\}$ so that, $B_g \ll \min(B_X, B_Y) \ll T_0$, and write, for any frequency ω_0,

$$\begin{aligned} U_x(t, \omega_0) &= \int_{t-T_0}^{t} g(u)\, X(t-u) \exp\{-i\omega_0(t-u)\}\, du \\ &\doteq \int_{-\infty}^{\infty} g(u)\, X(t-u) \exp\{-i\omega_0(t-u)\}\, du \end{aligned} \tag{5.3}$$

and

$$U_y(t,\omega_0) = \int_{t-T_0}^{t} g(u)\, Y(t-u)\exp\{-i\omega_0(t-u)\}\,du$$

$$\doteqdot \int_{-\infty}^{\infty} g(u)\, Y(t-u)\exp\{-i\omega_0(t-u)\}\,du. \tag{5.4}$$

Next let $w_{T'}(t)$ be a weight function, depending on the parameter T' (which characterizes its "width") and satisfying the usual condition (Priestley, 1965a). In particular, writing

$$W_{T'}(\lambda) = \int_{-\infty}^{\infty} \exp(-i\lambda t)\, w_{T'}(t)\,dt,$$

we assume that there exists a constant C such that

$$\lim_{T'\to\infty} \left\{T' \int_{-\infty}^{\infty} |W_{T'}(\lambda)|^2\,d\lambda\right\} = C.$$

Then we may estimate $f_{t,xy}(\omega_0)$ by

$$\hat{f}_{t,xy}(\omega_0) = \int_{t-T_0}^{t} w_{T'}(u)\,[U_x(t-u,\omega_0)\,U_y^*(t-u,\omega_0)]\,du. \tag{5.5}$$

The weight function $w_{T'}(u)$ is usually so chosen that the effective "width", T', of $\{w_{T'}(u)\}$ is *much larger than* B_g, the width of $\{g(u)\}$. We now assume further that $w_{T'}(u)$ decays sufficiently fast so that the *transient* effect may be neglected, and the limits of the integral in (5.5) replaced by $(-\infty,\infty)$. We then obtain

$$E[\hat{f}_{t,xy}(\omega_0)] \sim \bar{f}_{t,xy}(\omega_0), \tag{5.6}$$

where

$$\bar{f}_{t,xy}(\omega) = \int_{-\infty}^{\infty} w_{T'}(u)\, f_{t-u,xy}(\omega)\,du, \tag{5.7}$$

so that $\hat{f}_{t,xy}(\omega_0)$ is an (approximately) *unbiased estimate of the* (*weighted*) *average value of* $f_{t,xy}(\omega_0)$ *in the neighbourhood of* t.

An investigation of the sampling properties of $\hat{f}_{t,xy}(\omega_0)$ has been carried out. Although in many ways the method of investigation is similar to that of Priestley (1966), both the calculations and results are much more lengthy. We summarize here only the main results. (See Tong, 1972, for a full discussion of the sampling properties $\hat{f}_{t,xy}(\omega_0)$.) Making the usual assumptions concerning the "width" of the weight function $w_{T'}(u)$ (see, for example, Priestley, 1965a, pp. 219–220), we may show that the covariance between $\hat{f}_{t,xy}(\omega)$ and $\hat{f}_{s,xy}(\omega')$ is effectively zero if either

(i) $|\omega\pm\omega'|$ is sufficiently large such that $|\omega\pm\omega'| \gg$ bandwidth of $|\Gamma(\omega)|^2$, or
(ii) $|s-t|$ is sufficiently large, i.e. $|s-t| \gg$ "width" of the weight function $\{w_{T'}(u)\}$.

We may further show that the variance–covariance matrix of

$$\{\hat{f}_{t,xx}(\omega_0),\hat{f}_{t,yy}(\omega_0),\hat{c}_{t,xy}(\omega_0),\hat{q}_{t,xy}(\omega_0)\}$$

has a form similar to that arising in the case of bivariate stationary processes. Its detailed form is given in Subba Rao and Tong (1972), but for illustration we quote the following results

$$\operatorname{var} \hat{f}_{t,xx}(\omega) \sim K(\widetilde{f^2_{t,xx}(\omega)}),$$

$$\operatorname{var} \hat{c}_{t,xy}(\omega) \sim \tfrac{1}{2}K\{(\widetilde{f_{t,xx}(\omega)\, f_{t,yy}(\omega)}) + (\widetilde{c^2_{t,xy}(\omega)}) - (\widetilde{q^2_{t,xy}(\omega)})\},$$

where, for example,

$$(\widetilde{f^2_{t,xx}(\omega)}) = \int_{-\infty}^{\infty} f^2_{t-u,xx}(\omega)\{w_{T'}(u)\}^2\, du \Big/ \int_{-\infty}^{\infty} \{w_{T'}(u)\}^2\, du,$$

with similar definitions for the other terms, and

$$K = \frac{C}{T'} \int_{-\infty}^{\infty} |\Gamma(\theta)|^4\, d\theta.$$

In fact, the matrix reduces exactly to equation (42) of Jenkins (1963) in specializing to the stationary case.

(It is well known in classical cross-spectral analysis that an *alignment* of the cross-periodogram yields a more efficient cross-spectral estimate; see, for example, Priestley, 1965b, Jenkins and Watts, 1968. This device is certainly commendable in the stationary case, but in view of the fact that in our case the cross-spectrum is time dependent any alignment will involve a parameter which, in general, varies with t. This point no doubt deserves further study.)

Finally, we may note that an obvious estimate of the coherency spectrum, W_{xy}, is given by replacing $f_{t,xy}, f_{t,xx}, f_{t,yy}$ by their respective estimates in (4.10), giving

$$\hat{W}_{xy}(\omega_0) = \frac{|\hat{f}_{t,xy}(\omega_0)|}{\{\hat{f}_{t,xx}(\omega_0)\, \hat{f}_{t,yy}(\omega_0)\}^{\frac{1}{2}}} \tag{5.8}$$

provided $\hat{f}_{t,xx}(\omega_0) \neq 0$ and $\hat{f}_{t,yy}(\omega_0) \neq 0$.

6. LINEAR TRANSFORMATION

Suppose that a semi-stationary process $\{X(t)\}$ is transformed into a process $\{Y(t)\}$ via a time-dependent, linear transformation, viz.

$$Y(t) = \int_{-\infty}^{\infty} d_t(u)\, X(t-u)\, du. \tag{6.1}$$

It may be shown (Tong, 1972; Subba Rao and Tong, 1972) that under suitable conditions relating to the "width" and the time variation of $d_t(u)$, we obtain

$$f_{t,yy}(\omega) \doteq |D_t(\omega)|^2 f_{t,xx}(\omega), \tag{6.2}$$

where

$$D_t(\omega) = \int_{-\infty}^{\infty} d_t(u) \exp(-iu\omega)\, du, \quad \text{each } t, \tag{6.3}$$

is termed the *time-dependent transfer function* of the transformation (6.1). Equation (6.2) provides a natural generalization of the well-known filter relationship for the case of time-invariant transformation of stationary processes.

Suppose now that $\{Y(t)\}, \{X(t)\}$ are two general non-stationary processes, and we wish to "fit" a linear time-dependent relationship of the form (6.1). If we adopt a least-squares approach, then we would choose $\{d_t(u)\}$ so that, for each t,

$$V(t) = E\left| Y(t) - \int_{-\infty}^{\infty} d_t(u)\, X(t-u)\, du \right|^2$$

is minimized. It may be shown (Tong, 1972) that $V(t)$ is minimized, for each t by choosing $\{d_t(u)\}$ so that

$$D_t(\omega) \doteq f_{t,yx}(\omega)/f_{t,xx}(\omega). \tag{6.4}$$

We then have

$$\min_{\{d_t(u)\}} V(t) \doteq \int_{-\infty}^{\infty} f_{t,yy}(\omega)\{1 - |W_{yx}(\omega)|^2\}\, d\omega. \tag{6.5}$$

The expression on the right-hand side of the above equation may be termed the *residual variance bound* and its magnitude provides a measure of the degree of linear association between the two processes $\{Y(t)\}$ and $\{X(t)\}$. If $|W_{yx}(\omega)|^2$ is near to unity at all frequencies (recall that our definition of W_{yx} is independent of t), we would expect to obtain a close linear fit between the two processes (cf. Priestley, 1971b).

7. ESTIMATION OF THE TRANSFER FUNCTION OF A TIME-DEPENDENT OPEN LOOP SYSTEM

In many practical applications, the processes $\{X(t)\}$ and $\{Y(t)\}$ may be thought of as the *input* and *output* of an *open loop system*, with the associated *block diagram* given in Fig. 1. (See, for example, Jenkins, 1963 and Priestley, 1969). Typically, this situation would be described by a linear *time-invariant* model of the form

$$Y(t) = \int_{-\infty}^{\infty} d(u)\, X(t-u)\, du + e(t), \tag{7.1}$$

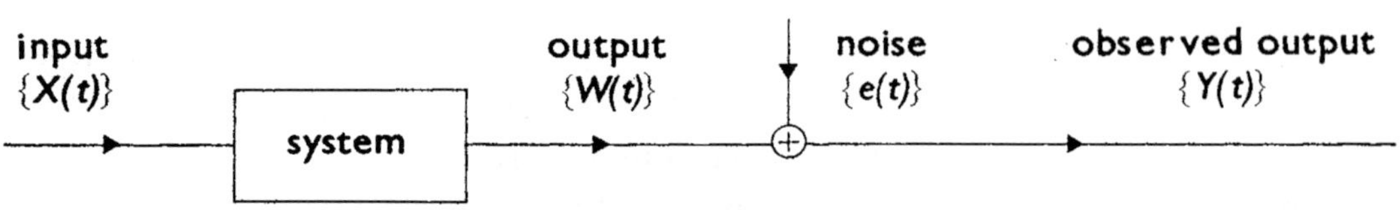

FIG. 1.

where $\{X(t)\}$, $\{Y(t)\}$ and $\{e(t)\}$ are stationary processes. However, a more general description is given by a time-dependent model of the form

$$Y(t) = \int_{-\infty}^{\infty} d_t(u)\, X(t-u)\, du + e(t), \tag{7.2}$$

where now $\{X(t)\}$, $\{Y(t)\}$ and $\{e(t)\}$ are semi-stationary processes and $\{d_t(u)\}$ describes a "slowly changing" filter. Suppose that we have available operating records of $\{X(t)\}$ and $\{Y(t)\}$ over the interval $0 \leqslant t \leqslant T$, and wish to estimate the time-dependent transfer function, $D_t(\omega)$, $0 \leqslant t \leqslant T$. From the discussion in Section 6, for each value of t in the interval $(0, T)$, a natural estimate of $D_t(\omega)$ is given by

$$\hat{D}_t(\omega) = \hat{f}_{t,yx}(\omega)/\hat{f}_{t,xx}(\omega), \quad \text{provided } \hat{f}_{t,xx}(\omega) \neq 0, \tag{7.3}$$

where $\hat{f}_{t,yx}$, $\hat{f}_{t,xx}$ denote the estimated cross and auto evolutionary spectral density functions computed from the given sample records. If we write

$$D_t(\omega) = G_t(\omega) \exp\{-i\phi_t(\omega)\}, \tag{7.4}$$

where $G_t(\omega)$ and $\phi_t(\omega)$ are termed respectively the *evolutionary gain-spectrum* and the *evolutionary phase-spectrum*, we may estimate $G_t(\omega)$ and $\phi_t(\omega)$ respectively by

$$\hat{G}_t(\omega) = \{\hat{c}^2_{t,yx}(\omega)+\hat{q}^2_{t,yx}(\omega)\}^{\frac{1}{2}}/\hat{f}_{t,xx}(\omega) \tag{7.5}$$

and

$$\hat{\phi}_t(\omega) = \tan^{-1}\{\hat{q}_{t,yx}(\omega)/\hat{c}_{t,yx}(\omega)\}. \tag{7.6}$$

The sampling properties of $\hat{G}_t(\omega), \hat{\phi}_t(\omega)$ are straightforward but lengthy and we shall briefly describe only the main results here. For details see Tong (1972).

Following Priestley and Subba Rao (1969), we assume (i) the "bandwidth" of $|\Gamma(\theta)|^2$ is small compared with the "frequency domain bandwidth" of $f_{t,xx}(\omega)$, $f_{t,yy}(\omega)$, etc. and (ii) the "bandwidth" of $w_{T'}(t)$ is small compared with the "time-domain bandwidth" of $f_{t,xx}(\omega)$, $f_{t,yy}(\omega)$, etc. In this case, equation (5.6) may be approximated by

$$E[\hat{f}_{t,xy}(\omega_0)] \sim f_{t,xy}(\omega_0) \tag{7.7}$$

and

$$E\{\hat{G}_t(\omega)\} \sim G_t(\omega), \quad E\{\hat{\phi}_t(\omega)\} \sim \phi_t(\omega), \tag{7.8}$$

$$\operatorname{var}\{\hat{G}_t(\omega)\} \sim \frac{CG_t^2(\omega)}{2T'}\left\{\int_{-\infty}^{\infty} |\Gamma(\theta)|^4 \, d\theta\right\}\{W_{xy}^{-2}(\omega)-1\}, \tag{7.9}$$

$$\operatorname{var}\{\hat{\phi}_t(\omega)\} \sim \frac{C}{2T'}\left\{\int_{-\infty}^{\infty} |\Gamma(\theta)|^4 \, d\theta\right\}\{W_{xy}^{-2}(\omega)-1\} \tag{7.10}$$

and

$$\operatorname{cov}\{\hat{\phi}_t(\omega), \hat{G}_t(\omega)\} \sim 0. \tag{7.11}$$

It is encouraging to note that on specializing to the stationary case, equations (7.3)–(7.11) reduce to similar equations obtained by Jenkins (1963).

We would mention that a test for time-dependence of the open loop system has recently been proposed by Subba Rao and Tong (1972), based on a two-factor MANOVA analysis.

8. NUMERICAL ILLUSTRATIONS

We consider the following slowly changing time-dependent open loop system in discrete parameter

$$Y(t+1) = X(t) + \tfrac{1}{2}\cos(t/100)\, X(t-1) + e(t+1) \quad (t = 1, 2, \ldots), \tag{8.1}$$

where $\{X(t)\}$ is a "uniformly modulated" process given by

$$X(t) = [\exp\{-(t-500)^2/2(200)^2\}]\, X_0(t) \quad (t = 0, 1, 2, \ldots), \tag{8.2}$$

with $\{X_0(t)\}$ being a second-order auto-regressive process given by

$$X_0(t+2) - 0{\cdot}8 X_0(t+1) + 0{\cdot}4 X_0(t) = Z(t+2) \quad (t = 0, 1, 2, \ldots), \tag{8.3}$$

in which the $\{Z(t)\}$, as independent random variables, each has the distribution $N(0, 100^2)$. The noise process $\{e(t)\}$ is independent of $\{X(t)\}$ and is itself a "uniformly modulated" process given by

$$e(t) = \{\exp(-0{\cdot}001|\, t\,|)\}\, e_0(t) \quad (t = 0, 1, 2, \ldots), \tag{8.4}$$

with

$$e_0(t+2) - 0{\cdot}8 e_0(t+1) + 0{\cdot}16 e_0(t) = Z'(t+2) \quad (t = 0, 1, 2, \ldots), \tag{8.5}$$

in which the $\{Z'(t)\}$ as independent random variables, each has the distribution $N(0, 10^2)$. Artificial realizations of the processes were constructed, and estimates of $f_{t,xx}(\omega)$, $f_{t,yy}(\omega)$ and $f_{t,yx}(\omega)$ were obtained for

$$t = 108\,(100)\,808, \quad \omega = \frac{\pi}{20}\left(\frac{\pi}{20}\right)\frac{19\pi}{20},$$

by using the discrete analogue of equation (5.5), in which $w_{T'}(u)$ is given by

$$w_{T'}(t) = \begin{cases} 1/T', & -\tfrac{1}{2}T' \leqslant t \leqslant \tfrac{1}{2}T', \\ 0, & \text{otherwise,} \end{cases} \tag{8.6}$$

with $T' = 200$, and $g(u)$ has the form

$$g(u) = \begin{cases} 1/\{2\sqrt{(h\pi)}\}, & |u| \leqslant h, \\ 0 & |u| > h, \end{cases} \tag{8.7}$$

with $h = 7$. Also, the window $|\Gamma(\omega)|^2$ has a bandwidth of approximately $\pi/h = \pi/7$, and the window $\{W_{T'}(u)\}$ has width $T' = 200$.

It may easily be verified that the transfer function $D_t(\omega)$ is given by

$$D_t(\omega) = 1 + \tfrac{1}{2}\exp(-i\omega)\cos(t/100). \tag{8.8}$$

For illustrations the resulting sample estimates of $G_t(\omega)$ and $\phi_t(\omega)$, for $0 \leqslant \omega \leqslant \pi$ and $t = 708$ are plotted in Figs. 2 and 3. (For estimates at other time points and

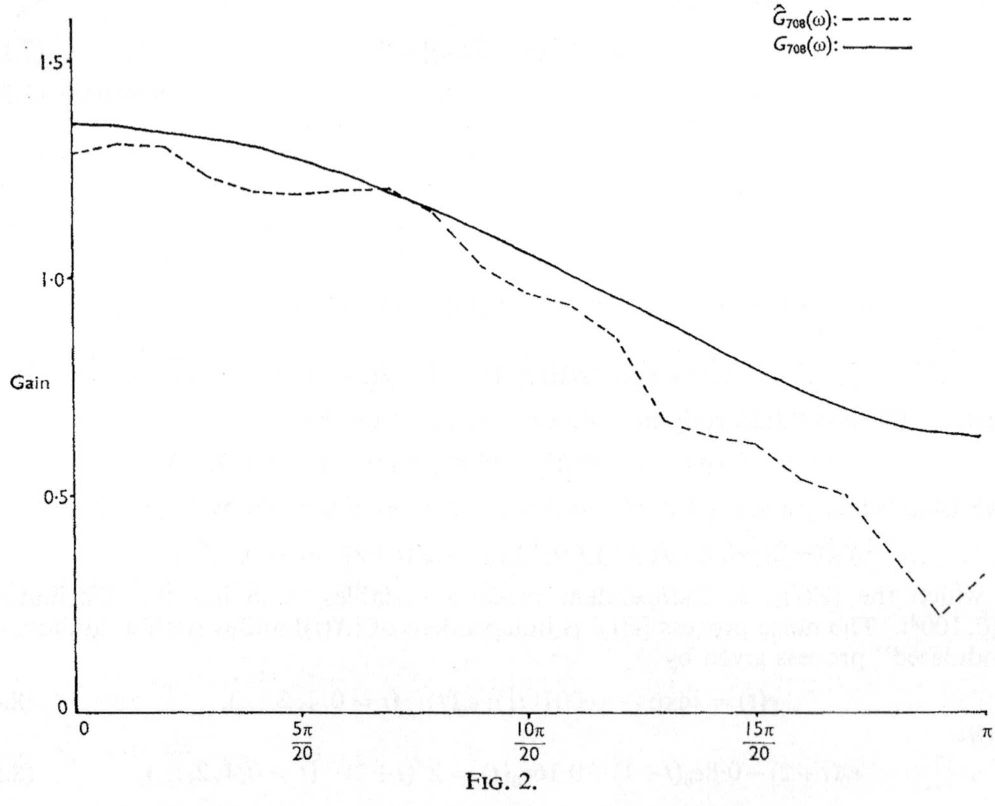

FIG. 2.

estimates of the coherency see Tong, 1972.) The estimate of $G_t(\omega)$ turns out to be reasonably accurate and, in fact, it exhibits the variation over t quite accurately. However, it does tend to somewhat underestimate the true value, which may possibly be explained by the fact that estimates of the auto-spectra, $\hat{f}_{t,xx}(\omega)$ and $\hat{f}_{t,yy}(\omega)$, tend

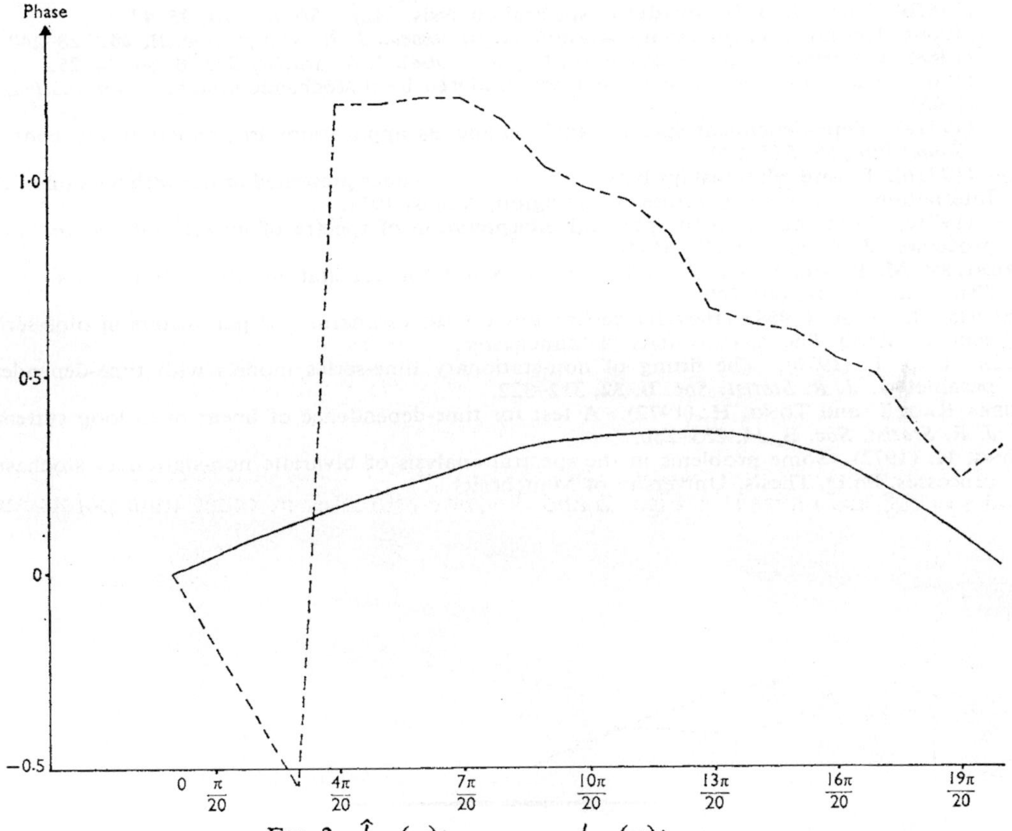

FIG. 3. $\hat{\phi}_{708}(\omega)$:– – – –. $\phi_{708}(\omega)$:———.

to overestimate the true values. (See Priestley, 1965a, p. 235.) The estimate of $\phi_t(\omega)$ turns out to be rather biased. This probably indicates the necessity for alignment of the cross periodogram.

REFERENCES

ABDRABBO, N. A. and PRIESTLEY, M. B. (1967). On the prediction of non-stationary processes. *J. R. Statist. Soc.* B, **29**, 570–585.

—— (1969). Filtering non-stationary signals. *J. R. Statist. Soc.* B, **31**, 150–159.

GRENANDER, U. and ROSENBLATT, M. (1957). *Statistical Analysis of Stationary Time Series*. New York: Wiley.

HAMMOND, J. K. (1968). On the response of single and multi-degree of freedom systems to non-stationary random excitations. *J. Sound Vib.*, **7**, 393–416.

JENKINS, G. M. (1963). Cross-spectral analysis and the estimation of linear open loop transfer functions. In *Symposium on Time Series Analysis* (M. Rosenblatt, ed.), pp. 267–276. New York: Wiley.

JENKINS, G. M. and WATTS, D. G. (1968). *Spectral Analysis and its Applications*. San Francisco: Holden-Day.

LOYNES, R. M. (1968). On the concept of the spectrum for non-stationary processes (with Discussion). *J. R. Statist. Soc.* B, **30**, 1–30.
MARK, W. D. (1970). Spectral analysis of the evolution and filtering of non-stationary stochastic processes. *J. Sound Vib.* **11**, 19–63.
PAGE, C. H. (1952). Instantaneous power spectra. *J. Appl. Phys.*, **23**, 103–106.
PRIESTLEY, M. B. (1965a). Evolutionary spectra and non-stationary processes. *J. R. Statist. Soc.* B, **27**, 204–237.
—— (1965b). The role of bandwidth in spectral analysis. *Appl. Statist.*, **14**, 33–47.
—— (1966). Design relations for non-stationary processes. *J. R. Statist. Soc.* B, **28**, 228–240.
—— (1968). Contribution to discussion on Loynes (1968), *J. R. Statist. Soc.* B, **30**, 24–25.
—— (1969). Estimation of transfer functions in closed loop stochastic systems. *Automatica*, **5**, 623–632.
—— (1971a). Time-dependent spectral analysis and its applications in prediction and control. *J. Sound Vib.*, **17**, 517–534.
—— (1971b). Fitting relationships between time series. Paper presented at the 38th Session of the International Statistical Institute, Washington, August 1971.
—— (1971c). Some notes on the physical interpretation of spectra of non-stationary stochastic processes. *J. Sound Vib.*, **17**, 51–54.
PRIESTLEY, M. B. and SUBBA RAO, T. (1969). A test for non-stationarity of time-series. *J. R. Statist. Soc.* B, **31**, 140–149.
REVFEIM, K. J. A. (1969). Iterative techniques for the estimation of parameters in time-series models. Ph.D. Thesis, University of Manchester.
SUBBA RAO, T. (1970). The fitting of non-stationary time-series models with time-dependent parameters. *J. R. Statist. Soc.* B, **32**, 312–322.
SUBBA RAO, T. and TONG, H. (1972). A test for time-dependence of linear open loop systems. *J. R. Statist. Soc.* B, **34**, 235–250.
TONG, H. (1972). Some problems in the spectral analysis of bivariate non-stationary stochastic processes. Ph.D. Thesis, University of Manchester.

J. R. Statist. Soc. B (1990)
52, *No.* 3, *pp.* 469–476

On Likelihood Ratio Tests for Threshold Autoregression

By K. S. CHAN and H. TONG†

University of Chicago, USA *University of Kent at Canterbury, UK*

[Received September 1988. Final revision August 1989]

SUMMARY

This paper addresses the null distribution of the likelihood ratio statistic for threshold autoregression with normally distributed noise. The problem is non-standard because the threshold parameter is a nuisance parameter which is absent under the null hypothesis. We reduce the problem to the first-passage probability associated with a Gaussian process which, in some special cases, turns out to be a Brownian bridge. It is also shown that, in some specific cases, the asymptotic null distribution of the test statistic depends only on the 'degrees of freedom' and not on the exact null joint distribution of the time series.

Keywords: BROWNIAN BRIDGE; CHANGEPOINT; FIRST-PASSAGE PROBABILITY; GAUSSIAN PROCESS; NON-LINEARITY; NON-STANDARD PROBLEM; ORNSTEIN–UHLENBECK PROCESS; THRESHOLD AUTOREGRESSION

1. INTRODUCTION

Threshold time series models, introduced by one of us, have been used for almost 10 years. (See, for example, Tong (1983) for a systematic account of the initial development.) To date, we have a fairly extensive range of results concerning the ergodicity and stationarity of the models, stationary probability distribution of the models, estimates of the unknown parameters and their sampling properties, graphical methods and diagnostics. (See, for example, Tong (1987) for a survey of these and other non-linear time series developments.) However, the likelihood ratio approach to the testing problems in threshold autoregression has been singularly incomplete, because the threshold parameter is a nuisance parameter which is *absent* under the null hypothesis of linearity (cf. Davies (1977, 1987)). To avoid this difficulty, Petruccelli and Davies (1986) have adopted a non-likelihood-ratio approach and developed a cumulative sum test for threshold-type non-linearity. However, empirical implementations of the likelihood ratio approach were reported in Chan and Tong (1986) and an unpublished technical report of Petruccelli in 1987. Ergodicity and stationarity of the threshold model will be assumed without further mention. (See, for example, Chan and Tong (1985).)

2. LIKELIHOOD RATIO TEST FOR THRESHOLD AUTOREGRESSION

The self-exciting threshold autoregressive model under consideration may be defined by

$$X_t - \theta_0 - \theta_1 X_{t-1} - \ldots - \theta_p X_{t-p} - I(X_{t-d} \leqslant r)\,(\phi_0 + \phi_1 X_{t-1} + \ldots + \phi_q X_{t-q}) = \epsilon_t, \tag{2.1}$$

†*Address for correspondence*: Institute of Mathematics, Cornwallis Building, University of Kent at Canterbury, Canterbury, Kent, CT2 7NF, UK.

 0035–9246/90/52469 $2.00

where the ϵ_t are independent identically distributed $N(0, \sigma^2)$ random variables $(0 < \sigma^2 < \infty)$, ϵ_t is independent of $X_{t-1}, X_{t-2}, \ldots$ and I is the indicator function with r the threshold parameter. The non-negative integers p, q and d are assumed known and are such that $0 \leqslant q \leqslant p$ and $1 \leqslant d \leqslant p$.

It is also assumed that the threshold parameter r belongs to a known bounded subset $\tilde{R}$ of R. In general, $\tilde{R}$ is in the form of a finite interval. More about the choice of $\tilde{R}$ will be given in Section 4. Also, all the roots of the characteristic equation

$$x^p - \theta_1 x^{p-1} - \ldots - \theta_p = 0$$

lie inside the unit circle. For convenience, d is assumed to be less than, or equal to, p. This is not always essential. In principle, the case with $d > p$ can be handled but the results are not as practically useful and will not be given here. Given observations $X_0, X_1, X_2, \ldots, X_N$, consider testing the null hypothesis

$$H_0\colon\ \phi_0 = \phi_1 = \ldots = \phi_q = 0. \tag{2.2}$$

Under H_0, the nuisance parameter r is absent. Let $\mathrm{RSS}_{\mathrm{AR}}$ and $\mathrm{RSS}_{\mathrm{TAR}(r)}$ denote the residual sums of squares under H_0 and model (2.1) respectively after a least squares fit. We study the asymptotic null distribution of the likelihood ratio test statistic

$$\lambda = \sup_{r\in\tilde{R}} (\mathrm{RSS}_{\mathrm{AR}} - \mathrm{RSS}_{\mathrm{TAR}(r)})/\hat{\sigma}^2, \tag{2.3}$$

$$\hat{\sigma}^2 = \inf_{r\in\tilde{R}} \mathrm{RSS}_{\mathrm{TAR}(r)}/n,$$

$$n = N-p+1, \text{ the effective number of observations.}$$

Let E denote the expectation under H_0 and $I_r(X_{t-d})$ denote the indicator function $I(X_{t-d} \leqslant r)$. Let

$$\Sigma_r = \left(\begin{array}{c|c} E\{I_r(X_{t-d})\} & E\{I_r(X_{t-d})X_{t-1}\}\ldots E\{I_r(X_{t-d})X_{t-q}\} \\ \hline E\{I_r(X_{t-d})X_{t-1}\} & \\ \vdots & S_r \\ E\{I_r(X_{t-d})X_{t-q}\} & \end{array}\right), \tag{2.4}$$

$$S_r = (E\{I_r(X_{t-d})X_{t-i}X_{t-j}\}), \qquad i = 1,\ldots,q, \quad j = 1,\ldots,q, \tag{2.5}$$

$$\Lambda_r = \left(\begin{array}{c|c} E\{I_r(X_{t-d})\} & E\{I_r(X_{t-d})X_{t-1}\}\ldots E\{I_r(X_{t-d})X_{t-p}\} \\ \hline E\{I_r(X_{t-d})X_{t-1}\} & \\ \vdots & L_r \\ E\{I_r(X_{t-d})X_{t-q}\} & \end{array}\right), \tag{2.6}$$

$$L_r = (E\{I_r(X_{t-d})X_{t-i}X_{t-j}\}), \qquad i = 1,\ldots,q, \quad j = 1,\ldots,p, \tag{2.7}$$

$$\Sigma = \left(\begin{array}{c|c} 1 & E(X_{t-1})\ldots E(X_{t-p}) \\ \hline E(X_{t-1}) & \\ \vdots & S \\ E(X_{t-p}) & \end{array}\right), \tag{2.8}$$

$$S = (E(X_{t-i}X_{t-j})), \qquad i = 1, \ldots, p, \quad j = 1, \ldots, p. \tag{2.9}$$

2.1. *Basic Result*

The asymptotic null distribution of the likelihood ratio test statistic λ is the same as the distribution of $\sup_{r \in \tilde{R}} \xi_r'(\Sigma_r - \Lambda_r \Sigma^{-1} \Lambda_r')^{-1} \xi_r$ where $\{\xi_r: r \in R\}$ is a $(q+1)$-dimensional Gaussian process with zero mean function and covariance kernel $\Sigma_{\min(r,s)} - \Lambda_r \Sigma^{-1} \Lambda_s'$.

An outline proof of this result is given in Appendix A. A rigorous demonstration under more general assumptions on ϵ_t is given in Chan (1988). Trivially, for each fixed r, $\xi_r'(\Sigma_r - \Lambda_r \Sigma^{-1} \Lambda_r')^{-1} \xi_r$ is asymptotically distributed as χ^2_{q+1}.

3. SPECIAL CASES

We state the results of two special cases. Details are given in the appendixes. Further discussion, including some other special cases, is available in an unpublished technical report of Chan and Tong (1988).

(a) The general hypothesis is

$$X_t - \theta_d X_{t-d} - \phi_d X_{t-d} I(X_{t-d} \leqslant r) = \epsilon_t, \tag{3.1}$$

and the null hypothesis is H_0: $\phi_d = 0$. In this case the asymptotic null distribution of λ reduces to the distribution of

$$\sup_s B_s^2/(s - s^2), \tag{3.2}$$

where $\{B_s: 0 \leqslant s \leqslant 1\}$ is a one-dimensional Brownian bridge and s ranges over the image of $\tilde{R}$ to be defined in Appendix B.

(b) The general hypothesis is

$$X_t - \theta_0 - \theta_1 X_{t-1} - \ldots - \theta_p X_{t-p} - I_r(X_{t-d})(\phi_0 + \phi_1 X_{t-1} + \ldots + \phi_p X_{t-p}) = \epsilon_t \tag{3.3}$$

and the null hypothesis is H_0: $\phi_i = 0, i = 0, \ldots, p$.

There exists a non-singular transformation Q such that $Q\xi_r$ is a Gaussian process, whose last $p - 1$ components are independent Brownian bridges and independent of the first two components which are correlated. The asymptotic null distribution of λ only depends on p, $\tilde{R}$, μ and σ_X.

4. FIRST-PASSAGE PROBABILITY

To calculate the tail probability of λ under H_0 for special case (a), we need to study

$$P\left(\sup_{a \leqslant s \leqslant b} \frac{B_s^2}{s - s^2} > z^2 \right) \qquad (0 < a < b < 1). \tag{4.1}$$

Setting

$$\alpha = \frac{1}{2} \ln \left\{ \frac{b(1-a)}{a(1-b)} \right\},$$

we may use the well-known result that expression (4.1) is the same as

$$P\left(\sup_{0\leqslant t\leqslant \alpha} |U(t)| > z\right), \tag{4.2}$$

where $\{U(t)\}$ is a stationary Ornstein–Uhlenbeck process with $E\{U(t)\} = 0$ and $E\{U(s)U(t)\} = \exp(-|t-s|)$. (See, for example, Anderson and Darling (1952).) From Dirkse (1975), for $z \to \infty$,

$$P\left(\sup_{0\leqslant t\leqslant \alpha} |U(t)| > z\right) \sim \left(\frac{2}{\pi}\right)^{1/2} \exp\left(-\frac{z^2}{2}\right)\left(\alpha z - \frac{\alpha}{z} + \frac{1}{z}\right). \tag{4.3}$$

Formula (4.3) gives us a convenient way of obtaining approximate significance points for the null distribution of λ, *provided* that $z \gg \alpha$. For example, with $\alpha = 1.1$ so that (a, b) is approximately the interquartile range, an approximate 5% point for λ is 7.84, which is remarkably close to the 5% point of the distribution of χ_3^2, the chi-square random variable on three degrees of freedom. In a different context, Hinkley (1969) has commented on a similar observation based on simulation results. An explanation of the χ_3^2 phenomenon in our case lies in that

$$P(\chi_3^2 > z^2) \sim (2/\pi)^{1/2} \exp(-z^2/2)(z + 1/z), \tag{4.4}$$

which may be compared with formula (4.3). The assumption $z \gg \alpha$ implies a practical bound on α and thus $\tilde{R}$. Indeed, if $\alpha = \infty$, then expression (4.2) is readily seen to be identically equal to unity for any finite z. In practice, we have found the above choice of the interquartile range quite convenient, and we adopt it henceforth in our discussion.

Unfortunately, for the more general cases such as special case (b), we are not aware of any generalization of the discussion leading to formula (4.3). Nevertheless, prompted by the χ_3^2-phenomenon, Moeanaddin and Tong (1988) have explored significance points via simulations in the neighbourhoods of χ_6^2 and χ_9^2 for the cases $p = 1 = d$ and $p = 2$, $d = 1$ or $d = 2$, respectively, and have recommended 12.6 and 15.6 as the respective empirical 5% points. On the basis of these values, they have reported the results in Table 1 with real data. (Data listings are given in Tong (1983).)

Two of the three sets of data in Table 1 are classic and need no further description. The blowfly data set was the result of an experiment by the Australian entomologist

TABLE 1

Data	*Data transformation*	*Conclusion of likelihood ratio test*†	*Order and delay*‡
Canadian lynx (1821–1934)	Raw	NL	$p = 1, d = 1$
	$\log_{10}$	NL	$p = 2, d = 1$
Sunspot numbers (1700–1955)	Raw	NL	$p = 2, d = 1$
	Square root	NL	$p = 2, d = 2$
Australian blowfly (first year)	Raw	NL	$p = 2, d = 1$
	Square root	NL	$p = 2, d = 1$
	$\log_{10}$	NL	$p = 1, d = 1$
Australian blowfly (second year)	Raw	L	
	Square root	L	
	$\log_{10}$	L	

†NL, non-linear; L, linear (nominal 5% significance used throughout).
‡p and d parameters with which the test rejects the hypothesis of linearity.

A. J. Nicholson. It is believed that the flies change their physiology after a year in captivity, thereby losing the cycle-generating mechanism which is non-linear. (Further details are given in Tong (1983).)

Our likelihood ratio results are asymptotic. An exact finite sample result based on a Bayesian approach may yield a more direct solution and avoid the arbitrariness of $\tilde{R}$ through the prior of r. For example, we may order our sample $X_{(0)} < X_{(1)} < X_{(2)} < \ldots < X_{(N)}$ and consider the odds ratio as in Smith (1975). We are studying this problem and expect some intricate numerical integration problems on the way.

We also stress that model selection is beyond the scope of the present paper. However, it is hoped that the results presented here (especially in Section 4) may help towards a more systematic study of the wider problem of linear models versus non-linear models.

ACKNOWLEDGEMENTS

We thank the referees and the Editor for their constructive comments.

APPENDIX A: OUTLINE PROOF OF BASIC RESULT

We treat model (2.1) as a 'regression model' with the 'added variables' $I_r(X_{t-d})$, $I_r(X_{t-d})X_{t-1}, \ldots, I_r(X_{t-d})X_{t-q}$. Employing the mechanics of added variables in regression, we readily obtain

$$\mathrm{RSS}_{\mathrm{AR}} - \mathrm{RSS}_{\mathrm{TAR}(r)} = T_r' \left\{ \frac{Y_r'Y_r}{n} - \frac{Y_r'X}{n} \left(\frac{X'X}{n} \right)^{-1} \frac{X'Y_r}{n} \right\}^{-1} T_r, \tag{A.1}$$

where

$$T_r = n^{-1/2}\{Y_r' - Y_r'X(X'X)^{-1}X'\}\epsilon, \tag{A.2}$$

$$Y_r = \begin{pmatrix} I_r(X_{p-d}) & X_{p-1}I_r(X_{p-d}) & \ldots & X_{p-q}I_r(X_{p-d}) \\ \vdots & \vdots & \vdots & \vdots \\ I_r(X_{N-d}) & X_{N-1}I_r(X_{N-d}) & \ldots & X_{N-q}I_r(X_{N-d}) \end{pmatrix}, \tag{A.3}$$

$$X = \begin{pmatrix} 1 & X_{p-1} & \ldots & X_0 \\ 1 & X_p & \ldots & X_1 \\ \vdots & \vdots & \vdots & \vdots \\ 1 & X_{N-1} & \ldots & X_{N-p} \end{pmatrix}, \tag{A.4}$$

and

$$\epsilon = (\epsilon_p, \epsilon_{p+1}, \ldots, \epsilon_N)'. \tag{A.5}$$

Now, by ergodicity it holds that as $N \to \infty$

$$\left(\frac{X'X}{n} \right)^{-1} \to \Sigma^{-1} \qquad \text{in probability,} \tag{A.6}$$

and, uniformly for $-\infty < -b \leqslant r \leqslant b < \infty$, b being any positive number,

$$\frac{Y_r'X}{n} \to \Lambda_r \qquad \text{in probability} \tag{A.7}$$

and

$$\frac{Y_r' Y_r}{n} \to \Sigma_r \qquad \text{in probability.} \tag{A.8}$$

Also, it is shown in Chan (1988) that $\hat{\sigma}^2$ is a consistent estimator of σ^2 and that $\Sigma_r - \Lambda_r \Sigma^{-1} \Lambda_r'$ is positive definite. Thus, the asymptotic null distribution of λ is given by that of

$$\sup_{r \in R} \left(\frac{T_r}{\sigma}\right)' (\Sigma_r - \Lambda_r \Sigma^{-1} \Lambda_r')^{-1} \left(\frac{T_r}{\sigma}\right). \tag{A.9}$$

$\{T_r: r \in R\}$ is a real parameter $(q+1)$-dimensional vector stochastic process. Now, from equations (A.2), (A.6) and (A.7), uniformly for $-b \leqslant r \leqslant b$,

$$T_r = n^{-1/2} Y_r' \epsilon - n^{-1/2} \Lambda_r \Sigma^{-1} X' \epsilon + O_p(1). \tag{A.10}$$

Applying the Cramer–Wold device (see, for example, Billingsley (1968), p. 48, or Brockwell and Davis (1987)), and a martingale central limit theorem (see, for example, theorem 23.1 in Billingsley (1968)), we obtain for each r

$$T_r \sim > N(0, \sigma^2(\Sigma_r - \Lambda_r \Sigma^{-1} \Lambda_r')), \tag{A.11}$$

with (T_r, T_s) converging weakly to the joint normal distribution with

$$\text{covariance} = \sigma^2(\Sigma_{\min(r,s)} - \Lambda_r \Sigma^{-1} \Lambda_s'). \tag{A.12}$$

Hence, on taking care of the tightness of the distributions and the topology of the function space (see Chan (1988)), $\{T_r\}$ is asymptotically a $(q+1)$-dimensional Gaussian process indexed by the threshold parameter $r \in R$. Denoting the limiting Gaussian process by $\{\xi_r\}$, we have completed our outline proof.

APPENDIX B: RESULTS FOR SPECIAL CASE (a)

$$\Sigma_r = \Lambda_r = E\{X_{t-d}^2 I(X_{t-d} \leqslant r)\}, \tag{B.1}$$

and

$$\Sigma = E(X_{t-d}^2). \tag{B.2}$$

Define s: $R \to [0, 1]$ by

$$s(r) = E\{X_{t-d}^2 I(X_{t-d} \leqslant r)\}/\sigma_X^2, \tag{B.3}$$

where $\sigma_X^2 = \text{var}\, X_t$. Then

$$\Sigma_r - \Lambda_r \Sigma^{-1} \Lambda_r' = \sigma_X^2\{s(r) - s(r)^2\}. \tag{B.4}$$

Let ρ: $[0, 1] \to R$ denote the inverse of s. Abusing the notation of s, let

$$B_s = \frac{\xi_{\rho(s)}}{\sigma_X}.$$

Then $\{B_s: 0 \leqslant s \leqslant 1\}$ is the one-dimensional Brownian bridge used in expression (3.2).

APPENDIX C: RESULTS FOR SPECIAL CASE (b)

In special case (b) $\Sigma_r = \Lambda_r$ and is as defined in equation (2.4). It is clear that $I_r(X_t) = I_{r-c}(X_t - c)$ for any real constant c. Thus, in the calculation of the covariance kernel of the Gaussian process, we may set $\mu = 0$ without loss of generality. Furthermore, let $d = p$. The proof for the case $d < p$ is similar and hence omitted. Now,

$$\Sigma = \left(\begin{array}{c|c} 1 & 0 \dots 0 \\ \hline 0 & \\ \vdots & S \\ 0 & \end{array}\right) \tag{C.1}$$

where $S = (E(X_{t-i} X_{t-j}))$ is $p \times p$. Partition S as

$$S = \begin{pmatrix} S_{11} & S_{12} \\ S_{21} & S_{22} \end{pmatrix} \tag{C.2}$$

where S_{11} is $(p-1) \times (p-1)$. Define $s_1 = E\{I(X_t \leqslant r)\}$, $s_3 = E\{X_t I(X_t \leqslant r)\}/\sigma_X$ and $u' = s_3(S_{21} S_{22}^{-1/2},\ S_{22}^{1/2})$. Then it can be shown that

$$\Sigma_r = \Lambda_r = \begin{pmatrix} s_1 & u' \\ u & S_r \end{pmatrix} \tag{C.3}$$

and $S_r = E\{X_{t-i} X_{t-j} I_r(X_{t-p})\}$ is $p \times p$.

Let G be the $(p+1) \times (p+1)$ matrix which, when premultiplied to another matrix, say A, permutes the second and the last rows of A. Let

$$Q = G \left(\begin{array}{c|c} 1 & 0 \dots 0 \\ \hline 0 & \\ \vdots & P \\ 0 & \end{array}\right), \tag{C.4}$$

where

$$P = P_3 P_2 P_1, \tag{C.5}$$

$$P_3 = \operatorname{diag}(1, 1, \dots, 1, (1-\Delta^2)^{-1/2}, 1), \tag{C.6}$$

$$P_1 = \left(\begin{array}{c|c} S_{11}^{-1/2} & \begin{matrix} 0 \\ \vdots \\ 0 \end{matrix} \\ \hline 0 \dots 0 & S_{22}^{-1/2} \end{array}\right), \tag{C.7}$$

$$P_2 = \left(\begin{array}{c|c} T & \begin{matrix} 0 \\ \vdots \\ 0 \\ -\Delta \end{matrix} \\ \hline 0 \dots 0 & 1 \end{array}\right). \tag{C.8}$$

$$\Delta^2 = S_{22}^{-1} S_{21} S_{11}^{-1} S_{12}, \qquad 0 \leqslant \Delta^2 < 1, \tag{C.9}$$

and T is a $(p-1) \times (p-1)$ orthogonal matrix such that

$$T S_{11}^{-1/2} S_{12} S_{22}^{-1/2} = \Delta(0, \dots, 0, 1)'. \tag{C.10}$$

Then it can be verified that

$$Q \Sigma Q' = I$$

and

$$Q\Sigma_r Q' = \begin{pmatrix} s_1 & s_3 & & & \\ s_3 & s_2 & & & \\ & & s_1 & & \\ & & & \ddots & \\ & & & & s_1 \end{pmatrix} \tag{C.11}$$

where

$$s_2 = s_2(r) = E\{X_t^2 I(X_t \leqslant r)/\sigma_X^2\}. \tag{C.12}$$

REFERENCES

Anderson, T. W. and Darling, D. A. (1952) Asymptotic theory of certain goodness of fit criterion based on stochastic processes. *Ann. Math. Statist.*, **23**, 193–211.

Billingsley, P. (1968) *Convergence of Probability Measures*. New York: Wiley.

Brockwell, P. J. and Davis, R. A. (1987) *Time Series: Theory and Methods*, p. 197. New York: Springer.

Chan, K. S. (1988) Testing for threshold autoregression. *Technical Report*. Department of Statistics, University of Chicago.

Chan, K. S. and Tong, H. (1985) On the use of the deterministic Lyapunov functions for the ergodicity of stochastic difference equations. *Adv. Appl. Probab.*, **17**, 666–678.

—— (1988) On likelihood ratio tests for threshold autoregression. *Technical Report*. Department of Statistics, University of Chicago.

Chan, W. S. and Tong, H. (1986) On tests for non-linearity in time series analysis. *J. Forecast.*, **5**, 217–228.

Davies, R. B. (1977) Hypothesis testing when a nuisance parameter is present only under the alternative. *Biometrika*, **64**, 247–254.

—— (1987) Hypothesis testing when a nuisance parameter is present only under the alternative. *Biometrika*, **74**, 33–43.

Dirkse, J. P. (1975) An absorption probability for the Ornstein–Uhlenbeck Process. *J. Appl. Probab.*, **12**, 595–599.

Hinkley, D. (1969) Inference about the intersection in two-phase regression. *Biometrika*, **56**, 495–504.

Moeanaddin, R. and Tong, H. (1988) Addendum and corrigendum for A comparison of the likelihood ratio test and CUSUM test for threshold autoregression. *Statistician*, **37**, 493–494.

Petruccelli, J. D. and Davies, N. (1986) A portmanteau test for self exciting threshold autoregressive-type nonlinearity in time series. *Biometrika*, **73**, 687–694.

Smith, A. F. M. (1975) A Bayesian approach to inference about a change-point in a sequence of random variables. *Biometrika*, **62**, 407–416.

Tong, H. (1983) Threshold models in non-linear time series analysis. *Lect. Notes Statist.*, **21**.

—— (1987) Non-linear time series models of regularly sampled data: a review. *Proc. 1st World Congr. Bernoulli Society, Tashkent*, vol. 2, pp. 355–367. Amsterdam: VNU.

Statistica Sinica 1(1991), 361-369

STRONG CONSISTENCY OF THE LEAST SQUARES ESTIMATOR FOR A NON-ERGODIC THRESHOLD AUTOREGRESSIVE MODEL

Dinh Tuan Pham, K. S. Chan and Howell Tong

Laboratory of Modelling and Computation, Grenoble, France,
University of Chicago and University of Kent

Abstract: **We have shown that the least squares estimator for a non-ergodic, first order, self-exciting, threshold autoregressive model is strongly consistent under quite general conditions.**

Key words and phrases: **Least squares estimator, martingale, nonlinear unit root, stationarity, strong consistency, threshold autoregressive model.**

1. Introduction

The class of self-exciting, threshold, autoregressive models (SETAR) has proved to be quite useful in nonlinear time series modelling. This class was introduced by Tong (1978) and has been studied by various authors (see Tong (1990) for references). In particular, it has been shown (Chan (1988)) that the least squares estimators (LSE) of the parameters of the model (including the thresholds and delay parameters) are strongly consistent. This result, however, depends crucially on the fact that the model is stationary and ergodic.

In this paper, we shall relax the above stationarity and ergodicity condition in the case of a simple model. Consider the first order SETAR model with only one threshold:

$$X_t = \begin{cases} \alpha X_{t-1} + \gamma + e_t, & \text{if } \ X_{t-1} \le r, \\ \beta X_{t-1} + \delta + e_t, & \text{if } \ X_{t-1} > r, \end{cases} \tag{1.1}$$

where e_t is a sequence of independent identically distributed (i.i.d.) random variables with zero mean and variance σ^2. The above model is stationary and geometrically ergodic if and only if $\alpha < 1$, $\beta < 1$ and $\alpha\beta < 1$, and is transient (as a Markov chain) if (α, β) is in the exterior of this region, but the situation is rather complex when (α, β) lies on the boundary (see Chan et al. (1985) and Guo and Petruccelli (1990)). Here, we are interested in the case where ergodicity may not hold. To simplify matters we shall assume that the threshold r is known, so that by simple substraction we may take $r = 0$. However, the above 4-parameter model is still too difficult to analyze (in the non-ergodic case). Therefore we assume further that the autoregression function is continuous at 0 and its value at this point is known, i.e. δ and γ are equal and known). For this simple model we shall show that the LSE of the parameters are strongly consistent if and only if $\alpha \le 1$ (< 1 if $\gamma < 0$) and $\beta \le 1$ (< 1 if $\gamma > 0$). The boundary case of $\alpha\beta = 1$ is also considered. This may be called the nonlinear unit root problem by analogy with the linear case.

2. Strong Consistency of the LSE

The least squares estimators of the parameters α, β of the model based on a sample $X_1, \dots, X_n$, say, are obtained by minimizing the sum of squares

$$Q_n(\alpha,\beta) = \sum_{t=2}^{n} [X_t - g(X_{t-1},\alpha,\beta)]^2 \tag{2.1}$$

where $g(x,\alpha,\beta) = [\alpha I(x \le 0) + \beta I(x > 0)]x + \gamma$ and $I(\cdot)$ denotes the set indicator function. Simple computation shows that these estimators are given by

$$\widehat{\alpha}_n = \left[\sum_{t=1}^{n-1} I(X_t \le 0) X_t (X_{t+1} - \gamma)\right] \Big/ \left[\sum_{t=1}^{n-1} I(X_t \le 0) X_t^2\right],$$

$$\widehat{\beta}_n = \left[\sum_{t=1}^{n-1} I(X_t > 0) X_t (X_{t+1} - \gamma)\right] \Big/ \left[\sum_{t=1}^{n-1} I(X_t > 0) X_t^2\right].$$

Let α_0, β_0 denote the true values of the parameters α, β. Then from (1.1) with $\alpha = \alpha_0$, $\beta = \beta_0$, $\gamma = \delta$ and $r = 0$, we get

$$\widehat{\alpha}_n - \alpha_0 = M_n^- / S_n^-, \quad \widehat{\beta}_n - \beta_0 = M_n^+ / S_n^+, \tag{2.2}$$

where

$$M_n^- = \sum_{t=1}^{n-1} I(X_t \le 0) X_t e_{t+1}, \quad S_n^- = \sum_{t=1}^{n-1} I(X_t \le 0) X_t^2,$$

$$M_n^+ = \sum_{t=1}^{n-1} I(X_t > 0) X_t e_{t+1}, \quad S_n^+ = \sum_{t=1}^{n-1} I(X_t > 0) X_t^2.$$

From (2.2), it is clear that the LSE is strongly consistent if and only if the following conditions hold:

(C0) $\lim_{n\to\infty} M_n^- / S_n^- = 0, \quad \lim_{n\to\infty} M_n^+ / S_n^+ = 0,$ almost surely.

Observe that for $n > 1$, M_n^- and M_n^+ are martingales adapted to the σ-field $\mathcal{a}_n$ generated by $X_1, \ldots ; X_n$. That is, they are $\mathcal{a}_n$-measurable and $E(M_n^- \mid \mathcal{a}_{n-1}) = M_{n-1}^-$, $E(M_n^+ \mid \mathcal{a}_{n-1}) = M_{n-1}^+$. Now let $M_n, n > 1$, be any martingale adapted to some σ-field $\mathcal{a}_n$ and denote by D_n the sum of the conditional variances $\sum_{t=2}^n E[(M_t - M_{t-1})^2 \mid \mathcal{a}_{t-1}]$. Then we have the following result (see e.g. Neveu (1965), p. 150): On the set $\{\lim_{n\to\infty} D_n = \infty\}$,

$$M_n / [D_n^{1/2} (\log D_n)^{(1/2)+\varepsilon}] \to 0, \quad \text{as } n \to \infty,$$

almost surely for every $\varepsilon > 0$, while on the set $\{\lim_{n\to\infty} D_n < \infty\}$, M_n converges almost surely to a finite limit. It is easy to see that the sum of the conditional variances for the martingales M_n^- and M_n^+ are precisely $\sigma^2 S_n^-$ and $\sigma^2 S_n^+$, respectively. Therefore the conditions (C0) are equivalent to

(C1) $\lim_{n\to\infty} S_n^- = \infty, \quad \lim_{n\to\infty} S_n^+ = \infty,$ almost surely.

We now proceed to study the behaviour of S_n^- and S_n^+. To this end we shall establish some results concerning the sample path behaviour of the process X_t. For ease of reading, the proofs are relegated to the Appendix. In the sequel we shall make the following assumptions:

(A1) $P(e_1 + \gamma > 0) > 0$,

(A2) $P(e_1 + \gamma < 0) > 0$.

Note that if we exclude the case $\sigma^2 = 0$, which corresponds to a deterministic process and is without interest, then from $E(e_1) = 0$, we have $P(e_1 > 0) > 0$ and $P(e_1 < 0) > 0$. Thus (A1) holds trivially when $\gamma \geq 0$ and so does (A2) when $\gamma \leq 0$. In particular, for $\gamma = 0$, both (A1) and (A2) hold. In any case, these assumptions hold if the distribution of e_1 has infinite positive and negative tails. The latter is a mild condition.

Lemma 2.1. *If $\alpha_0 < 1$ or $\alpha_0 = 1$, $\gamma \geq 0$, there exists a positive number c such that for all x,*

$$P(X_t > c \text{ for some } t > 0 \,|\, X_0 = x) = 1,$$

or equivalently, for any initial distribution on X_0, the Markov chain X_t enters the interval $[c, \infty)$ infinitely often almost surely.

Similarly, if $\beta_0 < 1$ or $\beta_0 = 1$, $\gamma \leq 0$, there exists a negative number d such that for all x,

$$P(X_t < d \text{ for some } t > 0 \,|\, X_0 = x) = 1,$$

or equivalently, for any initial distribution on X_0, X_t enters $(-\infty, d]$ infinitely often almost surely.

Note. Assumptions (A1) and (A2) cannot be relaxed, at least in the case $\alpha \geq 0$ and $\beta \geq 0$. Indeed, for $x < 0$, $P(X_t < 0 \mid X_{t-1} = x) \leq P(e_1 + \gamma \leq 0)$. Hence, if $P(e_1 + \gamma > 0) = 0$, we have $P(X_t < 0 \mid X_{t-1} = x) = 1$, which implies that starting at a point $x < 0$, the process remains indefinitely negative. Thus when $\alpha \geq 0$ and (A1) fails, the first conclusion of the Lemma would not hold. Similarly, $P(e_1 + \gamma < 0) = 0$ implies that starting at a point $x > 0$, the process remains indefinitely positive. Therefore when $\beta \geq 0$ and (A2) fails, the second conclusion of the Lemma would not hold.

Lemma 2.2. *If $\alpha_0 > 1$ or $\alpha_0 = 1$, $\gamma < 0$, then for any $x < 0$, $P(X_t \leq x$ for all $t > 0 \mid X_0 = x) > 0$. Similarly, if $\beta_0 > 1$ or $\beta_0 = 1$, $\gamma > 0$, then for any $x > 0$, $P(X_t \geq x$ for all $t > 0 \mid X_0 = x) > 0$.*

Lemma 2.1 shows that if α_0, β_0 and γ satisfy the conditions of this Lemma then (C1) holds since $S_n^+ \geq c^2 \sum_{t=2}^n I(X_{t-1} \geq c)$ and $S_n^- \geq d^2 \sum_{t=2}^n I(X_{t-1} \leq d)$. Lemmas 2.1 and 2.2 show that if these conditions are not satisfied then (C1) will not hold. Indeed, when α_0, γ satisfy the first condition of Lemma 2.2 and β_0, γ satisfy the second condition of Lemma 2.1, then by Lemma 2.1, X_t will almost surely eventually become negative and by Lemma 2.2, there is a positive probability that it remains negative indefinitely. Thus with positive probability, X_t enters the positive real line finitely often. The same is true when the positive real line is replaced by the negative one, when β_0, γ satisfy the second condition of Lemma 2.2 and when α_0, γ satisfy the first condition of Lemma 2.1. If α_0, β_0 and γ satisfy the conditions of Lemma 2.2, then depending on the starting value, there is a positive probability that X_t will be always positive or always negative. Therefore in all cases, at least one of the sequences of random variables S_n^- and S_n^+ is bounded with positive probability. This yields the following Proposition.

Proposition 1. *The* LSE *estimator* $(\hat{\alpha}, \hat{\beta})$ *is strongly consistent if and only if one of the following sets of conditions holds:* (i) $\alpha_0 \leq 1$, $\beta_0 \leq 1$ *and* $\gamma = 0$, (ii) $\alpha_0 < 1$, $\beta_0 \leq 1$ *and* $\gamma < 0$ *and* (iii) $\alpha_0 \leq 1$, $\beta_0 < 1$ *and* $\gamma > 0$.

We now consider the boundary case, that is when it is *known* that the point (α, β) lies on the boundary of the stationary region. This boundary is formed by three curves: $\{\alpha \leq 1,\ \beta = 1\}$, $\{\alpha = 1,\ \beta \leq 1\}$ and $\{\alpha = 1/\beta < 0\}$. For the case $\alpha \leq 1$, $\beta = 1$, the LSE of α is the same as in the general case. A similar remark holds for the case $\alpha = 1$, $\beta \leq 1$. However, in case $\alpha = 1/\beta < 0$, the least squares method would estimate α by minimizing $Q_n(\alpha, 1/\alpha)$, where Q_n is as in (2.1), leading to a different estimator. The following Proposition shows that the resulting estimator is also strongly consistent.

Proposition 2. *Suppose that $\alpha_0 = 1/\beta_0 < 0$; then the estimator of α obtained*

by minimizing $Q_n(\alpha, 1/\alpha)$ is strongly consistent.

Proof. We have

$$\begin{aligned}
&Q_n(\alpha,1/\alpha) - Q_n(\alpha_0,1/\alpha_0)\\
&= \sum_{t=2}^{n}\{[e_t + g(X_{t-1},\alpha_0,1/\alpha_0) - g(X_{t-1},\alpha,1/\alpha)]^2 - e_t^2\}\\
&= (\alpha-\alpha_0)^2 S_n^- - 2(\alpha-\alpha_0)M_n^- + (1/\alpha - 1/\alpha_0)S_n^+ - 2(1/\alpha - 1/\alpha_0)M_n^+.
\end{aligned}$$

Let δ be an arbitrary positive number. Observe that $x^2 - 2bx > \delta^2 - 2|b|\delta$ for $|x| \geq \delta \geq |b|$ and $|\alpha - \alpha_0| > \delta$ implies $|1/\alpha - 1/\alpha_0| > \delta/[|\alpha_0|(|\alpha_0| + \delta)]$. Therefore when $|M_n^-|/S_n^- \leq \delta$ and $|M_n^+|/S_n^+ \leq \delta/[|\alpha_0|(|\alpha_0|+\delta)]$, we have

$$\begin{aligned}
&\inf_{\alpha:|\alpha-\alpha_0|>\delta}[Q_n(\alpha,1/\alpha) - Q_n(\alpha_0,1/\alpha_0)]\\
&\geq S_n^+\delta^2 - 2|M_n^-|\delta + S_n^+\delta^2/[|\alpha_0|(|\alpha_0|+\delta)]^2 - 2|M_n^+|\delta/[|\alpha_0|(|\alpha_0|+\delta)]. \quad (2.3)
\end{aligned}$$

But we have already shown that (C1) holds and hence (C0) also holds, which implies that almost surely for n large enough, $|M_n^-|/S_n^- \leq \delta$, $|M_n^+|/S_n^+ \leq \delta/[|\alpha_0|(|\alpha_0| + \delta)]$ and hence (2.3) is satisfied. Further, from (C0) and (C1), the right hand side of (2.3) tends to infinity almost surely as n goes to infinity. Therefore

$$\lim_{n\to\infty}\ \inf_{\alpha:|\alpha-\alpha_0|>\delta}[Q_n(\alpha,1/\alpha) - Q_n(\alpha_0,1/\alpha_0)] = \infty, \quad \text{almost surely.}$$

Thus there exists, for almost all sample paths of the X_t process, an integer N such that for all $n > N$ and for all α satisfying $|\alpha - \alpha_0| > \delta$, we have $Q_n(\alpha,1/\alpha) - Q_n(\alpha_0,1/\alpha_0) > 0$. This implies that for all $n > N$, the function $Q_n(\alpha,1/\alpha)$ cannot attain its minimum outside the closed interval $[\alpha_0 - \delta, \alpha_0 + \delta]$; and since this function is continuous, its restriction on the compact set $[\alpha_0 - \delta, \alpha_0 + \delta]$ always admits a minimum. Thus for all $\delta > 0$, there exists, for almost all sample paths, an integer N such that for all $n > N$, the function $Q_n(\alpha,1/\alpha)$ admits a minimum which is in the interval $[\alpha_0 - \delta, \alpha_0 + \delta]$. This completes the proof of our Proposition.

Remark. A close look at the above proof reveals that the LSE of α, β under the constraints $\alpha\beta = c$, $\alpha < 0$, $\beta < 0$, where c is a given positive number, are also strongly consistent provided that the true values satisfy the same constraints. However, there is no reason to consider the constraint $\alpha\beta = c$ except for $c = 1$.

3. Some Open Problems

Our work is still incomplete in that some important questions have not been addressed. The first one concerns the more general model in which the threshold is unknown and/or the autoregression function takes unknown value at this threshold, with or without discontinuity there. Again, the least squares method may be used to estimate the model parameters. Strong consistency of the estimators has been established for the ergodic case provided that the invariant probability measure of the Markov chain X_t admits a strictly positive density (see Chan (1988)). In the general, not necessarily ergodic case, we conjecture that the last condition needs to be replaced by a recurrence property of any open interval for the Markov chain X_t. The second open problem concerns the convergence rate of the estimators. This rate may be faster than the usual rate $n^{-1/2}$ since, intuitively, the process would explode or drift to infinity with alternating signs, according to $\alpha_0\beta_0 > 1$ or $\alpha_0\beta_0 = 1$ ($\alpha_0 < 0$, $\beta_0 < 0$). The third open problem is the limiting distribution of the estimators.

Appendix : Proof of Lemmas 2.1 and 2.2

Proof of Lemma 2.1. We shall prove only the first part of the Lemma since the proof for the second part is similar.

We begin by establishing the equivalence between the two statements of the Lemma. Suppose that X_t enters $[c,\infty)$ infinitely often, regardless of the initial distribution of X_0. Then, taking this distribution to be the Dirac distribution with mass at x, we get

$$P(X_t > c \ \text{ for some } \ t > 0 \,|\, X_0 = x) = 1.$$

Conversely, if the above equality holds for all x, then from the homogeneity of the Markov chain X_t, we also have for all $k > 0$ and all x,

$$P(X_{k+t} > c \ \text{ for some } \ t > 0 \,|\, X_k = x) = 1.$$

By integration with respect to the distribution of X_k, we get $P(X_t > c$ for some $t > k) = 1$, and since this is true for all k, this means that the process enters $[c,\infty)$ infinitely often almost surely.

We now show the validity of the Lemma. Let c be a positive number. We have for all $x \in (0,c)$

$$\begin{aligned} P(X_1 \geq c \,|\, X_0 = x) &= P(\beta_0 x + e_1 + \gamma \geq c) = P(e_1 + \gamma \geq c - \beta_0 x) \\ &\geq P\{e_1 + \gamma \geq c(1 + |\beta_0|)\}. \end{aligned}$$

Choose c small enough such that $P\{e_1 + \gamma \geq (1 + |\beta_0|)c\} > 0$, which is possible because of (A1). Then

$$\inf_{x\in[0,c)} P(X_t \geq c \text{ for some } t > 0 \,|\, X_0 = x) > 0.$$

Hence, by Proposition 5.1 in Orey (1971), for any initial distribution on X_0,

$$\{X_t \in [0,c) \text{ infinitely often}\} \subseteq \{X_t \geq c \text{ infinitely often}\}$$

almost surely. But clearly,

$$\begin{aligned}&\{X_t \geq 0 \text{ infinitely often}\}\\ &=\{X_t \in [0,c) \text{ infinitely often}\} \cup \{X_t \geq c \text{ infinitely often}\}.\end{aligned}$$

Therefore we get the result of the Lemma: $P\{X_t \geq c \text{ infinitely often}\} = 1$, if we can prove that the event on the above left hand side has probability one, or equivalently (by the same argument as the beginning of this proof)

$$P(X_t < 0 \text{ for all } t > 0 \,|\, X_0 = x) = 0,$$

for all x. For this purpose, it is enough to show that for all $y < 0$,

$$P(X_t < 0 \text{ for all } t > 1 \,|\, X_1 = y) = 0,$$

which would yield the desired result by integration with respect to the conditional distribution of X_1 given $X_0 = x$.

We now prove the last equality. Note that the left hand side of this equality is the same as the probability that a first order autoregressive process with parameter α_0 and constant term γ, starting at $y < 0$, remains indefinitely negative. For $\alpha_0 = 1$, this process reduces to a random walk with increments having mean γ and hence the corresponding probability, for $\gamma \geq 0$, is zero (see Feller (1966), pp. 395, 396). The same is true for $0 < \alpha_0 < 1$, since then the above process is a stationary Markov chain with invariant probability measure having support not contained in $(-\infty, 0]$. To see this, note that this measure is the probability distribution of $\sum_{k=0}^{\infty} \alpha_0^k(e_{-k} + \gamma)$. By (A1) there exists $\epsilon > 0$ such that $P(e_{-k} + \gamma > \epsilon) > 0$. Thus $\sum_{k=0}^{n-1} \alpha_0^k(e_{-k} + \gamma)$ is greater than $\epsilon(1 - \alpha_0^n)/(1 - \alpha_0)$ with positive probability. Now, choose n large enough such that $(1 - \alpha_0^n)\epsilon + \alpha_0^n\gamma$ is positive (if $\gamma \geq 0$, any value of n would do). Then the random variable

$$\sum_{k=n}^{\infty} \alpha_0^k(e_{-k} + \gamma) + \epsilon(1 - \alpha_0^n)/(1 - \alpha_0)$$

is positive with positive probability since it has positive expectation. Thus $\sum_{k=0}^{\infty} \alpha_0^k(e_{-k} + \gamma)$ can be expressed as the sum of two random variables greater than $\epsilon(1 - \alpha_0^n)/(1 - \alpha_0)$ and $-\epsilon(1 - \alpha_0^n)/(1 - \alpha_0)$ with positive probability, respectively, yielding the desired result.

Finally, for $\alpha_0 \leq 0$, one obtains the conclusion of the Lemma by taking the limit, as k goes to infinity, of the extreme sides of

$$P(X_t < 0, t = 2, \ldots, k \,|\, X_1 = y) \leq P(e_t + \gamma < 0, t = 2, \ldots, k) \\ = [P(e_1 + \gamma < 0)]^{k-1},$$

and noting that $P(e_1 + \gamma < 0) < 1$ by (A1).

Proof of Lemma 2.2. Again, we prove only the first part of the Lemma since the proof for the second part is similar.

Suppose that $\alpha_0 \geq 1$. Then $X_{t-1} \leq x < 0$ implies

$$X_t - x = \alpha_0(X_{t-1} - x) + e_t + \gamma + (\alpha_0 - 1)x \\ \leq X_{t-1} - x + e_t + \gamma + (\alpha_0 - 1)x.$$

Define the random walk Y_t by $Y_0 = X_0 - x$, $Y_t = Y_{t-1} + e_t + \gamma + (\alpha_0 - 1)x$. Then by induction, it is easily seen that $Y_t \leq 0$ for all $t \geq 0$ implies $X_t - x \leq Y_t$. For $\alpha_0 = 1$ and $\gamma < 0$, or $\alpha_0 > 1$ and $x < -\gamma/(\alpha_0 - 1)$, the random variable $(\alpha_0 - 1)x + e_t + \gamma$ has negative expectation, which implies that the random walk Y_t drifts to minus infinity (Feller (1966), pp. 395, 396) or equivalently

$$P(Y_t \leq 0 \text{ for all } t > 0 \,|\, Y_0 = 0) > 0,$$

giving

$$P(X_t \leq x \text{ for all } t > 0 \,|\, X_0 = x) > 0.$$

To complete the proof of the Lemma, we need only show that for $\alpha_0 > 1$, there is a positive probability that starting at $X_0 = x < 0$, the X_t process becomes less than $-\gamma/(\alpha_0 - 1)$ after a finite number of steps. (Here $\gamma > 0$, otherwise there is nothing to prove.) That this is true follows from the fact that for all $y < 0$,

$$P(X_t < y \,|\, X_{t-1} = y) = P\{(\alpha_0 - 1)y + e_t + \gamma < 0\} \\ \geq P(e_1 + \gamma < 0) > 0.$$

Acknowledgement

This paper was presented at the Edinburgh International Workshop on Non-linear Time Series under the auspices of the Science and Engineering Research Council (UK) in July 1989. The research of Chan was partially supported by National Science Foundation Grant DMS-9006.

References

Chan, K. S. (1988). Consistency and limiting distribution of the least squares estimator of a threshold autoregressive model. Technical Report 245, University of Chicago.

Chan, K. S., Petruccelli, J. D., Tong, H. and Woolford, S. W. (1985). A multiple-threshold AR(1) model. *J. Appl. Probab.* **22**, 267-279.

Feller, W. (1966). *An Introduction to Probability Theory and Its Applications II*, 2nd edition. John Wiley, New York.

Guo, M. and Petruccelli, J. D. (1990). On the null recurrence and transience of a first order SETAR model. Unpublished manuscript.

Neveu, J. (1965). *Mathematical Foundations of the Calculus of Probability.* Holden-Day, San Francisco.

Orey, S. (1971). *Lecture Notes on Limit Theorems for Markov Chain Transition Probabilities.* Van Nostrand and Reinhold Company, London.

Tong, H. (1978). On a threshold model. In *Pattern Recognition and Signal Processing* (Edited by C. H. Chen). Sijthoff and Noordhoff, Holland.

Tong, H. (1990). *Non-linear Time Series: A Dynamical System Approach.* Oxford University Press, Oxford.

Laboratory of Modelling and Computation, CNRS, Grenoble, B. P. 53x, 38041 Grenoble Cedex, France.

Department of Statistics, University of Chicago, Illinois 60637, U.S.A.

Institute of Mathematics and Statistics, University of Kent at Canterbury, Canterbury CT2 7NF, U.K.

(Received November 1989; accepted January 1991)

Some Remarks on Professor Tong's Two Papers

SHIQING LING*

Department of Mathematics, Hong Kong University of Science and Technology
Clear Water Bay, Hong Kong, P. R. China
E-mail: maling@ust.hk

This note makes some remarks on the two papers, "Strong Consistency of the Least Squares Estimator for a Non-Ergodic Threshold Autoregressive Model" in *Statistica Sinica* (1991) and "On Likelihood Ratio Tests for Threshold Autoregression" in *Journal of the Royal Statistical Society, Series B* (1990), by Professor Tong and his collaborators. This note also discusses the limiting distribution of the LSE for the unit root TAR(1) model and gives the rate of convergence of the LSE for an explosive TAR(1).

1. On "Strong Consistency of the Least Squares Estimator for a Non-Ergodic Threshold Autoregressive Model" by Pham, Chan and Tong (1991)

Tong's (1978) threshold autoregressive (TAR) models have been extensively investigated in the literature. The standard TAR(1) model can be written as follows:

$$y_t = \begin{cases} \gamma + \alpha y_{t-1} + \varepsilon_t \text{ if } y_{t-1} \le r, \\ \delta + \beta y_{t-1} + \varepsilon_t \text{ if } y_{t-1} > r, \end{cases} \tag{1}$$

where $\{\varepsilon_t\}$ is a sequence of i.i.d. random variables with mean zero and variance σ^2. Petruccelli and Woolford (1984) and Chan, Petruccelli, Tong and Woolford (1985) showed that the necessary and sufficient condition for the strictly stationary and geometrically ergodic solution to model (1) is

$$\alpha < 1, \ \beta < 1 \text{ and } \alpha\beta < 1, \tag{2}$$

see also Chen and Tsay (1991). The properties of the least squares estimator (LSE) of model (1) were established in the stationary case by Chan (1993) and later by Chan and Tsay (1996) for the continuous case. When (α, β) does not lie in the stationary region (2), the estimation theory of the LSE of model (1) has not been fully developed.

The paper by Pham, Chan and Tong (1991) was the first one to consider the non-stationary case of model (1). Since model (1) is too complicated when (α, β) does not lie in the stationary region (2), they focus on the following special case:

$$\gamma = \delta$$

and assume that δ and r are known parameters. Without loss of generality, they assume

*The author thanks Professors K.S. Chan and H. Tong for their comments and the Hong Kong Research Grants Commission (601607 and 602609) for financial support.

that $r = 0$. They considered the LSE of (α, β), which are

$$\hat{\alpha}_n = \frac{\sum_{t=1}^{n-1} I(y_t \le 0) y_t (y_{t+1} - \gamma)}{\sum_{t=1}^{n-1} I(y_t \le 0) y_t^2},$$
$$\hat{\beta}_n = \frac{\sum_{t=1}^{n-1} I(y_t > 0) y_t (y_{t+1} - \gamma)}{\sum_{t=1}^{n-1} I(y_t > 0) y_t^2}.$$

Let (α_0, β_0) be the true value of (α, β). The main results in the paper are that

$$(\hat{\alpha}_n, \hat{\beta}_n) \to (\alpha_0, \beta_0) \text{ almost surely (a.s.)}$$

if and only if the one of the following conditions holds:

$$\alpha_0 \le 1, \ \beta_0 \le 1, \text{ and } \gamma = 0, \tag{3}$$
$$\alpha_0 < 1, \ \beta_0 \le 1, \text{ and } \gamma < 0, \tag{4}$$
$$\alpha_0 \le 1, \ \beta_0 < 1, \text{ and } \gamma > 0. \tag{5}$$

They also showed that, when $\alpha_0 = 1/\beta_0$, i.e., (α_0, β_0) is on the boundary with $\alpha\beta = 1$, the estimator of α_0 is strongly consistent. However, the rate of convergence and the limiting distribution of LSE are not clear when (α_0, β_0) lies in the non-ergodic region. They identified these as two open problems.

I would like to discuss two special cases, which may provide some useful insights for the previous problems.

A. The case when $\alpha_0 = 1$, $\beta_0 = -1$ and $\gamma = 0$. Let $I_t = I(y_t \le 0) - I(y_t > 0)$. Then, we can write y_t in the following form:

$$\begin{aligned} y_t &= \varepsilon_t + I_{t-1} y_{t-1} \\ &= \sum_{i=0}^{t-1} \prod_{j=1}^{i} I_{t-j} \varepsilon_{t-i} + \prod_{j=1}^{t} I_{t-j} y_0 \\ &= \sum_{i=0}^{t-1} \prod_{j=t-i}^{t-1} I_j \varepsilon_{t-i} + \prod_{j=0}^{t-1} I_j y_0 \\ &= \sum_{k=1}^{t} \prod_{j=k}^{t-1} I_j \varepsilon_k + \prod_{j=0}^{t-1} I_j y_0, \end{aligned}$$

where $\prod_{j=1}^{-1} I_{t-j} = 1$. Let $A_k = \prod_{j=0}^{k-1} I_j$. We have

$$A_t^{-1} y_t = \sum_{k=1}^{t} A_k^{-1} \varepsilon_k + y_0. \tag{6}$$

The first term is a martingale in terms of $\mathcal{F}_t \equiv \sigma\{y_0, \varepsilon_1, \cdots, \varepsilon_t\}$. Note that $E[A_k^{-2} \varepsilon_k^2 | \mathcal{F}_{t-1}] = \sigma^2$. It is not hard to see that

$$\frac{1}{\sqrt{n}} A_{[n\tau]}^{-1} y_{[n\tau]} \longrightarrow_{\mathcal{L}} \sigma B(\tau) \text{ in } D[0,1], \tag{7}$$

where $\to_{\mathcal{L}}$ denotes the convergence in distribution and $B(\tau)$ is the standard Brownian motion on $D[0, 1]$ which is the Skorokhod space. Note that $I(y_t \le 0) A_{t+1} = I(y_t \le 0) A_t$

and $A_t^{-2} = 1$. It is possible to show that

$$\begin{aligned} n(\hat{\alpha}_n - 1) &= \frac{n\sum_{t=1}^{n-1} I(y_t \le 0) y_t \varepsilon_{t+1}}{\sum_{t=1}^{n-1} I(y_t \le 0) y_t^2}, \\ &= \frac{\sum_{t=1}^{n-1} I(A_t^{-1} y_t/\sqrt{n} \le 0)(A_t^{-1} y_t/\sqrt{n})(A_{t+1}^{-1}\varepsilon_{t+1}/\sqrt{n})}{\sum_{t=1}^{n-1} I(A_t^{-1} y_t/\sqrt{n} \le 0)[A_t^{-1} y_t/\sqrt{n}]^2}, \\ &\longrightarrow_{\mathcal{L}} \frac{\int_0^1 I(B(\tau) \le 0) B(\tau) dB(\tau)}{\int_0^1 I(B(\tau) \le 0) B^2(\tau) d\tau}. \end{aligned}$$

But we can not prove that $I(y_t \le 0)$ and $I(A_t^{-1} y_t/\sqrt{n} \le 0)$ have the same limiting distribution. Thus, the limiting distributions of $n(\hat{\alpha}_n - 1)$ and $n(\hat{\beta}_n + 1)$ remain open problems. Similar situation happens when $\alpha_0 = -1$ and $\beta_0 = 1$.

B. The case when $\alpha_0 > 1$ and $\beta_0 > 1$. Let $I_t = \alpha_0 I(y_t \le 0) + \beta_0 I(y_t > 0)$. Similar to (6), we can write y_t in the following form:

$$A_t^{-1} y_t = \sum_{k=1}^{t} A_k^{-1}(\gamma + \varepsilon_k) + y_0. \tag{8}$$

Since $A_k^{-1} \le \rho^k$ with $\rho = \max\{1/\alpha_0, 1/\beta_0\} < 1$, we have

$$A_t^{-1} y_t \longrightarrow \xi \equiv \sum_{k=1}^{\infty} A_k^{-1}(\gamma + \varepsilon_k) + y_0 \text{ a.s. and } L^2, \tag{9}$$

when $t \to \infty$. Furthermore, since $A_t/A_n \le \rho^{n-t}$ and $xI(x \le 0)$ is a continuous function, we can show that

$$\sum_{t=1}^{n-1} \left(\frac{A_t}{A_n}\right)^2 [I(y_t \le 0) A_t^{-2} y_t^2 - I(\xi \le 0)\xi^2] = o(1).$$

By the previous equation, we have

$$\begin{aligned} \frac{1}{A_n^2} \sum_{t=1}^{n-1} I(y_t \le 0) y_t^2 &= I(\xi \le 0)\xi^2 \sum_{t=1}^{n-1} \left(\frac{A_t}{A_n}\right)^2 + o(1) \\ &\ge cI(\xi \le 0)\xi^2 + o(1), \end{aligned} \tag{10}$$

where $c > 0$ is some constant. Note that

$$E[\max_t |A_t^{-1} y_t|] \le E[\sum_{k=1}^{\infty} \rho^k |\gamma + \varepsilon_k| + |y_0|] < \infty.$$

Furthermore, by (9), the continuity of the function $xI(x \le 0)$ and dominated convergence, we can show that

$$\begin{aligned} &E\Big|\sum_{t=1}^{n-1} \frac{A_t}{A_n} [I(y_t \le 0) A_t^{-1} y_t - I(\xi \le 0)\xi]\varepsilon_{t+1}\Big| \\ &\le \sigma \sum_{t=1}^{n-1} \rho^{n-t} E\Big| I(y_t \le 0) A_t^{-1} y_t - I(\xi \le 0)\xi\Big| = o(1). \end{aligned}$$

Thus, it follows that

$$\frac{1}{A_n}\sum_{t=1}^{n-1} I(y_t \le 0)y_t\varepsilon_{t+1} = I(\xi \le 0)\xi \sum_{t=1}^{n-1}\left(\frac{A_t}{A_n}\right)\varepsilon_{t+1} + o_p(1) = O_p(1). \tag{11}$$

By (10)-(11), we have

$$A_n(\hat{\alpha}_n - \alpha_0) = O_p(1).$$

Similarly, we can show that

$$A_n(\hat{\beta}_n - \beta_0) = O_p(1).$$

This means that the rate of convergence of the LSE of (α_0, β_0) is exponential and much faster than the $\sqrt{n}$−convergent rate in the stationary case. It is similar to the rate of convergence of the LSE for the explosive AR(1) model. The results also indicate that the conditions (3)-(5) may not be necessary for weak consistency.

2. On "On Likelihood Ratio Tests for Threshold Autoregression" by Chan and Tong (1990)

Consider the general TAR(p) model:

$$y_t = \phi_0 + \phi_1 y_{t-1} + \cdots + \phi_p y_{t-p} + I(y_{t-d} \le r)(\psi_0 + \psi_1 y_{t-1} + \cdots + \psi_p y_{t-p}) + \varepsilon_t,$$

where ε_t are i.i.d white noises with variance $\sigma^2 > 0$. It is obvious that the TAR(p) model nests the linear AR(p) model. The paper by Chan and Tong (1990) considered the likelihood ratio test for the null hypothesis:

$$H_0 : \phi_0 = \phi_1 = \cdots = \phi_p = 0$$

against the alternative TAR(p) model. A special feature is that the threshold parameter r is absent under H_0. For a given r, the log-likelihood ratio is

$$LR_n(r) = \tilde{T}_n'(r)(\Sigma_{nr} - \Sigma_{nr}\Sigma_n^{-1}\Sigma_{nr})^{-1}\tilde{T}_n(r),$$

where $\Sigma_{nr} = \sum_{t=1}^n X_{t-1}X_{t-1}' I\{y_{t-d} \le x\}/n$, $\Sigma_n = \Sigma_{n,\infty}$, $X_t = (y_t, \cdots, y_{t-p+1})'$ and

$$T_n(r) = \frac{1}{\sqrt{n}}\sum_{t=1}^n X_{t-1}\varepsilon_t I\{y_{t-d} \le r\},$$

$$\widetilde{T}_n(r) = T_n(r) - \frac{\Sigma_{nr}\Sigma_n^{-1}}{\sqrt{n}}\sum_{t=1}^n X_{t-1}\varepsilon_t.$$

They considered the following test statistic:

$$LR_n = \max_{r\in[a,b]} LR_n(r),$$

where $-\infty < a < b < \infty$. They showed that the limiting distribution of LR_n is the maxima of a Gaussian process ξ_r on $[a, b]$. This involves the weak convergence of the empirical process of $\{T_n(r)\}$ on $D[a, b]$ which was established by Chan (1990).

To use the test statistic LR_n, one needs to obtain its critical values. For the special case,

$$y_t = \phi_1 y_{t-1} + I(y_{t-d} \le r)\psi_1 y_{t-1} + \varepsilon_t,$$

Chan and Tong (1990) showed that the limiting distribution of LR_n reduces to that of

$$\sup_{\tau \in [s_a, s_b]} \frac{B(\tau)}{s - s^2},$$

where $s_r = E[y_{t-d}^2 I(y_{t-d} \leq r)]/var(y)$ and $\{B(\tau) : s \in [0,1]\}$ is the one-dimensional Brownian bridge. After connecting this to a stationary Ornstein-Uhlenbeck process, they provided a simple way to obtain the approximating critical values of LR_n. For the general case, they showed that there exists a non-singular transformation Q such that the last $p-1$ components of $Q\xi_r$ are independent Brownian bridges and independent of its first two components. But it needs to use the Poisson clumping heuristic developed by Aldous (1989) to obtain the approximating critical values of LR_n, which was discussed by Chan (1991).

The paper by Chan and Tong (1990) together with Chan (1990, 1991) fully developed the theory and procedure for testing AR(p) against TAR(p) models [see Tsay (1998) for the vector case]. One may be interested in testing the following general null and alternative models:

$$H_0 : y_t = \mu_t(\theta_0) + \eta_t \sqrt{h_t(\theta_0)},$$
$$H_1 : y_t = \mu_t(\theta_0) + \mu_t(\theta_1) I\{y_{t-1} \leq r\} + \eta_t \sqrt{h_t(\theta_0) + h_t(\theta_1) I\{y_{t-1} \leq r\}},$$

where $\theta_0 \neq \theta_1$. Some special models were studied in the literature, e.g., testing the AR-ARCH model against threshold AR-ARCH models by Wong and Li (1997, 2000) and testing the linear MA model against the threshold MA models by Ling and Tong (2006). Under suitable conditions and under H_0, the log-likelihood ratio test statistic for H_0 against H_1 can be approximated generally by

$$\begin{aligned} S_n(r) &\equiv T_n'(r, \hat{\theta}_n)(\Sigma_r - \Sigma_r \Sigma^{-1} \Sigma_r)^{-1} T_n(r, \hat{\theta}_n) \\ &= \widetilde{T}_n'(r, \theta_0)(\Sigma_r - \Sigma_r \Sigma^{-1} \Sigma_r)^{-1} \widetilde{T}_n(r, \theta_0) + o_p(1), \end{aligned}$$

uniformly in $r \in R$, where $\hat{\theta}_n$ is a suitable estimator of θ_0,

$$T_n(r, \theta) = \frac{1}{\sqrt{n}} \sum_{t=1}^{n} D_t(\theta) I\{y_{t-1} \leq r\}, \tag{12}$$

$$\widetilde{T}_n(r, \theta_0) = T_n(r, \theta_0) - \frac{\Sigma_r \Sigma^{-1}}{\sqrt{n}} \sum_{t=1}^{n} D_t(\theta_0), \tag{13}$$

$D_t(\theta)$ is the score function under H_0, $\Sigma = \Sigma_\infty$, and

$$\Sigma_r = E[D_t(\theta_0) D_t'(\theta_0) I\{y_{t-1} \leq r\}].$$

Following Chan and Tong (1990), it is natural to consider the test statistic

$$LR_n = \max_{r \in [a,b]} S_n(r).$$

Unlike Chan (1991) and Chan and Tong (1990), we have to use the simulation method to find the critical values case by case, see Wong and Li (1997) and Ling and Tong (2005). This greatly hinders research progress in this direction.

To avoid the previous difficulty, Ling and Tong (2007) considered a linear transformation of $T_n(r, \hat{\theta}_n)$. Let β be a nonzero $p \times 1$ constant vector and

$$\sigma_r = \frac{\beta'(\Sigma_r^{-1} - \Sigma^{-1})\beta}{\beta'(\Sigma_a^{-1} - \Sigma^{-1})\beta}.$$

They showed that $\beta'\hat{\Sigma}_{nr}^{-1}T_n(r,\hat{\theta}_n)/\sqrt{\beta'(\hat{\Sigma}_{na}^{-1}-\Sigma^{-1})\beta}$ converges weakly to a Gaussian process with covariance kernel $\sigma_x \wedge \sigma_y$, where $\hat{\Sigma}_{nx}$ is a consistent estimator of Σ_x uniformly in $r \in R$. Let $b > a$ be a constant. Define $T_n^a = T_n^a(\infty)$ and

$$T_n^a(b) = \max_{a \le r \le b} \frac{[\beta'\hat{\Sigma}_{nx}^{-1}T_n(r,\hat{\theta}_n)]^2}{\beta'(\hat{\Sigma}_{na}^{-1}-\Sigma^{-1})\beta}.$$

Here, T_n^a is the maxima of the linear combination of the score marked empirical process on $r \in [a,\infty)$. They showed that

$$P\left[T_n^a(b) \le x\right] \approx P\left[\max_{\tau\in[0,1]} B^2(\tau) \le x\right]$$

for any $x \in R$, where $B(\tau)$ is a standard Brownian motion on $C[0,1]$. The constant C_α such that $P[\max_{\tau\in[0,1]} B^2(\tau) \ge C_\alpha] = \alpha$ can be used as an approximating critical value of T_n^a for rejecting the null H_0 at the significance level α. From Shorack and Wellner (1986, p.34), it follows that

$$P\Big[\max_{\tau\in[0,1]} B^2(\tau) \ge x\Big] = 1 - \frac{4}{\pi}\sum_{k=0}^{\infty}\frac{(-1)^k}{2k+1}\exp\Big[-\frac{(2k+1)^2\pi^2}{8x}\Big],$$

for all $x > 0$, and $C_{0.1} = 3.83$, $C_{0.05} = 5.00$ and $C_{0.01} = 7.63$.

The choice of a will depend on the data, but the choice should ensure that $\hat{\Sigma}_{na}^{-1}$ exists. T_n^a will be invariant in terms of $\|\beta\|$ and a natural choice for β is $(1,\cdots,1)'$. It is not really a likelihood ratio test. Rather it is a portmanteau test with non-specific alternatives. The simulation in Ling and Tong (2007) showed that this choice with a being the 5% quantile of data S_n works well in terms of both size and power when the null models are the ARMA and GARCH models, while the alternatives are the threshold ARMA and threshold GARCH models, respectively.

References

1. Aldous, D. (1989) Probability approximations via the Poisson clumping heuristic. *Applied Mathematical Sciences*, **77**. Springer-Verlag, New York.
2. Chan, K.S. (1990) Testing for threshold autoregression. *Ann. Statist.* **18**, 1886-1893.
3. Chan, K.S. (1991) Percentage points of likelihood ratio tests for threshold autoregression. *J. Roy. Statist. Soc. Ser. B* **53**, 691-696.
4. Chan, K.S. (1993) Consistency and limiting distribution of the least squares estimator of a threshold autoregressive model. *Ann. Statist.* **21**, 520-533.
5. Chan, K.S. and Tong, H. (1990) On likelihood ratio tests for threshold autoregression. *J. Roy. Statist. Soc. Ser. B* **52** (1990), 469-476.
6. Chan, K.S. and Tsay, R.S. (1998) Limiting properties of the least squares estimator of a continuous threshold autoregressive model. *Biometrika* **85** (1998), 413-426.
7. Chan, K.S., Petruccelli, J.D., Tong, H. and Woolford, S.W. (1985) A multiple-threshold AR(1) model. *J. Appl. Probab.* **22**, 267-279.
8. Chen, R. and Tsay, R.S. (1991) On the ergodicity of TAR(1) processes. *Ann. Appl. Probab.* **1**, 613-634.
9. Ling, S. and Tong, H. (2006) Testing a linear MA model against threshold MA models. *Ann. Statist.* **34**, 994-1012.
10. Ling, S. and Tong, H. (2007) A general approach to goodness-of-fit tests for time series models. Revised for *JRSSB*.
11. Petruccelli, J.D. and Woolford, S.W. (1984) A threshold AR(1) model. *J. Appl. Probab.* **21**, 270–286

12. Pham, D.T., Chan, K.S. and Tong, H. (1991) Strong consistency of the least squares estimator for a nonergodic threshold autoregressive model. *Statist. Sinica* **1**, 361-369.
13. Shorack, G.R. and Wellner, J.A. (1986) *Empirical processes with applications to statistics.* John Wiley & Sons, Inc., New York.
14. Tong, H. (1978) On a threshold model. *Pattern recognition and signal processing.* (ed. 6.H.Chen). Sijthoff and Noordhoff, Amsterdam.
15. Tong, H. (1990) *Nonlinear time series. A dynamical system approach.* The Clarendon Press, Oxford University Press, New York.
16. Tsay, R.S. (1998) Testing and modeling multivariate threshold models. *J. Amer. Statist. Assoc.* **93**, 1188-1202.
17. Wong, C.S. and Li, W.K. (1997) Testing for threshold autoregression with conditional heteroscedasticity. *Biometrika* **84**, 407-418.
18. Wong, C.S and Li, W.K. (2000) Testing for double threshold autoregressive conditional heteroscedastic model. *Statist. Sinica* **10**, 173-189.

With George Tiao and Rong Chen at NSF-NBER Time Series meeting in Iowa, 2007

With Nils Christian Stenseth in Hong Kong, 2008

With friends at the 40th Anniversary celebration of the Statistics Department, University of Hong Kong, in 2008 (Fred Ho, Wai-Keung Li, John Aitchitson, HT, Yeh Lam, Shiqing Ling)

J. R. Statist. Soc. B (2002)
64, *Part* 3, *pp.* 363–410

An adaptive estimation of dimension reduction space

Yingcun Xia,

University of Cambridge, UK, and Jinan University, People's Republic of China

Howell Tong,

University of Hong Kong, People's Republic of China, and London School of Economics and Political Science, UK

W. K. Li

University of Hong Kong, People's Republic of China

and Li-Xing Zhu

University of Hong Kong and Chinese Academy of Sciences, Beijing, People's Republic of China

[*Read before* The Royal Statistical Society *at a meeting organized by the* Research Section *on Wednesday, February 13th, 2002,* Professor D. Firth *in the Chair*]

Summary. Searching for an effective dimension reduction space is an important problem in regression, especially for high dimensional data. We propose an adaptive approach based on semiparametric models, which we call the (conditional) minimum average variance estimation (MAVE) method, within quite a general setting. The MAVE method has the following advantages. Most existing methods must *undersmooth* the nonparametric link function estimator to achieve a faster rate of consistency for the estimator of the parameters (than for that of the nonparametric function). In contrast, a faster consistency rate can be achieved by the MAVE method even without undersmoothing the nonparametric link function estimator. The MAVE method is applicable to a wide range of models, with fewer restrictions on the distribution of the covariates, to the extent that even time series can be included. Because of the faster rate of consistency for the parameter estimators, it is possible for us to estimate the dimension of the space consistently. The relationship of the MAVE method with other methods is also investigated. In particular, a simple outer product gradient estimator is proposed as an initial estimator. In addition to theoretical results, we demonstrate the efficacy of the MAVE method for high dimensional data sets through simulation. Two real data sets are analysed by using the MAVE approach.

Keywords: Average derivative estimation; Dimension reduction; Generalized linear models; Local linear smoother; Multiple time series; Non-linear time series analysis; Nonparametric regression; Principal Hessian direction; Projection pursuit; Semiparametrics; Sliced inverse regression estimation

1. Introduction

Let y and X be respectively $\mathbb{R}$-valued and $\mathbb{R}^p$-valued random variables. Without prior knowledge about the relationship between y and X, the regression function $g(x) = E(y|X = x)$ is often

Address for correspondence: Howell Tong, Department of Statistics, London School of Economics and Political Science, Houghton Street, London, WC2A 2AE, UK.
E-mail: h.tong@lse.ac.uk

 1369–7412/02/64363

modelled in a flexible nonparametric fashion. When the dimension of X is high, recent efforts have been expended in finding the relationship between y and X efficiently. The final goal is to approximate $g(x)$ by a function having simplifying structure which makes estimation and interpretation possible even for moderate sample sizes. There are essentially two approaches: the first is largely concerned with function approximation and the second with dimension reduction. Examples of the former are the additive model approach of Hastie and Tibshirani (1986) and the projection pursuit regression proposed by Friedman and Stuetzle (1981); both assume that the regression function is a sum of univariate smooth functions. Examples of the latter are the dimension reduction of Li (1991) and the regression graphics of Cook (1998).

A regression-type model for dimension reduction can be written as

$$y = g(B_0^{\mathrm{T}} X) + \varepsilon, \tag{1.1}$$

where g is an unknown smooth link function, $B_0 = (\beta_1, \ldots, \beta_D)$ is a $p \times D$ orthogonal matrix $(B_0^{\mathrm{T}} B_0 = I_{D\times D})$ with $D < p$ and $E(\varepsilon|X) = 0$ almost surely. The last condition allows ε to be dependent on X. When model (1.1) holds, the projection of the p-dimensional covariates X onto the D-dimensional subspace $B_0^{\mathrm{T}} X$ captures all the information that is provided by X on y. We call the D-dimensional subspace $B_0^{\mathrm{T}} X$ the effective dimension reduction (EDR) space. Li (1991) introduced the EDR space in a similar but more general context; the difference disappears for the case of additive noise as in model (1.1). See also Carroll and Li (1995), Chen and Li (1989) and Cook (1994). Note that the space spanned by the column vectors of B_0 is uniquely defined under some mild conditions (given in Section 3) and is our focus of interest. For convenience, we shall refer to these column vectors as EDR directions, which are unique up to orthogonal transformations. The estimation of the EDR space includes the estimation of the directions, namely B_0, and the corresponding dimension of the EDR space. For specific semiparametric models, methods have been introduced to estimate B_0. Next, we give a brief review of these methods.

One of the important approaches is the projection pursuit regression proposed by Friedman and Stuetzle (1981). Huber (1985) has given a comprehensive discussion. Chen (1991) has investigated a projection pursuit type of regression model. The primary focus of projection pursuit regression is more on the approximation of $g(x)$ by a sum of ridge functions $g_k(\cdot)$, namely

$$g(X) \approx \sum_{k=1}^{D} g_k(\beta_k^{\mathrm{T}} X),$$

than on looking for the EDR space.

A simple approach that is directly related to the estimation of EDR directions is the average derivative estimation (ADE) proposed by Härdle and Stoker (1989). For the single-index model $y = g_1(\beta_1^{\mathrm{T}} X) + \varepsilon$, the expectation of the gradient $\nabla g_1(X)$ is a scalar multiple of β_1. A nonparametric estimator of $\nabla g_1(X)$ leads to an estimator of β_1. There are several limitations of ADE.

(a) To estimate β_1, the condition $E\{g_1'(\beta_1^{\mathrm{T}} X)\} \neq 0$ is needed. This condition is violated when $g_1(\cdot)$ is an even function and X is symmetrically distributed.
(b) As far as we know, there is no successful extension to the case of more than one EDR direction.

The sliced inverse regression (SIR) method proposed by Li (1991) is perhaps up to now the most powerful method for searching for EDR directions and dimension reduction. However, the SIR method imposes some strong probabilistic structure on X. Specifically, the method requires that, for any constant vector $b^{\mathrm{T}} = (b_1, \ldots, b_p)$, there are constants c_0 and $c^{\mathrm{T}} = (c_1, \ldots, c_D)$ depending on b such that, for the directions B_0 in model (1.1),

$$E(b^{\mathrm{T}} X | B_0^{\mathrm{T}} X) = c_0 + c^{\mathrm{T}} B_0^{\mathrm{T}} X. \tag{1.2}$$

As pointed out by Cook and Weisberg in their discussion of Li (1991), the most important family of distributions satisfying condition (1.2) is that of elliptically symmetric distributions. Now, in time series analysis we typically set $X = (y_{t-1}, \ldots, y_{t-p})^{\mathrm{T}}$, where $\{y_t\}$ is a time series. Then it is easy to prove that elliptical symmetry of X for all p with (second-order) stationarity of $\{y_t\}$ implies that $\{y_t\}$ is time reversible, a feature which is the exception rather than the rule in time series analysis. (For a discussion of time reversibility, see, for example, Tong (1990).)

Another aspect of searching for the EDR space is the determination of the corresponding dimension. The method proposed by Li (1991) can be applied to determine the dimension of the EDR space in some cases but for reasons mentioned above it is typically not relevant for time series data.

In this paper, we shall propose a new method to estimate the EDR directions. We call it the (conditional) minimum average variance estimation (MAVE) method. Our approach is inspired by the SIR method, the ADE method and the idea of local linear smoothers (see, for example, Fan and Gijbels (1996)). It is easy to implement and needs no strong assumptions on the probabilistic structure of X. Specifically, our methods apply to model (1.1) including its generalization within the additive noise set-up. The joint density function of covariate X is needed if we search for the EDR space globally. However, if we have some prior information about the EDR directions and we look for them locally, then existence of density of X in the directions around EDR directions will suffice. These cases include those in which some of the covariates are categorical or functionally related. The observations need not be independent, e.g. time series data. On the basis of the properties of the MAVE method, we shall propose a method to estimate the dimension of the EDR space, which again does not require strong assumptions on the design X and has wide applicability.

Let Z be an $\mathbb{R}^q$-valued random variable. A general semiparametric model can be written as

$$y = G\{\phi(B_0^{\mathrm{T}} X), Z, \theta\} + \varepsilon, \tag{1.3}$$

where G is a *known* smooth function up to a parameter vector $\theta \in \mathbb{R}^l$, $\phi(\cdot)$: $\mathbb{R}^D \mapsto \mathbb{R}^{D'}$ is an *unknown* smooth function and $E(\varepsilon|X, Z) = 0$ almost surely. Special cases are the generalized partially linear single-index model of Carroll *et al.* (1997) and the single-index functional coefficient model in Xia and Li (1999). Searching for the EDR space $B_0^{\mathrm{T}} X$ in model (1.3) is of theoretical as well as practical interest. However, the existing methods are not always appropriate for this model. An extension of our method to handle this model will be discussed.

The rest of this paper is organized as follows. Section 2 describes the MAVE procedure and gives some results. Section 3 discusses some comparisons with existing methods and proposes a simple average outer product of gradients (OPG) estimation method and an inverse MAVE method. To check the feasibility of our approach, we have conducted many simulations, typical ones of which are reported in Section 4. In Section 5 we study the circulatory and respiratory data of Hong Kong and the hitters' salary data of the USA using the MAVE methodology. In practice, we standardize our observations. Appendix A establishes the efficiency of the algorithm proposed. Some of our theoretical proofs are very lengthy and not included here. However, they are available on request from the authors. Finally, the programs are available at

```
http://www.blackwellpublishers.co.uk/rss/
```

2. Estimation of effective dimension reduction space

2.1. The estimation of effective dimension reduction directions

Let us denote the working dimension by d with $1 \leqslant d \leqslant p$. Therefore, we need to estimate only a set of orthogonal vectors. There are many related methods for this and similar purposes. Most of the existing methods adopt two separate cost functions. The first is used to estimate the link function and the second the directions based on the estimated link function. See, for example, Hall (1989), Härdle and Stoker (1989) and Carroll *et al.* (1997). It is therefore not surprising that the performance of the direction estimator suffers from the bias problem in nonparametric estimation. Härdle *et al.* (1993) noticed this and overcame the problem for a single-index model by minimizing a cross-validation-type sum of squares of the residuals *simultaneously* with respect to the bandwidth and the directions. However, the cross-validation-type sum of squares of residuals affects the performance of estimation. See Xia *et al.* (1999). Moreover, the minimization is not trivial. Härdle *et al.* (1993) used the grid search method in their simulations, which is quite inefficient when the dimension is high.

Consider the simple regression model (1.1). The direction B_0 is the solution of

$$\min_B [E\{y - E(y|B^T X)\}^2]. \tag{2.1}$$

For any orthogonal matrix $B = (\beta_1, \ldots, \beta_d)$, the conditional variance given $B^T X$ is

$$\sigma_B^2(B^T X) = E[\{y - E(y|B^T X)\}^2 | B^T X]. \tag{2.2}$$

It follows that

$$E\{y - E(y|B^T X)\}^2 = E\{\sigma_B^2(B^T X)\}.$$

Therefore, minimizing expression (2.1) is equivalent to minimizing, with respect to B,

$$E\{\sigma_B^2(B^T X)\} \qquad \text{subject to } B^T B = I. \tag{2.3}$$

We shall call this MAVE. Suppose that $\{(X_i, y_i)\ i = 1, 2, \ldots, n\}$ is a sample from (X, y). Let

$$g_B(v_1, \ldots, v_d) = E(y|\beta_1^T X = v_1, \ldots, \beta_d^T X = v_d).$$

For any given X_0, a local linear expansion of $E(y_i|B^T X_i)$ at X_0 is

$$E(y_i|B^T X_i) \approx a + b^T B^T (X_i - X_0), \tag{2.4}$$

where $a = g_B(B^T X_0)$ and $b^T = (b_{(1)}, \ldots, b_{(d)})$ with

$$b_{(k)} = \left. \frac{\partial g_B(v_1, \ldots, v_d)}{\partial v_k} \right|_{v_1 = \beta_1^T X_0, \ldots, v_d = \beta_d^T X_0}, \qquad k = 1, \ldots, d.$$

Note that the right-hand side of approximation (2.4) is the tangent plane of g_B at $B^T X_0$. The residuals are then

$$y_i - g_B(B^T X_i) \approx y_i - \{a + b^T B^T (X_i - X_0)\}.$$

Following the idea of local linear smoothing estimation, we can estimate $\sigma_B^2(B^T X_0)$ by exploiting the approximation

$$\sum_{i=1}^n \{y_i - E(y_i|B^T X_i)\}^2 w_{i0} \approx \sum_{i=1}^n [y_i - \{a + b^T B^T (X_i - X_0)\}]^2 w_{i0}, \tag{2.5}$$

where $w_{i0} \geqslant 0$ are some weights with $\Sigma_{i=1}^{n} w_{i0} = 1$ and typically centred at $B^{\mathrm{T}} X_0$. The choice of the weights w_{i0} plays a key role in searching for the EDR directions. We shall discuss this issue in detail later. Usually,

$$w_{i0} = K_h\{B^{\mathrm{T}}(X_i - X_0)\} \Big/ \sum_{l=1}^{n} K_h\{B^{\mathrm{T}}(X_l - X_0)\},$$

where $K_h(\cdot) = h^d K(\cdot/h)$ and d is the dimension of $K(\cdot)$. For ease of exposition, $K(\cdot)$ denotes different kernel functions at different places. The estimators of a and b are just the minimum point of approximation (2.5). Therefore, the estimator of σ_B^2 at $B^{\mathrm{T}} X_0$ is just the minimum value of expression (2.5), namely

$$\hat{\sigma}_B^2(B^{\mathrm{T}} X_0) = \min_{a,b} \left(\sum_{i=1}^{n} [y_i - \{a + b^{\mathrm{T}} B^{\mathrm{T}}(X_i - X_0)\}]^2 w_{i0} \right). \tag{2.6}$$

Under some mild conditions, we have $\hat{\sigma}_B^2(B^{\mathrm{T}} X_0) - \sigma_B^2(B^{\mathrm{T}} X_0) = o_P(1)$. On the basis of expressions (2.1), (2.3) and (2.6), we can estimate the EDR directions by solving the minimization problem

$$\min_{B:B^{\mathrm{T}}B=I} \left\{ \sum_{j=1}^{n} \hat{\sigma}_B^2(B^{\mathrm{T}} X_j) \right\} = \min_{\substack{B:B^{\mathrm{T}}B=I \\ a_j, b_j, j=1,\ldots,n}} \left(\sum_{j=1}^{n} \sum_{i=1}^{n} [y_i - \{a_j + b_j^{\mathrm{T}} B^{\mathrm{T}}(X_i - X_j)\}]^2 w_{ij} \right), \tag{2.7}$$

where $b_j^{\mathrm{T}} = (b_{j1}, \ldots, b_{jd})$. The MAVE method or the minimization in problem (2.7) can be seen as a combination of nonparametric function estimation and direction estimation, which is executed simultaneously with respect to the directions and the nonparametric link function. As we shall see, we benefit from this simultaneous minimization.

If the weights depend on B, the implementation of the minimization in problem (2.7) is non-trivial. The weight w_{i0} in approximation (2.5) should be chosen such that *the value of* w_{i0} *is a function of the distance between* X_i *and* X_0. Next, we give two choices of w_{i0}.

2.1.1. *Multidimensional kernel weight*

To simplify problem (2.7), a natural choice is

$$w_{i0} = K_h(X_i - X_0) \Big/ \sum_{l=1}^{n} K_h(X_l - X_0).$$

This kind of weight can be used as an initial step of estimation. Given d, we obtain a set of directions $\hat{B}$ via the minimization in problem (2.7). Let $\mathcal{S}(\hat{B})$ denote the subspace spanned by the column vectors of $\hat{B}$. The distance between the space $\mathcal{S}(B_0)$, the space spanned by the column vectors of B_0, and the space $\mathcal{S}(\hat{B})$ can be measured by $\|(I - B_0 B_0^{\mathrm{T}}) \hat{B}\|$ if $d < D$ and $\|(I - \hat{B}\hat{B}^{\mathrm{T}}) B_0\|$ if $d \geqslant D$. Here and later, obvious augmentations by zero vectors are understood and the distance is denoted by $m(\hat{B}, B_0)$.

Theorem 1. Suppose that conditions 1–6 (in Appendix A) hold, model (1.1) is true and as $n \to \infty$ both $nh^p/\log(n) \to \infty$ and $h \to 0$. If $d < D$, then

$$m(\hat{B}, B_0) = O_P(h^2 + h^{-1}\delta_n^2),$$

where $\delta_n = \{\log(n)/nh^p\}^{1/2}$. If $d \geqslant D$, then

$$m(\hat{B}, B_0) = O_P(h^3 + h^{-1}\delta_n^2).$$

Provided that the dimension is chosen correctly, the rate of consistency for $\hat{B}$ is $O_P\{h_{\text{opt}}^3 \log(n)\}$ if we use the optimal bandwidth h_{opt} of the regression function estimation in the sense of minimizing the mean integrated squared errors. This is faster than the rate that is achieved by the other methods, which is $O_P(h_{\text{opt}}^2)$. Note that the consistency rate for the local linear estimator of the link function is also $O_P(h_{\text{opt}}^2)$. The faster rate is due to minimizing the average (conditional) variance with respect to both directions and the local linearization of the link function. Moreover, if we extend the idea to higher order local polynomial smoothers, root n consistency for the estimator of B_0 can be achieved; see the discussion in Section 6.

2.1.2. *Refined kernel weight*

If we know the dimension of the EDR space, which is usually less than p, we can then search for the EDR directions in a lower dimensional space, thereby reducing the effect of high dimension and improving the accuracy of the estimation. Suppose that we have an initial estimator of B_0, say $\hat{B}$. Let

$$\tilde{w}_{ij} = K_h\{\hat{B}^{\mathrm{T}}(X_i - X_j)\} \Big/ \sum_{l=1}^{n} K_h\{\hat{B}^{\mathrm{T}}(X_l - X_j)\}. \tag{2.8}$$

Re-estimate B_0 by the minimization in problem (2.7) with weights $\tilde{w}_{ij}$ replacing w_{ij}. By an abuse of notation, we denote the new estimator of B_0 by $\hat{B}$ also. Replace $\hat{B}$ in equation (2.8) by the latest $\hat{B}$ and estimate B_0. Repeat this procedure until $\hat{B}$ converges; we call the limit the refined MAVE (RMAVE) estimator. Results similar to those of theorem 1 can be obtained. We here use a lower dimensional kernel and the bandwidth now is smaller than that used in the multidimensional w_{ij}, leading to a faster rate of consistency.

One of the referees has drawn our attention to an unpublished paper by W. H. Wong and X. Shen, who have been working on a similar problem. They have proposed the nearest neighbour method and used the weights

$$w_{ij} = \frac{1}{n} \mathbf{1}_{\{X_i \text{ is one of the } \mathcal{N} \text{ nearest observations to } X_j\}},$$

where $\mathcal{N} < n$ is a suitable integer and $\mathbf{1}_A$ denotes the indicator function of the set A.

2.2. *Dimension of effective dimension reduction space*

Methods have been proposed for the determination of the number of the EDR directions. See, for example, Li (1992), Schott (1994) and Cook (1998). Their approaches tend to be based on similar probabilistic assumptions on the covariates X imposed by SIR. We now propose an alternative approach within our set-up. It is well known that a cross-validation approach penalizes the complexity of the model. See, for example, Stone (1974). We now extend the cross-validation method of Cheng and Tong (1992) and Yao and Tong (1994) to solve the above problem. A similar extension may be effected by using the approach of Auestad and Tjøstheim (1990), which is asymptotically equivalent to the cross-validation method.

Suppose that $\beta_1, \ldots, \beta_D$ are the EDR directions, i.e. $y = g(\beta_1^{\mathrm{T}} X, \ldots, \beta_D^{\mathrm{T}} X) + \varepsilon$ with $E(\varepsilon|X) = 0$ almost surely. If $D < p$, we can nominally extend the number of directions to p, say $\{\beta_1, \ldots, \beta_D, \ldots, \beta_p\}$, such that they are perpendicular to one another. Now, the problem becomes the selection of the covariates among $\{\beta_1^{\mathrm{T}} X, \ldots, \beta_p^{\mathrm{T}} X\}$. However, because $\beta_1, \ldots, \beta_p$ are unknown, we must replace β_ks by their estimators $\hat{\beta}_k$s. As we have proved that the rate of consistency of the $\hat{\beta}_k$s is faster than that of the nonparametric link function estimators, the replacement is justified. Let

$$\hat{a}_{d0,j} = \sum_{i=1, i\neq j}^{n} K_{h_d}^{(i,j)} y_i \Big/ \sum_{i=1,i\neq j}^{n} K_{h_d}^{(i,j)},$$

where $K_{h_d}^{(i,j)} = K_{h_d}\{\hat{\beta}_1^{\mathrm{T}}(X_i - X_j), \ldots, \hat{\beta}_d^{\mathrm{T}}(X_i - X_j)\}$. Here, we use the suffix d to highlight the fact that the bandwidth depends on the working dimension d. Let

$$\mathrm{CV}(d) = n^{-1} \sum_{j=1}^{n} (y_j - \hat{a}_{d0,j})^2, \qquad d = 1, \ldots, p.$$

Suppose that model (1.1) holds and $B_d^{\mathrm{T}} X$ has a density $f_d(v_1, \ldots, v_d)$ with compact support, where $B_d = (\beta_1, \ldots, \beta_d)$. For ease of exposition, we temporarily abbreviate $g(v_1, \ldots, v_D)$ to $g(v)$ and $f_d(v_1, \ldots, v_d)$ to $f_d(v)$. When $d \geqslant D$, we have

$$\mathrm{CV}(d) = \sigma^2 + h_d^4 \int \left[\frac{1}{2}\mathrm{tr}\{\nabla^2 g(v)\} + f_d^{-1}(v)\, \nabla^{\mathrm{T}} g(v)\, \nabla\, f_d(v)\right]^2 f_d(v)\, \mathrm{d}v_1 \ldots \mathrm{d}v_d$$
$$+ \frac{\alpha_d}{n h_d^d}\{1 + o_P(1)\} + O_P(n^{-1/2} + h_d^5),$$

where $\sigma^2 = \mathrm{var}(\varepsilon)$,

$$\alpha_d = E\{E(\varepsilon^2 | B^{\mathrm{T}} X)/f_d(B^{\mathrm{T}} X)\} \int K^2(v_1, \ldots, v_d)\, \mathrm{d}v_1 \ldots \mathrm{d}v_d$$

and $\nabla^2 g(v)$ is a $d \times d$ matrix whose (i, j)th element is $\partial^2 g(v)/\partial v_i \partial v_j$. If h_d is monotonic increasing such that $h_{d+1}^{d+1} = o(h_d^d)$, then $\mathrm{CV}(d)$ increases with d. Note that the optimal bandwidth $h_d \sim n^{-1/(d+4)}$ satisfies this requirement. When $d < D$, it is not difficult to see that $\mathrm{CV}(d) > \mathrm{CV}(D)$ because of the lack of fit. To include the case that y and X are independent, we define

$$\mathrm{CV}(0) = n^{-1} \sum_{i=1}^{n} (y_i - \bar{y})^2.$$

It is easy to see that $\mathrm{CV}(0) = \sigma^2 + O_P(n^{-1/2})$. Thus, we estimate the dimension of EDR space as

$$\hat{d} = \arg \min_{0 \leqslant d \leqslant p} \{\mathrm{CV}(d)\}.$$

Theorem 2. Suppose that the assumptions 1–6 (in Appendix A) hold. Under model (1.1) with X having a density with compact support, we have

$$\hat{d} \to D \qquad \text{in probability}$$

If X is not bounded, we may consider only a compact domain over which the density is positive. Then we have a small probability of overestimating the dimension (Cheng and Tong, 1992; Yao and Tong, 1994). Note that $a_{d0,j}$ is the Nadaraya–Watson estimator of a. We can use alternatively the local linear estimator for $a_{d0,j}$, which also leads to a consistent $\hat{d}$. However, the local linear estimator involves more complicated computation. Moreover, as far as cross-validatory determination of the dimension is concerned, our experience shows that using the local linear estimator tends to lead to a poorer performance in comparison with using the Nadaraya–Watson estimator. Empirical evidence suggests that using the latter tends to incur a smaller bandwidth and to lead to a heavier penalty for overfitting.

2.3. Bandwidth and algorithm

An important feature of the MAVE method is that we do not need to undersmooth the link function estimator for the EDR direction estimator to achieve a higher rate of consistency than the former. Therefore, the optimal bandwidth in the sense of mean integrated squared error can be used and, in practice, a variable bandwidth is normally recommended, e.g. (in obvious notation)

$$K_{\mathbf{h}}(u_1, \ldots, u_d) = \frac{K(u_1/h_{(1)}, \ldots, u_d/h_{(d)})}{\prod_{k=1}^{d} h_{(k)}},$$

$\mathbf{h} = (h_{(1)}, \ldots, h_{(d)})$ and d is the dimension of $K(\cdot)$. There are many ways to obtain such a bandwidth $\mathbf{h}$. See, for example, Fan and Gijbels (1996) and Yang and Tschernig (1999).

Our search procedure is as follows.

Step 1 (directions): for each d, $1 \leqslant d \leqslant p$, we search for the d directions as follows.

(a) Initial value: use the multidimensional kernel weight to obtain an initial estimate of possible EDR directions $\hat{\beta}_1, \hat{\beta}_2, \ldots, \hat{\beta}_d$ by minimizing problem (2.7).
(b) Refined estimation: let $\hat{B} = (\hat{\beta}_1, \ldots, \hat{\beta}_d)$ constitute the latest estimator of B. Therefore we obtain refined kernel weights by using equation (2.8). We refine the estimator via expression (2.7) using the refined kernel weights. Continue this procedure until convergence. The CV(d) values can be obtained by using the final estimators of the directions.

Step 2 (dimension and output results): compare the CV(d), $0 \leqslant d \leqslant p$. The d with the smallest CV(d) value is the estimated dimension. The corresponding estimator of B in step 1(b) gives the estimated EDR directions.

Let $\hat{B}_a$ and $\hat{B}_b$ be the estimators of B in two adjacent iterations in step 1(b). A suggested stopping rule for step 1(b) is when the distances $m(\hat{B}_a, \hat{B}_b)$ in several adjacent iterations are each less than a pre-set tolerance. Next, we describe one method to implement the minimization in problem (2.7). For any d, let $B = (\beta_1, \ldots, \beta_d)$ be the initial value (set $\beta_1 = \beta_2 = \ldots = \beta_d = 0$ in step 1(a)). $B_{l,k} = (\beta_1, \ldots, \beta_{k-1})$ and $B_{r,k} = (\beta_{k+1}, \ldots, \beta_d)$, $k = 1, 2, \ldots, d$. Minimize

$$S_{n,k} = \sum_{j=1}^{n} \sum_{i=1}^{n} \left\{ y_i - a_j - (X_i - X_j)^{\mathrm{T}} (B_{l,k}, \beta, B_{r,k}) \begin{pmatrix} c_j \\ d_j \\ e_j \end{pmatrix} \right\}^2 w_{ij}$$

$$\text{subject to } B_{l,k}^{\mathrm{T}} \beta = 0 \text{ and } B_{r,k}^{\mathrm{T}} \beta = 0,$$

where c_j is a $(k-1) \times 1$ vector, d_j a scalar and e_j a $(d-k) \times 1$ vector. This is a typical constrained quadratic programming problem. See, for example, Rao (1973), page 232. Let

$$C_j = \sum_{i=1}^{n} w_{ij}(X_i - X_j),$$

$$D_j = \sum_{i=1}^{n} w_{ij}(X_i - X_j)(X_i - X_j)^{\mathrm{T}},$$

$$E_j = \sum_{i=1}^{n} w_{ij} y_i,$$

$$F_j = \sum_{i=1}^{n} w_{ij}(X_i - X_j) y_i.$$

With β given, the (a_j, c_j, d_j, e_j) which minimizes $S_{n,k}$ is given by

$$\begin{pmatrix} a_j \\ c_j \\ d_j \\ e_j \end{pmatrix} = \begin{pmatrix} 1 & C_j^{\mathrm{T}}(B_{l,k}, \beta, B_{r,k}) \\ (B_{l,k}, \beta, B_{r,k})^{\mathrm{T}} C_j & (B_{l,k}, \beta, B_{r,k})^{\mathrm{T}} D_j (B_{l,k}, \beta, B_{r,k}) \end{pmatrix}^{-1} \times \begin{pmatrix} E_j \\ (B_{l,k}, \beta, B_{r,k})^{\mathrm{T}} F_j \end{pmatrix},$$

$j = 1, \ldots, n$. If a_j, c_j, d_j and e_j are given, then the β which minimizes $S_{n,k}$ is given by

$$\begin{pmatrix} \beta \\ \lambda \end{pmatrix} = \begin{pmatrix} \sum_{j=1}^{n} d_j^2 D_j & \tilde{B}_k \\ \tilde{B}_k^{\mathrm{T}} & 0 \end{pmatrix}^{+} \begin{pmatrix} \sum_{j=1}^{n} d_j \left\{ F_j - a_j C_j - D_j \tilde{B}_k \begin{pmatrix} c_j \\ e_j \end{pmatrix} \right\} \\ 0 \end{pmatrix},$$

where $\tilde{B}_k = (B_{l,k}, B_{r,k})$ and A^+ denotes the Moore–Penrose inverse of a matrix A. Here λ is the usual Lagrangian multiplier for the constraint minimization. Finally, we normalize β.

3. Links with other methods and generalization

3.1. *Outer product of gradients estimation*

Suppose that $y = \tilde{g}(X) + \varepsilon$ with $E(\varepsilon|X) = 0$ almost surely. Consider the minimization in problem (2.6). Under assumptions 1–6 (in Appendix A) and

$$w_{i0} = K_h(X_i - X_0) \Big/ \sum_{l=1}^{n} K_h(X_l - X_0),$$

we have

$$\min_{a,b} \left(\sum_{i=1}^{n} [y_i - \{a + b^{\mathrm{T}} B^{\mathrm{T}} (X_i - X_0)\}]^2 w_{i0} \right) = \hat{\sigma}^2(X_0) + h^2 \nabla^{\mathrm{T}} \tilde{g}(X_0) \times (I_{p \times p} - BB^{\mathrm{T}}) \nabla \tilde{g}(X_0) + o_P(h^2),$$

where $\hat{\sigma}^2(X_0) = \Sigma_{i=1}^{n} \varepsilon_i^2 w_{i0}$ does not depend on B. Thus, the minimization problem (2.7) depends mainly on

$$E\{\nabla^{\mathrm{T}} \tilde{g}(X)(I_{p \times p} - BB^{\mathrm{T}}) \nabla \tilde{g}(X)\} = \mathrm{tr}[(I_{p \times p} - BB^{\mathrm{T}}) E\{\nabla \tilde{g}(X) \nabla^{\mathrm{T}} \tilde{g}(X)\}] = \mathrm{tr}[E\{\nabla \tilde{g}(X) \nabla^{\mathrm{T}} \tilde{g}(X)\}] - \mathrm{tr}[B^{\mathrm{T}} E\{\nabla \tilde{g}(X) \nabla^{\mathrm{T}} \tilde{g}(X)\} B].$$

Therefore, the B which minimizes this equation is the first d eigenvectors corresponding to the d largest eigenvalues of $E\{\nabla \tilde{g}(X) \nabla^{\mathrm{T}} \tilde{g}(X)\}$, which is the average OPG of $\tilde{g}(\cdot)$.

Lemma 1. Suppose that $\tilde{g}(\cdot)$ is differentiable. If model (1.1) is true, then B_0 is in the space spanned by the first D eigenvectors of $E[\nabla \tilde{g}(X) \nabla^{\mathrm{T}} \tilde{g}(X)]$ corresponding to the largest D eigenvalues.

This relationship was also noticed in Li (1991). By lemma 1, it is easy to see that the EDR space is unique up to orthogonal transformations if the density function of X has a compact support. We may use lemma 1 and propose the following estimation procedure. First, estimate the gradients by local polynomial smoothing. Specifically, we consider the local linear fitting in the form of the minimization problem

$$\min_{a_j, b_j}\left(\sum_{i=1}^{n}[\{y_i - a_j - b_j^{\mathrm{T}}(X_i - X_j)\}^2 w_{ij}]\right). \tag{3.1}$$

We then estimate $E\{\nabla\tilde{g}(X)\,\nabla^{\mathrm{T}}\,\tilde{g}(X)\}$ by

$$\hat{\Sigma} = \frac{1}{n}\sum_{j=1}^{n}\hat{b}_j\hat{b}_j^{\mathrm{T}},$$

where $\hat{b}_j$ is the minimizer from expression (3.1). Finally, we estimate the EDR directions by the first d eigenvectors of $\hat{\Sigma}$. We call this method the method of OPG estimation.

Theorem 3. Let $\hat{\beta}_1, \ldots, \hat{\beta}_d$ be the first d eigenvectors of $\hat{\Sigma}$ corresponding to the largest d eigenvalues, and $\hat{B} = (\hat{\beta}_1, \ldots, \hat{\beta}_d)$. Suppose that conditions 1–6 (in Appendix A) hold and model (1.1) is true. If $nh^p/\log(n) \to \infty$ and $h \to 0$, then

$$m(\hat{B}, B_0) = O_P(h^2 + \delta_n^2 h^{-1}).$$

Unlike the ADE method, the OPG method still works even if $E\{\nabla\tilde{g}(X)\} = 0$. Moreover, the OPG method can handle multiple EDR directions simultaneously whereas the ADE method can only handle the first EDR direction (i.e. the single-index model). We can further refine the OPG estimator using refined weights as in the RMAVE method. Compared with the MAVE method, the OPG method still suffers from the effect of the bias term in nonparametric function estimation. Therefore, the rate of consistency is slower than that of the MAVE method when the dimension is chosen correctly. However, the OPG method is easy to implement and can be used as an initial value of other estimation methods. Li (1992) proposed the principal Hessian directions (PHD) method by estimating the Hessian matrix of $g(\cdot)$. Similarly to the OPG method, the directions are the eigenvectors of the Hessian matrix. For a normally distributed design X, the Hessian matrix can be properly estimated simply by Stein's lemma. However, the PHD method assumes some probabilistic structure on design X which is frequently violated in time series analysis. More fundamentally, the PHD method involves estimators of second derivatives whereas the OPG method involves only the first derivatives, which are considerably simpler and easier to estimate.

3.2. *Inverse regression minimum average (conditional) variance estimation*

We start with

$$\bar{w}_{ij} = K_h(y_i - y_j)\Big/\sum_{l=1}^{n}K_h(y_l - y_j). \tag{3.2}$$

Now, with this weight function, the minimization in equation (2.6) becomes the minimization of

$$\sum_{i=1}^{n}[y_i - \{a + b\beta^{\mathrm{T}}(X_i - X_0)\}]^2\bar{w}_{i0},$$

and the MAVE method involves the minimization of

$$\sum_{i=1}^{n}\sum_{j=1}^{n}[y_i - \{a_j + b_j\beta^{\mathrm{T}}(X_i - X_j)\}]^2\bar{w}_{ij}.$$

A 'dual' of this is the minimization of

$$\sum_{j=1}^{n}\sum_{i=1}^{n}\{\beta^{\mathrm{T}}X_i - c_j - d_j(y_i - y_j)\}^2\bar{w}_{ij}. \tag{3.3}$$

This may be considered an alternative derivation of the SIR method. The extension of expression (3.3) to more than one direction can be stated as follows. Suppose that the first k directions have been calculated and are denoted by $\hat{\beta}_1, \ldots, \hat{\beta}_k$ respectively. To obtain the $(k+1)$th direction, we need to perform

$$\min_{\alpha_{1j},\ldots,\alpha_{kj},c_j,d_j,\beta}\left[\sum_{j=1}^{n}\sum_{i=1}^{n}\{\beta^{\mathrm{T}}X_i + \alpha_{1j}\hat{\beta}_1^{\mathrm{T}}X_i + \ldots + \alpha_{kj}\hat{\beta}_k^{\mathrm{T}}X_i - c_j - d_j(y_i - y_j)\}^2\bar{w}_{ij}\right]$$

$$\text{subject to } \beta^{\mathrm{T}}(\hat{\beta}_1, \ldots, \hat{\beta}_k) = 0 \text{ and } \|\beta\| = 1. \tag{3.4}$$

We call the estimation method based on minimizing expression (3.3) with $\bar{w}_{ij}$ as defined in equation (3.2) the inverse MAVE (IMAVE) method. The IMAVE method is in line with the most predictable variate (Hotelling, 1935). The minimizations in expressions (3.3) and (3.4) can be seen as looking for linear combinations of X that are most predictable from y. Under a similar assumption on X as in SIR, we have the following result.

Theorem 4. Suppose that equation (1.2) and assumptions 1, 2(b), 3(b), 4, 5(b) and 6 (in Appendix A) hold. Let $\hat{b} = (\hat{\beta}_1, \ldots, \hat{\beta}_d)$. If $h \to 0$ and $nh/\log(n) \to \infty$, then

$$m(\hat{B}, B_0) = O_P\{h^2 + \log(n)/nh + n^{-1/2}\}.$$

This result is similar to that of Zhu and Fang (1996). As noted previously, the assumption on the design X can be a handicap as far as applications of the IMAVE method are concerned. Interestingly, simulations show that the SIR method and the IMAVE method can sometimes produce useful results in the case of independent data even when this assumption is mildly violated. However, for time series data, we find that this is often not so.

3.3. *Semiparametric multi-index models*

Consider the general model (1.3). Suppose that $G(v, Z, \theta)$ is differentiable. For ease of exposition we set $D' = 1$. Let $G'(v, Z, \theta) = \partial G(v, Z, \theta)/\partial v$. For $B^{\mathrm{T}}X_i$ close to $B^{\mathrm{T}}X_0$ we have

$$G\{\phi(B^{\mathrm{T}}X_i), Z_i, \theta\} \approx G\{\phi(B^{\mathrm{T}}X_0), Z_i, \theta\} + G'\{\phi(B^{\mathrm{T}}X_0), Z_i, \theta\}\, \nabla^{\mathrm{T}}\phi(B^{\mathrm{T}}X_0)B^{\mathrm{T}}(X_i - X_0).$$

To estimate B, we minimize

$$\sum_{j=1}^{n}\sum_{i=1}^{n}\{y_i - G(a_j, Z_i, \theta) - G'(a_j, Z_i, \theta)b_j^{\mathrm{T}}B^{\mathrm{T}}(X_i - X_j)\}^2 w_{ij}$$

with respect to $a_j, b_j, j = 1, \ldots, n$, θ and B. Similarly, we may first use the multidimensional kernel weight to obtain an initial estimate and then repeatedly use the refined kernel weight.

Model (1.3) includes many models with a fixed dimension of EDR space. Examples are the single-index model of Ichimura and Lee (1991), the generalized partially linear single-index model of Carroll *et al.* (1997) and Xia *et al.* (1999) and the single-index coefficient regression model of Xia and Li (1999). Here the estimation of the unknown function is also important. An obvious question is whether we can estimate both the function and the directions (multi-indices) with their optimal rates of consistency simultaneously. This problem has attracted much attention. See, for example, Härdle *et al.* (1993), Severini and Wong (1992) and Carroll *et al.* (1997).

For most methods, the estimator of the direction suffers from the effect of the bias in the estimator of the unknown link function. Therefore, undersmoothing the estimator of the link function is necessary for the estimator of the direction to achieve its optimal rate of consistency. We are not aware of any recommended method to select the undersmooth bandwidth. By minimizing a cross-validation-type sum of squares of residuals simultaneously with respect to both the bandwidth and the direction, Härdle *et al.* (1993) have given a positive answer to the question raised in the previous paragraph. However, we have discussed the problems with this approach in Section 2. In contrast, the MAVE-type methods can handle all the models mentioned above effectively. Specifically, when $D'=1$, the root n rate of consistency for the direction estimator can be obtained and at the same time the optimal rate of consistency for the nonparametric function estimator can be achieved.

3.4. Discrete or functionally related covariates

Generally, dimension reduction methods cannot be applied to models with discrete or functionally related covariates because they are not estimable, in the sense that there can be more than one dimension reduction space up to orthogonal transformations.

We believe that, provided that the link function can be approximated locally by 'tangent' planes, the MAVE method can still be practically useful for discrete or functionally related covariates. The limiting accuracy will, of course, depend on the accuracy of the tangent plane approximation. We must keep in mind two points:

(a) the bandwidth cannot be selected to be smaller than a critical value because we must use adjacent points to estimate the 'tangent' plane and
(b) if none of the X design points has repeated measurements then bandwidth selection methods based on cross-validation may be considered. If the latter methods are ruled out, a feasible alternative may be one based on the idea of the nearest neighbours as follows. For any point x_k, we choose a nearest neighbour of x_k which includes observations $(\tilde{x}_1, \tilde{y}_1), \ldots, (\tilde{x}_{\tilde{p}}, \tilde{y}_{\tilde{p}})$, such that the plane $y = a + b^{\mathrm{T}}X$ is estimable, i.e. there is a unique solution of (a, b) to $\min_{a,b}\{\Sigma_{i=1}^{\tilde{p}}(\tilde{y}_i - a - b^{\mathrm{T}}\tilde{X}_i)^2\}$; cf. the nearest neighbour method due to Wong and Shen (unpublished) mentioned in Section 2.

If X includes continuous covariates as well as categorical or functionally related covariates, then the RMAVE method still applies with appropriate initial values. If we carry out a global search for the EDR directions, the procedure may be trapped by directions with positive probability due to the categorical data. If we have some prior information about the EDR directions such that we only need to search for the directions locally, then the density requirement can be relaxed, namely the density function of $B^{\mathrm{T}}X$ exists for all $B \in \mathcal{B} = \{B\colon B^{\mathrm{T}}B = I_D$ and $\|B - B_0\| < c\}$ for some $c > 0$. Suppose further that $E(XX^{\mathrm{T}}|B^{\mathrm{T}}X = v)$ and $E(X|B^{\mathrm{T}}X = v)$ exist and have continuous second-order derivatives. Then the RMAVE method in our paper applies with appropriate initial values in $\mathcal{B}$ and the search for the directions conducted within the same region.

4. Simulations

In this section, we carry out simulations to check the performance of the proposed OPG method and the MAVE-type methods. We shall use the square-distance function m^2, where m was defined in Section 2, to measure the error of estimation when we compare our method with others.

4.1. Example 1

We first adopt the examples used in Li (1991). Let $p=10$ and $\varepsilon, x_1, x_2, \ldots, x_{10}$ be independent random variables each with a standard normal distribution. Consider two regression models:

$$y = x_1(x_1 + x_2 + 1) + 0.5\varepsilon, \tag{4.1}$$

$$y = x_1/\{0.5 + (x_2 + 1.5)^2\} + 0.5\varepsilon. \tag{4.2}$$

The sample size is set at $n=200$ or $n=400$ and 100 replications are drawn in each case. Let $\beta_1 = (1, 0, \ldots, 0)^{\mathrm{T}}$, $\beta_2 = (0, 1, \ldots, 0)^{\mathrm{T}}$ and $B_0 = (\beta_1, \beta_2)$. Fig. 1 shows the means of the estimation errors $m^2(\hat{\beta}_1, B_0)$ and $m^2(\hat{\beta}_2, B_0)$; they are labelled '1' and '2' for β_1 and β_2 respectively. In our simulations, the IMAVE method outperforms the SIR method but is outperformed by the MAVE method. The RMAVE method performs best of all the methods. Zhu and Fang (1996) proposed a kernel smooth version of the SIR method. However, their method does not show a significant improvement over that of the original SIR method.

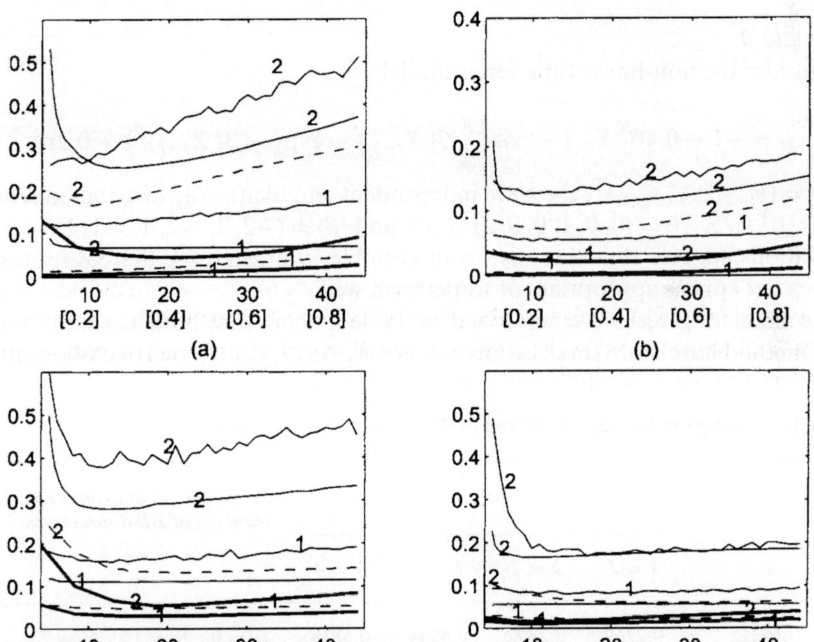

Fig. 1. Means of $m^2(\hat{\beta}_1, B_0)$ (labelled 1) and $m^2(\hat{\beta}_2, B_0)$ (labelled 2) (broken curves are based on the MAVE method; full curves are based on the IMAVE method; wavy curves are based on the SIR method; bold curves are based on the RMAVE method; the horizontal axes give the numbers of slices or the bandwidth (in square brackets) for the SIR method or IMAVE method respectively): (a) model (4.1), sample size 200, bandwidths 1–3 (MAVE method) and 0.1–1 (RMAVE method); (b) model (4.1), sample size 400, bandwidths 1–2 (MAVE method) and 0.1–1 (RMAVE method); (c) model (4.2), sample size 200, bandwidths 1–3 (MAVE method) and 0.1–1 (RMAVE method); (d) model (4.2), sample size 400, bandwidths 1–2 (MAVE method) and 0.1–1 (RMAVE method)

4.2. Example 2

Consider the model

$$y = X^{\mathrm{T}}\beta_1(X^{\mathrm{T}}\beta_2)^2 + (X^{\mathrm{T}}\beta_3)(X^{\mathrm{T}}\beta_4) + 0.5\varepsilon, \tag{4.3}$$

where $X \sim N(0, I_{10})$ and $\varepsilon \sim N(0, 1)$ and they are independent. In model (4.3), the coefficients $\beta_1 = (1, 2, 3, 4, 0, 0, 0, 0, 0, 0)^{\mathrm{T}}/\sqrt{30}$, $\beta_2 = (-2, 1, -4, 3, 1, 2, 0, 0, 0, 0)^{\mathrm{T}}/\sqrt{35}$, $\beta_3 = (0, 0, 0, 0, 2, -1, 2, 1, 2, 1)^{\mathrm{T}}/\sqrt{15}$, $\beta_4 = (0, 0, 0, 0, 0, 0, -1, -1, 1, 1)^{\mathrm{T}}/2$ and there are four EDR directions. Let $B_0 = (\beta_1, \beta_2, \beta_3, \beta_4)$. In our simulations, the SIR method and the IMAVE method perform quite poorly for this model. Next, we use this model to check the OPG method and the MAVE method.

With sample size $n = 100$, 200, 400, 200 independent samples are drawn in each case. The average distance from the estimated EDR directions to $\mathcal{S}(B_0)$ is calculated for the PHD method (Li, 1992), the OPG method, the MAVE method and the RMAVE method. The results are listed in Table 1. The results suggest that the MAVE method performs better than the OPG method, which performs better than the PHD method, whereas the RMAVE method shows a significant improvement over the MAVE method. Our method for the estimation of the number of EDR directions also gives satisfactory results.

4.3. Example 3

We next consider the non-linear time series model

$$y_t = -1 + 0.4\beta_1^{\mathrm{T}}X_{t-1} - \cos\left(\frac{\pi}{2}\beta_2^{\mathrm{T}}X_{t-1}\right) + \exp\{-(\beta_3^{\mathrm{T}}X_{t-1})^2\} + 0.2\varepsilon_t, \tag{4.4}$$

where $X_{t-1} = (y_{t-1}, \ldots, y_{t-6})^{\mathrm{T}}$, the ε are independent and identically distributed $N(0, 1)$, $\beta_1 = (1, 0, 0, 2, 0, 0)^{\mathrm{T}}/\sqrt{5}$, $\beta_2 = (0, 0, 1, 0, 0, 2)^{\mathrm{T}}/\sqrt{5}$ and $\beta_3 = (-2, 2, -2, 1, -1, 1)^{\mathrm{T}}/\sqrt{15}$. Fairly large simulations suggest that there is no discernible symmetry for the covariates; the SIR method does not appear appropriate or to perform well.

Now, the simulation results summarized in Table 2 show that both the OPG method and the MAVE method have quite small estimation errors. As expected, the RMAVE method works

Table 1. Average $m^2(\hat\beta_k, B_0)$ for model (4.3) by using different methods

n	*Method*	$m^2(\hat\beta_k, B_0)$				*Frequencies of estimated numbers of EDR directions*
		k = 1	*k = 2*	*k = 3*	*k = 4*	
100	PHD	0.2769	0.2992	0.4544	0.5818	$f_1 = 0,\ f_2 = 10,\ f_3 = 23,$
	OPG	0.1524	0.2438	0.3444	0.4886	$f_4 = 78,\ f_5 = 44,\ f_6 = 32,$
	MAVE	0.1364	0.1870	0.2165	0.3395	$f_7 = 11,\ f_8 = 1,\ f_9 = 1,$
	RMAVE	0.1137	0.1397	0.1848	0.3356	$f_{10} = 0$
200	PHD	0.1684	0.1892	0.3917	0.6006	$f_1 = 0,\ f_2 = 0,\ f_3 = 5,$
	OPG	0.0713	0.1013	0.1349	0.2604	$f_4 = 121,\ f_5 = 50,\ f_6 = 16,$
	MAVE	0.0710	0.0810	0.0752	0.1093	$f_7 = 8,\ f_8 = 0,\ f_9 = 0,$
	RMAVE	0.0469	0.0464	0.0437	0.0609	$f_{10} = 0$
400	PHD	0.0961	0.1151	0.3559	0.6020	$f_1 = 0,\ f_2 = 0,\ f_3 = 0,$
	OPG	0.0286	0.0388	0.0448	0.0565	$f_4 = 188,\ f_5 = 16,\ f_6 = 6,$
	MAVE	0.0300	0.0344	0.0292	0.0303	$f_7 = 0,\ f_8 = 0,\ f_9 = 0,$
	RMAVE	0.0170	0.0119	0.0116	0.0115	$f_{10} = 0$

Table 2. Average $m^2(\hat{\beta}_k, B_0)$ for model (4.4) by using different methods

n	*Method*	$m^2(\hat{\beta}_k, B_0)$ $k=1$	$k=2$	$k=3$	*Frequency of estimated number of EDR directions*
100	PHD	0.1582	0.2742	0.3817	$f_1=3,\ f_2=73,$
	OPG	0.0427	0.1202	0.2803	$f_3=94,\ f_4=25,$
	MAVE	0.0295	0.1201	0.2924	$f_5=4,\ f_6=1$
	RMAVE	0.0096	0.1712	0.2003	
200	PHD	0.1565	0.2656	0.3690	$f_1=0,\ f_2=34,$
	OPG	0.0117	0.0613	0.1170	$f_3=160,\ f_4=5,$
	MAVE	0.0059	0.0399	0.1209	$f_5=1,\ f_6=0$
	RMAVE	0.0030	0.0224	0.0632	
300	PHD	0.1619	0.2681	0.3710	$f_1=0,\ f_2=11,$
	OPG	0.0076	0.0364	0.0809	$f_3=185,\ f_4=4,$
	MAVE	0.0040	0.0274	0.0666	$f_5=0,\ f_6=0$
	RMAVE	0.0017	0.0106	0.0262	

better than the MAVE method, which outperforms the OPG method. The PHD method does not fare very well. The number of the EDR directions is also estimated correctly most of the time.

5. Examples

5.1. Circulatory and respiratory problems in Hong Kong

Consider the effect of the levels of pollutants and weather on the total number y_t of daily hospital admissions of patients suffering from circulatory and respiratory problems. The pollutant and weather data are the daily average levels of sulphur dioxide (x_{1t} (μg m^{-3})), nitrogen dioxide (x_{2t} (μg m^{-3})), respirable suspended particulates (x_{3t} (μg m^{-3})), ozone (x_{4t} (μg m^{-3})), temperature (x_{5t} (°C)) and relative humidity (x_{6t} (%)). The data were collected daily in Hong Kong from January 1st, 1994, to December 31st, 1995, and are shown in Fig. 2. The basic question is this: are the prevailing levels of the pollutants a cause for concern?

A naïve approach may be to start with a simple linear regression model such as

$$\begin{array}{l} y_t = 255.45 - 0.55x_{1t} + 0.58x_{2t} + 0.18x_{3t} - 0.33x_{4t} - 0.12x_{5t} - 0.16x_{6t}. \\ \qquad (20.64) \quad (0.18) \qquad (0.17) \qquad (0.13) \qquad (0.11) \qquad (0.46) \qquad (0.23) \end{array} \tag{5.1}$$

Note that the coefficients of x_{3t}, x_{5t} and x_{6t} are not significantly different from 0 (at the 5% level of significance) by reference to their standard errors shown inside the parentheses and the negative and significant coefficients of x_{1t} and x_{4t} are difficult to interpret. Refinements of this model are, of course, possible within the linear framework but are unlikely to throw much light with respect to the opening question because, as we shall see, the situation is quite complex. Previous analyses, such as Fan and Zhang (1999) and Cai *et al.* (2000), have not included the weather effect. However, it turns out that the weather has an important role to play.

The daily admissions shown in Fig. 2(a) suggest non-stationarity in the form of almost a level shift taking place in early 1995 although none of the covariates seems to show a similar level shift. Now, a trend was also observed by Smith *et al.* (1999) in their study of the effect of particulates on human health. They conjectured that the trend was due to the epidemic effect. In our case, we understand from our data provider that additional hospital beds were released to accommodate circulatory and respiratory patients in the course of his joint project. As a

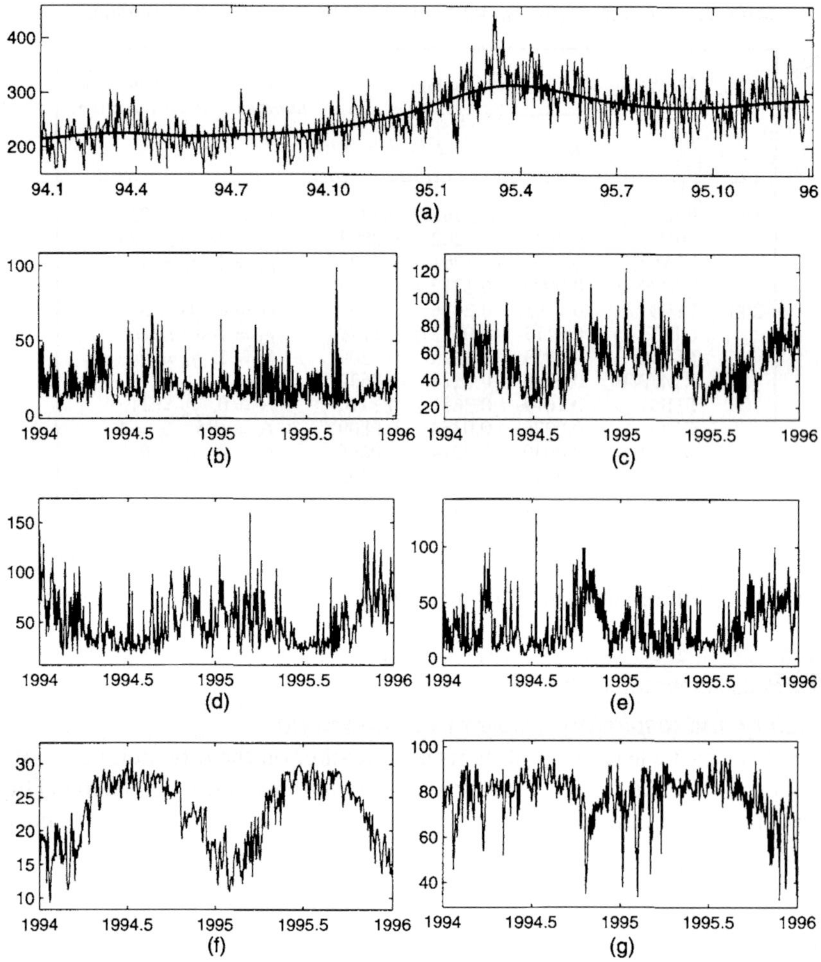

Fig. 2. (a) Total number of daily hospital admissions of circulatory and respiratory patients (——, time trend) and average levels of (b) sulphur dioxide, (c) nitrogen dioxide, (d) respirable suspended particulates, (e) ozone, (f) temperature and (g) humidity

result, we estimate the time dependence by a simple kernel method and the result is shown in Fig. 2(a). Another factor is the day of the week effect, presumably due to the hospital booking system. The day of the week effect can be estimated by a simple regression method using dummy variables. To assess the effect of pollutants better, we remove these two factors first. By an abuse of notation, we shall continue to use y_t to denote the 'filtered' data, now shown in Fig. 3.

As the pollutant-based and weather-based covariates may affect the circulatory and respiratory system with a time delay, we consider the six covariates in the last 7 days (1 week). Altogether, we have 42 covariates:

$$X_t = (x_{1,t-1}, x_{1,t-2}, \dots, x_{1,t-7}, x_{2,t-1}, x_{2,t-2}, \dots, x_{2,t-7}, \dots, x_{6,t-1}, x_{6,t-2}, \dots, x_{6,t-7})^{\mathrm{T}}.$$

Now, using the RMAVE method and with a cross-validation bandwidth, we have the results in Table 3. The cross-validation choice of the dimension is 3. The corresponding direction estimates are listed in Table 4.

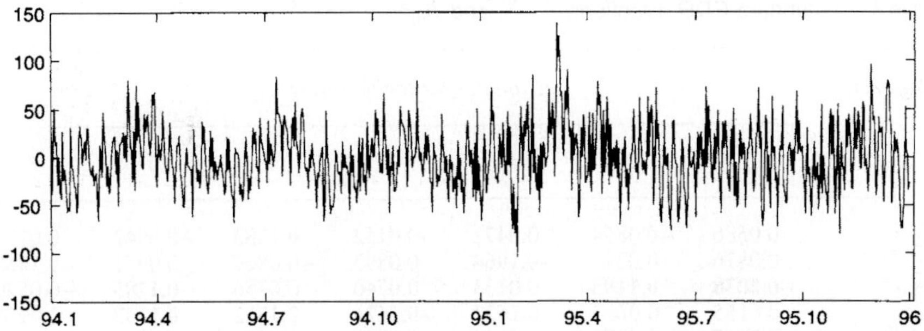

Fig. 3. 'Filtered' number of daily hospital admissions of circulatory and respiratory patients by removing the time trend and the day of the week effect

Figs 4(a)–4(c) show y_t plotted against the respective EDR directions. These plots and Table 4 suggest the following features.

(a) Rapid temperature changes play an important role. (Note the dominant coefficients for temperature in the two recent past days in $\hat{\beta}_{\rm I}^{\rm T} X$.)

(b) Of the pollutants, the most influential seems to be the particulates (note the large coefficient for particulates at lag 5 in $\hat{\beta}_{\rm II}^{\rm T} X$) and the least influential seems to be sulphur dioxide.

(c) The weather covariates are influential. (Note the many large coefficients for the weather covariates in all the three $\hat{\beta}$s.)

Comparing the levels of the individual pollutants in Hong Kong against the national ambient quality standard of the USA lends further support to feature (b).

Bearing these features in mind, we may explore further by focusing on the suspended particulates (x_3), the ozone level (x_4), the temperature (x_5) and its variation, and the relative humidity (x_6). First, we define the variation of temperature as

$$v_t = \text{std}(x_{5,t-k}, k = 1, 2, 3, 4, 5).$$

Further simplification is obtained by selecting only one lag for each covariate. For this, we use the method of Yao and Tong (1994). The lagged covariates selected are $x_{3,t-2}$, $x_{4,t-6}$, $x_{5,t-4}$ and $x_{6,t-2}$. Let $Z_t = (x_{3,t-2}, x_{4,t-6}, v_t, x_{5,t-4}, x_{6,t-2})^{\rm T}$. We then consider a model of the form

$$y_t = g(Z_t) + \varepsilon_t.$$

Table 3. Results of the CV method

Dimension	*Bandwidth*	*CV(d) value*
1	0.10	0.33
2	0.13	0.28
3	0.16	0.27
4	0.20	0.29
5	0.21	0.29
6	0.24	0.31
7	0.24	0.34
8	0.29	0.31
9	0.31	0.34
10	0.31	0.37

Table 4. Estimated EDR directions $\hat{\beta}_{\mathrm{I}}$, $\hat{\beta}_{\mathrm{II}}$ and $\hat{\beta}_{\mathrm{III}}$†

Parameter	*Estimates for the following lags:*						
	1	*2*	*3*	*4*	*5*	*6*	*7*
x_1	0.0586	−0.0854	0.0472	−0.0152	**0.1083**	−0.0942	0.0734
x_2	0.0876	0.0313	**−0.1964**	0.0893	−0.0867	0.0951	**−0.1068**
x_3	**−0.2038**	**0.1103**	0.0153	0.0740	−0.0756	**0.1283**	−0.0520
x_4	0.0155	0.0692	**0.1622**	**−0.2624**	**0.1312**	**0.1342**	0.0976
x_5	**0.5065**	**−0.4079**	0.0743	0.0859	**−0.3024**	−0.1734	−0.0302
x_6	−0.0294	−0.0610	0.0129	−0.0392	−0.0075	**0.2850**	0.0513
x_1	**−0.1525**	0.0962	**−0.1112**	**0.1170**	−0.0388	−0.0605	−0.0326
x_2	−0.0029	**0.1614**	−0.0955	**−0.1160**	**−0.2185**	0.0826	**0.1696**
x_3	−0.0096	**−0.1874**	**0.2422**	−0.0047	**0.3272**	**−0.2646**	−0.0041
x_4	−0.0013	**−0.1162**	0.0673	**0.2113**	**−0.2193**	**0.1235**	**−0.1282**
x_5	**0.1410**	**0.1193**	**−0.1425**	**0.1819**	**−0.2793**	−0.0880	−0.0325
x_6	−0.0345	**−0.1479**	−0.0400	**0.4033**	0.0474	0.0899	**0.1336**
x_1	0.0701	0.0065	−0.0535	**−0.1570**	−0.0553	−0.0091	−0.0363
x_2	−0.0529	**0.1360**	0.0723	**0.1045**	−0.0045	−0.0200	0.0221
x_3	−0.0121	**−0.1189**	0.0715	−0.0814	0.0112	0.0155	**0.1214**
x_4	**0.2215**	0.0103	**−0.3304**	**0.1028**	0.0160	**−0.1805**	**0.1341**
x_5	**0.2909**	**−0.2372**	0.0621	−0.0211	0.0950	−0.0954	**0.2507**
x_6	**0.2797**	**−0.1094**	**−0.3038**	0.0452	**0.1754**	**−0.3937**	**0.2597**

†Entries in bold have relatively large absolute values.

The above proposed procedure yields the results in Table 5.

On the basis of Table 5 the dimension of EDR space is chosen to be 3 with the following estimated basis vectors for the space:

$$\begin{aligned}
\hat{\beta}_1 &= (-0.1317 \quad -0.0772 \quad 0.5256 \quad -0.8366 \quad -0.0235)^{\mathrm{T}},\\
\hat{\beta}_2 &= (0.4809 \quad 0.3154 \quad -0.6414 \quad -0.5078 \quad 0.0018)^{\mathrm{T}},\\
\hat{\beta}_3 &= (0.0101 \quad 0.3815 \quad 0.1345 \quad 0.0734 \quad -0.9115)^{\mathrm{T}}.
\end{aligned}$$

Figs 4(d)–4(f) show y_t plotted against the three directions. The 'price' of using the reduced set with five covariates instead of the original set with 42 covariates is, loosely speaking, an increase in the percentage of unexplained variation from about 27% to about 34%. (As we use standardized observations, we may interpret the CV(d) value as a percentage of unexplained variation.) In return, we can gain further insight.

(a) The first EDR direction is $-0.1317x_{3,t-2} - 0.0772x_{4,t-6} + 0.5256v_t - 0.8366x_{5,t-4} - 0.0235x_{6,t-2}$, with temperature and temperature variation being the two dominant

Table 5. Results of the cross-validation method

Dimension	*Bandwidth*	*CV(d) value*
1	0.325	0.3593
2	0.325	0.3516
3	0.325	0.3435
4	0.325	0.3523
5	0.475	0.3450

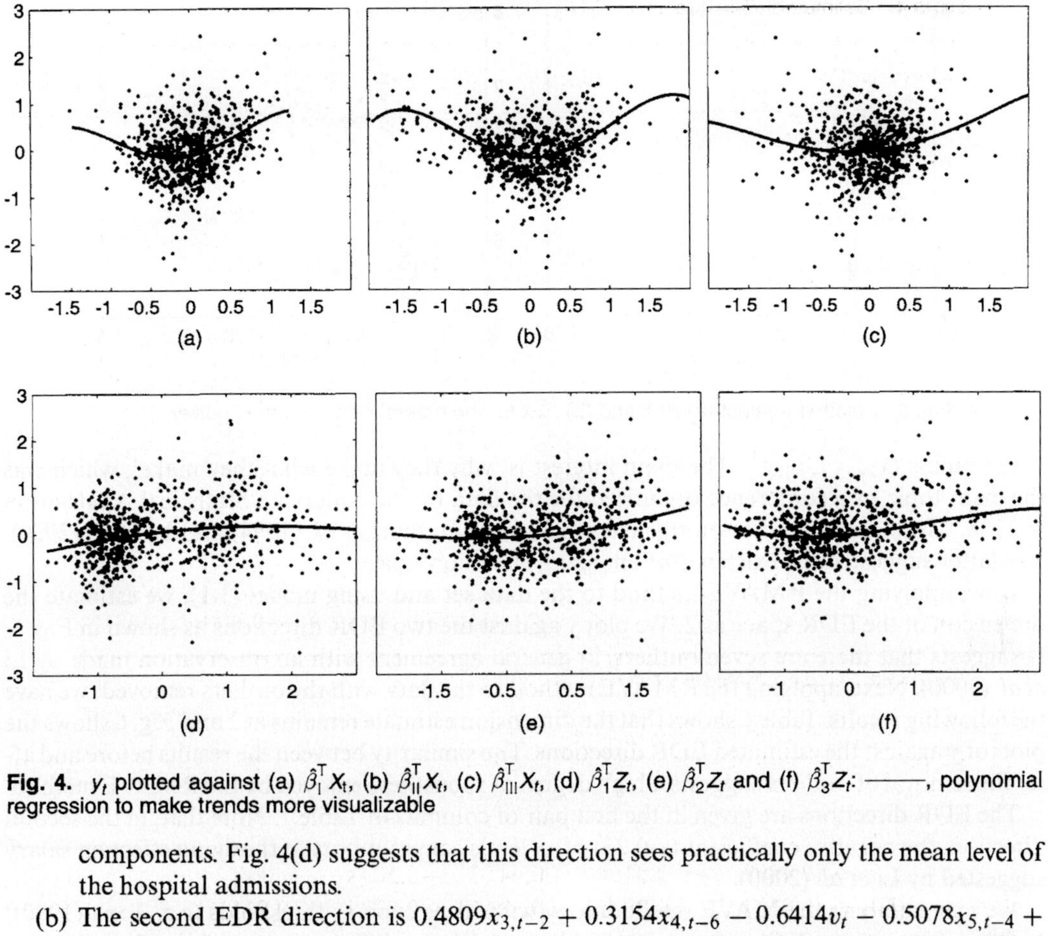

Fig. 4. y_t plotted against (a) $\hat{\beta}_{\mathrm{I}}^{\mathrm{T}} X_t$, (b) $\hat{\beta}_{\mathrm{II}}^{\mathrm{T}} X_t$, (c) $\hat{\beta}_{\mathrm{III}}^{\mathrm{T}} X_t$, (d) $\hat{\beta}_1^{\mathrm{T}} Z_t$, (e) $\hat{\beta}_2^{\mathrm{T}} Z_t$ and (f) $\hat{\beta}_3^{\mathrm{T}} Z_t$: ——, polynomial regression to make trends more visualizable

components. Fig. 4(d) suggests that this direction sees practically only the mean level of the hospital admissions.

(b) The second EDR direction is $0.4809x_{3,t-2} + 0.3154x_{4,t-6} - 0.6414v_t - 0.5078x_{5,t-4} + 0.0018x_{6,t-2}$, which, together with Fig. 4(e), suggests that high levels of suspended particulates and/or high levels of ozone during cold spells tend to cause high admissions.

(c) The third EDR direction is $0.0101x_{3,t-2} + 0.3815x_{4,t-6} + 0.1345v_t + 0.0734x_{5,t-4} - 0.9115x_{6,t-2}$, which, together with Fig. 4(f), suggests that high ozone levels on extremely dry days tend to cause high admissions.

This analysis suggests that pollutants have reached such a level in Hong Kong that it only takes the weather to enter the right regime to exacerbate the circulatory and respiratory problems there.

5.2. *Hitters' salary data*

The hitters' salary data set has attracted much attention among statisticians. The data consist of times at bat (x_1), hits (x_2), home runs (x_3), runs (x_4), runs batted in (x_5) and walks (x_6) in 1986, years in major leagues (x_7), times at bat (x_8), hits (x_9), home runs (x_{10}), runs (x_{11}), runs batted in (x_{12}) and walks (x_{13}) during their entire career up to 1986, annual salary (y) in 1987, put-outs (x_{14}), assistances (x_{15}) and errors (x_{16}). For ease of exposition, we abuse the notation and set y as the logarithm of annual salary in 1987, x_j the standardized x_j ($j = 1, \ldots, 16$) and

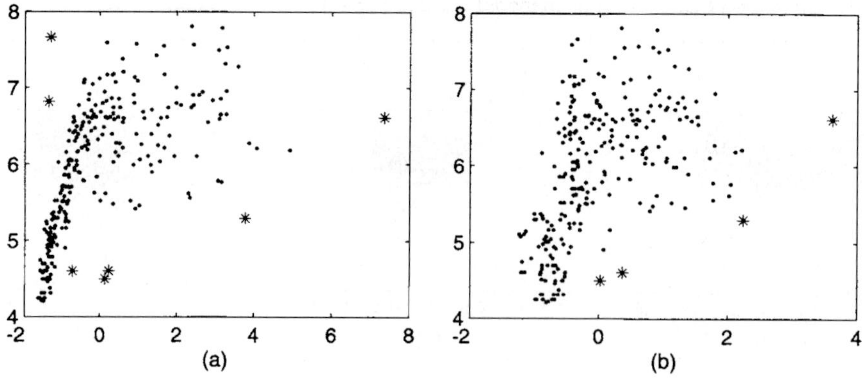

Fig. 5. y plotted against (a) $\hat{\beta}_1^{\mathrm{T}}X$ and (b) $\hat{\beta}_2^{\mathrm{T}}X$ for the hitters' salary data: *, outlier

X the vector $(x_1, \ldots, x_{16})^{\mathrm{T}}$. The main interest is 'why they make what they make', which was the main topic of a conference organized for the data by the American Statistical Association in 1988. More recent studies on this data include Chaudhuri *et al.* (1994) and Li *et al.* (2000). The latter suggested the existence of an 'aging effect' on salary.

Now, applying the RMAVE method to the data set and using model (1.1), we estimate the dimension of the EDR space as 2. We plot y against the two EDR directions as shown in Fig. 5. It suggests that there are seven outliers, in general agreement with an observation made by Li *et al.* (2000). Next, applying the RMAVE method to the data with the outliers removed, we have the following results. Table 6 shows that the dimension estimate remains at 2 and Fig. 6 shows the plots of y against the estimated EDR directions. The similarity between the results before and after the removal of outliers suggests a high degree of robustness enjoyed by the RMAVE method.

The EDR directions are given in the first pair of columns of Table 7. Note that, in the second direction, the negative coefficient (-0.23) of x_7 lends some support to the aging effect on salary suggested by Li *et al.* (2000).

We may combine the MAVE methodology with ideas such as thresholds (e.g. Tong (1990)) and regression trees to fit different regression models to different parts of the data set. For regression trees, we may mention the classification and regression trees method of Breiman *et al.* (1984), the SUPPORT algorithm of Chaudhuri *et al.* (1994) and the PHDRT algorithm of Li *et al.* (2000) and others. As an illustration, the left-hand 'regime' in Fig. 6(a) can be fitted

Table 6. Results of the cross-validation method (with the outliers removed)

Dimension	*Bandwidth*	*CV(d) value*
1	0.148	0.265
2	0.395	0.118
3	0.473	0.134
4	0.609	0.158
5	0.544	0.139
6	0.596	0.135
7	0.662	0.146
8	0.572	0.133
9	0.655	0.132
10	0.927	0.178

Table 7. Estimated EDR directions in model (1.1) and (5.2)†

$\hat\beta_1$	$\hat\beta_2$	$\hat\beta_1$	$\hat\beta_2$	$\hat\beta$	$\hat\theta$
−0.25 (x_1)	0.08 (x_1)	−0.05 (x_1)	0.14 (x_1)	−0.01 (x_1)	**−0.27** (x_1)
0.24 (x_2)	0.04 (x_2)	0.04 (x_2)	**−0.20** (x_2)	0.15 (x_2)	0.17 (x_2)
0.09 (x_3)	−0.01 (x_3)	−0.03 (x_3)	−0.09 (x_3)	0.02 (x_3)	0.09 (x_3)
0.00 (x_4)	0.07 (x_4)	0.03 (x_4)	**0.40** (x_4)	0.01 (x_4)	−0.04 (x_4)
−0.01 (x_5)	−0.04 (x_5)	−0.03 (x_5)	−0.03 (x_5)	−0.06 (x_5)	0.01 (x_5)
0.05 (x_6)	0.04 (x_6)	−0.09 (x_6)	**−0.29** (x_6)	0.06 (x_6)	0.04 (x_6)
0.52 (x_7)	**−0.23** (x_7)	0.18 (x_7)	0.02 (x_7)	0.01 (x_7)	**0.51** (x_7)
0.55 (x_8)	**−0.49** (x_8)	**0.51** (x_8)	**−0.57** (x_8)	0.03 (x_8)	**0.74** (x_8)
0.37 (x_9)	**0.75** (x_9)	**−0.81** (x_9)	**−0.26** (x_9)	**0.90** (x_9)	−0.11 (x_9)
0.10 (x_{10})	0.15 (x_{10})	0.00 (x_{10})	−0.08 (x_{10})	**0.24** (x_{10})	0.03 (x_{10})
0.23 (x_{11})	0.12 (x_{11})	0.08 (x_{11})	**0.27** (x_{11})	0.14 (x_{11})	0.16 (x_{11})
0.08 (x_{12})	0.17 (x_{12})	−0.10 (x_{12})	0.08 (x_{12})	0.04 (x_{12})	−0.13 (x_{12})
0.30 (x_{13})	**0.22** (x_{13})	0.07 (x_{13})	**0.43** (x_{13})	**0.27** (x_{13})	0.08 (x_{13})
−0.01 (x_{14})	0.09 (x_{14})	−0.06 (x_{14})	0.09 (x_{14})	0.08 (x_{14})	−0.04 (x_{14})
0.04 (x_{15})	−0.03 (x_{15})	0.03 (x_{15})	0.12 (x_{15})	−0.03 (x_{15})	0.06 (x_{15})
0.00 (x_{16})	0.04 (x_{16})	−0.05 (x_{16})	−0.08 (x_{16})	0.05 (x_{16})	−0.02 (x_{16})

†Entries in bold have relatively large absolute values.

by a simple straight line, say

$$y = 7.24 + 0.09x_3 + 0.38x_7 + 1.49x_9 + 0.83x_{13}.$$
$$(0.07)\quad (0.02)\quad (0.07)\quad (0.15)\quad (0.15)$$

The standard deviation of the fitted residuals, $\hat\sigma$, is 0.26 and $R^2 = 0.865$. The threshold is set at −0.47. The right-hand regime is much more volatile and we may return to the RMAVE method. The estimated dimension is still 2 and the estimated directions are given in the second pair of columns in Table 7. Let $z_1 = \hat\beta_1^{\mathrm{T}} X$ and $z_2 = \hat\beta_2^{\mathrm{T}} X$. We may fit to the right-hand regime a polynomial regression such as

$$y = 6.61 - 1.86z_1 + 0.21z_2 - 1.19z_2^2.$$
$$(0.03)\quad (0.11)\quad (0.09)\quad (0.19)$$

For this model, $\hat\sigma = 0.28$ and $R^2 = 0.714$. The overall $\hat\sigma$ is 0.27. A simple calculation shows

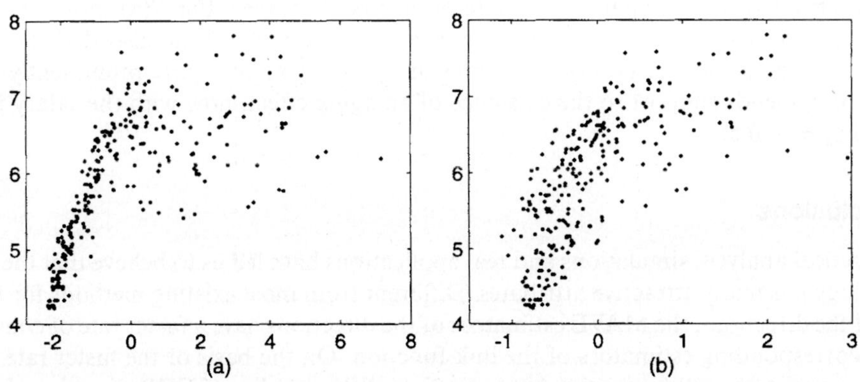

Fig. 6. y plotted against (a) $\hat\beta_1^{\mathrm{T}} X$ and (b) $\hat\beta_2^{\mathrm{T}} X$ for the hitters' salary data with the outliers removed

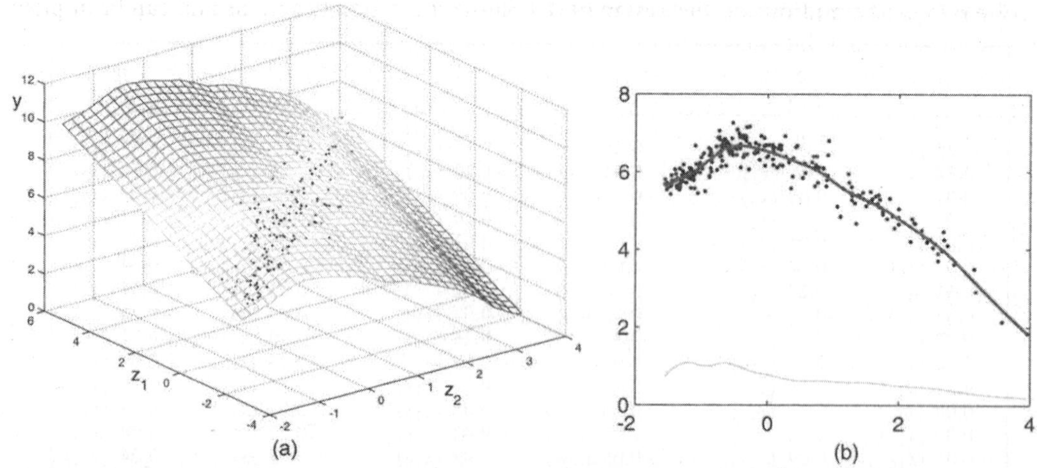

Fig. 7. (a) Estimated regression surface of model (5.2) (•, observations) and (b) estimated regression function of g (———) and estimate of the density function along the direction (·······) (•, residuals after removing the linear part in model (5.2))

that the coefficient of x_7 in the model for the right-hand regime is again negative, with the implication mentioned previously. As a comparison with the regression tree results obtained by Li *et al.* (2000), we quote $\hat{\sigma}=0.422$ for the classification and regression trees method with five bases, 0.33 for the multivariate adaptive regression splines method with 13 bases, 0.44 for the SUPPORT algorithm with two bases and 0.35 for the PHDRT algorithm. For our simple-minded hybrid, the overall $\hat{\sigma}=0.27$ with two bases.

Finally, we may consider the model

$$y = a\beta^{\mathrm{T}}X + g(\theta^{\mathrm{T}}X) + \varepsilon, \tag{5.2}$$

where $\beta \perp \theta$ with $\|\theta\| = \|\beta\| = 1$. This is a special case of model (1.3). See Xia *et al.* (1999) for details. Using the method described in Section 3, we obtain estimates of β and θ as listed in the third pair of columns in Table 7, $\hat{a}=0.75$, $\hat{\sigma}=0.26$ and the estimate of the function g as shown in Fig. 7(b). (Because the density of $\hat{\theta}^{\mathrm{T}}X$ is not so uniform, a variable bandwidth is used. See Fan and Gijbels (1996), page 152.) The dominant covariates in $z_1 = \hat{\beta}^{\mathrm{T}}X$ are x_2, x_9, x_{10}, x_{11} and x_{13}, all with positive coefficients. Four out of these five covariates measure past performance and so we may interpret z_1 as principally a measure of past performance. Fig. 7(a) shows that, along the z_1-axis, players with better past performance are paid better. Note also that the number of years in the major league (x_7) only features in z_2, i.e. $\hat{\theta}^{\mathrm{T}}X$, and quite prominently so. The estimated $g(z_2)$ lends support to the existence of an aging effect, now with the salary peaking at around $z_2 = -0.5$.

6. Conclusions

Our theoretical analysis, simulations and real applications have led us to believe that the MAVE methodology has many attractive attributes. Different from most existing methods for the estimation of the directions, the MAVE estimators of the directions have a faster rate of consistency than the corresponding estimators of the link function. On the basis of the faster rate of consistency, a consistent method for the determination of the number of EDR directions has been proposed. The MAVE method can easily be extended to more complicated models. It does not require strong assumptions on the design of X and the regression functions and can be applied to both independent data and dependent data.

As a by-product, we have extended the ADE method of Härdle and Stoker (1989) to the case of more than one EDR direction, resulting in the OPG method. This method has wider applicability with respect to designs for X and regression functions. Our basic idea has also led to the IMAVE method, which is closely related to the SIR method and the most predictable problem of Hotelling (1935), but in our simulations IMAVE seems to enjoy a better performance than SIR. The refined kernel based on the determination of the number of the directions can further improve the accuracy of estimation of the directions. Our simulations show that substantial improvements can be achieved.

Theoretical improvements on the MAVE method and the OPG method can be made by using higher order local polynomial smoothing. For example, we may replace expressions (2.7) and (3.1) by

$$\min_{\substack{B:B^{\mathrm{T}}B=I\\ a_j,b_j,c_j}} \left[\sum_{j=1}^{n} \sum_{i=1}^{n} \{y_i - a_j - b_j B^{\mathrm{T}}(X_i - X_j) - \sum_{1<k\leqslant r} \sum_{i_1+\ldots+i_p=k} (c_{j,i_1,i_2,\ldots,i_p} \{X_i - X_j\}_1^{i_1} \{X_i - X_j\}_2^{i_2} \ldots \{X_i - X_j\}_p^{i_p})\}^2 w_{ij} \right],$$

where $c_j = \{c_{j,i_1,i_2,\ldots,i_p}, i_1 + \ldots + i_p = k, 1 < k \leqslant r\}$, and

$$\min_{a_j,b_j,c_j} \left[\sum_{i=1}^{n} \{y_i - a_j - b_j^{\mathrm{T}}(X_i - X_j) - \sum_{1<k\leqslant r} \sum_{i_1+\ldots+i_p=k} (c_{j,i_1,i_2,\ldots,i_p} \{X_i - X_j\}_1^{i_1} \{X_i - X_j\}_2^{i_2} \ldots \{X_i - X_j\}_p^{i_p})\}^2 K_h(X_i - X_j) \right]$$

respectively. Higher rates of consistency can then be obtained.

Unlike the SIR method, the MAVE method is well adapted to time series; our experience suggests that the MAVE method is also robust against outliers. Furthermore, all our simulations show that the MAVE method has a much better performance than the SIR method (and OPG method). Although theorem 2 furnishes a partial explanation, we are still intrigued because SIR uses the one-dimensional kernel (for the kernel version) whereas the MAVE method uses a multidimensional kernel. However, because the SIR method uses y to produce the kernel weight, its efficiency will suffer from fluctuations in the link function. The gain by using the y-based one-dimensional kernel does not seem to be sufficient to compensate for the loss in efficiency caused by these fluctuations, but further research is needed here.

Acknowledgements

We thank the Biotechnology and Biological Science Research Council and Engineering and Physical Sciences Research Council of the UK, the Research Grants Council of Hong Kong, the Committee on Research and Conference Grants of the University of Hong Kong, the Friends of London School of Economics (Hong Kong) and the Wellcome Trust for partial support. We are most grateful to two referees for constructive comments. We thank Professor Wing Hung Wong and Professor X. Shen for making available to us their unpublished work and Professor T. S. Lau for providing the Hong Kong data and some background information.

Appendix A

A.1. Assumptions and remarks

The observations of X should be standardized before the analysis. Define the generalized conditional density

$$p_{\xi|\zeta}(u|v) = \lim_{du\to 0, dv\to 0} \left\{ \frac{P(\xi \in [u, u+du), \zeta \in [v, v+dv))}{du\, P(\zeta \in [v, v+dv))} \right\},$$

and we define $0/0=0$. In our proofs, we need the following conditions. (In all our theorems, weaker conditions can be adopted at the expense of much lengthier proofs.)

Condition 1: $\{(X_\iota, y_i)\}$ is a stationary (with the same distribution as (X, y)) and absolutely regular sequence, i.e.

$$\beta(k) = \sup_{i\geqslant 1} \left[E\left\{ \sup_{A\in\mathcal{F}_{i+k}^{\infty}} |P(A|\mathcal{F}_1^i) - P(A)| \right\} \right] \to 0 \qquad \text{as } k \to \infty,$$

where $\mathcal{F}_i^k$ denotes the σ-field generated by $\{(X_l, y_l) : i \leqslant l \leqslant k\}$. Further, $\beta(k)$ decreases at a geometric rate.

Condition 2:

(a) $E|y|^k < \infty$ for all $k > 0$;
(b) $E\|X\|^k < \infty$ for all $k > 0$.

Condition 3:

(a) the density function f of X has bounded fourth derivative and is bounded away from 0 in a neighbourhood $\mathcal{D}$ around 0;
(b) the density function f_y of y has bounded derivative and is bounded away from 0 on a compact support.

Condition 4: the generalized conditional densities $p_{X|y}(x|y)$ of X given y and $p_{(X_0, X_l)|(y_0, y_l)}$ of (X_0, X_l) given (y_0, y_l) are bounded for all $l \geqslant 1$.

Condition 5:

(a) g has bounded, continuous third derivatives;
(b) $E(X|y)$ and $E(XX^{\mathrm{T}}|y)$ have bounded, continuous third derivatives.

Condition 6: $K(\cdot)$ is a spherical symmetric density function with a bounded derivative. All the moments of $K(\cdot)$ exist.

Condition 1 is made only for the purpose of simplicity of proof. It can be weakened to $\beta(k)=O(k^{-\iota})$ for some $\iota > 0$. Many time series models, including the autoregressive single-index model (Xia and An, 1999), satisfy assumption 1. Condition 2(a) is also made for simplicity of proof. See, for example, Härdle *et al.* (1993). The existence of finite moments is sufficient. Condition 3(a) is needed for the uniform rate of consistency of the kernel smoothing methods. Condition 4 is needed for kernel estimation of dependent data. Condition 5(a) is made to meet the continuous requirement for kernel smoothing. The kernel assumption 6 is satisfied by most of the commonly used kernel functions. For ease of exposition, we further assume that

$$\int UU^{\mathrm{T}} K(U)\, dU = I.$$

A.2. The efficiency of the algorithm

To explain the mechanism of the MAVE method, we here consider only the single-index model, i.e.

$$y = g(\beta_0^{\mathrm{T}} X) + \varepsilon.$$

We estimate β_0 by minimizing

$$\sum_{j=1}^{n} \sum_{i=1}^{n} \{y_i - a_j - b_j \beta^{\mathrm{T}}(X_i - X_j)\}^2 w_{ij} \tag{A.1}$$

iteratively with respect to (a_j, b_j) and β. Let

$$s_{n,0}(x) = n^{-1} \sum_{i=1}^{n} K_{h,i}(x),$$

$$s_{n,1}(x) = n^{-1} \sum_{i=1}^{n} K_{h,i}(x)((X_i - x)/h),$$

$$s_{n,2}(x) = n^{-1} \sum_{i=1}^{n} K_{h,i}(x)((X_i - x)/h)((X_i - x)/h)^{\mathrm{T}}.$$

Then $w_{ij} = n^{-1} K_{h,i}(X_j)/s_{n,0}(X_j)$. According to our estimation procedure, if we begin with any unit norm vector β, we have by minimizing expression (A.1)

$$\hat{a}_j = \frac{n^{-1} \sum_{i=1}^{n} \{\beta^{\mathrm{T}} s_{n,2} \beta K_{h,i}(X_j) - \beta^{\mathrm{T}} s_{n,1} \beta^{\mathrm{T}} K_{h,i}(X_j)((X_i - X_j)/h)\} y_i}{\beta^{\mathrm{T}}(s_{n,0}(X_j) s_{n,2}(X_j) - s_{n,1}(X_j) s_{n,1}^{\mathrm{T}}(X_j))\beta},$$

$$\hat{b}_j h = \frac{n^{-1} \sum_{i=1}^{n} \{s_{n,0}(X_j) \beta^{\mathrm{T}} K_{h,i}(X_j)((X_i - x)/h) - \beta^{\mathrm{T}} s_{n,1} K_{h,i}(X_j)\} y_i}{\beta^{\mathrm{T}}(s_{n,0}(X_j) s_{n,2}(X_j) - s_{n,1}(X_j) s_{n,1}^{\mathrm{T}}(X_j))\beta}.$$

After one step of iteration, we obtain the estimate of β_0 as

$$\tilde{\beta} = \left\{ \sum_{j=1}^{n} h^2 \hat{b}_j^2 \sum_{i=1}^{n} \frac{K_{h,i}(X_j)((X_i - X_j)/h)((X_i - X_j)/h)^{\mathrm{T}}}{s_{n,0}(X_j)} \right\}^{-1} \times \sum_{j=1}^{n} h \hat{b}_j \sum_{i=1}^{n} \frac{K_{h,i}(X_j)((X_i - X_j)/h)(y_i - \hat{a}_j)}{s_{n,0}(X_j)}.$$

If β is not perpendicular to β_0, we have

$$\tilde{\beta} = \{1 + (1 - \beta^{\mathrm{T}} \beta_0) + o_P(1)\}\beta_0 + O_P[(h^2 + \delta_n)\{h + m(\beta, \beta_0)\} + h^{-1}\delta_n^2]\beta_0^{\perp}, \tag{A.2}$$

where $\beta_0^{\perp}$ is a vector perpendicular to β_0. Equation (A.2) means that the effect of the initial value is quite small. Note that $\delta_n \sim h^2 \log(n)^{1/2}$ if we use the optimal bandwidth of the estimation of the regression function, i.e. $h \sim n^{-1/(p+4)}$. Suppose that we start with an initial estimator of β_0 which has a consistency rate of $O_P\{h^2 \log(n)\}$. Then $m(\beta, \beta_0) = O_P\{h^2 \log(n)\}$ and we have

$$\tilde{\beta} = \{1 + (1 - \beta^{\mathrm{T}} \beta_0) + o_P(1)\}\beta_0 + O_P(h^3 + h\delta_n + h^{-1}\delta_n^2)\beta_0^{\perp}.$$

Therefore,

$$m(\tilde{\beta}, \beta_0) = O_P(h^3 + h\delta_n + h^{-1}\delta_n^2).$$

This estimation procedure is very efficient in that, in theory, after two steps the estimate from our procedure can achieve the final consistency rate.

A similar result was discovered in a different context by Hannan (1969). Specifically, he developed an estimation procedure for the parameters of autoregressive moving average processes. Starting with arbitrary consistent estimators of the parameters, a modification by one step of the Newton–Raphson-type iteration can make the estimators asymptotically efficient. In the MAVE method, the first step is to find a consistent 'initial' estimator. The second step is to modify the 'initial' estimator, which can also make the estimate asymptotically efficient. In spite of the asymptotic efficiency, the iterative application of the procedure beyond the two steps was suggested by Hannan (1969) as a way of further improving the estimator. For the MAVE method, our simulation also suggests that further iterations are beneficial.

References

Auestad, B. and Tjøstheim, D. (1990) Identification of nonlinear time series: first order characterization and order determination. *Biometrika*, **77**, 669–688.
Breiman, L., Friedman, J. H., Olshen, R. A. and Stone, C. J. (1984) *Classification and Regression Trees.* Belmont: Wadsworth.
Cai, Z., Fan, J. and Yao, Q. (2000) Functional-coefficient regression models for nonlinear time series. *J. Am. Statist. Ass.*, **95**, 941–956.
Carroll, R. J., Fan, J., Gijbels, I. and Wand, M. P. (1997) Generalized partially linear single-index models. *J. Am. Statist. Ass.*, **92**, 477–489.
Carroll, R. J. and Li, K. C. (1995) Binary regressors in dimension reduction models: a new look at treatment comparisons. *Statist. Sin.*, **5**, 667–688.
Chaudhuri, P., Huang, M. C., Loh, W. Y. and Yao, R. (1994) Piecewise-polynomial regression trees. *Statist. Sin.*, **4**, 143–167.
Chen, C.-H. and Li, K. C. (1989) Can SIR be as popular as multiple linear regression? *Statist. Sin.*, **8**, 289–316.
Chen, H. (1991) Estimation of a projection-pursuit type regression model. *Ann. Statist.*, **19**, 142–157.
Cheng, B. and Tong, H. (1992) On consistent nonparametric order determination and chaos (with discussion). *J. R. Statist. Soc.* B, **54**, 427–449.
Cook, R. D. (1994) On the interpretation of regression plots. *J. Am. Statist. Ass.*, **89**, 177–189.
——(1998) Principal Hessian directions revisited (with discussions). *J. Am. Statist. Ass.*, **93**, 85–100.
Fan, J. and Gijbels, I. (1996) *Local Polynomial Modeling and Its Applications.* London: Chapman and Hall.
Fan, J. and Zhang, W. Y. (1999) Statistical estimation in varying coefficient models. *Ann. Statist.*, **27**, 1491–1518.
Friedman, J. H. and Stuetzle, W. (1981) Projection pursuit regression. *J. Am. Statist. Ass.*, **76**, 817–823.
Hall, P. (1989) On projection pursuit regression. *Ann. Statist.*, **17**, 573–588.
Hannan, E. J. (1969) The estimation of mixed moving average autoregressive system. *Biometrika*, **56**, 579–593.
Härdle, W., Hall, P. and Ichimura, H. (1993) Optimal smoothing in single-index models. *Ann. Statist.*, **21**, 157–178.
Härdle, W. and Stoker, T. M. (1989) Investigating smooth multiple regression by method of average derivatives. *J. Am. Statist. Ass.*, **84**, 986–995.
Hastie, T. J. and Tibshirani, R. (1986) Generalized additive models (with discussion). *Statist. Sci.*, **1**, 297–318.
Hotelling, H. (1935) The most predictable criterion. *J. Educ. Psychol.*, **26**, 139–142.
Huber, P. J. (1985) Projection pursuit (with discussion). *Ann. Statist.*, **13**, 435–525.
Ichimura, H. and Lee, L. (1991) Semiparametric least squares estimation of multiple index models: single equation estimation. In *Nonparametric and Semiparametric Methods in Econometrics and Statistics* (eds W. Barnett, J. Powell and G. Tauchen). Cambridge: Cambridge University Press.
Li, K. C. (1991) Sliced inverse regression for dimension reduction (with discussion). *J. Am. Statist. Ass.*, **86**, 316–342.
——(1992) On principal Hessian directions for data visualisation and dimension reduction: another application of Stein's Lemma. *Ann. Statist.*, **87**, 1025–1039.
Li, K. C., Lue, H. H. and Chen, C. H. (2000) Interactive tree-structured regression via principal Hessian directions. *J. Am. Statist. Ass.*, **95**, 547–560.
Rao, C. R. (1973) *Linear Statistical Inference and Its Applications.* New York: Wiley.
Schott, J. R. (1994) Determining the dimensionality in sliced inverse regression. *J. Am. Statist. Ass.*, **89**, 141–148.
Severini, T. A. and Wong, W. H. (1992) Profile likelihood and conditionally parametric models. *Ann. Statist.*, **20**, 1768–1802.
Smith, R. L., Davis, J. M. and Speckman, P. (1999) Assessing the human health risk of atmospheric particles. In *Environmental Statistics: Analysing Data for Environmental Policy.* New York: Wiley.
Stone, M. (1974) Cross-validatory choice and assessment of statistical predictions (with discussion). *J. R. Statist. Soc.* B, **36**, 111–147.
Tong, H. (1990) *Nonlinear Time Series Analysis: a Dynamical System Approach.* Oxford: Oxford University Press.
Xia, Y. and An, H.-Z. (1999) Projection pursuit autoregression in time series. *J. Time Ser. Anal.*, **20**, 693–714.
Xia, Y. and Li, W. K. (1999) On single-index coefficient regression models. *J. Am. Statist. Ass.*, **94**, 1275–1285.
Xia, Y., Tong, H. and Li, W. K. (1999) On extended partially linear single-index models. *Biometrika*, **86**, 831–842.
Yang, L. and Tschernig, R. (1999) Multivariate bandwidth selection for local linear regression. *J. R. Statist. Soc.* B, **61**, 793–815.
Yao, Q. and Tong, H. (1994) On subset selection in nonparametric stochastic regression. *Statist. Sin.*, **4**, 51–70.
Zhu, L. X. and Fang, K.-T. (1996) Asymptotics for kernel estimate of sliced inverse regression. *Ann. Statist.*, **24**, 1053–1068.

Discussion on the paper by Xia, Tong, Li and Zhu

J. T. Kent (*University of Leeds*)

The paper is an ambitious attempt to tackle high dimensional regression problems. There are connections to several areas of statistics, including multivariate analysis, nonparametric regression and linear regression. I would like to direct some comments to each area in turn.

Multivariate analysis

A standard model in multivariate analysis of variance involves k groups of p-dimensional observations X with different means. The group membership can be represented in terms of a random variable y taking integer values $j = 1, \ldots, k$, with probabilities π_j. Conditional on $y = j$, the distribution of X is modelled by $N_p(\mu_j, \Sigma)$, $j = 1, \ldots, k$. Let $\bar{\mu}$ denote the average of these mean values. Canonical variate analysis is a tool for improving the interpretability in this setting via dimension reduction. It is assumed that these means lie on a lower dimensional plane of dimension D, say, where $D < \min(k-1, p)$, i.e. we assume that the $\{\mu_j - \bar{\mu}\}$ span a subspace of dimension D. Let B $(p \times D)$ be a matrix whose columns span this subspace and let C $(p \times (p-D))$ be a complementary matrix so that (B, C) is non-singular. Reversing the conditioning yields the logistic-type regression model

$$P(y = j|X) \propto \pi_j \exp\{(\mu_j - \bar{\mu})^{\mathrm{T}}\Sigma^{-1}(X - \bar{\mu}) - \tfrac{1}{2}(\mu_j - \bar{\mu})^{\mathrm{T}}\Sigma^{-1}(\mu_j - \bar{\mu})\}$$

in which the exponent is a linear function of X with different coefficients for each j.

It can be checked that this conditional probability in fact depends only on $B^{\mathrm{T}}X$, not on all of X, and so yields the conditional independence statement

$$(y \perp C^{\mathrm{T}}X)|B^{\mathrm{T}}X.$$

Thus this model can be regarded as a discrete and parametric version of the authors' model (1.1). In passing, note that similar conditional independence statements form the building-blocks of graphical models, except that in our setting B is unknown.

In the k-groups model, the marginal distribution of X is a mixture of p-variate normals. However, when attention is focused on the conditional distribution of $y|X$ in the logistic-type regression model, it is usual to allow more general possibilities for the marginal distribution of X. The k-groups model can be viewed as a motivating example for the sliced inverse regression approach to nonparametric multiple regression, whereas the logistic-type regression model better matches the tone of the current paper.

Nonparametric regression

A generalized additive model takes the form $y = \Sigma_{j=1}^{D} g_j(\beta_j^{\mathrm{T}}X) + \varepsilon$. The ridge terms $g_j(\beta_j^{\mathrm{T}}X)$ can be viewed as 'main effects' in the directions β_j. In contrast, the more general model (1.1), $y = g(B_0^{\mathrm{T}}X) + \varepsilon$, which forms the foundation of the paper, also allows 'interaction terms'. However, I am concerned that there is a tendency in practice to interpret the columns of B_0 as main effects and to ignore possible interactions. For example, consider the plots of y *versus* $\hat{\beta}_1^{\mathrm{T}}X$ and y *versus* $\hat{\beta}_2^{\mathrm{T}}X$ in Fig. 5. There are two related problems with these plots. First any possible interactions are ignored; it might be better to represent the whole response surface. The second problem is that these two directions $\hat{\beta}_1$ and $\hat{\beta}_2$ have no preferred status. It is possible to take any other basis of their column space without affecting the validity of the model.

Linear regression

Reduced rank models are also of interest in linear regression analysis. Of course the ordinary least squares regression model is a special case of model (1.1) with $D = 1$ and g linear. However, when p is large, it is well known that the least squares estimator can be unstable, so attempts are often made to reduce the dimensionality of X. One class of methods involves variable selection. However, a class of methods that is more in keeping with the current paper involves the construction of new linear composite variables from X. One of the simplest such methods is principal components regression in which X is replaced by its first few dominant principal components. Unfortunately, this method is rather unsatisfactory since the dominant principal components depend just on the X-variability and not on the relationship to y. A hybrid approach between ordinary least squares and principal components regression is partial least squares; see Stone and Brooks (1990) for a unified treatment. Of course these methods of dimension reduction (including variable selection methods as well) depend heavily on the covariance structure of X.

Are there any lessons from this methodology for this paper? In particular, what happens when there is very high correlation between the X-variables or, more generally, when the X-variables become nearly collinear? My concern is that the estimate of the column space of B will become unstable and that problem (2.7) might have multiple solutions.

I have found the paper tremendously stimulating, and it gives me great pleasure to propose the vote of thanks.

Adrian Bowman (*University of Glasgow*)

It is a great pleasure to add my thanks for this paper. I enjoyed both its reading and its presentation. Over the past few years there has been a considerable amount of work in the dimension reduction area. Regression used to be a topic which we thought we understood. Now we are not so sure. It is one of the merits of this paper that it brings together a variety of approaches in this area and synthesizes them into a simple but potentially powerful idea. Direct and simultaneous estimation of both the nonparametric and the directional components of the model brings some significant benefits. These include an avoidance of some of the usual difficulties with bias incurred by smoothing, a weakening of assumptions, the ability to handle the special but important case of time series, some impressively strong supporting asymptotics and evidence of good behaviour in numerical work. However, it is difficult to believe that these properties are not bought at some price and I would like to explore one or two aspects of where the costs may lie.

The first relevant feature is that, although the central idea is attractively simple, the implementation is necessarily more sophisticated. It involves a variety of steps. The first is smoothing in, possibly high dimensional, covariate space. Most people feel comfortable when applying smoothing in one, two or occasionally three dimensions. The authors have been courageous in going rather beyond that. In the hospital admissions data courage gives way to heroism by smoothing in 42 dimensions. Of course, the refinements introduced by the authors quickly reduce attention to the much smaller dimensional space defined by the current effective dimension reduction (EDR) directions where smoothing can be applied without difficulty. At the same time, there is a high dimensional minimization in operation to identify the EDR directions. Beyond this lies a cross-validation step to compare the EDR dimensions. Finally, there is some mention in the paper of the possibility of using a data-dependent bandwidth choice, although the authors wisely do not routinely incorporate this. The end result is a set of EDR directions which have been produced by a set of complex operations on the data. However, there is no difficulty in principle with that. Complex data may require complex methods of analysis and if the end result brings insight then it has been worthwhile.

On the question of insight, I would like to use the hospital admissions data as a means of raising some practical issues. The first concerns the *robustness* and *sensitivity* of the procedure. A scatterplot matrix reveals a variety of features in the covariates. One is the presence of substantial skewness. The sulphur dioxide variable is a good example of this and it includes in particular two very large observations. Since the sulphur dioxide, nitrogen dioxide and particulates covariates are all concentrations, it would be natural to take a log-transformation of each. Ozone, although also skewed, contains observations at or close to zero and so it may be best left unaltered, along with temperature. Humidity is a percentage, with many observations at high values and so the logistic transformation would be natural here. The question is whether the broad qualitative conclusions of the analysis will remain unchanged when repeated using the variables on these, arguably more natural, scales. The assumptions of the model are weak but one can only feel that there will be greater stability if the variables exhibit approximately normal variation. A second issue arises from the scatterplot of log(nitrogen dioxide) and log(particulates) which shows a strong linear relationship between these two variables. This is exactly the situation assumed by the model. However, it then seems surprising that particulates feature strongly in the conclusions whereas nitrogen dioxide does not. This raises the question of whether the decisions being made by the procedure on the weights to assign to variables are ones which we shall always feel comfortable with.

An issue of the *appropriateness* of the model is raised by the scatterplot of nitrogen dioxide against temperature. This shows a clear non-linear pattern which will be obscured by the linear combinations around which the model is built. Of course, a second dimension will, in this case, allow the full relationship between the covariates to be expressed. However, it would seem more appropriate to incorporate specific non-linear relationships into the model in a more direct way, where these are appropriate.

Finally, some important issues arise under the heading of *interpretation*. The first derives from the fact that EDR delivers a subspace, not a co-ordinate system. The same subspace can be represented by EDR directions which are rotated in different ways. This makes the interpretation of specific elements of the EDR direction vectors rather difficult. The nonparametric surface g has an unspecified shape, built from all EDR directions simultaneously. The marginal space may change radically as the EDR co-ordinate system is rotated. An interpretation can therefore only be made from the entire collection of EDRs and this is not an easy task. In addition, if we simulate data where y is unrelated to x we are still likely to identify EDR directions of apparent meaning. This highlights the need for some statistical methods of model comparison, beyond CV(d), to ensure that the results of EDR can safely be attributed to meaningful structure rather than to noise.

When the authors have come so far, it may seem churlish to ask them to go yet further. However, I raise these issues in the hope that the authors will be able to devote their considerable powers to addressing them. To return to the original remarks, this is clearly a simple but potentially powerful idea which deserves to be considered carefully. I have great pleasure again in congratulating the authors on their paper and in warmly seconding the vote of thanks.

The vote of thanks was passed by acclamation.

Santiago Velilla (*Universidad Carlos III de Madrid, Getafe*)
In developing minimum average variance estimation (MAVE) the authors seem to have in mind a first-order regression problem in which all the information that X carries on the response y is captured by the conditional expectation $E(y|X)$. In this sense, the populational objective function (2.1) and its sample version (2.7) seem to be appropriate when the error ε in model (1.1) not only satisfies the condition $E(\varepsilon|X) = 0$ but also $\text{var}(\varepsilon|X) = \sigma^2$. If the conditional variance is not constant, expressions (2.1) and (2.7) should perhaps be modified accordingly.

In comparing the four new methods proposed in this interesting paper, I find that both the outer product of gradients method, in Section 3.1, and inverse MAVE, in Section 3.2, have a natural *nested* character. Once a decision has been taken on the value of the dimension of the effective dimension reduction space, directions are determined sequentially. In contrast, both MAVE and refined MAVE seem to require specific computation in each step $d = 1, 2, \ldots$. Moreover, as indicated in the algorithm of Section 2.3, computation is required for all $1 \leqslant d \leqslant p$. In view of the pattern of Tables 3, 5 and 6 in the examples in Sections 5.1 and 5.2, where the change in the CV(d) value is 'small' when spurious directions are considered, for 'large' values of d the algorithm could be initialized using the results for $d-1$ making it 'nested', i.e. looking only for $\hat{\beta}_d$, once $\hat{\beta}_1, \hat{\beta}_2, \ldots, \hat{\beta}_{d-1}$ have been determined. Of course, this is just a suggestion based on the pattern of the tables in the examples, but this simplified scheme for *spurious* values of d might save some computational time.

Finally, in connection with condition (1.2), in Velilla (1998), section 4.1, I proposed a method for generating regressors X satisfying condition (1.2) that are not necessarily elliptical. This method has been applied, for example, in Bura and Cook (2001a, b) for assessing by simulation the performance of some methods for testing for dimension.

Wenyang Zhang (*University of Kent at Canterbury*)
I have two comments to make on this interesting paper.

Shannon's entropy
A measure of uncertainty, Shannon's entropy, was introduced by Shannon (1948), which is extremely useful in communication theory. It also can be used to reduce dimension in regression to avoid the 'curse of dimensionality'.

Let ξ and η be two random variables with joint density function $f(x, y)$. $p(x)$ is the density of ξ, the entropy of ξ is defined as

$$H(\xi) = -\int p(x) \log\{p(x)\}\, \mathrm{d}x$$

and the conditional entropy of ξ given η is

$$H_\eta(\xi) = -\int\int f(x, y)[\log\{f(x, y)\} - \log\{q(y)\}]\, \mathrm{d}x\, \mathrm{d}y$$

where $q(y)$ is the density of η. The information contained in η about ξ is

$$I(\xi, \eta) = H(\xi) - H_\eta(\xi).$$

Let Y be the response, X be the covariate with high dimension p and $(X_i, Y_i), i = 1, \ldots, n$, be a sample from (X, Y). For any fixed β, the estimate $\hat{I}(Y, \beta^\tau X)$ of $I(Y, \beta^\tau X)$ can be obtained by standard density estimation; see Fan and Gijbels (1996). An alternative dimension reduction procedure is maximize $\hat{I}(Y, \beta^\tau X)$ subject to $||\beta|| = 1$, to find the maximizer β_1 and maximum I_1, then maximize $\hat{I}(Y, \beta^\tau X)$, subject to $\beta^\tau \beta_1 = 0$ and $||\beta|| = 1$, to find the maximizer β_2 and maximum I_2, and continue this exercise until I_q is less than a selected critical value c which may be obtained by cross-validation. $(\beta_1, \ldots, \beta_q)$ forms the efficient directions to reduce the dimension. It would be very interesting to compare this approach with that in the paper.

Curse of dimensionality
In Section 2.1.1, the initial B is obtained based on

$$w_{i0} = K_h(X_i - X_0) \Big/ \sum_{l=1}^{n} K_h(X_l - X_0).$$

If the dimension p of X is very large, it would be impossible to obtain an initial B with small bias owing to the 'curse of dimensionality'. My question is does this bias matter in your procedure? If not, why could we not take the whole range of X_i as the initial bandwidth?

Frank Critchley (*The Open University, Milton Keynes*)
In welcoming the faster rate of consistency and time series extensions afforded by the paper, I would like to make the following points in which $Y_x := (Y|X = x)$ and $\varepsilon_x := (\varepsilon|X = x)$.

(a) I was somewhat surprised not to find fuller reference to the important body of work by Cook and co-workers, surveyed to that date in Cook (1998). Among other attractive features, such as its graphical emphasis, this approach examines how the whole distribution of Y_x—not just, as here, its mean $g(x)$—varies with x. Again, it exploits a conditional independence formulation throughout, that is both logically cogent and statistically intuitive. I would also like to draw attention to two forthcoming papers, available on the *Annals of Statistics* Web site and directly relevant to this paper: Cook and Li (2002), which addresses dimension reduction for $g(x)$, and Chiaromonte *et al.* (2002), which overlaps with Section 3.4.
(b) There are two apparent significant errors of omission.
 (i) In the sentence two after equation (1.1), a simple counter-example is

$$X \sim N_2(0, I),$$

$$Y = g(X_1) + \varepsilon$$

and

$$\varepsilon_x \sim N(0, \sigma^2 x_2^2).$$

The omission appears to be that model (1.1) should be augmented by the *location regression* requirement $Y \perp\!\!\!\perp X | E(Y|X)$ (Cook (1998), page 111); a similar remark applies to model (1.3).
 (ii) In the sentence including expression (2.1), additional conditions—such as constancy of $\mathrm{var}(\varepsilon_x)$ over x—apparently are required.
(c) The benefits of this paper—including relaxation of condition (1.2) on X—come at the price of other non-trivial restrictions to its applicability: in particular, to additive error models that are special cases of location regression and in which certain additional conditions hold.
(d) In unpublished preliminary discussions with Cook, it was noted that the conditional independence approach seems natural in a variety of time series contexts, autoregressive processes being obvious examples. This would seem a promising line of enquiry.
(e) In view of the quadratic nature of the criterion minimized, I was somewhat surprised by the robustness to outliers claim (Section 6) and would value further details.
(f) Concerning Section 2.1.2, under what conditions is convergence (to a unique solution) guaranteed?

Anthony Atkinson (*London School of Economics and Political Science*)
I congratulate the authors on an interesting paper which stimulated an excellent discussion. I have five points.

(a) John Kent placed the authors' proposal in the context of other dimension reduction methods, including partial least squares. This method is often used with p close to n. Is this likely to cause any problems? Partial least squares is also often used with $p \gg n$, e.g. in the spectroscopic data set analysed again by Brown *et al.* (2001). Can the authors' method be extended to this important class of problems?
(b) The interpretation of results like those of Table 4 seems beset with difficulties, since the directions

can be rotated in the D-dimensional subspace. Basilevsky (1994), section 6.10, discussed the similar problem of rotation and interpretation in factor analysis.

(c) On pages 378–379 the data have the effects of two factors removed, so that the y_t are indeed notationally abused, being residuals. The method of added variables (e.g. Atkinson and Riani (2000), section 2.2) indicates that the same regression should be performed on the explanatory variables as on the response, so that the analysis becomes one of residuals on residuals. Incidentally, this use of only one set of residuals is a frequent occurrence in time series analysis, where a series is 'pre-whitened', but the regressors left untouched.

(d) Some dicussants have mentioned robustness. It has been the experience of Marco Riani and myself that use of the forward search (Atkinson and Riani, 2000) reveals masked outliers and their effects in a way that is impossible by looking at a fit to all the data. The data are fitted to subsets of increasing size and parameter estimates, residuals and other quantities monitored. The starting-point for the searches is a robustly chosen subset of p, or a few more, observations. Could relatively small subsets of the data be used here to start such a process?

(e) Many statistical methods, including, I suspect, that described here, tend to work better if the data are approximately normal. In applications of inverse regression for dimension reduction, the data are sometimes transformed to approximate multivariate normality by using a multivariate Box–Cox transformation. An example is the analysis of data on New Zealand mussels in chapters 10 and 11 of Cook and Weisberg (1994). A robust version of this transformation using the forward search is illustrated in Riani and Atkinson (2001). What is the effect here of such transformations both on computation time and on the conclusions drawn from Tables 4 and 7?

Qiwei Yao (*London School of Economics and Political Science*)
The authors should be congratulated for making a further contribution along their impressive list of publications on nonparametric multivariate regression—a very important and immensely difficult topic.

Theorem 1 may be presented in a slightly stronger form by defining the weights w_{ij} in terms of $\{B^{\mathrm{T}}X_i\}$ instead of $\{X_i\}$. This effectively changes a p-dimensional smoothing problem into a d-dimensional one. The gain in convergence rate would now be $h_{\mathrm{opt}}\log(n) = O\{n^{-1/(d+4)}\log(n)\}$ at the price of the added computational complication in the minimization of problem (2.7).

As B_0 is only defined up to any orthogonal transforms, will the alternating iteration between refined kernel weights and estimating β_j in step 1(b) lead to stable $\hat{\beta}_j$? The use of refined kernel weights only makes sense if such a stable solution is guaranteed.

An alternative version for the distance measure would be

$$m(\hat{B}, B_0) = ||(I - B_0B_0^{\mathrm{T}})\hat{B}|| + ||(I - \hat{B}\hat{B}^{\mathrm{T}})B_0||.$$

Then $m(\hat{B}, B_0) \rightarrow 0$ in probability if and only if $\hat{B}$ estimates B_0 'correctly'.

Finally the method proposed is most useful when D is small such as 2 or 3, as we still need to estimate the link function even if we have the right effective dimension reduction. If model (1.1) does not hold, will the procedure lead to a 'good' approximation for the conditional expectation of y given X?

A. H. Welsh (*University of Southampton*)
Comparisons of minimum average variance estimation (MAVE) with sliced average variance estimation (SAVE) proposed by Cook and Weisberg (1991) (see Cook and Yin (2001) for recent references) in addition to sliced inverse regression may be interesting and more insightful. Robustness issues in sliced inverse regression and SAVE were raised at the 2000 Australian conference in a presentation by Ursula Gather and the discussion to Cook and Yin (2001). The issues are subtle so the claim that MAVE has good robustness properties needs a proper investigation.

In the single-index model, the asymptotic distribution of $\tilde{\beta}$ is essentially determined by

$$\sum_{i=1}^{n}\sum_{j=1}^{n} \hat{b}_j\, w_{ij}(X_i - X_j)\{\varepsilon_i + g(X_i^{\mathrm{T}}\beta_0) - \hat{a}_j\},$$

the 'numerator' in $\tilde{\beta}$. The approach in which we estimate g and g' by smoothing (as in the present paper) but estimate β_0 by standard maximum likelihood (Brillinger, 1992; Weisberg and Welsh, 1994) seems rather different. However, it is important to centre X_i about an estimate of $E(X|X^{\mathrm{T}}\beta_0 = X_i^{\mathrm{T}}\beta_0)$ and, under the simplifying conditions of the present paper and using local linear smoothing (Ruckstuhl and Welsh, 1999), the equivalent expression for this estimator is

$$\sum_{i=1}^{n}\sum_{j=1}^{n}\hat{b}_i\tilde{w}_{ji}(X_i - X_j)\{\varepsilon_i + g(X_i^{\mathrm{T}}\beta_0) - \hat{a}_i\}.$$

Whereas we usually use undersmoothing, higher order kernels or higher order polynomials in local polynomial smoothing to increase the rate of convergence of $\hat{a}_i - g(X_i^{\mathrm{T}}\beta_0)$ so that it is asymptotically negligible, MAVE estimates integrals of g rather than g so we can use optimal bandwidths for g while estimating β_0. If the above expressions are correct, MAVE should have the same asymptotic distribution (possibly up to centring of the covariates) as the maximum likelihood estimator but this needs to be checked carefully. Finally, MAVE should also be extended to other distributions, presumably by maximizing the average local log-likelihood.

Hengjian Cui (*Beijing Normal University*) **and Guoying Li** (*Academy of Mathematics and System Sciences, Beijing*)

This paper is very interesting and very provocative! The authors give us new ideas to search the effective dimension reduction (EDR) space in nonparametric regression settings.

The minimum average variance estimation (MAVE) is effective provided that model (1.1) is correct. It is different from projection pursuit (PP) (Huber, 1985; Li and Cheng, 1993), which assumes that the link function is a sum of several ridge functions; we call it the PP regression (PPR) model here. If model (1.1) is true, the first PP approximation is $E(y|\beta_1^{\mathrm{T}}X)$. However, β_1 is not necessarily in the space spanned by B_0 although $E(y|\beta_1^{\mathrm{T}}X)$ is the first-order optimal PP approximation of $g(B_0^{\mathrm{T}}X)$. If the PPR model is true and the number of ridge functions is less than p, model (1.1) holds obviously. However, MAVE concentrates on finding the EDR directions whereas the PP approach provides estimators for both the directions and the link function. Another point is that MAVE uses a high dimensional kernel whereas PP needs only a one-dimensional kernel. To simplify computation in MAVE, we may use the following iterative algorithm to search the EDR directions one by one:

$$\min_{\substack{\beta_d \perp \hat{\beta}_1, \ldots, \hat{\beta}_{d-1} \\ \|\beta_d\|=1}} \left\{ \sum_{j=1}^{n} \hat{\sigma}^2_{\hat{B}_{d-1},\beta_d}(\hat{\beta}_1^{\mathrm{T}}X_j, \ldots, \hat{\beta}_{d-1}^{\mathrm{T}}X_j, \beta_d^{\mathrm{T}}X_j) \right\}$$

where $\hat{B}_{d-1} = (\hat{\beta}_1, \ldots, \hat{\beta}_{d-1})$. Then, the associated p-dimensional kernel can be taken as a product of p one-dimensional kernels. This intuitively makes sense by theorem 1 and lemma 1. Also, we may refine the kernel weights and determine the number D by the procedures described in Sections 2.1.2 and 2.2 respectively.

The example in Section 5.2 shows that the (refined) MAVE method is robust. It seems to us that it is robust against outliers in X-space because the local smoother puts lower weights on further X_js. If the outliers occur in Y-space the story may be different.

There are at least two obvious questions. One is the inference of the EDR directions, which involves the asymptotic normality of the $\hat{B}$. This is true for single-index models (Härdle *et al.*, 1993; Xia and Li, 1999). We believe that the $\hat{B}$ obtained by (refined) MAVE has $\sqrt{n}$-consistency and asymptotic normality under some regular conditions. The expression of the asymptotic covariance matrix of $\hat{B}$ could be complicated, and its consistent estimator is needed. This may be given by, say, a bootstrap method. Moreover, the estimation of the link function is also important. In particular, we may first ask whether the link function is additive (Cui *et al.*, 2001). Also, it is expected that the MAVE method may be extended to the case that X includes continuous as well as categorical (or, generally, discrete) or functionally related covariates, as mentioned in Section 3.4. Further work is definitely needed in this area.

Vladimir Spokoiny (*Weierstrass Institute and Humboldt University, Berlin*)

The authors discuss an excellent idea for solving the dimension reduction problem by minimizing the sum

$$\sum_{j=1}^{n}\sum_{i=1}^{n}[y_i - \{a_j + b_j^{\mathrm{T}}B^{\mathrm{T}}(X_i - X_j)\}]^2 w_{ij}$$

over all $p \times D$ matrices B fulfilling $B^{\mathrm{T}}B = 1$. Here w_{ij} are non-negative weights. The approach has genuine benefits compared with the existing methods like sliced inverse regression or average derivative estimation. The choice of the weights w_{ij} plays the central role in this method. The authors discuss two possibilities. The first is to apply the usual multidimensional kernel weights

$$w_{ij} = K_h(X_i - X_j) \Big/ \sum_l K_h(X_l - X_j).$$

This approach, similarly to the average derivative estimation or outer product of gradients methods, suffers from the *curse of dimensionality* problem. Indeed, even for the optimal choice of the bandwidth h, the accuracy of estimation of the effective dimension reduction space is very low if the dimensionality p is large. The refined weights

$$\tilde{w}_{ij} = K_h\{\hat{B}^{\mathrm{T}}(X_i - X_j)\} \Big/ \sum_l K_h\{\hat{B}^{\mathrm{T}}(X_l - X_j)\}$$

are based on the knowledge of the structure of the model and they allow us to obtain better accuracy of estimation corresponding to the problem of the reduced dimension. However, the refined weights proposal utilizes the estimator $\hat{B}$ which comes from the first-step estimation with the multidimensional weights. If this first-step estimator is not sufficiently precise then the advantage of using the refined weights disappears and the whole procedure may fail in estimating the true effective dimension reduction. Hristache, Juditski and Spokoiny (2001) and Hristache, Juditski, Polzehl and Spokoiny (2001) proposed another way of selecting the refined weights w_{ij} based on the idea of structural adaptation. The idea is to pass progressively from multidimensional weights w_{ij} to the low dimensional weights of type $\tilde{w}_{ij}$. In this context, an interesting question is the possibility of joining the proposal of this paper (to estimate the index space by minimizing the mean average squared error) with the structural adaptation method.

The following contributions were received in writing after the meeting.

K. S. Chan (*University of Iowa, Iowa City*) **and Ming-Chung Li** (*EMMES Corporation, Rockville*)
We congratulate the authors for their masterly piece of work that will certainly stimulate much research on semiparametric modelling and non-linear time series.

The authors considered the case of univariate responses. Interestingly, we have independently done some related work with multivariate responses. Li and Chan (2001) (and also Li (2000)) proposed the semiparametric reduced rank regression model

$$Y_t = C\, f(BX_t) + \varepsilon_t,$$

where Y_t and X_t are m- and n-dimensional componentwise standardized random vectors, ε_t is of zero mean and identical variance given the current X and past Xs and Ys, C and B are $m \times r_1$ and $r_2 \times n$ coefficient matrices and r_1 and r_2 are the ranks of the model. The unknown (link) function f maps from R^{r_2} to R^{r_1}. The model is unaltered on replacing C, $f(\cdot)$ and B by CP, $P^{-1}\ f(Q^{-1}\cdot)$ and QB for any two invertible matrices P and Q. So, identification requires constraining, for example, the leading subsquare matrices of C and B as identity matrices, after suitable permutations of the variables. We may interpret the r_1 components of $f(BX_t) = (f_1(U_{1,t}, \ldots, U_{r_2,t}), \ldots, f_{r_1}(U_{1,t}, \ldots, U_{r_2,t}))^{\mathrm{T}}$ as non-linear principal components which depend on the indices $BX_t = (U_{1,t}, \ldots, U_{r_2,t})^{\mathrm{T}}$. Li and Chan (2001) proposed an estimation procedure that resembles the minimum average variance estimation method for $m = 1$.

We now use the respiratory problem data to illustrate the semiparametric reduced rank regression model with some preliminary analysis of the dynamic structure of air pollution in Honk Kong. Let Y consist of (log-transformed) sulphur dioxide (S), nitrogen dioxide (N), (log-transformed) respirable suspended particulates (P) and (square-root-transformed) ozone (O); X consists of lags 1, 2 and 7 of the Y-variable and lags 0 and 1 of temperature (T) and humidity (H). From cross-validation, $r_1 = r_2 = 2$. B is estimated to equal (standard errors are given in parentheses; NA denotes 'not applicable')

S_{t-1}	N_{t-1}	P_{t-1}	O_{t-1}	S_{t-2}	N_{t-2}	P_{t-2}	O_{t-2}
−0.617	1	−0.011	0.523	0.038	−0.033	0.046	−0.146
(0.085)	(NA)	(0.104)	(0.122)	(0.087)	(0.117)	(0.099)	(0.083)
0.510	0	0.159	−0.121	−0.110	−0.057	0.036	0.034
(0.064)	(NA)	(0.076)	(0.079)	(0.070)	(0.087)	(0.074)	(0.061)

S_{t-7}	N_{t-7}	P_{t-7}	O_{t-7}	T_t	T_{t-1}	H_t	H_{t-1}
−0.136	0.036	0.104	0.084	0	0.210	−1.145	0.071
(0.064)	(0.085)	(0.087)	(0.060)	(NA)	(0.075)	(0.177)	(0.112)
0.120	0.018	−0.038	−0.047	1	−1.167	0.349	−0.179
(0.049)	(0.067)	(0.060)	(0.048)	(NA)	(0.056)	(0.093)	(0.074)

Here, the subsquare matrix corresponding to N_{t-1} and T_t is normalized as the identity matrix. Fig. 8 displays the smoothed graphs of the non-linear principal components $\hat{f}_i$ *versus* the indices u_1 and u_2. Whereas $\hat{f}_1$ seems linear, $\hat{f}_2$ appears to be piecewise linear. Below is the estimate of C and that after transformation that renders the two non-linear principal components uncorrelated and of unit variance:

$$\hat{C} = \begin{pmatrix} 0.000 & 1.000 \\ (\mathrm{NA}) & (\mathrm{NA}) \\ 1.000 & 0.000 \\ (\mathrm{NA}) & (\mathrm{NA}) \\ 1.065 & -0.189 \\ (0.038) & (0.063) \\ 0.977 & -0.890 \\ (0.050) & (0.097) \end{pmatrix}; \qquad \hat{C}_{\mathrm{rotated}} = \begin{pmatrix} 0.124 & 0.526 \\ (\mathrm{NA}) & (\mathrm{NA}) \\ 0.729 & 0.124 \\ (\mathrm{NA}) & (\mathrm{NA}) \\ 0.753 & 0.033 \\ (0.029) & (0.034) \\ 0.601 & -0.347 \\ (0.036) & (0.050) \end{pmatrix}$$

The Euclidean distance between any two rows of the rotated C measures the dissimilarity in the dynamics of the corresponding variables. The rotated $\hat{C}$ suggests that the sulphur dioxide variable enjoyed different dynamics from other variables whereas the suspended particulates and nitrogen dioxide variables shared similar dynamics, over the study period; see also Fig. 8.

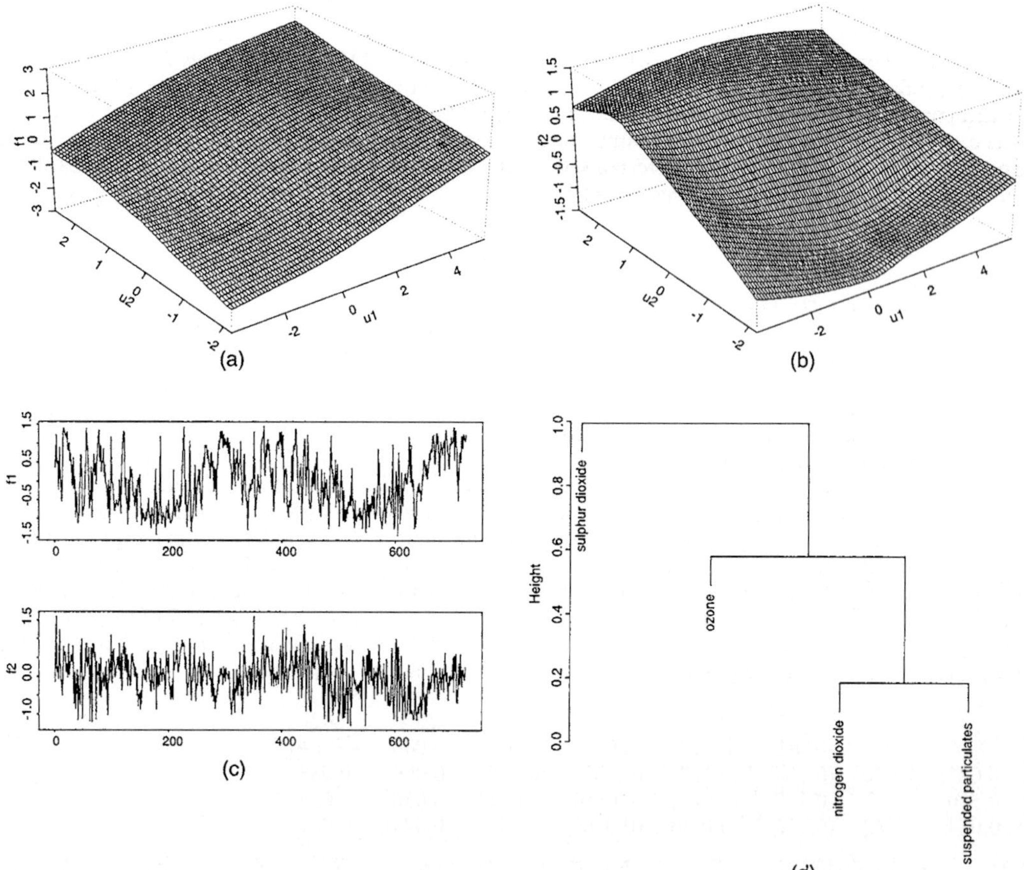

Fig. 8. (a) Smoothed graph of $\hat{f}_1$, (b) smoothed graph of $\hat{f}_2$, (c) time series plots of the two non-linear principal components and (d) dendrogram from a cluster analysis of the dynamics of the four pollution variables, based on $\hat{C}_{\mathrm{rotated}}$

Pavel Čížek and Wolfgang Härdle (*Humboldt University, Berlin*) **and Lijian Yang** (*Michigan State University, East Lansing*)

This paper addresses the challenging problem of dimension reduction and we congratulate the authors for this new insight into modelling high dimensional data. They provide the new minimum average variance estimation (MAVE) approach that creates a variety of semiparametric modelling strategies. The technical treatment is excellent and the algorithms derived are directly implementable. From a practitioner's point of view, there are probably questions about the performance of the method in non-standard situations.

For an assumed number of directions, the MAVE method is based on the local linear approximation of a regression function. The main idea is to use this approximation (conditionally on yet unknown indices) directly in the local linear smoothing procedure by using a multidimensional kernel. This is just a simultaneous minimization with respect to function and direction estimates, which is broader than the usual methods that estimate only function values or only directions. According to theorem 1, this makes undersmoothing of the bandwidth selection unnecessary. Additionally, MAVE together with a cross-validation procedure can be used to estimate the effective dimension reduction (EDR) dimension.

On the basis of MAVE, the authors design generalizations of several existing methods (e.g. the outer product of gradients (OPG) method is a generalization of additive derivatives estimation by Härdle and Stoker (1989)). Additionally, these extensions even outperform the original methods. However, we must keep in mind that these generalizations are valid only under assumptions on the smoothness of all the variables and cannot therefore replace the corresponding single- and multi-index methods that can also handle discrete variables (e.g. semiparametric least squares by Ichimura (1993)).

Finally, the MAVE method is claimed to be robust against outliers, supposedly in the space of explanatory variables. We examined the robustness of the choice of the EDR dimension and the OPG and MAVE methods to outliers and random noise in more detail. In the first case, our simulations regarding the cross-validation procedure in the presence of a single outlier show two main effects: the outlier results generally in an upwardly biased estimate of the EDR dimension, and additionally, in most cases, model estimates under contamination do not reduce the variance of the dependent variable conditionally on the regression function. In the second case, we studied the behaviour of MAVE and OPG under contamination. The most interesting result is that OPG, which for clean data is always worse than MAVE, can keep up with or even outperform MAVE when applied to contaminated data. We achieved similar results also under no contamination and a high variance of the error term.

R. D. Cook (*University of Minnesota, St Paul*)

The authors refer to span(B_0) from model (1.1) as the effective dimension reduction (EDR) subspace, but I find this characterization to be incorrect. Li (1991) defined the EDR subspace as the span(B) in the representation $y = g(B^{\mathrm{T}}X, \delta)$, where the error $\delta \perp\!\!\!\perp X$ and $B = (b_1, \ldots, b_k)$. Because ε may depend on X, equation (1.1) permits a model with $\varepsilon = \sigma(C_0^{\mathrm{T}}X)\delta$, where $\sigma(C_0^{\mathrm{T}}X) \geqslant 0$. For this version of model (1.1), the EDR subspace is span(B_0) + span(C_0), not span(B_0) as the paper implies. This confusion is unfortunate but perhaps understandable because published descriptions of the EDR subspace are not explicitly constructive.

A mean subspace is any subspace span(B) of $\mathbb{R}^p$ such that $y \perp\!\!\!\perp E(y|X)|B^{\mathrm{T}}X$. If the intersection of all mean subspaces is itself a mean subspace it is called the central mean subspace (CMS) and may be taken as the subject of a regression inquiry. Recently introduced by Cook and Li (2002), the CMS seems to be the subspace pursued in this paper.

A dimension reduction subspace (DRS) is any subspace span(B) such that $y \perp\!\!\!\perp X|B^{\mathrm{T}}X$. When the intersection of all DRSs is itself a DRS it is called the central subspace (CS; Cook (1996a, b, 1998)), which is a metaparameter for dimension reduction. The CS may not exist when the EDR subspace does exist. And the CS may exist straightforwardly when the construction of the EDR subspace is problematic (e.g. binary responses). I find the CS to be *much* easier to handle in theory and widely applicable in practice. The CMS is contained in the CS. The CS is invariant under strictly monotonic transformations of Y, whereas the CMS and span(B_0) are not. Compactness of the support of X is not required for the CMS or the CS (see the discussion following lemma 1).

I do not regard sliced inverse regression (SIR) and refined minimum average variance estimation (RMAVE) to be direct competitors. SIR estimates directions in the CS, whereas RMAVE apparently estimates the CMS. The authors demonstrate that RMAVE does better than SIR in some situations that RMAVE was designed to handle. I wonder how RMAVE would perform across the many situations where SIR, sliced average variance estimation and related methods have apparently uncovered key regression structures.

The fact that SIR will not perform well in models like model (4.3) is known (Cook and Weisberg, 1991). Does the performance of RMAVE degrade when there are strong non-linear relationships among the predictors, the kind that would render SIR ineffective?

I found this paper interesting because of the suggestion that local methods might mitigate the need for restrictions on the predictors.

Jianqing Fan (*University of North Carolina at Chapel Hill*)

Model parameters and identifiability

The basic assumption of the paper is that model (1.1) holds. In practice, it is at best an approximation. In general, following Fan *et al.* (2001), the parameters $\mathbf{B}_0$ and the function g can be defined as the minimizer of

$$\min_{g\mathbf{B}:\mathbf{B}^{\mathrm{T}}\mathbf{B}=I_D} [E\{Y - g(\mathbf{B}^{\mathrm{T}}\mathbf{X})\}^2] = \min_{\mathbf{B}:\mathbf{B}^{\mathrm{T}}\mathbf{B}=I_D} [E\{Y - E(Y|\mathbf{X}^{\mathrm{T}}\mathbf{B})\}^2].$$

This is the same as expression (2.1). Hence, the model assumption (2.1) is not needed as far as the procedure for estimating $\mathbf{B}_0$ and g is concerned. Under what conditions does the optimization problem (2.1) have a unique solution, namely when is the parameter $\mathbf{B}_0$ identifiable? (Indeed, only the space spanned by the columns of $\mathbf{B}_0$ is possibly identifiable.)

The identifiability condition is necessary for asymptotic results to hold. To elaborate the identifiability issue, consider the model studied by Fan *et al.* (2001):

$$Y = \sum_{j=0}^{p} g_i(\mathbf{B}^{\mathrm{T}}\mathbf{X})X_j + \varepsilon$$

with $X_0 = 1$. Consider the specific case where $D = 1$ and write $\mathbf{B} = \boldsymbol{\beta}$. When $g_j(x) = \alpha_j x$ with $\alpha_0 = 0$, this model becomes

$$Y = (\boldsymbol{\alpha}^{\mathrm{T}}\mathbf{X})(\boldsymbol{\beta}^{\mathrm{T}}\mathbf{X}) + \varepsilon,$$

where $\boldsymbol{\alpha} = (\alpha_1, \ldots, \alpha_p)^{\mathrm{T}}$. When they are not parallel, the parameters $\boldsymbol{\alpha}$ and $\boldsymbol{\beta}$ are not identifiable for $D = 1$. This is the only case where the parameters are not identifiable for $D = 1$, following theorem 1 of Fan *et al.* (2001). This case does not appear in model (1.1), since the authors implicitly assume that $g(B_0^{\mathrm{T}}\mathbf{X}) = E(Y|B_0^{\mathrm{T}}\mathbf{X})$.

Minimum average variance estimation and profile likelihood

The profile likelihood is commonly used to estimate parameters and nonparametric functions in semiparametric models. The basic idea, in the current context, is to estimate the function g for a given $\mathbf{B}$ by using a nonparametric approach, resulting in an estimator $\hat{g}_{\mathbf{B}}(\cdot)$. Now, find the parameter $\mathbf{B}$ to minimize

$$\sum_{i=1}^{n} \{Y_i - \hat{g}_{\mathbf{B}}(\mathbf{B}^{\mathrm{T}}\mathbf{X}_i)\}^2.$$

The fully iterated procedure in Carroll *et al.* (1997) used this idea. Minimum average variance estimation is a nice variation of the profile likelihood method. It is motivated from estimating the conditional variance by a kernel estimator rather than minimizing directly the mean-square errors. As a result, it has the nice expression (2.7) which facilitates theoretical studies but involves an extra loop of summation in computation. The merits of both approaches are worth exploring further. However, it is worthwhile to mention that the profile likelihood method generally gives semiparametric efficient estimators (see, for example, Carroll *et al.* (1997) and Murphy and van der Vaart (2000)). Whether minimum average variance estimation has this kind of optimality remains to be seen. Two procedures share at least one merit in common: no undersmoothing is needed for estimating parametric components (Carroll *et al.* (1997) and theorem 1 of the present paper). In fact, the criteria that the two procedures optimize are approximately the same.

Expression (2.7) is somewhat informal, since its minimization with respect to $\mathbf{B}$ is not unique though its effective dimension reduction is. Could the authors therefore explain how problem (2.7) is minimized and clarify the convergence criterion in Section 2.3?

L. Ferré (*University of Toulouse le Mirail*)
The paper is interesting since it substitutes local linear smoothing for inverse regression for estimating the effective dimension space. The main advantage of the method over inverse regression is that condition (1.2) is relaxed, allowing applications to time series. Even if my own experience of the application of sliced inverse regression in times series is quite positive, time reversibility is indeed an awkward condition derived from equation (1.2). However, an argument in favour of inverse regression is simplicity: estimates of the effective dimension reduction space are deduced from a simple eigenvalue decomposition of a matrix independently from g. This feature allows in particular extensions to functional data (see for example Dauxois *et al.* (2001)). This necessary reduction of the dimension (recall the goal: overcome the 'curse of dimensionality') comes before (and independently of) the nonparametric estimation of g. For deriving this dimension, tests have been proposed, relying, in the original papers, on distributional assumptions. These assumptions can be removed since recent unpublished work has shown that the existence of the first four moments is sufficient. An alternative is to use a model selection approach based on the distance between $S(B_0)$ and $S(\hat{B}_d)$ by letting d vary (Ferré, 1998). The main idea is that a working dimension that is lower than the 'true' dimension D can be preferable and the distance between $S(\hat{B}_d)$ and a d-subspace of the unknown $S(B_0)$ is finally used. Simple estimates of this criterion have been proposed for elliptically distributed explanatory variates but also for the general case by using the bootstrap or jackknife (see Ferré (1997, 1998)). Local linear smoothing intends to estimate at the same time the regression function and the effective dimension reduction space. The price to pay is that more local linear smoothing is needed than covariates are included in the model. For the dimensionality a global model selection approach is considered, but cross-validation, in addition to the high computational cost, does not avoid the curse of dimensionality. Indeed, $\hat{a}_{d0,j}$ is the Nadaraya–Watson estimator which may perform poorly for large values of d and my feeling is that overparameterization is to be feared.

Ker-chau Li (*University of California at Los Angeles*)
The dramatic improvement of the methods proposed over sliced inverse regression (SIR) and the principal Hessian directions method for the three examples deserves some non-asymptotic explanations. For $n = 200$ and $p = 10$, it is difficult to tell why the nice asymptotic theorems are relevant. For the first two examples, a simple explanation goes like this. First, least squares regression is known to be consistent in finding an effective dimension reduction direction (Brillinger, 1983; Li and Duan, 1989) under condition (1.2). It is straightforward to extend this result to weighted least squares regression provided that the weight function depends on (y, x) only through $(y, B_0^{\mathrm{T}} X)$. Now because equation (2.6) is basically a weighted least squares regression, one can prove that, for the population version of equation (2.6), $b^{\mathrm{T}} B^{\mathrm{T}}$ should be in the effective dimension reduction space. If condition (1.2) does not hold, then the result may be biased and an upper bound of bias can be evaluated (Duan and Li, 1991; Li, 1997). Problem (2.7) amounts to averaging over a number of weight functions. Averaging may help the cancellation of bias in the time series context.

For fairness, I would like to point out that weighted versions of SIR and similar procedures have been proposed before to temper the bias problem; see the discussion and rejoinder in Li (1991). It is worth pointing out the difference between condition (1.2) and elliptical symmetry (Hall and Li, 1993). Also SIR and principal Hessian directions can be applied to residuals after deterministic components have been taken out. Iteration does improve the results. However, the issue of non-linear confounding (Li, 1997) sets a limitation that is difficult to bypass by any procedure. It is not clear to me whether the new approach can do anything about it.

For brevity, I shall not go over the long list of clever ideas that I found interesting in this path breaking work by the authors. Let me close by noting that they did not compare their procedure with projection pursuit regression. A dozen years ago when I submitted my SIR paper to the *Journal of the American Statistical Association*, the Associate Editor recommended rejection because he or she thought that SIR was not as good as projection pursuit regression. Luckily my paper was salvaged by the Editor, who allowed me to explain the difference between the two approaches. Apparently the authors have done more than enough to convince the reviewers just as they have convinced me!

Lexin Li (*University of Minnesota, St Paul*)
Adopting the notation in model (1.1) and following the definitions of the central mean subspace (CMS) (Cook and Li, 2002), the minimum average variance estimation (MAVE) methods seem to pursue the CMS only. To confirm this, simulations were done on models of the form $y = g(B_1^{\mathrm{T}} X) + h(B_2^{\mathrm{T}} X)\varepsilon$, where g and h are both unknown functions, ε is independent of X and $E(\varepsilon) = 0$. My results indicate that MAVE

methods can successfully estimate B_1 in the mean structure $E(y|X)$, whereas they always miss B_2 in the error structure.

Refined MAVE (RMAVE) does not require sliced inverse regression's (Li, 1991) linearity condition. Simulations were done to examine the performance of RMAVE when there are strong non-linear relationships among the predictors X. I considered one-dimensional models only, where $B \in \Re^p$. The results show that RMAVE has good performance for one-dimensional models when the non-linearity in X is strong.

Under the assumption $D = 1$, however, there is still room for improvement, compared with RMAVE, to estimate the underlying true direction without the requirement of the linearity condition. Cook and Nachtsheim (1994) suggested a co-ordinatewise reweighting approach to remove the non-linearity in X and to make X elliptically contoured. I have been investigating the possibility of extending the idea of removing the non-linearity in X by clustering on X-space as the first step. An ordinary least squares (OLS) estimate is obtained from each cluster, and all those estimates are combined to estimate the true direction. Intuitively, the clusterwise OLS method works because non-linearity in X is broken and within each cluster the linearity condition should hold approximately. Then the Li–Duan proposition (Li and Duan (1989), theorem 2.1, and Cook (1998), proposition 8.1) is applicable within each cluster. I also consider an iterative version of the algorithm, which obtains the estimate by iteratively clustering on $\hat{B}_i^{\mathrm{T}} X$, where $\hat{B}_i$ is the estimate from the ith iteration. Simulations show that the OLS estimate with clustering achieves a better performance than RMAVE. As an example, consider the model $x_1 \sim$ uniform(0,1) and $x_2 = \log(x_1) + e$, where $e \sim$ uniform$(-0.3, 0.3)$, and $y = \log(x_1) + \varepsilon$, where $\varepsilon \sim N(0, 0.01)$. The actual direction is $B = (1, 0)^{\mathrm{T}}$. With 100 observations, RMAVE gives an estimate of $\hat{B} = (0.991, 0.133)^{\mathrm{T}}$ with the angle to B equal to 7.626°, whereas OLS with five clusters produces $\hat{B} = (0.999, 0.038)^{\mathrm{T}}$ with the angle to B equal to 2.196°. Here the number of clusters, 5, is chosen before we see the computational results, to make the comparison fair. Details of this work will be reported elsewhere.

Oliver Linton (*London School of Economics and Political Science*)
This is a comprehensive paper. I shall just focus on the new implementation of Ichimura's semiparametric least squares method for estimating index models. In expression (A.1) the authors sequentially minimize

$$\sum_{j=1}^{n}\sum_{i=1}^{n} \{y_i - a_j - b_j\beta^{\mathrm{T}}(X_i - X_j)\}^2 w_{ij}$$

with respect to (a, b, β) holding w_{ij} constant and starting from some initial consistent estimator $\tilde{\beta}_0$. The Ichimura (1993) procedure involves sequential minimization with the difference that he uses only local constant but also includes the dependence of w_{ij} on β; this leads to a nasty non-linear optimization problem, whereas the authors' procedure is just bilinear least squares, and so is conditionally linear. They apparently prove that after two iterations their $\tilde{\beta}$ behaves as if (a, b) were known in expression (A.1). I think that this is an important idea that will make estimation of these models much easier. The authors develop many useful tools and apply them impressively. I have some comments and questions.

The initial consistent estimator that lurks in Appendix A.2 is either the average derivative estimator (in which case the criticisms in (a) and (b) of the second page apply) or some non-linear least squares estimator, which itself will be heavily computational.

I suppose that the authors' estimator achieves the semiparametric efficiency bound in for example the special case of Appendix A.2 with independent and identically distributed ε, but it is not so clear to me.

In time series, we come across special sorts of indices like $\Sigma_{k=0}^{\infty} \beta^k X_{t-k}$, where β is unknown; this would generalize the linear model $y_t = \beta y_{t-1} + \gamma X_t + \varepsilon_t$ that is widely used. Have the authors thought about this case?

I do not think that the optimal amount of smoothing for the function will always be the same as the optimal amount of smoothing for the parameter. Generally speaking it seems that in 'adaptive' cases the optimal bandwidth for the parameter and the function have the same magnitude, although not the same constant. See for example Carroll and Härdle (1989). In non-adaptive cases this is not usually so. In the partially linear model $y = \beta_x + g(z) + e$, Linton (1995) showed that the Robinson (1998) estimator $\hat{\beta}$ for β has expansion $\hat{\beta} - \beta = O_p(n^{-1/2}) + O_p(h^4) + O_p(n^{-1}h^{-1/2})$ under twice continuous differentiability of g, which suggests an optimal bandwidth rate of $h \propto n^{-1/9}$, i.e. it is optimal to undersmooth. Although maybe the authors can find an estimator of β that has the optimal bandwidth rate of $h \propto n^{-1/5}$.

Liqiang Ni (*University of Minnesota, St Paul*)
I applaud the authors for the promising refined minimum average variance estimation (RMAVE) algorithm and the intriguing idea of determining the dimension in a cross-validation approach. Many methods have been proposed to estimate directions in the effective dimension reduction space (Li, 1991), or the central subspace (Cook, 1996). Sliced inverse regression (SIR) can discover directions of linear terms in mean functions but fails in symmetric situations like $y = (\beta^T X)^2 + \varepsilon$ with X normal, $E(X) = 0$ and $\varepsilon \perp\!\!\!\perp X$, where the direction can be detected by sliced average variance estimation (Cook and Weisberg, 1991). In my experience, RMAVE can estimate both linear and quadratic terms well.

Suppose that we have a continuous predictor $X \in R^p$ and a categorical predictor $C \in R$ representing different subpopulations. If the mean function of Y does have a form as

$$g\left\{\beta^T \begin{pmatrix} X \\ C \end{pmatrix}\right\},$$

which may indicate shifts between subpopulations, RMAVE can be practically useful under the circumstances described by the authors. However, when $y = G_C(\beta_C^T X) + \varepsilon$, so each subpopulation may have its own unique directions and functions, mixing continuous and categorical predictors may be inappropriate. Partial SIR (Chiaromonte *et al.*, 2002) directly addresses this issue. In the same spirit, we may consider 'partial RMAVE'. One way to do this may be simply to let the weight w_{ij} in expression (3.8) multiply an indicator function $I(C_i = C_j)$ and modify the cross-validation (CV) function as well. Details of this approach, which seems to work quite well, will be reported elsewhere.

The selection of the bandwidth seems tricky. The estimation of dimension is much more stable when CV adopts the Nadaraya–Watson estimator than when using a local linear estimator. Neverthless, it is still sensitive to the bandwidth. I applied RMAVE to the AIS data (Chiaromonte *et al.*, 2002) which consist of a mixture of two linear regressions determined by the only categorical predictor—gender. Considering only continuous predictors, the Nadaraya–Watson CV values suggested two dimensions with larger bandwidth and only one dimension with smaller bandwidth. The partial RMAVE method described as above, however, suggested one dimension consistently, which confirmed that both linear regressions associate with the same direction, $y = G_C(\beta^T X) + \varepsilon$.

I have a question about inverse MAVE. The essence of SIR is that, under the linearity condition (1.2), the space spanned by $E(Z|Y)$ where $E(Z) = 0$ and $\text{cov}(X) = I$ is a subset of the EDR space. To estimate this space, Li (1991) proposed slicing on Y, and Zhu and Fang (1996) proposed kernel methods. I am not sure whether inverse MAVE is intended to estimate span$\{E(Z|Y)\}$ also.

Megu Ohtaki and Yasunori Fujikoshi (*Hiroshima University*)
We praise the authors of this paper, which has a highly original and fascinating content. The paper is sure to be one of the monumental works in the field of multivariate analysis.

In the paper it is clearly shown that the minimum average variance estimation (MAVE) method and its algorithm have many advantages over existing methods for searching an effective dimension reduction (EDR) space. Just like the sliced inverse regression method, however, no description for the reduction in the number of the original covariables was given. It is also important to consider selection of the original variables as well as the covariables $\beta_1^T X, \ldots, \beta_p^T X$. In practical situations of data analysis, a model with a small number of original covariables is preferable while the bias is negligible. This problem may be formulated mathematically as below.

Suppose, for example, in model (1.1)

$$y = g(B_0^T X) + \varepsilon,$$

where B_0 and X are decomposed as

$$B_0 = \begin{pmatrix} B_{01} \\ B_{02} \end{pmatrix}_{p\times D}, \qquad X = \begin{pmatrix} X_1 \\ X_2 \end{pmatrix}_{p\times 1},$$

and hence $B_0^T X = B_{01}^T X_1 + B_{02}^T X_2$. If $B_{02} = O$, then it is expected by analogy (Akaike, 1973; Mallows, 1973) for cases of linear regression that we shall be able to have a more efficient EDR.

For not only such a mathematical background but also economical reasons, those covariables which have no effect on the response should not be used in regression analysis. Therefore, we propose the regression model

$$y = g(B^T X) + \varepsilon,$$

where $Q \subset \{1, \ldots, p\}$ and $D_Q = \text{diag}(q_1, \ldots, q_p)_{p \times p}$, with $q_i = 1$ if $i \in Q$ and $q_i = 0$ otherwise, for selecting the optimal model, and to choose a model attaining $\min_{d,Q}\{\text{CV}(d, Q)\}$ that will be constructed by modifying the cross-validation criterion, CV(d), which is given in the paper. Thus the MAVE method may be extended easily to reduce the number of the original covariables as well as the dimension of an EDR space simultaneously. Furthermore, the MAVE method has the advantage that it may be generalized to multivariate regression.

In linear statistical inference, it has been reported that the model selection method using Akaike's information criteria AIC is not consistent for estimating the true model (see, for example, Shibata (1976) and Fujikoshi (1985)). Stone (1974) showed that the cross-validation criterion and AIC are asymptotically equivalent for model selection. Given these results, we wonder whether theorem 2 is consistent with the classical results.

James R. Schott (*University of Central Florida, Orlando*)

Over the past decade, there has been a considerable amount of work on dimensionality reduction techniques in the regression setting. This paper represents a substantial contribution to that area. I have just a couple of minor comments relating to the sliced inverse regression (SIR) procedure of Li (1991) and subsequent similar types of procedure such as the sliced average variance estimate of Cook and Weisberg (1991).

The linear condition given in equation (1.2) is a fundamental requirement for most of these procedures. Additional assumptions may be needed; for instance, sliced average variance estimation requires a constant variance assumption, and inferential methods, associated with these procedures, for determining the correct dimension often require stronger conditions. These additional assumptions are certainly restrictive, but it is important to note that equation (1.2) is a fairly mild condition. It is weaker than elliptical symmetry because it only has to hold for the directions B_0. Thus, we may not have elliptical symmetry but be sufficiently lucky still to have condition (1.2) hold. In fact, Hall and Li (1993) have shown that, loosely speaking, if the dimension of X is high, then it is likely that condition (1.2) holds at least approximately.

A further point to note is that procedures like SIR estimate a space that may be a proper subspace of the space spanned by the columns of B_0. Have we missed any important directions? If so, how do we recover them? These are questions that may need to be answered when using SIR. However, they are not relevant questions for the adaptive procedures proposed here since they directly estimate the space spanned by the columns of B_0.

C. M. Setodji (*University of Minnesota, St Paul*)

We have been presented with a constructive and useful paper and the authors are to be congratulated. Minimum average variance estimation (MAVE) seems to be an interesting and intriguing method for dimension reduction estimation. Equation (1.1) is applicable to any regression problem since, for any Y and X, we can always define $\varepsilon = Y - E(Y|X)$ which depends on X and satisfies the conditions in the paper. I have applied MAVE to three well-known sets of data that have been studied in the dimension reduction literature, and the optimal bandwidth was used throughout. Background on the examples was given by Cook and Critchley (2000). In all three examples, MAVE fails to produce the directions obtained by other methods.

First the methods proposed were applied to the bank-note data. With a binary response (the bank-note's authenticity) and six predictors, all the information in the regression is contained in the mean function. The refined MAVE method gave $\hat{d} = 21$, which is the same as the result produced by sliced average variance estimation (SAVE) (Cook and Critchley, 2000; Chiaromonte *et al.*, 2002) and projection pursuit analysis (Posse, 1995). Whereas the first MAVE and SAVE directions are essentially the same, the second directions are quite different. The second SAVE direction shows two kinds of forged notes, but the role of the second MAVE direction is unclear. It misses the clustering in the counterfeit notes.

We also applied MAVE to the Hawkins data, designed to challenge traditional and robust regression methods with outliers. Although the data with four covariates and a continuous response have two directions in the mean function, refined MAVE and inverse MAVE suggest independence whereas the outer products of gradients method suggests only one direction. SAVE correctly identifies the regression structure. Lastly, the method was applied to the AIS data, a data set with mixtures. MAVE gave $\hat{d} = 1$, suggesting one direction, whereas sliced inverse regression infers $\hat{d} = 2$. MAVE evidently missed the 'joining information' for males and females.

Many regression problems are filled with 'mixtures' which is the one thing that all these data sets have in common. Mixtures increase the dimension of the mean function. My experience suggests that the MAVE methods fail to detect mixture regressions. Is it possible to enhance the proposed method to face such an issue?

Finally, for me, one of the weaknesses of the method proposed is the fact that it is not invariant under linear transformations. Using (x_1, x_2) or $(x_1 + x_2, x_2)$ as predictors may yield different first directions when $d = 1$. More developments need to be pursued for these methods.

Nils Chr. Stenseth and Ole Chr. Lingjærde (*University of Oslo*)

Lynx populations undergo regular density cycles all across the boreal forest of Canada (see, for example, Stenseth *et al.* (1998)). In a previous analysis of the lynx dynamics (Stenseth *et al.*, 1999) two competing hypotheses were put forward regarding the spatial structure of the dynamics. One predicts that the dynamical structure clusters into groups defined according to ecological-based features, whereas the other predicts that it clusters into groups according to climatic-based features. On the basis of an analysis of 21 time series from 1821 onwards, Stenseth *et al.* (1999) found evidence in support of the latter hypothesis, assuming a piecewise linear autoregressive model for each population. However, their model did not explicitly include any climatic effects.

Here, we propose to use the authors' minimum average variance estimation (MAVE) methodology to study the spatial structure of the Canadian lynx populations, on the basis of a more general nonparametric model of the dynamics that includes as a covariate the potentially important climatic variable known as the North Atlantic oscillation winter index. Specifically, let L_t^s denote the natural logarithm of the abundance of lynx in region s in year t, and let NAO_t denote the North Atlantic oscillation winter index in year t. For each s and t define the response $y_t^s = L_t^s$ and the vector of covariates

$$\mathbf{X}_t^s = (L_{t-1}^s, L_{t-2}^s, L_{t-3}^s, L_{t-4}^s, \mathrm{NAO}_t, \mathrm{NAO}_{t-1}, \mathrm{NAO}_{t-2})^{\mathrm{T}}.$$

For each region s we assume the model

$$y_t^s = g_s(\mathbf{B}_{s,0}^{\mathrm{T}} \mathbf{X}_t^s) + \varepsilon_{s,t} = g_s(\beta_{s,1}^{\mathrm{T}} \mathbf{X}_t^s, \ldots, \beta_{s,d}^{\mathrm{T}} \mathbf{X}_t^s) + \varepsilon_{s,t}$$

where g_s is an unknown smooth link function, $\mathbf{B}_{s,0} = (\beta_{s,1}, \beta_{s,2}, \ldots, \beta_{s,d}) \in \mathbb{R}^{7,d(s)}$ is an orthogonal matrix and $E(\varepsilon_{s,t}|\mathbf{X}_s^t) = 0$ almost surely. Using refined MAVE and cross-validation, we estimated $d(s)$ and $\mathbf{B}_{s,0}$ for each s. To compare the dynamics in two regions s and s' we considered the largest principal angle $\varphi(s, s')$ between the subspaces spanned by the columns of $\mathbf{B}_{s,0}$ and $\mathbf{B}_{s',0}$ respectively. This angle can be determined from the relationships $0 \leqslant \varphi(s, s') \leqslant \pi/2$ and $\sin\{\varphi(s, s')\} = ||\mathbf{B}_{s,0}\mathbf{B}_{s,0}^{\mathrm{T}} - \mathbf{B}_{s',0}\mathbf{B}_{s',0}^{\mathrm{T}}||_2$. See Fig. 9 for results when $d(s)$ is estimated by cross-validation. Note that rows and columns are permuted to obtain coherent blocks of similar dynamics.

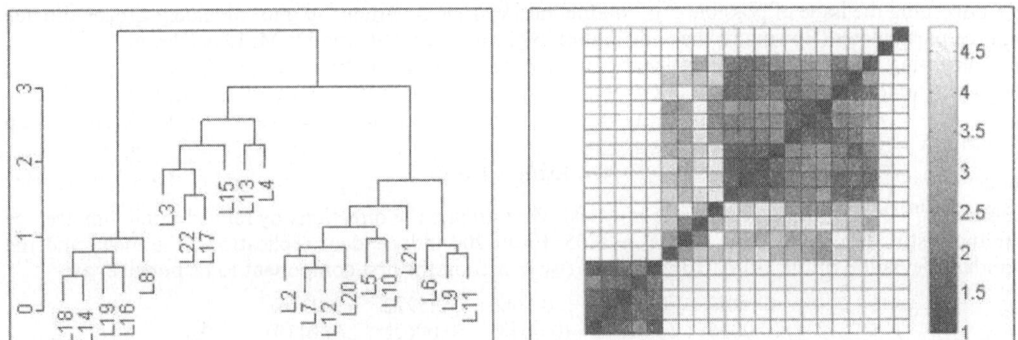

Fig. 9. Comparison of dynamic structures across Canada, using cross-validation estimates for the orders $d(s)$ (the comparison is based on the largest principal angles between the estimated reduction subspaces for each region): (a) average linkage hierarchical clustering of the 21 time series; (b) pseudocolour checker-board plot of distances (the plotted values are non-linearly scaled as $\exp\{\varphi(s, s')\}$ to accentuate the regions of similar dynamics; order of regions (from left to right) with two major clusters emphasized, **L18, L19, L16, L8, L14**, L3, L22, L17, L15, **L2, L7, L12, L20, L6, L9, L11, L5, L10, L21**, L13, L4)

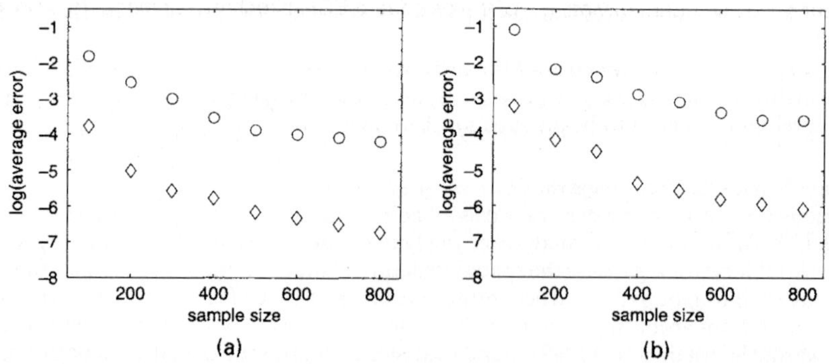

Fig. 10. (a) Parametric estimation and (b) nonparametric estimation (◇, results with uncorrelated design; ○, results for designs with functional relationships)

The results are strikingly similar to what we proposed as the ecological region structuring, and there is no strong support for the climatic region structuring, the latter of which was concluded to be the most appropriate region by Stenseth *et al.* (1999). To understand the underlying reasons for these differences certainly requires further work, both on the ecological and on the statistical side—work that we would like to pursue.

The **authors** replied later, in writing, as follows.

The extraordinarily kind words from so many distinguished discussants have overwhelmed us. We thank all the discussants for their constructive remarks and stimulating questions. Limitations of time and space prevent us from answering every question raised. Moreover, some of the suggestions will keep us busy for a while!

We thank Professor Kent for pointing out possible connections with other areas. His point regarding reduced rank models is clearly related to Chan and M. Li's important contribution. Turning to partial least squares, one of us has studied a nonparametric partial least squares regression after transformation. For data (y, X), a spline transformation $G(\cdot)$ of the response y is carried out so that the partial least squares regression can be modelled without knowing the exact form of $G(\cdot)$. Readers can refer to Zhu (2002) for more details. The basic idea is to 'linearize' a smooth function $G(\cdot)$ of the response y by $\pi(\cdot)^{\mathrm{T}}\theta$, where $\pi(\cdot)$ is a vector of B-spline basis functions of y and θ is an unknown projection parameter.

Concerning the issue of possible confounding between the covariates sulphur dioxide, nitrogen dioxide and the particulates (Bowman), the contribution by Professor Chan and Dr M. Li is relevant.

Concerning the challenging non-linear confounding problem mentioned by Professor K. C. Li, let us study the model used in Li (1997). Let $u_1 \sim \text{uniform}(0, 1)$, $u_2 = \log(u_1) + e$ with $e \sim \text{uniform}(-0.5, 0.5)$; $u_3, u_4, u_5 \sim^{\mathrm{IID}} N(0, 1)$ and $x_1 = u_1 + u_3$, $x_2 = u_2 + u_4 + u_5$, $x_3 = u_3 - u_4$, $x_4 = u_4$ and $x_5 = u_5$. A relationship of y with $X = (x_1, \ldots, x_5)^{\mathrm{T}}$ via u_1 is

$$y = \log(u_1) + 0.1\varepsilon, \tag{1}$$

where $\varepsilon \sim^{\mathrm{IID}} N(0, 1)$. The sample size $n = 100$. We estimate the directions by refined minimum average variance estimation (RMAVE) with $h = 0.05$. From 200 independent replications, the mean and the standard deviation of the estimated directions (we constrain the first component to be positive) are

$$\begin{array}{ccccc} (0.5662 & 0.0311 & -0.5660 & -0.5972 & -0.0316)^{\mathrm{T}}. \\ (0.0046) & (0.0107) & (0.0043) & (0.0067) & (0.0119) \end{array}$$

Because $u_1 = (1, 0, -1, -1, 0)^{\mathrm{T}} X$, the true direction is $(0.5774, 0, -0.5774, -0.5774, 0)^{\mathrm{T}}$. Our estimation results are quite encouraging especially since the structure of model (1) can hardly be detected by any of the other procedures. See for example Li (1997).

We agree with Professor Kent and Professor Bowman that the issue of collinearity is important. With a large set of near collinear covariates, some prescreening is recommended using such devices as principal components and others. Our limited simulations suggest that the MAVE method can still give some useful information when there is strong collinearity of functional relationships between covariates. Here we report the simulations for the model

Table 8. Means (and standard deviations) of the estimated EDR directions for model (5) with a sample size 200 and 100 replications†

β_1	(0.4538	0.4387	0.4467	0.4443	−0.0041	0.0013	0.0242	0.0067	0.0193	$0.0128)^T$
Standard deviation	(0.0985	0.0988	0.1089	0.0969	0.0765	0.1025	0.1973	0.1900	0.1815	$0.1961)^T$
β_2	(0.0106	−0.0033	0.0223	0.0127	0.0016	0.0071	0.3983	0.3833	0.3765	$0.3428)^T$
Standard deviation	(0.2290	0.2300	0.2435	0.2520	0.1339	0.1511	0.2063	0.1963	0.1869	$0.2223)^T$

†$h = 0.6$ was used.

$$y = (\beta_1^T X)^2 (a + \beta_2^T X) + \varepsilon, \tag{2}$$

where $a = 1$, $\beta_1 = (-\frac{2}{3}, \frac{1}{3}, \frac{2}{3}, 0)^T$, $\beta_2 = (1, 0, 1, 0)^T/\sqrt{2}$ and $X = (x_1, x_2, x_3, x_4)^T$. Two cases are considered:

(a) an uncorrelated design, $x_1, x_2, x_3, x_4, \varepsilon \sim^{IID} N(0, 1)$, and
(b) a design with functional relationships, $x_3 = (2x_1 + 2x_2 + \varepsilon_1)/3$, $x_4 = \{\text{sgn}(x_1)|x_1|^2 + \varepsilon_2\}/2$ and $x_1, x_2, \varepsilon_1, \varepsilon_2, \varepsilon \sim^{IID} N(0, 1)$.

We estimate model (2) under respectively the nonparametric setting and the non-linear parametric setting. With different sample sizes and bandwidths 0.6, 0.5, 0.45, 0.4, 0.35, 0.3, 0.28 and 0.25, results for the parametric estimators (obtained with the SAS software) and RMAVE estimators are shown in Fig. 10, where the error is defined as $m^2(\hat{\beta}_1, B_0) + m^2(\hat{\beta}_2, B_0)$ with $B_0 = (\beta_1, \beta_2)$. It is clear that both methods suffer from functional relationships between covariates. The relative degradation of efficiency for RMAVE due to collinearity and functional relationships between covariates is similar to that for the parametric case.

Our remark on the apparent robustness, based on our experience with MAVE, has somewhat to our surprise aroused substantial interest among the discussants (Critchley, Atkinson, Cui, G. Li, Yao, Čížek, Härdle, Yang and Welsh). The issue is important but we have as yet no theoretical results to offer.

We take Professor Cook's point about effective dimension reduction (EDR), a name which we adopted only after a suggestion from a referee. We also thank Professor Cook (and Professor Critchley) for clarifying the differences between the central subspace and the central mean subspace and their roles in the sliced inverse regression and the RMAVE methods. Professor Cook, Professor Critchley, Dr L. Li, Dr Schott and Dr Velilla raise concerns about heteroscedastic variance and wonder whether RMAVE can detect directions in the variance specification. If the conditional mean and the (not necessarily homogeneous) noise are additive, a two-step procedure may be adopted as follows. MAVE is first used to search the directions in the conditional mean and then applied to the squares of the residuals to look for the other directions. An alternative approach is as follows. Suppose that

$$y = g(B_0^T X, \varepsilon). \tag{3}$$

Some of the EDR directions will be ignored if only the usual conditional mean is investigated. For any values δ and Δ, the data $(X_j + \delta, |y_j - \Delta|)$ are from the following model which has the same EDR space:

$$|y - \Delta| = E[|g^{\delta,\Delta}\{B_0^T(X + \delta), \varepsilon\}| |B_0^T X] + \eta^{\delta,\Delta} \tag{4}$$

with $g^{\delta,\Delta}$ denoting some measurable function, where

$$\eta^{\delta,\Delta} = |g^{\delta,\Delta}\{B_0^T(X + \delta), \varepsilon\}| - E[|g^{\delta,\Delta}\{B_0^T(X + \delta), \varepsilon\}| |B_0^T X]$$

with $E(\eta^{\delta,\Delta}|X) = 0$. By choosing Δ appropriately, the conditional mean of model (4) can detect the other EDR directions. To avoid the difficulty of choosing Δ, we may use several of them together. For the following model, we consider three different pairs of (δ, Δ) and then we have four samples $\{(X_i, y_i)\}$, $\{(X_i + \delta_k, |y_i - \Delta_k|)\}$, $k = 1, 2, 3$. We re-denote them as $\{(X_{ki}, y_{ki})\}$, $k = 1, 2, 3, 4$. Using MAVE, for each sample we have from problem (2.7) a double summation to look for B:

$$S_k(B) = \sum_{j=1}^{n} \sum_{i=1}^{n} \{y_{ki} - b_{kj}^T B^T (X_{ki} - X_{kj})\}^2 w_{k,ij},$$

$k = 1, 2, 3, 4$. The common thing in these double summations is the direction matrix B. To find B, we can minimize $S_1(B) + S_2(B) + S_3(B) + S_4(B)$. We illustrate this approach with the model

Table 9. Means of estimation error $m^2\{(\beta_1, \beta_2), \hat{\beta}_1\}, m^2\{(\beta_1, \beta_2), \hat{\beta}_2\}$ for model (6) based on different algorithms†

Method	*Means of estimation errors for various bandwidths or spans*		
PPR (S-PLUS)	[0.4] (0.3459, 0.2876)	[0.5] (0.2997, 0.2613)	[0.6] (0.3707, 0.2776)
RMAVE (additive)	[0.2] (0.0415, 0.0355)	[0.3] (0.0088, 0.0170)	[0.4] (0.0214, 0.0212)
RMAVE (non-additive)	[0.3] (0.0305, 0.0516)	[0.4] (0.0481, 0.0731)	[0.5] (0.1104, 0.0586)

†Bandwidths or spans are given in square brackets.

$$y = \exp\{-2(\beta_1^{\mathrm{T}} X)^2\} + 0.5(\beta_2^{\mathrm{T}} X)\varepsilon, \tag{5}$$

where $X = (x_1, \ldots, x_{10})^{\mathrm{T}}$ with $\varepsilon, x_j, j = 1, \ldots, 10 \sim^{\mathrm{IID}} N(0, 1), \beta_1 = (0.5, 0.5, 0.5, 0.5, 0, \ldots, 0)^{\mathrm{T}}$ and $\beta_2 = (0, \ldots, 0, 0.5, 0.5, 0.5, 0.5)^{\mathrm{T}}$. The simulation results are reported in Table 8. And Fig. 11 shows that by using the conditional expectation $E(|y - \Delta_k||X)$ we can capture all the EDR directions.

To answer questions concerning the minimization of problem (2.7) raised by the following discussants in this paragraph, we state some additional properties of RMAVE here. First, the estimation error for RMAVE is

$$m(\hat{B}, B_0) = O_p\left\{h_d^3 + \frac{\log(n)}{nh_d^{d+1}} + n^{-1/2}\right\},$$

provided that $d \geqslant D$. The estimation error depends only on d (and not on p). When d is small, root n consistency can be achieved (similar results were obtained by Hristache *et al.* (2002) from an approach that is analogous to the outer product of gradients method using refined weights). This answers the question of Professor K. C. Li and Professor Yao and gives an intuitive reason why our simulation works well. Secondly, the MAVE method can be applied easily to semiparametric models such as the model given in Professor Fan's comments. For all the single-index type of models that we have investigated (e.g. the single-index model and the generalized partially linear single-index model; see Xia *et al.* (2002)), the estimators are efficient in the semiparametric sense (Bickel *et al.*, 1993), and undersmoothing is unnecessary. This addresses Professor Linton's question.

We welcome the mention of projection pursuit regression (PPR) by Dr Cui, Dr G. Li, Professor K. C. Li and Dr Zhang, who have reiterated the differences between MAVE and PPR. Consider the PPR model

$$y = g_1(\beta_1^{\mathrm{T}} X) + \ldots + g_D(\beta_D^{\mathrm{T}} X) + \varepsilon, \tag{6}$$

where $E(\varepsilon|X) = 0$ and $(\beta_1^{\mathrm{T}}, \ldots, \beta_D^{\mathrm{T}})$ spans an EDR space. In the absence of extra conditions, we cannot ensure that the directions searched by PPR are in the EDR space. We compare the RMAVE algo-

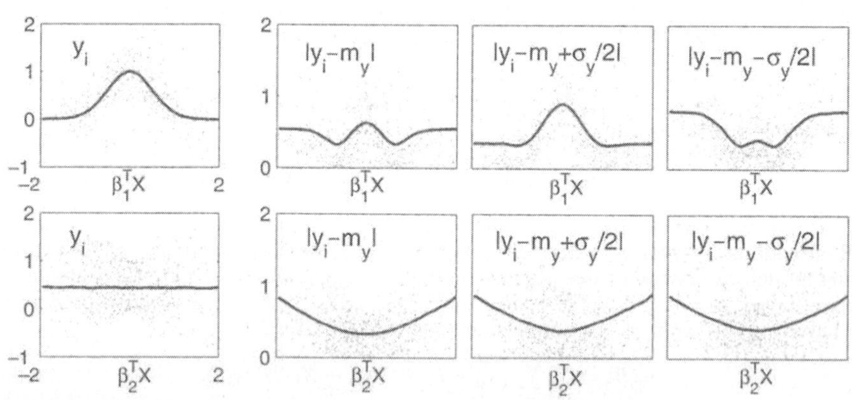

Fig. 11. 1000 observations from model (3) (·) and conditional expectations (———), based on kernel regression from 1 million observations

Table 10. Results of the CV methods

Data	*Method†*	*Results for the following dimensions:*						
		0	*1*	*2*	*3*	*4*	*5*	*6*
Bank-note	Bandwidth	—	0.1	0.3	0.5	0.6	0.7	0.7
	LL–CV value	0.2525	**0.0016**	0.0029	0.0045	0.0061	0.0093	0.0153
	NW–CV value	0.2525	**0.0036**	0.0049	0.0047	0.0079	0.0078	0.0085
AIS	Bandwidth	—	0.6	0.7	0.9			
	LL–CV value	150.5675	13.7718	**12.4450**	12.9045			
	NW–CV value	150.5675	20.2026	**19.8053**	27.1200			
Hawkins	Bandwidth	—	0.28	0.26	0.28	1		
	LL–CV value	9.2133	**7.8666**	8.8623	10.4843	18.5332		
	NW–CV value	9.2133	**7.6566**	9.0900	11.1208	11.6386		

†LL, local linear; NW, Nadaraya–Watson.

rithm with the PPR program in S-PLUS by reference to the distances from the EDR space based on the estimated directions. In our simulations, we take $D = 2, g_1(v) = \exp(-2v^2), g_2(v) = -\cos(2v), X = (x_1, x_2, \ldots, x_{15})^T, \beta_1 = (1, 2, 3, 4, 5, 6, 7, 8, 7, 6, 5, 4, 3, 2, 1)^T/\sqrt{344}, \beta_2 = (-7, -6, -5, -4, -3, -2, -1, 0, 1, 2, 3, 4, 5, 6, 7)^T/\sqrt{280}$ and $x_1, \ldots, x_{15}, 2\varepsilon \sim^{\text{IID}} N(0, 1)$. With a sample size of 200 and 200 independent replications, the estimated errors are listed in Table 9. The PPR algorithm in S-PLUS performs much worse than the MAVE algorithm; even without the benefit of the additive noise structure, the RMAVE method still outperforms the PPR algorithm in S-PLUS.

We refer Professor Ohtaki and Professor Fujikoshi to Cheng and Tong (1992), which establishes consistency of the cross-validation (CV) estimate, and to Professor Ferré's contribution.

We now consider Professor Setodji's examples. Because of the estimation of the remainder term, we have fewer problems to face than undersmoothing. It allows us to use the optimal bandwidth chosen by data-driven methods. For example, the CV method for the local linear smoothing of y_i on $X_i^T \hat{B}$ can be applied to step 1(b) of our algorithm to choose the bandwidth that is used for the next iteration of estimation. Using this kind of bandwidth, we have re-examined the data sets cited by Professor Setodji. As usual we standardize each covariate before applying the RMAVE method. Table 10 shows our results with the smallest CV values highlighted in bold.

For the bank-note data, the dimension is estimated by CV to be 1 (instead of Setodji's 2). The corresponding direction is estimated as $\beta_1 = (-0.0521, 0.1438, -0.2036, 0.8103, 0.2242, -0.4779)^T$. On the basis of this direction, we further have the following fit, which turns out to be practically deterministic:

$$y_i = f(\beta_1^T X_i), \qquad \text{where } f(v) = 1 \text{ if } v \leqslant -0.2 \text{ or } f(v) = 0 \text{ otherwise.}$$

See also Fig. 12(a). With this simple deterministic single-index relationship, it seems difficult to believe that the efficient dimension is 2 as suggested by the sliced average variance estimation (SAVE) method in Cook and Critchley (2000). One possible explanation for suggesting a second dimension is that, if we classify $\{\beta_1^T X_i, i = 1, \ldots, n\}$ into two groups then one of the notes might be in the wrong group on the basis of the SIR (or SAVE) direction as shown in Fig. 12(c). However, on the basis of the RMAVE direction above there is no such 'outlier'. See Fig. 12(b). For the AIS data, the CV estimated dimension is 2, which is the same as that suggested by SAVE. The results are shown in Figs 12(d) and 12(e). It seems to us that RMAVE has not missed any information. For the Hawkins data, the dimension is estimated to be 1. The model seems to give a reasonable fit to the data although the estimated dimension is lower than 2; see Fig. 12(f). Since the data set was generated from two regression models, we have also explored RMAVE with dimension 2 (and bandwidth 0.2). The directions are estimated as $\beta_1 = (0.0326, 0.7432, -0.2440, 0.6221)^T$ and $\beta_2 = (0.7139, -0.1634, 0.5653, 0.3796)^T$. The difference between these directions and the directions β_{01} and β_{02} that are estimated on the basis of the two regressions above is very small. See also Figs 12(g)–12(j). Fig. 12(g) can distinguish the observations by their models. The rotation in Figs 12(g) and 12(h) is useful for interpretation purposes and is related to questions about the effect of rotation raised by Professor Bowman, Professor Atkinson, Professor Chan and Dr M. Li, and Professor Yao.

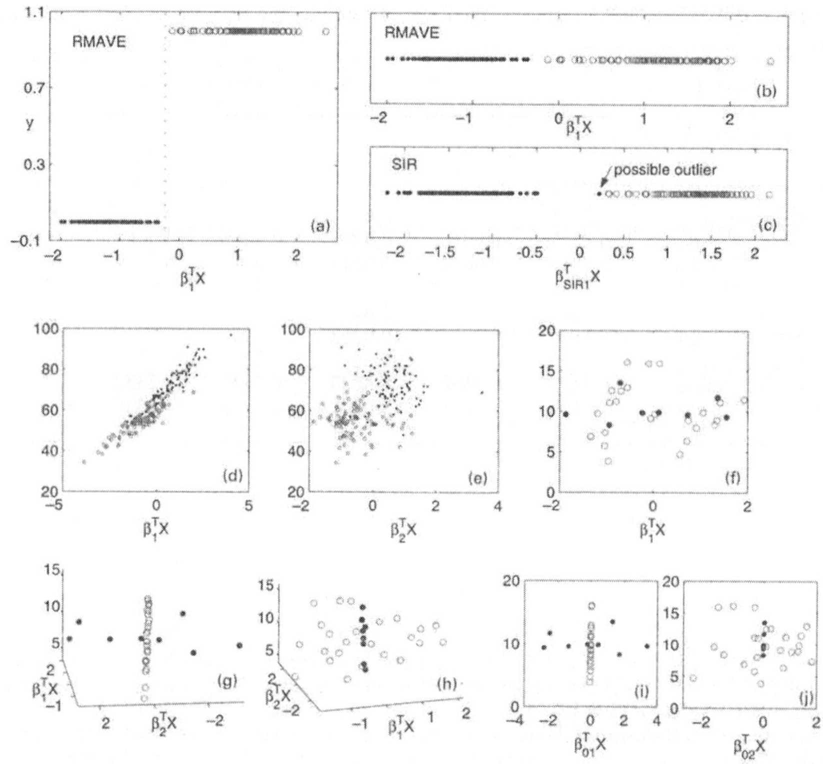

Fig. 12. Calculations for (a), (b), (c) the bank-note data (○, y = '1'; •, y = '0'), (d), (e) the AIS data (○, females; •, males) and (f), (g), (h), (i), (j) the Hawkins data (○, primary regression; •, second regression)

Professor Stenseth and Dr Lingjærde's application of the RMAVE method to the Canadian lynx populations is clearly very interesting. We also look forward to using the partial RMAVE method suggested by Professor Ni.

Concerning Professor Spokoiny's question, a further improvement on MAVE can be made. For example, we can improve the stability of the algorithm along the lines suggested by him.

References in the discussion

Akaike, H. (1973) Information theory and an extension of the maximum likelihood principle. In *Proc. 2nd Int. Symp. Information Theory* (eds B. N. Petrov and F. Csáki), pp. 267–281. Budapest: Akadémiai Kiadó.

Atkinson, A. C. and Riani, M. (2000) *Robust Diagnostic Regression Analysis*. New York: Springer.

Basilevsky, A. (1994) *Statistical Factor Analysis and Related Methods*. New York: Wiley.

Bickel, P. J., Klaassen, A. J., Ritov, Y. and Wellner, J. A. (1993) *Efficient and Adaptive Inference in Semiparametric Models*. Baltimore: Johns Hopkins University Press.

Brillinger, D. R. (1983) A generalized linear model with "Gaussian" regressor variables. In *A Festschrift for Erich L. Lehmann* (eds P. J. Bickel, K. A. Doksum and J. L. Hodges, Jr), pp. 97–114. Belmont: Wadsworth.

——(1992) Nerve cell spike train data analysis: a progression of technique. *J. Am. Statist. Ass.*, **87**, 260–271.

Brown, P. J., Fearn, T. and Vannucci, M. (2001) Bayesian wavelet regression on curves with application to a spectroscopic calibration problem. *J. Am. Statist. Ass.*, **96**, 398–408.

Bura, E. and Cook, R. D. (2001a) Extending sliced inverse regression: the weighted chi-squared test. *J. Am. Statist. Ass.*, **96**, 990–1003.

——(2001b) Estimating the structural dimension of regressions via parametric inverse regression. *J. R. Statist. Soc.* B, **63**, 393–410.

Carroll, R. J., Fan, J., Gijbels, I. and Wand, M. P. (1997) Generalized partially linear single-index models. *J. Am. Statist. Ass.*, **92**, 477–489.

Carroll, R. J. and Härdle, W. (1989) Second order effects in semiparametric weighted least squares regression. *Statistics*, **2**, 179–186.
Cheng, B. and Tong, H. (1992) On consistent nonparametric order determination and chaos (with discussion). *J. R. Statist. Soc.* B, **54**, 427–449, 451–474.
Chiaromonte, F., Cook, R. D. and Li, B. (2002) Sufficient dimension reduction in regressions with categorical predictors. *Ann. Statist.*, **30**, in the press.
Cook, R. D. (1996a) Graphics for regressions with a binary response. *J. Am. Statist. Ass.*, **91**, 983–992.
——(1996b) *Regression Graphics: Ideas for Studying Regressions through Graphics*. New York: Wiley.
——(1998) *Regression Graphics*. New York: Wiley.
Cook, R. D. and Critchley, F. (2000) Identifying regression outliers and mixtures graphically. *J. Am. Statist. Ass.*, **95**, 781–794.
Cook, R. D. and Li, B. (2002) Dimension reduction for conditional mean in regression. *Ann. Statist.*, **30**, in the press.
Cook, R. D. and Nachtsheim, C. J. (1994) Re-weighting to achieve elliptically contoured covariates in regression. *J. Am. Statist. Ass.*, **89**, 592–600.
Cook, R. D. and Weisberg, S. (1991) Discussion on 'Sliced inverse regression' (by K. C. Li). *J. Am. Statist. Ass.*, **86**, 316–342.
——(1994) *An Introduction to Regression Graphics*. New York: Wiley.
Cook, R. D. and Yin, X. (2001) Dimension reduction and visualization in discriminant analysis (with discussion). *Aust. New Z. J. Statist.*, **43**, 147–200.
Cui, H., He, X. and Liu, L. (2001) Testing additivity with regression splines. Submitted to *Ann. Statist.*
Dauxois, J., Ferré, L. and Yao, A. F. (2001) Un modèle semi-paramétrique pour variables hilbertiennes. *C. R. Acad. Sci.*, **333**, 947–952.
Duan, N. and Li, K. C. (1991) A bias bound for least squares linear regression. *Statist. Sin.*, **1**, 127–136.
Fan, J. and Gijbels, I. (1996) *Local Polynomical Modelling and Its Applications*. London: Chapman and Hall.
Fan, J., Yao, Q. and Cai, Z. (2001) Adaptive varying-coefficient linear models. Submitted to *J. R. Statist. Soc.* B.
Ferré L. (1997) Dimension choice for Sliced Inverse Regression based on ranks. *Student*, **2**, 95–108.
——(1998) Determination of the dimension in SIR and related methods. *J. Am. Statist. Ass.*, **93**, 132–140.
Fujikoshi, Y. (1985) Selection of variables in two-group discriminant analysis by error rate and Akaike's information criteria. *J. Multiv. Anal.*, **17**, 27–37.
Hall, P. and Li, K. C. (1993) On almost linearity of low dimensional projections from high dimensional data. *Ann. Statist.*, **21**, 867–889.
Härdle, W., Hall, P. and Ichimura, H. (1993) Optimal smoothing in single-index models. *Ann. Statist.*, **21**, 157–178.
Härdle, W. and Stoker, T. M. (1989) Investigating smooth multiple regression by method of average derivatives. *J. Am. Statist. Ass.*, **84**, 986–995.
Hristache, M., Juditsky, A., Polzehl, J. and Spokoiny, V. (2001) Direct estimation of the index coefficients in a single-index model. *Ann. Statist.*, **29**, 1537–1566.
——(2002) Structure adaptive approach for dimension reduction. *Ann. Statist.*, to be published.
Hristache, M., Juditsky, A. and Spokoiny, V. (2001) Direct estimation of the index coefficients in a single-index model. *Ann. Statist.*, **29**, 595–623.
Huber, P. J. (1985) Projection pursuit (with discussion). *Ann. Statist.*, **13**, 435–525.
Ichimura, H. (1993) Semiparametric least squares (SLS) and weighted SLS estimation of single index models. *J. Econometr.*, **58**, 71–120.
Li, G. and Cheng, P. (1993) Some recent developments in projection pursuit in China. *Statist. Sin.*, **3**, 35–51.
Li, K. C. (1991) Sliced inverse regression for dimension reduction (with discussion). *J. Am. Statist. Ass.*, **86**, 316–342.
——(1997) Nonlinear confounding in high dimensional regression. *Ann. Statist.*, **57**, 577–612.
Li, M.-C. (2000) Multivariate nonlinear time series modeling. *PhD Thesis*. University of Iowa, Iowa City.
Li, M.-C. and Chan, K. S. (2001) Semiparametric reduced-rank regression. *Technical Report 310*. Department of Statistics and Actuarial Science, University of Iowa, Iowa City. (Available from `http://www.stat.uiowa.edu/techrep/`.)
Li, K. C. and Duan, N. (1989) Regression analysis under link violation. *Ann. Statist.*, **17**, 1009–1052.
Linton, O. B. (1995) Second order approximation in the partially linear regression model. *Econometrica*, **63**, 1079–1112.
Mallows, C. L. (1973) Some comments on C_p. *Technometrics*, **15**, 661–671.
Murphy, S. A. and van der Vaart, A. W. (2000) On profile likelihood (with discussion). *J. Am. Statist. Ass.*, **95**, 449–485.
Posse, C. (1995) Projection pursuit exploratory data analysis. *Comput. Statist. Data Anal.*, **20**, 669–687.
Riani, M. and Atkinson, A. C. (2001) A unifed approach to outliers, influence, and transformations in discriminant analysis. *J. Comput. Graph Statist.*, **10**, 513–544.
Robinson, P. M. (1988) Root-N-consistent semiparametric regression. *Econometrica*, **56**, 931–954.
Ruckstuhl, A. F. and Welsh, A. H. (1999) Reference band for nonparametrically estimated link function. *J. Comput. Graph. Statist.*, **8**, 699–714.

Shannon, C. E. (1948) A mathematical theory of communication. *Bell Syst. Tech. J.*, **27**, 379–423, 623–656.
Shibata, R. (1976) Selection of the order of an autoregressive model by Akaike's information criterion. *Biometrika*, **63**, 117–126.
Stenseth, N. C., Chan, K.-S., Tong, H., Boonstra, R., Boutin, S., Krebs, C. J., Post, E., O'Donoghue, M., Yoccoz, N. G., Forchhammer, M. C. and Hurrell, J. W. (1999) Common dynamic structure of Canada lynx populations within three climatic regions. *Science*, **285**, 1017–1073.
Stenseth, N. C., Falck, W., Chan, K.-S., Bjørnstad, O. N., O'Donoghue, M., Tong, H., Boonstra, R., Boutin, S., Krebs, C. J. and Yoccoz, N. G. (1998) From patterns to processes: phases and density dependencies in the Canadian lynx cycle. *Proc. Natn. Acad. Sci. USA*, **95**, 15430–15435.
Stone, M. (1974) Cross-validatory choice and assessment of statistical predictions (with discussion). *J. R. Statist. Soc.* B, **36**, 111–147.
Stone, M. and Brooks, R. J. (1990) Continuum regression: cross-validated sequentially constructed prediction embracing ordinary least squares, partial least squares and principal components regression (with discussion). *J. R. Statist. Soc.* B, **52**, 237–269.
Velilla, S. (1998) Assessing the number of linear components in a general regression problem. *J. Am. Statist. Ass.*, **93**, 1008–1098.
Weisberg, S. and Welsh, A. H. (1994) Adapting for the missing link. *Ann. Statist.*, **22**, 1674–1700.
Xia, Y. and Li, W. K. (1999) On single-index coefficient regression models. *J. Am. Statist. Ass.*, **94**, 1275–1285.
Xia, Y., Tong, H. and Li, W. K. (2002) Single index volatility model and its estimation. *Statist. Sin.*, to be published.
Zhu, L. X. (2002) Transforming a response variable for partial least squares regression. *Technical Report*. Department of Statistics and Actuarial Science, Univesity of Hong Kong, Hong Kong.
Zhu, L. X. and Fang, K.-T. (1996) Asymptotics for kernel estimate of sliced inverse regression. *Ann. Statist.*, **24**, 1053–1068.

An Adaptive Estimation Method for Semiparametric Models and Dimension Reduction

CHENLEI LENG*, YINGCUN XIA† AND JINFENG XU‡

Department of Statistics and Applied Probability
National University of Singapore, Singapore
E-mails: **stalc@stat.nus.edu.sg*, *†staxyc@stat.nus.edu.sg*, *‡staxj@stat.nus.edu.sg*

Xia, Tong, Li and Zhu (2002) proposed a general estimation method termed minimum average variance estimation (MAVE) for semiparametric models. The method has been found very useful in estimating complicated semiparametric models (Xia, Zhang and Tong, 2004; Xia and Härdle, 2006) and general dimension reduction (Xia, 2008; Wang and Xia, 2008). The method is also convenient to combine with other methods in order to incorporate additional statistical requirements (Wang and Yin, 2007). In this paper, we give a general review on the method and discuss some issues arising in estimating semiparametric models and dimension reduction (Li, 1991 and Cook, 1998) when complicated statistical requirements are imposed, including quantile regression, sparsity of variables and censored data.

Keywords: censored data; dimension reduction; Lasso; minimum average variance estimation; nonparametric quantile regression; semiparametric models.

1. Introduction

Suppose Y is a response and $X = (\mathbf{x}_1, ..., \mathbf{x}_p)$ are covariates. One of the basic statistical goals is to estimate the conditional mean function $m(x) = E(Y|X = x)$ or model $Y = m(X) + \epsilon$ with $E(\epsilon|X) = 0$ almost surely. More generally, statisticians are interested in function

$$m(x) = \sup_{m \in \mathcal{C}} E\{\rho(Y - m(x))|X = x\},$$

where $\rho(.)$ is a loss function and $\mathcal{C}$ is a class of (smooth) function. As it is well known, without any information about the structure of function $m(.)$, it is difficult to estimate the function well when the dimension p is greater than 1, due to the so called "curse of dimensionality". As a consequence, many parametric and semiparametric models were proposed in the last decades by imposing structure or special functional forms. Write $X^\top = (X_{[1]}^\top, X_{[2]}^\top)$ where $X_{[1]}^\top = (\mathbf{x}_1, ..., \mathbf{x}_q)$ and $X_{[2]}^\top = (\mathbf{x}_{q+1}, ..., \mathbf{x}_p)$ with $q < p$. Here are some examples of the popular semiparametric models.

- Partially linear model (e.g. Speckman, 1988): $Y = \alpha^\top X_{[1]} + g(X_{[2]}) + \varepsilon$.
- Single-index model (e.g. Ichimura, 1993): $Y = g(\beta_0^\top X) + \varepsilon$.
- Semi-varying coefficient model (e.g. Zhang et al, 1999; Xia, Zhang and Tong, 2004): $Y = g_1(\mathbf{x}_1)\mathbf{x}_1 + g_2(\mathbf{x}_1)\mathbf{x}_2 + ... + g_q(\mathbf{x}_1)\mathbf{x}_q + a_{q+1}\mathbf{x}_{q+1} ... + a_p\mathbf{x}_p + \varepsilon$.
- Single-indexing varying coefficient model (Xia and Li, 1999; Fan, Yao and Cai, 2003): $Y = g_0(\beta_0^\top X) + g_1(\beta_0^\top X)\mathbf{x}_1 + ... + g_p(\beta_0^\top X)\mathbf{x}_p + \varepsilon$.
- Dimension reduction model (Li, 1991): $Y = g(\beta_1^\top X, ..., \beta_d^\top X, \varepsilon)$, where $d < p$.

Developing estimation method for the parameters in the semiparametric models has a long history. It was well understood from the very beginning that the root-n consistency for the estimator of parameters can be achieved, but was generally believed that the under-smoothing is necessary. The under-smoothing approach utilizes a smaller bandwidth (than the optimal bandwidth for the estimation of nonparametric functions in the model); see Robinson (1988). It was found later that many semiparametric models do not need the under-smoothing (Speckman, 1988; Härdle, Hall and Ichimura, 1993). Applying MAVE, it was well demonstrated that for a large set of semiparametric models, under-smoothing is also unnecessary for the parameters to achieve the root-n consistency rate (Xia, Zhang and Tong, 2004; Xia, 2007).

Another important feature for MAVE is its easy implementation and availability of algorithms. The estimation of the semiparametric models, especially those that contain single indices, needs to solve a complicated nonlinear minimization problem, which can be difficult. A naive approach is the Newton-Raphson method, in which evaluations of the derivatives or the Hessian matrix of the unknown link function are needed. However, it is well known that the estimation of the derivatives and the Hessian matrix can be complicated. As a consequence, the Newton-Raphson method does not work well. Instead, MAVE provides a very simple way for the computation by a local linear approximation such that the calculation is eventually converted to problems of "linear minimization". For the latter the calculation is much easier and many efficient algorithms are available. In this review, we will give a few examples for which MAVE can be conveniently used in calculation.

Lastly, we will discuss the application of the method in dimension reduction (Li, 1991 and Cook, 1998). Xia et al (2002) proposed the MAVE method for dimension reduction in the conditional mean function. Xia (2007) and Wang and Xia (2008) generalized the idea to general dimension reduction problem. In this review, we consider the dimension reduction problem for survival data, which are often subject to censoring. When censoring occurs, the incompleteness of the observed data may induce a substantial bias in the sample. A number of approaches have been suggested to overcome the associated difficulties in regression with pre-specified model assumptions, including the censored linear regression model, the Cox proportional hazard model and many others. It is interesting to consider the problem without model specification, leading to dimension reduction in censored data. We shall show how censored data can still be easily analyzed by applying MAVE.

2. Estimating Semiparametric Regression Models with Nonsmooth Loss Functions

In this section, we consider the estimation of the semiparametric models with nonsmooth loss functions, including the quantile regression and estimation with L_1 penalty. The latter is now a popular choice used for variable selection, known as Lasso (Tibshirani, 1996). For simplicity, let us focus on the single-index model. Its extension to other models is not difficult. In the last decade or so, a series of papers (e.g. Powell, Stock and Stoker, 1989; Härdle and Stoker, 1989; Ichimura, 1993; Klein and Spady, 1993; Härdle, Hall and Ichimura, 1993; Horowitz and Härdle, 1996; Hristache, Juditski and Spokoiny, 2001; Xia, Tong, Li and Zhu, 2001) have considered the estimation of the parametric index and the nonparametric link function (i.e. the function g), focusing on the root-n consistency of the former; efficiency issues have also been studied. Amongst the various methods of estimation, the more popular ones are the average derivative estimation (ADE) method investigated by Härdle and Stoker (1989), the sliced inverse regression (SIR) method proposed by Li (1989), and the simultaneous minimization method of Härdle, Hall and Ichimura (1993).

The basic algorithm for estimating the parameters in the single-index model is based on observing that

$$\theta_0 = \arg\min_{\theta} E\Big[y - g(\theta^\top X)\Big]^2 \tag{1}$$

subject to $\theta^\top\theta = 1$. By conditioning on $\xi = \theta^\top X$, we see that (1) equals $E_\xi \sigma_\theta^2(\xi)$ where

$$\sigma_\theta^2(\xi) = E\Big[\Big(y - g(\xi)\Big)^2 \Big| \theta^\top X = \xi\Big].$$

It follows that

$$E\Big[y - g(\theta^\top X)\Big]^2 = E_\xi \sigma_\theta^2(\theta^\top X).$$

Therefore, minimization (1) is equivalent to ,

$$\theta_0 = \arg\min_{\theta} E_\xi \sigma_\theta^2(\xi) \tag{2}$$

subject to $\theta^\top\theta = 1$. Let $\{(X_i, y_i)\ i = 1, 2, \cdots, n\}$ be a sample from (X, y). The conditional expectation in (2) is now approximated by the sample analogue. For X_i close to x, we have the following local linear approximation

$$y_i - g(\theta_0^\top X_i) \approx y_i - g(\theta_0^\top x) - g'(\theta_0^\top x) X_{ix}^\top \theta_0,$$

where $X_{ix} = X_i - x$. Following the idea of local linear smoothing, we may estimate $\sigma_\theta^2(\theta^\top x)$ by

$$\hat\sigma_\theta^2(\theta^\top x) = \min_{a,d} \sum_{i=1}^n \Big\{y_i - a - dX_{ix}^\top\theta\Big\}^2 w_{i0}. \tag{3}$$

Here, $w_{i0} \geq 0, i = 1, 2, \cdots, n$, are some weights with $\sum_{i=1}^n w_{i0} = 1$, typically centering at x. Let $X_{ij} = X_i - X_j$. By (2) and (3), our estimation procedure is to minimize

$$\frac{1}{n}\sum_{j=1}^n \sum_{i=1}^n \Big\{y_i - a_j - d_j X_{ij}^\top\theta\Big\}^2 w_{ij} \tag{4}$$

with respect to (a_j, d_j) and θ. If the kernel smoothing is used with kernel function $H(.)$ and bandwidth h, then the weight functions $w_{ij} = H_h(X_{ij})$, where $H_h(.) = H(./h)/h^p$. We call the estimation procedure the minimum average (conditional) variance estimation (MAVE) method; see Xia et al (2002) for more discussions.

2.1. *Semiparametric estimation with Lasso*

When p is large, the coefficients are usually sparse. As a consequence, many of the coefficients are zeros. To automatically select the variables with nonzero coefficients and to estimate the model, Tibishrani (1996) proposed to use the L_1 penalty. For the single-index model, we can also implement the variable selection and model estimation simultaneously by imposing the L_1 penalty. Following the MAVE idea and Lasso, we need to estimate the single-index by

$$\hat\theta = \arg \min_{\substack{\theta:||\theta||=1,\\ a_j, d_j, j=1,\ldots,n}} \sum_{j=1}^n \sum_{i=1}^n \{Y_i - a_j - d_j\theta^\top X_{ij}\}^2 w_{ij} + \lambda|\theta|, \tag{5}$$

where $||.||$ stands for Euclidean norm and $|\theta| = |\theta_1|+...+|\theta_p|$ is the L_1 norm, $\theta = (\theta_1, ..., \theta_p)^\top$. The calculation of the above minimization problem can be decomposed into two minimization problems as follows.

- Fixing $\vartheta = \theta$ and $w_{ij}^\vartheta = K_h(\vartheta^\top X_{ij})$, the solution to (5) of a_j and d_j are

$$\begin{pmatrix} a_j^\theta \\ d_j^\theta h \end{pmatrix} = \Big\{ \sum_{i=1}^n w_{ij}^\vartheta \begin{pmatrix} 1 \\ X_{ij} \end{pmatrix} \begin{pmatrix} 1 \\ \theta^\top X_{ij} \end{pmatrix}^\top \Big\}^{-1} \sum_{i=1}^n w_{ij}^\vartheta \begin{pmatrix} 1 \\ \theta^\top X_{ij} \end{pmatrix} Y_i,$$

- Fixing a_j and d_j, the minimizaiton in (5) with respect to θ can be done as follows. Let

$$Y_{ij}^\vartheta = Y_i(w_{ij}^\vartheta)^{1/2} - a_j(w_{ij}^\vartheta)^{1/2}, \quad X_{ij}^\vartheta = d_j X_{ij}(w_{ij}^\vartheta)^{1/2}.$$

  Then the problem becomes that of minimizing

$$\sum_{i,j=1}^n \{Y_{ij}^\vartheta - \theta^\top X_{ij}^\vartheta\}^2 + \lambda|\theta|. \tag{6}$$

Repeat the two steps with $\vartheta := \theta/||\theta||$ until convergence. Similar to the linear regression model, the model estimation and variable selection can be implemented simultaneously. Using a similar idea, a more general model was investigated in Wang and Yin (2007). Note that the above algorithm is based on Lasso. One can also use lars algorithm (Efron et al, 2004).

Two pilot parameters, bandwidth h and penalty λ, need to be selected. For the bandwidth, Xia (2002) gave a discussion. With θ fixed, the bandwidth is actually selected for a univariate local linear smoothing. Many existing methods can be used. In practice, the simple rule-of-thumb works well. More discussion is given later in section 3.2.

For the penalty parameter, we use the BIC criterion which is defined as

$$BIC(\lambda) = \log(RSS(\lambda)) + \frac{d_\lambda \log n}{n},$$

where d_λ is the number of nonzero entries in the estimator for the tuning parameter λ and $RSS(\lambda)$ is the residual sum of squares. It is remarkable that the BIC here is the one used for parametric models instead of nonparametric ones, since we are selecting number of parameters rather than nonparametric functions though it is under a semiparametric setting. We found that $BIC(\lambda)$ works well in simulations though rigorous justification is needed.

2.2. *Quantile regression*

Regression quantiles, along with the dual methods of regression rank scores, can be considered one of the major statistical breakthroughs of the past decades. Its advantages over the other estimation methods have been well investigated. Regression quantile methods provide a much more complete statistical analysis of the stochastic relationships among variables; in addition, they are more robust against possible outliers or extremely values, and can be computed via traditional linear programming methods. Although median regression ideas go back to the 18th century and the work of Laplace, regression quantile methods were first introduced by Koenker and Bassett (1978).

For a general loss function $\rho(.)$, we are interested in the following minimization problem

$$\min E\{\rho(Y - m(\theta^\top X))|X = x\} \tag{7}$$

with respect to $||\theta|| = 1$ and $m \in \mathcal{C}$, continuous functions. Suppose the minima is achieved by $\{\theta_0, m(.)\}$ and $\rho(.)$ is piece-wise differentiable with derivative $\varphi(.)$, then (7) leads to a single-index M-regression model

$$Y = m(\theta_0^\top X) + \varepsilon, \quad E(\varphi(\varepsilon)|X) = 0 \tag{8}$$

almost surely. An important special case for the loss function is

$$\rho(v) = \tau I(v > 0)v + (\tau - 1)I(v \leq 0)v,$$

where $0 < \tau < 1$ and $I(.)$ is the indicator function, leading to the quantile regression, see Koenker and Bassett (1978).

Our main focus is the estimation of θ_0. Suppose $\{X_i, Y_i\}_{i=1}^n$ are observations from underlying model (8). Following MAVE, we propose to estimate the index parameter θ_0 by

$$\hat{\theta} = \arg \min_{\substack{\theta:||\theta||=1,\\ a_j, d_j, j=1,\ldots,n}} \sum_{i=1}^n \sum_{j=1}^n w_{ij} \rho[Y_i - a_j - d_j \theta^\top X_{ij}]. \tag{9}$$

Again, the calculation of the above minimization problem can be decomposed into two minimization problems.

- Fixing $\theta = \vartheta$ and $w_{ij}^\vartheta = K_h(\vartheta^\top X_{ij})$, the estimation of a_j and d_j are

$$\sum_{i=1}^n \rho\{Y_i - a_j - d_j \vartheta^\top X_{ij}\} w_{ij}^\vartheta.$$

- Fixing a_j and d_j, the minimization respect to θ can be done as follows. Again, let

$$Y_{ij}^\vartheta = Y_i (w_{ij}^\vartheta)^{1/2} - a_j (w_{ij}^\vartheta)^{1/2}, \quad X_{ij}^\vartheta = d_j X_{ij} (w_{ij}^\vartheta)^{1/2}.$$

  Then the problem becomes

$$\min_\theta \sum_{i,j=1}^n \rho\{Y_{ij}^\vartheta - \theta^\top X_{ij}^\vartheta\}$$

  Suppose the solution to the above problem is θ. Standardize it to $\theta := \theta/||\theta||$.

Set $\vartheta = \theta$ and repeat the two steps until convergence. Note that both steps are simple linear quantile regression problems and that several efficient algorithms are available, see Koenker (2005). Kong and Xia (2008) proved that the above iteration converges, i.e. there exists a constant $0 < c < 1$ such that

$$||\theta - \theta_0|| = c||\vartheta - \theta_0|| + O_p(n^{-1/2}).$$

3. Dimension Reduction with Censored Data

Censoring occurs when the value of an observation is only partially known. This type of data happens quite often in the investigate of epidemiology. Let Y^o be the true but sometimes unobservable lifetime, and $X = (\mathbf{x}_1, ..., \mathbf{x}_p)^\top$ be the covariates. Let

$$C = \text{the censoring time,}$$
$$\delta = \text{the censoring indicator; } \delta = 1, \text{if } Y^o \text{ is not censored; } \delta = 0, \text{otherwise,}$$
$$Y = \begin{cases} Y^o, & \text{if } Y^o \text{ is not censored;} \\ \min\{Y^o, C\}, & \text{otherwise.} \end{cases}$$

There are two kinds of censoring

Type I: C is independent of X and Y^o.
Type II: Conditional on X, C is independent of Y^o.

Suppose $S_{Y^o|X}$ is the central space (CS, Cook, 1998) of Y on X. That is for a base $B = (\beta_1, ..., \beta_q)$ of $S_{Y|X}$ with $B^\top B = I_q$, we have

$$Y^o \perp\!\!\!\perp X | B^\top X,$$

which indicates that given $B^\top X$, X and Y^o are independent. Similarly, we can define $S_{Y|X}$ and $S_{C|X}$. See Li et al (1999).

For Type I censoring, the estimation is relative easy since the censoring does not effect the CS of Y on X. Let $S_{Y|X}$ be the CS of Y on X with type I censoring. For censoring type II, besides the dimension reduction for Y^o we also need to consider the dimension reduction for the censoring C. Let $S_{C|X}$ be the CS for C, i.e. $C \perp\!\!\!\perp X | B_c^\top X$, where B_c is a base of $S_{C|X}$. They have the following relationship.

Proposition 3.1. *For type I censoring, we have $S_{Y|X} = S_{Y^o|X}$. For type II censoring, we have $S_{Y|X} \subseteq S_{Y^o|X} \oplus S_{C|X}$.*

Next, we shall only develop estimation methods for the first type.

3.1. *Dimension reduction of complete data*

We first give a brief review on how MAVE method estimates the dimension reduction space when the data is complete.

Theorem 3.2. *For any matrix B, $Y \perp\!\!\!\perp X | B^\top X$ is equivalent to $P(Y \le y | X = x) = P(Y \le y | B^\top X = B^\top x)$ for all $y \in \mathcal{R}^1$ and $x \in \mathcal{R}^p$.*

The proof of Proposition 3.2 can be found in Cook (1998). Since $P(Y \le y|X) = E\{I(Y \le y)|X\}$, Proposition 3.2 implies that the CS of Y is closely related to the CMS of $I(Y \le y)$. Consequently, as long as the CMS of $I(Y \le y)$ can be estimated for all $y \in \mathcal{R}^1$, the CS of Y should be able to be recovered. Let $M(x|y) = E\{I(Y \le y)|X = x\}$ and $G(u|y) = E(I(Y < y)|B_0^\top X = u)$. Therefore, we need to consider the following regression model.Let $m(x|y) = E(I(Y \le y)|X = x)$. Based on the discussion, we have $m(x|y) = E(I(Y \le y)|B_0^\top X = B^\top x) := G(B^\top x|y)$, leading to model

$$I(Y \le y) = G(B_0^\top X|y) + \epsilon_y, \tag{10}$$

where $\epsilon_y = I(Y \le y) - E(I(Y \le y)|X) = I(Y \le y) - G(B_0^\top X)$.

Consider the gradients $\bigtriangledown M(x|y) = (\partial M(x|y)/\partial x_1, ..., \partial M(x|y)/\partial x_p)^\top$ and $\bigtriangledown G(u|y) = (\partial G(u|y)/\partial u_1, ..., \partial G(u|y)/\partial u_{d_0})^\top$, where $u = (u_1, \cdots, u_{d_0})^\top$. It follows that

$$\frac{\partial M(x|y)}{\partial x} = B_0 \bigtriangledown G_b(B_0^\top x|y). \tag{11}$$

Moreover, the following results indicate that the CS cannot be missed as long as y runs over its whole sample space.

Proposition 3.3. *Let* $\Omega(y) = E\{\bigtriangledown M(X|y) \bigtriangledown^\top M(X|y)\}$ *and* $\Lambda(y) = E\{\bigtriangledown G(B_0^\top X|y) \bigtriangledown^\top G(B_0^\top X|y)\}$. *If* B_0 *is a basis of the CS and that* $\bigtriangledown M(x|y)$ *is continuous in* x, *then (1)* $E\Omega(Y) = B_0 E\{\Lambda(Y)\} B_0^\top$, *and (2)* $E\{\Lambda(Y)\}$ *is of full rank.*

A proof of Proposition 3.3 can be found in Wang and Xia (2007).

3.2. *Dimension reduction for the censored data*

Let us first consider the distribution estimation of censored data. Suppose $Y_1, ..., Y_n$ are the randomly observed life times, amongst them $Y_{(1]}, ..., Y_{(n_c]}$ are censored and $Y_{(1)}, ..., Y_{(n_c')}$ are not censored, where $n_c + n_c' = n$. We propose to estimate the empirical distribution by

$$\hat{G}(y) = \frac{\#\{Y_{(i)} \le y\}}{n_c' + \#\{Y_{(i]} > y\}}$$

It is easy to see that $\hat{G}(y)$ is an increasing function, because $\#\{Y_{(i)} \le y\}$ is increasing and $\#\{Y_{(i]} > y\}$ is deceasing. Note that if the censored data is not used, a naive estimator of the empirical distribution is $\hat{F}(y) = \#\{Y_{(i)} \le y\}/n_c'$. As a comparison we have the following results.

Proposition 3.4. *Suppose* $Y_{(1]}, ..., Y_{(n_c]}$ *are randomly censored, i.e.* $Y_{(k]} = \min\{Y_k^*, C_k\}, k = 1, ..., n_c$. *Then* $E\{\hat{G}(y)\} = F(y)$ *and* $E\{\hat{G}(y) - F(y)\}^2 \le E\{\hat{F}(y) - F(y)\}^2$. *The equality holds only at* y *with* $\#\{Y_{(i]} > y\} = 0$.

Proposition 3.4 indicates that $\hat{G}(y)$ is a more efficient estimator of the distribution and thus the censored data is used more efficiently than discarding the incomplete data.

Initial estimator. Based on Proposition 3.3, we immediately have the following estimation method. Suppose that $\{(X_i, Y_i, \Delta_i), i = 1, 2, \cdots n\}$ is a random sample from (X, Y), where $\Delta_i = 1$ or 0 respectively for complete data or censored data. To estimate the gradient $\partial M(x|y)/\partial x$, we can use the nonparametric kernel smoothing methods. For simplicity, we adopt the following notation scheme. Let $K_0(v^2)$ be a univariate symmetric density function and define $K(v_1, \cdots, v_d) = K_0(v_1^2 + \cdots + v_d^2)$ for any integer d and $K_h(u) = h^{-d}K(u/h)$, where d is the dimension of u and $h > 0$ is a bandwidth. For any (x, y), the principle of the local linear smoother suggests minimizing

$$n^{-1} \sum_{\substack{\Delta_i = 1 \text{ or} \\ Y_i \ge y}} \Big\{I(Y_i \le y) - a - b^\top (X_i - x)\Big\}^2 K_h(X_{ix}) \tag{12}$$

with respect to a and b to estimate $M(x|y)$ and $\partial M(x|y)/\partial x$ respectively, where $X_{ix} = X_i - x$. See Fan and Gijbels (1996) for more details. For each pair of (X_j, Y_k), we consider the following minimization problem

$$(\hat{a}_{jk}, \hat{b}_{jk}) = \arg \min_{a_{jk}, b_{jk}} \sum_{\substack{\Delta_i=1 \text{ or} \\ Y_i \geq Y_k}} \left[I(Y_i \leq Y_k) - a_{jk} - b_{jk}^\top X_{ij} \right]^2 w_{ij}, \tag{13}$$

where $X_{ij} = X_i - X_j$ and $w_{ij} = K_h(X_{ij})$. We consider an average of their outer products

$$\hat{\Sigma} = n^{-2} \sum_{k=1}^{n} \sum_{j=1}^{n} \hat{\rho}_{jk} \hat{b}_{jk} \hat{b}_{jk}^\top,$$

where $\hat{\rho}_{jk}$ is a trimming function introduced for technical purpose to handle the notorious boundary points. In this paper, we adopt the following trimming scheme. For any given point (x, y), we use all observations to estimate its function value and its gradient as in (12). We then consider the estimates in a compact region of (x, y). Moreover, for those points with too few observations around, their estimates might be unreliable. They should not be used in the estimation of the CS directions and should be trimmed off. Let $\rho(\cdot)$ be any bounded function with bounded second order derivatives on $\mathbb{R}$ such that $\rho(v) > 0$ if $v > \omega_0$; $\rho(v) = 0$ if $v \leq \omega_0$ for some small $\omega_0 > 0$. We take $\hat{\rho}_{jk} = \rho(\hat{f}(X_j))$, where $\hat{f}(x)$ is estimator of the density functions of X. The CS directions can be estimated by the first q eigenvectors of $\hat{\Sigma}$. We call this estimator the outer product of gradient estimator (OPG); see also Xia et al (2002).

MAVE for censored data Note that if (10) holds, then the gradients $\partial M(x|y)/\partial x$ at all (x, y) are in a common q-dimensional subspace as shown in equation (11). To use this observation, we can replace b in (12), which is an estimate of the gradient, by $Bd(x, y)$ and have the following local linear approximation

$$n^{-1} \sum_{\substack{\Delta_i=1 \text{ or} \\ Y_i \geq Y_k}} \{I(Y_i \leq y) - a - d^\top B^\top (X_i - x)\}^2 K_h(B^\top X_{ix}),$$

where $d = d(x, y)$ is introduced to take the role of $\bigtriangledown G(B_0^\top x|y)$ in (11). Note that the above weighted mean of squares is the local approximation errors of $I(Y_i \leq y)$ by a hyperplane with the normal vectors in a common space spanned by B. Since B is common for all x and y, it should be estimated with aims to minimize the approximation errors for all possible X_j and Y_k. As a consequence, we propose to estimate B_0 by minimizing

$$n^{-3} \sum_{k=1}^{n} \sum_{j=1}^{n} \hat{\rho}_{jk} \sum_{\substack{\Delta_i=1 \text{ or} \\ Y_i \geq Y_k}} \{I(Y_i \leq Y_k) - a_{jk} - d_{jk}^\top B^\top X_{ij}\}^2 w_{ij} \tag{14}$$

with respect to $a_{jk}, d_{jk} = (d_{jk1}, \cdots, d_{jkq})^\top, j, k = 1, ..., n$ and $B : B^\top B = I_d$, where $\hat{\rho}_{jk}$ is defined above. Because the method is proposed for censored data using MAVE, we call it the minimum average (conditional) variance estimation for censored data (cMAVE).

The minimization problem in (14) can be solved by fixing $(a_{jk}, d_{jk}), j, k = 1, ..., n$, and B alternatively. As a consequence, we need to solve two quadratic programming problems which have simple analytic solutions. For any matrix $B = (\beta_1, \cdots, \beta_q)$, we define operators $\ell(.)$ and $\mathcal{M}(.)$ respectively as

$$\ell(B) = (\beta_1^\top, \cdots, \beta_q \top)^\top \qquad \text{and} \qquad \mathcal{M}(\ell(B)) = B.$$

We propose the following cMAVE algorithm to implement the estimation.

Step 0: Let $B_{(1)}$ be an initial estimator of the CS directions. Set $t = 1$.

Step 1: Let $B = B_{(t)}$, calculate the solutions of $(a_{jk}, d_{jk}), j, k = 1, ..., n$, to the minimization problem in (14)

$$\begin{pmatrix} a_{jk}^{(t)} \\ d_{jk}^{(t)} \end{pmatrix} = \Big\{ \sum_{\substack{\Delta_i=1 \text{ or} \\ Y_i \geq Y_k}} K_{h_t}(B_{(t)}^\top X_{ij}) \begin{pmatrix} 1 \\ B_{(t)}^\top X_{ij} \end{pmatrix} \begin{pmatrix} 1 \\ B_{(t)}^\top X_{ij} \end{pmatrix}^\top \Big\}^{-1}$$

$$\times \sum_{\substack{\Delta_i=1 \text{ or} \\ Y_i \geq Y_k}} K_{h_t}(B_{(t)}^\top X_{ij}) \begin{pmatrix} 1 \\ B_{(t)}^\top X_{ij} \end{pmatrix} I(Y_i \leq Y_k),$$

where h_t and b_t are two bandwidths (details are discussed below).

Step 2: Let $\rho_{jk}^{(t)} = \rho(\hat{f}_{B_{(t)}}(X_j))$ with $\hat{f}_{B_{(t)}}(x) = n^{-1} \sum_{\Delta_i=1 \text{ or } Y_i \geq Y_k} K_{h_t}(B_{(t)}^\top X_{ix})$. Fixing $a_{jk} = a_{jk}^{(t)}$ and $d_{jk} = d_{jk}^{(t)}$, calculate the solution of B or $\ell(B)$ to (14)

$$\mathbf{b}^{(t+1)} = \Big\{ \sum_{k,j=1}^{n} \sum_{\substack{\Delta_i=1 \text{ or} \\ Y_i \geq Y_k}} \rho_{jk}^{(t)} K_{h_t}(B_{(t)}^\top X_{ij}) X_{ijk}^{(t)} (X_{ijk}^{(t)})^\top \Big\}^{-1}$$

$$\times \sum_{k,j=1}^{n} \sum_{\substack{\Delta_i=1 \text{ or} \\ Y_i \geq Y_k}} \rho_{jk}^{(t)} K_{h_t}(B_{(t)}^\top X_{ij}) X_{ijk}^{(t)} \{I(Y_i \leq Y_k) - a_{jk}^{(t)}\},$$

where $X_{ijk}^{(t)} = d_{jk}^{(t)} \otimes X_{ij}$.

Step 3: Calculate $\Lambda_{(t+1)} = \{\mathcal{M}(\mathbf{b}^{(t+1)})\}^\top \mathcal{M}(\mathbf{b}^{(t+1)})$ and $B_{(t+1)} = \mathcal{M}(\mathbf{b}^{(t+1)}) \Lambda_{(t+1)}^{-1/2}$. Set $t := t + 1$ and go to Step 1.

Step 4: Repeat steps 1–3 until convergence. The final value of $B_{(t)}$ is our estimator of the direction, denoted by $\hat{B}$.

The cMAVE algorithm needs a consistent initial estimator in Step 0 to guarantee its theoretical justification. The OPG estimator above can be served as the initial estimator. In practice, we may need to standardize $X_i = (X_{i1}, \cdots, X_{ip})^\top$ by setting $X_i := S_X^{-1/2}(X_i - \bar{X})$ and standardize Y_i by setting $Y_i := (Y_i - \bar{Y})/\sqrt{s_Y}$, where $\bar{X} = n^{-1} \sum_{i=1}^n X_i$ and $S_X = n^{-1} \sum_{i=1}^n (X_i - \bar{X})(X_i - \bar{X})^\top$, $\bar{Y} = n^{-1} \sum_{i=1}^n Y_i$ and $s_Y = n^{-1} \sum_{i=1}^n (Y_i - \bar{Y})^2$. Then the estimated CS directions are the first q columns of $S_X^{-1/2} \hat{B}$.

Note that the estimation in the procedure is related with nonparametric estimations of conditional density functions. Several bandwidth selection methods are available for the estimation. See, e.g. Silverman (1986), Scott (1992) and Fan et al (1996). Our theoretical verification of the convergence for the algorithms requires some constraints on the bandwidths although we believe these constraints can be removed with more complicated technical proofs. To ensure the requirements on bandwidths can be satisfied, after standardizing the variables we use the following bandwidths in our calculations. In the first iteration, we use slightly larger bandwidths than the optimal ones in terms of MISE as

$$h_0 = c_0 n^{-\frac{1}{p_0+6}}, \qquad b_0 = c_0 n^{-\frac{1}{p_0+5}}, \tag{15}$$

where $p_0 = \max(p, 3)$. Then we reduce the bandwidths in each iteration as

$$h_{t+1} = \max\{r_n h_t, c_0 n^{-\frac{1}{q+4}}\}, \qquad b_{t+1} = \max\{r_n b_t, c_0 n^{-\frac{1}{q+3}}, c_0 n^{-\frac{1}{5}}\} \tag{16}$$

for $t \geq 0$, where $r_n = n^{-1/(2(p_0+6))}$, $c_0 = 2.34$ as suggested by Silverman (1986) if the Epanechnikov kernel is used. Here, the bandwidth b is selected smaller than h based on simulation comparisons.

3.3. *Theoretical results*

Assume that both B_0 and $\hat{B}$ have been standardized such that $B_0^\top B_0 = I_q$ and $\hat{B}^\top \hat{B} = I_q$. Furthermore, we use $||A||$ to denote the maximum singular value of an arbitrary matrix A, which is the Euclidean norm if A is a vector. Following Wang and Xia (2007), we have the following asymptotic results for the estimation.

Theorem 3.5. *Suppose conditions (C1)-(C5) in the appendix hold and the final bandwidth is $\hbar$, then there is a rotation matrix Q such that the estimator $\hat{B}$ is consistent with*

$$||\hat{B} - B_0 Q|| = O_p\{\hbar^4 + \log n/(n\hbar^d) + n^{-1/2}\}.$$

If we use higher order polynomial smoothing, it is possible to show that the root-n consistency can be achieved for any dimension q. See, e.g. Härdle and Stoker (1989) and Samarov (1993), where the higher order kernel, a counterpart of the higher order polynomial smoother, was used. However, using higher order polynomial smoothers increases the difficulty of calculations while the improvement of finite sample performance is not substantial.

4. Some Numerical Results

In this section we discuss some simulation results to check the numerical performance of the proposed methods.

Example 4.1 (Variable selection single-index via Lasso). *We consider in this example a linear regression model*

$$y = \theta_0^\top X + 0.5\varepsilon, \tag{17}$$

where $\varepsilon \sim N(0,1)$, $\theta_0 \in \mathbb{R}^{18}$ with $\theta_{0,1} = \theta_{0,7} = \theta_{0,13} = 1/\sqrt{3}$ and the rest entries are zero. We compare the proposed estimator with the unpenalized MAVE. We refer to our method as sparse MAVE (sMAVE), where we choose λ in (6) by minimizing the BIC criterion. The covariate vector are generated as $X \sim \Sigma_0^{1/2} X_0$ with $X_0 \sim N(0, I_5)$ and $\Sigma_0 = (\rho^{|i-j|})_{0\leq i,j\leq 5}$, where $\rho = 0, 0.5, 0.9$. We consider two sample sizes $n = 60, 120$ and repeat the simulation for 100 times. The results are summarized by two measures for an estimator θ: the correlation (CORR) between $X^\top\theta_0$ and $X^\top\theta$ and the mean square error $MSE = E(\theta-\theta_0)^\top XX^\top(\theta-\theta_0)$. Furthermore, we record the percentage of models which are correctly identified (CM). Table 1 clearly shows that sMAVE outperforms MAVE. At the same time, sMAVE can select the variables efficiently.

Example 4.2 (Median regression of single-index). *In this example we consider the following model*

$$y = \exp\{-5(\theta_0^\top X)^2\} + \varepsilon, \tag{18}$$

where $X \sim \Sigma_0^{1/2} X_0$ with $X_0 \sim N(0, I_5)$ and $\Sigma_0 = (0.5^{|i-j|})_{0\leq i,j\leq 5}$. For the noise term, we consider several distributions with both heavy tail and thin tails as well. For simplicity, we

ρ	size	method	CORR	MSE × 100	CM (%)
0	60	MAVE	0.960 (0.013)	10.87 (3.98)	0
		sMAVE	0.989 (0.010)	2.481 (2.35)	59
	120	MAVE	0.982 (0.006)	4.093 (1.25)	0
		sMAVE	0.997 (0.003)	0.618 (0.63)	78
0.5	60	MAVE	0.962 (0.014)	10.77 (4.73)	0
		sMAVE	0.991 (0.008)	2.118 (2.12)	68
	120	MAVE	0.982 (0.007)	4.270 (1.44)	0
		sMAVE	0.997 (0.004)	0.724 (0.76)	79
0.9	60	MAVE	0.979 (0.008)	23.56 (10.24)	0
		sMAVE	0.990 (0.009)	6.053 (5.69)	36
	120	MAVE	0.990 (0.003)	9.531 (5.59)	0
		sMAVE	0.998 (0.002)	1.597 (2.88)	74

consider the median regression only. As a comparison, we also run the MAVE where a least square type estimation is used. With different sample sizes $n = 100, 200$, *we carried out 100 replications. The calculation results are listed in Table 2.*

		Distribution of ε			
size	method	$0.05t(1)$	$0.1(N(0,1)^4 - 3)$	$\sqrt{5}t(5)/20$	N(0,1)/4
100	MAVE	0.3641(0.3526)	0.3530(0.3102)	0.0401(0.0182)	0.0581(0.0263)
	qMAVE	0.0902(0.1074)	0.1512(0.1957)	0.0833(0.0785)	0.1146(0.0651)
200	MAVE	0.3381(0.3389)	0.2859(0.2887)	0.0232(0.0091)	0.0373(0.0147)
	qMAVE	0.0681(0.1415)	0.0581(0.0698)	0.0402(0.0173)	0.0652(0.0272)

the MAVE method with quadratic loss function has very bad performance when the noise has heavy tail (e.g. $t(1)$*) or is highly asymmetric (e.g.* $N(0,1)^4$*). With the absolute value loss function, the performance is much better. Even in the situation when the noise has thin tail and symmetric, qMAVE still performance reasonably well.*

Example 4.3 (Dimension reduction for censored data). *In this example, we consider the following censored regression model*

$$Y^o = 4 - |\theta_0^T X - 1| + 0.1\varepsilon_1,$$

$$C = a + 0.1\varepsilon_2 \tag{19}$$

where $X \sim N(0, I_6), \varepsilon_1 \sim N(0,1), \varepsilon_2 \sim N(0,1)$, $C \perp (X, Y^o)$, $\theta_0 = (1,0,0,0,0,0)^T$, *and* a *is set to* 3.6 *or* 3.2 *resulting in* 20% *or* 40% *censoring respectively. The estimation error between the estimator* $\hat{\theta}$ *and* θ_0 *is measured by both MSVD and SQABD, where* $MSVD = \max(|svd(\hat{\theta}\hat{\theta}^T - \theta_0\theta_0^T)|)$ *and* $SQABD = \sqrt{1 - |\hat{\theta}^T\theta_0|}$. *With different sample sizes, 100 replications are drawn from the model and the comparison is made between a naive MAVE method discarding the censored observations (MAVE) and the one we proposed (cMAVE). The simulation results are summarized in Table 3. As expected, cMAVE performs more favorably than MAVE by both measures of the estimation error, especially in the heavier censoring (40%) case.*

censoring (%)	size	method	MSVD	SQABD
20	100	MAVE	0.0439(0.0161)	0.0310(0.0114)
		cMAVE	0.0419(0.0185)	0.0296(0.0131)
	200	MAVE	0.0277(0.0088)	0.0196(0.0062)
		cMAVE	0.0257(0.0080)	0.0182(0.0056)
40	100	MAVE	0.0827(0.0318)	0.0586(0.0225)
		cMAVE	0.0445(0.0174)	0.0315(0.0123)
	200	MAVE	0.0527(0.0231)	0.0373(0.0164)
		cMAVE	0.0280(0.0106)	0.0198(0.0075)

Example 4.4 (Primary biliary cirrhosis data). *The Mayo clinic has established a database of 424 patients having primary biliary cirrhosis (PBC). 276 of them have complete information of 18 variables.*

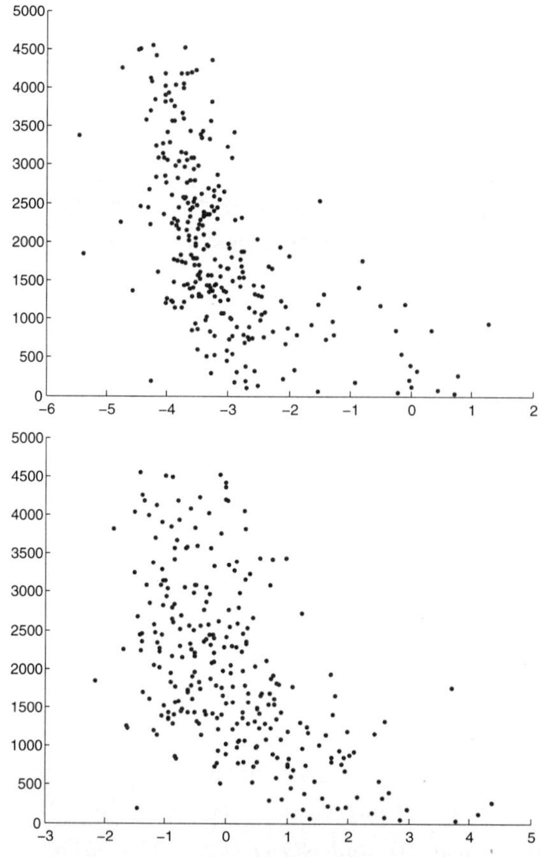

Fig. 1. The first panel is the plot of response gainst the dimension reduction direction estimated by cMAVE and the second that against the dimension reduction direction estimated by uMAVE.

Considering all 18 variables, we analyze it using the cMAVE and uMAVE methods and the estimated effective dimension reductions (with the number of dimension reduction set to

1) are (0.0043, -0.6152, -0.0001, 0.3802, 0.1674, -0.0001, -0.2972, 0.5421, 0.0215, 0.0001, -0.1882, -0.0015, -0.0010, 0.0639, 0.1565, -0.0466, -0.0009, 0.0027) and (0.0317, -0.7340, -0.0001, 0.2852, 0.0437 ,0.0003 0.1475 , 0.2563 , 0.2799,-0.0009 , -0.0387 , -0.1907 , 0.0025, 0.2266 , 0.1804 , 0.2991 , -0.0014 , 0.0044) which are plotted against the response variable in the two panels of Figure 1 respectively. Based on the plots, we can see that the proposed estimation that uses the censored observations can give clearer pattern than the method that ignores the censored observations.

5. Conclusion

In this review, we present some problems that MAVE can be applied easily. Discussions and simulations suggested that it is promising to apply the MAVE to complicated problems.

This work is a brief review of the the possible application of MAVE. More details related to the implementation, including how to select the pilot parameters h and λ, and how to determine the dimension of the central space, need to presented. Asymptotic theory is another topic which needs to be addressed in the future.

Acknowledgement

The authors thank two referees for their very valuable comments.

References

1. Cook, R.D. (1998). *Regression Graphics*, John Wiley, New York, NY.
2. Efron, B., Hastie, T., Johnstone, I., Tibshirani, R. (2004). Least angle regression. *Annals of Statistics* **32** 407-499.
3. Koenker, R. (2005). *Quantile regression.* Cambridge University Press, New York.
4. Fan, J. and Gijbels, I. (1996). *Local Polynomial Modelling and Its Applications*, Chapman and Hall, New York, NY.
5. Härdle, W. and Stoker, T. M. (1989). Investigating smooth multiple regression by method of average derivative, *Journal of the American Statistical Association*, 84, 986–995.
6. Li, L. and Li, H. (2004). Dimension reduction methods for microarrays with application to censored survival data, *Bioinformatics* **20** 3406-3412
7. Li, K. C. (1991). Sliced inverse regression for dimension reduction, *Journal of the American Statistical Association*, 86, 316–327.
8. Li, K. C., Wang J. L. and C. H. Chen (1999). Dimension reduction for censored regression data, *Annals of Statistics* **27** 1-23.
9. Robinson, Peter M (1988). Root-N-consistent semiparametric regression. *Econometrica*, 56(4), 931-954.
10. Silverman, B. W. (1986). *Density Estimation for Statistics and Data Analysis*, London: Chapman and Hall.
11. Speckman, P. (1988). Kernel smoothing in partial linear models. *J. Roy. Statist. Soc. Ser. B* 50, 413–436.
12. Tibshirani, R. (1996). Regression shrinkage and selection via the lasso. *J. Royal. Statist. Soc B.*, **58**,267-288.
13. Wang, Q. and Yin X. Y. (2007). A nonlinear multi-dimensional variable selection method for high dimensional data: Sparse MAVE. Manuscript, Department of Statistics, University of Georgia.
14. Wang, H. and Xia, Y. (2008). Sliced Regression for Dimension Reduction. Manuscript, Department of Statistics and Applied Probability, National University of Singapore.

15. Xia, Y. (2007). A constructive approach to the estimation of dimension reduction directions, *The Annals of Statistics*, To appear.
16. Xia, Y., Tong, H., Li, W. K., and Zhu, L. (2002). An adaptive estimation of dimension reduction space. *Journal of Royal Statistical Society, Series B.*, 64, 363–410.

Common Dynamic Structure of Canada Lynx Populations Within Three Climatic Regions

Nils Chr. Stenseth,[1,2]* Kung-Sik Chan,[3] Howell Tong,[4,5] Rudy Boonstra,[1,6] Stan Boutin,[7] Charles J. Krebs,[1,8] Eric Post,[2] Mark O'Donoghue,[8,9] Nigel G. Yoccoz,[1,10] Mads C. Forchhammer,[11,12] James W. Hurrell[13]

Across the boreal forest of Canada, lynx populations undergo regular density cycles. Analysis of 21 time series from 1821 onward demonstrated structural similarity in these cycles within large regions of Canada. The observed population dynamics are consistent with a regional structure caused by climatic features, resulting in a grouping of lynx population dynamics into three types (corresponding to three climatic-based geographic regions): Pacific-maritime, Continental, and Atlantic-maritime. A possible link with the North Atlantic Oscillation is suggested.

Periodic population fluctuations of the Canada lynx (*Lynx canadensis*) have greatly influenced both ecological theory and statistical time series modeling [(*1*, *2*); see (*3*) for a summary]. Recent analyses have focused on the extent of synchrony in population fluctuations, assessing the importance of external abiotic factors (such as weather) and internal biotic factors (such as dispersal among populations) in causing spatial patterns (*4*). Such empirical and theoretical approaches have, however, assumed that the populations were structurally similar [that is, the density-dependent relationships are identical among populations (*5*)]. This assumption has never been thoroughly evaluated. To do so requires determining whether the lynx populations display the same phase- and density-dependent structure (*3*) and then searching for similar underlying causes of the observed dynamics. Using new statistical methods developed for this purpose (*6*), we ask to what extent the time series on the Canada lynx (Fig. 1) compiled by the Hudson Bay Company for the period 1821 to 1939 (*7*) and the corresponding more modern time series compiled by Statistics Canada for the period 1921 to present (*8*), taken together, are structurally similar. Specifically, we ask whether the phase- and density-dependent structure of changes in lynx abundance cluster into groups defined according to ecological-based features (*9*) or according to climatic-based features (*10*, *11*).

The available time series (Fig. 1A) cover two ecosystems (referred to below as ecological regions): the northern, open boreal forest (Fig. 1B) and the southern, closed boreal forest. In western Canada, the mountainous topography adds complexity. Additionally, the series cover three climatic regions defined by the spatial influences of the North Atlantic Oscillation (NAO) [Fig. 1C; see (*12*)], which may contribute to spatial differences in trophic interactions (*13*).

Previously, we fitted a piecewise linear autoregressive model (*14*) to each of the series (*3*). A general hare-lynx model (*3*, *15*) may be expressed as an equivalent model in delay coordinates of the lynx (the species for which we have data). Here we check whether all the time series, or some subsets of these, display the same underlying phase- and density-dependent structure. For this purpose we use a piecewise linear model (*14*, *15*):

$$y_{s,t} = \begin{cases} \beta_{s,1,0} + \beta_{s,1,1}y_{s,t-1} + \beta_{s,1,2}y_{s,t-2} + \varepsilon_{s,1,t} & y_{s,t-d} \leq \theta_s \\ \beta_{s,2,0} + \beta_{s,2,1}y_{s,t-1} + \beta_{s,2,2}y_{s,t-2} + \varepsilon_{s,2,t} & y_{s,t-d} > \theta_s \end{cases} \quad (1)$$

where $y_{s,t}$ is the log-transformed abundance of lynx at site s and for year t [that is, $y_{s,t} = \log(Y_{s,t})$ where $Y_{s,t}$ is the abundance of lynx at site s and in year t, and where s = 1, 2, . . . , represent the sites corresponding to the individual time series; see Fig. 1A]; $\beta_{s,i,j}$ are the statistical parameters that determine the phase- and density-dependent structure of the system (i = 1 and 2 correspond to the lower and the upper regimes of the model; j = 0, 1, 2 correspond to the constant term, the first lag, and the second lag, respectively) at site s; $\varepsilon_{s,i,t}$ is normally distributed, time-independent noise [$N(0,\sigma^2_{s,i})$]; and θ_s is the threshold applicable to the log-transformed density d years earlier.

[1]Center for Advanced Study, The Norwegian Academy of Science and Letters, Drammensveien 78, N-0271 Oslo, Norway. [2]Division of Zoology, Department of Biology, University of Oslo, P.O. Box 1050 Blindern, N-0316 Oslo, Norway. [3]Department of Statistics and Actuarial Science, University of Iowa, Iowa City, IA 52242, USA. [4]Department of Statistics, London School of Economics, London WC2A 2AE, UK. [5]Department of Statistics, University of Hong Kong, Hong Kong. [6]Division of Life Sciences, University of Toronto at Scarborough, Scarborough, Ontario M1C 1A4, Canada. [7]Department of Biological Sciences, University of Alberta, Edmonton, Alberta T6G 2E9, Canada. [8]Department of Zoology, University of British Columbia, Vancouver, British Columbia V6T 1Z4, Canada. [9]Department of Renewable Resources, Fish and Wildlife Branch, P.O. Box 310, Mayo, Yukon Y0B 1M0, Canada. [10]Department of Arctic Ecology, Norwegian Institute for Nature Research, Polar Environmental Centre, N-9296 Tromsø, Norway. [11]Department of Zoology, University of Cambridge, Downing Street, Cambridge CB2 3EJ, UK. [12]Department of Landscape Ecology, National Environmental Research Institute, Kalø, Grenåvej 12, DK-8410 Rønde, Denmark. [13]National Center for Atmospheric Research, Climate Analysis Section, P. O. Box 3000, Boulder, CO 80307–3000, USA.

*To whom correspondence should be addressed. E-mail: n.c.stenseth@bio.uio.no

Such a threshold approach has several statistical advantages (*16*). The upper (respective lower) regime has been found to correspond to the decrease (respective increase) phase (*3*).

Because of varying carrying capacities and trapping efforts across sites, we expect series to have different means and standard deviations. Therefore, Chan *et al.* (*6*) considered the hypothesis of common structure that all series enjoy the same dynamics up to their means and standard deviations. The hypothesis of common structure is equivalent to two hypotheses: the hypothesis of common slopes $\beta_{s,i,j} = \beta_{1,i,j}$, and the hypothesis of common ratio of intercepts [(the intercept in the upper regime)/(the intercept in the lower regime)] takes the same value at the threshold across the different sites. Chan *et al.* (*6*) derived some test statistics for checking these hypotheses.

We also compared model fits between various possible groupings of the time series into subgroups. The SETAR (self-exciting threshold autoregression) models (Eq. 1) may be constrained to have some coefficients identical across series within a given subgroup. Different groupings can be compared in terms of their respective AICs (Akaike information criteria) (*14, 17*). For groupings involving different series to be comparable, each grouping will be defined for all series. A model with minimal AIC strikes a good balance between parsimony and goodness of fit to the data.

Time series coming from the same locations exhibit the same dynamic structure, indicating a common underlying dynamic model (*18*). The time series come from forested biomes across Canada and thus from areas with greatly different plant species composition and habitat structure. Focusing on the vegetation, we may classify the time series into two ecological-based groups (Fig. 1B): the northern forest tundra, which consists of shrub tundra and low-density trees (*19*), and the true boreal forest, which is a mixture of conifer and deciduous trees (*19*). [A western ecological-based group with heterogeneous topography and habitat, as well as climate (*20*) may also be identified.] However, treating the old and modern series separately, the ecological-based grouping represents no improvement over the baseline of no grouping [Table 1; see also (*21, 22*)].

As an alternative to this ecological-based grouping, the Canada lynx series may be grouped according to three major climate-based features: the Pacific-maritime region, the Continental region, and the Atlantic-maritime region (Fig. 1C). This grouping clearly provides a better description of the data [Table 1; see also (*22*)]. The similarity is particularly strong for the decrease phase. There is evidence in support of climate-based properties contributing to the structuring of the lynx dynamics.

Over much of central and western Canada, surface climate is most strongly influenced by the atmospheric circulation upstream over the North Pacific and in particular by a natural mode of large-scale atmospheric variability known as the Pacific–North American (PNA) teleconnection pattern (*23*). However, the influence of the PNA on Canadian surface temperature is spatially homogeneous. In contrast, the influence of the NAO on surface winter temperatures varies considerably from coast to coast (Fig. 1C) and shows spatial variation corresponding well to the best grouping of the lynx series (Table 1). Hence, it is the winter atmospheric circulation, for which the NAO may serve as a proxy, that probably contributes to making the nonlinear structure of the hare-lynx dynamics similar within each of the three Canadian groups. Although it is generally known that climate profoundly influences regional variation in vegetation [for example, see (*9*)], our results suggest that the spatiotemporal patterns of climatic variation also influence the trophic interaction between the lynx and its main prey, the snowshoe hare, differently across these regions.

Because the NAO may have a delayed effect on the lynx dynamics, we have to choose between using lag-0 or lag-1 NAO as the covariate. Several statistical techniques are available, including Cox's test of separate families of hypotheses, AIC or its many variants, and others (*17*). Even though the effect of the NAO on lynx abundance is not strong, the lynx series fall along an east-west gradient progressing from negative to positive and

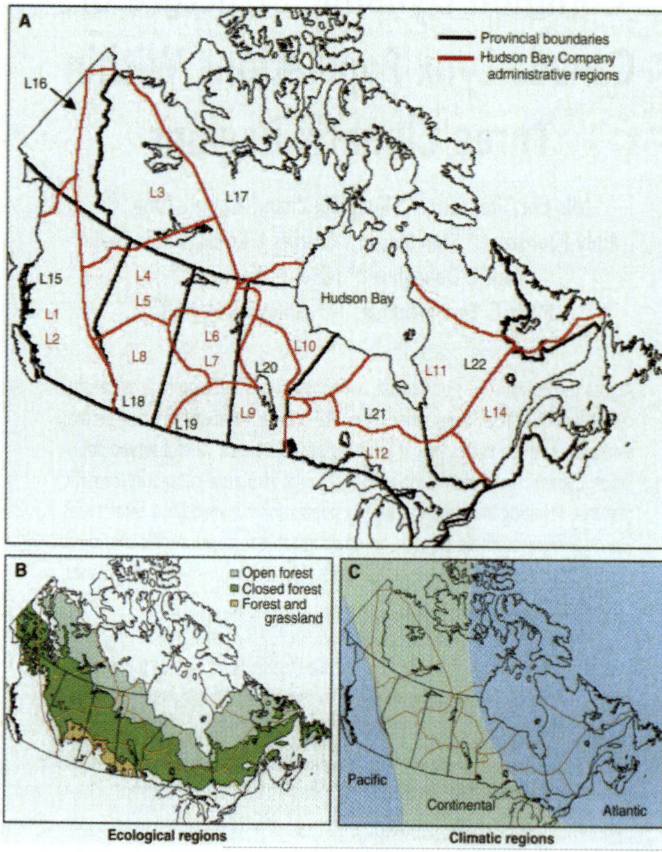

Fig. 1. Time series data studied. (**A**) Map of Canada with demarcations of the studied time series [red indicates the Hudson Bay Company time series (*7*) and black indicates the recent series (*8*)]. See (*3*) for definitions of names of the individual time series used. (**B**) Ecological regions of Canada (*24*). (**C**) Climatic regions of Canada (*10*). The NAO refers to a meridional oscillation in surface pressures with centers of action near Iceland and over the subtropical Atlantic. When surface pressures are lower than normal near Iceland and higher than normal over the subtropical Atlantic (the positive phase of the NAO), enhanced northerly flow over eastern Canada cools surface temperatures and enhanced southerly flow from the Gulf of Mexico into much of central Canada produces warm surface anomalies. Over the Pacific-maritime region, there is no significant NAO signature.

Table 1. Common structure in the underlying dynamics. Groupings with the smallest AIC value represent the best ones. (Only the AIC differences between the particular grouping and the baseline comparison are given [see (*22*), where the absolute AIC values are also given]. In some of the subgroups, the SETAR models are constrained to have some coefficients identical across series within the subgroup (see "Identical constraint structure" column). These constraints are suggested by the new statistical tests developed by Chan *et al.* (*6*). For different groupings involving different series to be comparable, each grouping is defined for all series in the panel; each series not in a grouping implicitly forms a singleton group. For definition of names of the series (L1, L2, . . . , L22), see (*3*).

Groupings	Group definitions	AIC difference*	Identical constraint structure†
Baseline comparison			
No groupings	(Individual series without any constraints) (L1), (L2), . . . (L14), (L15), . . . (L22)	0	
Old and modern as separate groupings	All Hudson Bay Co. series (L1, L2, . . ., L14)	−28.12	1
	All Statistics Canada series (L15, L16, . . ., L22)		1
Old and modern grouped together	All series together (L1, L2, L3, . . ., L22)	−21.36	1
Ecological-based grouping			
	Western (L1, L2)	−28.75	2
	(L15)	(−20.25)	
	Northern (L3, L10)		3
	(L16, L17)		1
	Southern (L4, L5, L6, L7, L9)		1
	(L18, L19, L20)		3
	Eastern (L11, L12, L14)		1
	(L21, L22)		1
Climate-based grouping (maritime vs. continental)			
	Pacific-maritime (L1, L2)	−32.14	2
	(L15)	(−23.39)	
	Continental (L3, L4, L5, L6, L7, L9, L10)		1
	(L16, L17, L18, L19, L20)		3
	Atlantic-maritime‡ (L11, L12, L14)		3
	(L21, L22)		3

*AIC differences given in parentheses correspond to assuming only constraints on lag-1 and lag-2 in the upper regime. †1, $\beta_{s,2,1}$ and $\beta_{s,2,2}$ each being common for different series (different s); 2, $\beta_{s,i,j}$ for each *i* and each *j* being common for different series (different s); 3, $\beta_{s,1,2}$, $\beta_{s,2,1}$, and $\beta_{s,2,2}$ each being common for different series (different s). ‡The old and modern series in the Eastern (Atlantic-maritime) seem to share the same lag-1 and lag-2 coefficients in the decrease phase, as imposing these constraints further decreases the AIC by 3.39 for both groupings.

finally to no effect of the NAO. The previously observed phase-dependent nature of the density-dependent structure (*3*) remains even after the NAO is included as a covariate. As a result, this study is consistent with earlier results but adds the geographic component to the structure of the lynx time series.

We can now reach a comprehensive synthesis of the time series of the Canadian lynx—namely, the lynx cycle is a direct result of trophic interactions varying structurally in three different regions of Canada, a grouping that is associated with the large-scale climatic effects known to be associated with the NAO. We argue that the extensive similarity during the decrease phase is to a large extent a result of region-specific winter conditions and suggest that these may be linked to the state of NAO. We do not yet know how these winter climatic events influence the lynx cycle, but we suggest that lynx hunting efficiency needs to be measured in the three climatic regions.

References and Notes

1. C. S. Elton, *Br. J. Exp. Biol.* **2**, 119 (1924); *Voles, Mice and Lemmings* (Clarendon, Oxford, 1942); see also (*3*).
2. See P. A. P. Moran [*Aust. J. Zool.* **1**, 163 (1953)], who fitted a linear autoregressive model of order two, which exhibits quasi-periodicity, but was well aware of its inadequacy. In particular, he pointed out the inhomogeneity of the fitted residuals, which violated the assumption of a common and constant variance for the white noise term in the fitted model. As an interesting historical point, it should be noted that Moran learned about the lynx data when he visited Charles Elton and Dennis Chitty in the Bureau of Animal Population at Oxford after World War II.
3. N. C. Stenseth *et al. Proc. Natl. Acad. Sci. U.S.A.* **95**, 15430 (1998).
4. E. Ranta, V. Kaitala, P. Lundberg, *Science* **278**, 1621 (1997); E. Ranta, V. Kaitala, J. Lindström, *Ecography* **20**, 454 (1997).
5. The focus on one particular lynx series has to some extent distracted both ecologists and statisticians up to the present from the fact that similar time series exist for the entire continent [but see T. Royama, *Analytical Population Dynamics* (Chapman and Hall, London, 1992)].
6. K. S. Chan, H. Tong, N. C. Stenseth, unpublished data.
7. C. S. Elton and M. Nicholson, *J. Anim. Ecol.* **11**, 215 (1942).
8. Statistics Canada; Dominion Bureau of Statistics 1965; Statistics Canada 1983–1995.
9. H. Walters, *Vegetation of the Earth and Ecological Systems of the Geobiosphere* (Springer-Verlag, Berlin, 1985); R. G. Bailey, *Ecoregions: The Ecosystem Geography of the Oceans and Continents* (Springer-Verlag, Berlin, 1998).
10. J. W. Hurrell and H. Van Loon, *Clim. Change* **36**, 301 (1997).
11. In principle there are more than 2^{21} classifications of the 21 lynx series. A "black-box" approach means doing an exhaustive search for the optimum combination in some sense—an approach troubled by multiple comparison in statistical modeling. Instead, we focus on two sets of combinations, one based on ecological features and the other on climatic condition.
12. J. W. Hurrell, *Geophys. Res. Lett.* **23**, 665 (1996).
13. E. Post and N. C. Stenseth, *J. Anim. Ecol.* **67**, 537 (1998); M. C. Forchhammer, N. C. Stenseth, E. Post, R. Langvatn, *Proc. R. Soc. London Ser. B* **265**, 341 (1998); E. Post and N. C. Stenseth, *Ecology* **80**, 1322 (1999).
14. H. Tong, *Non-Linear Time Series: A Dynamical System Approach* (Clarendon Press, Oxford, 1990); H. Tong, *Threshold Models in Non-Linear Time Series Analysis* (Springer-Verlag, Berlin, 1983). The model given by Eq. (1) is a SETAR(2;2,2) model. For a nontechnical presentation of the SETAR models, see Stenseth *et al.* [N. C. Stenseth, K.-S. Chan, E. Framstad, H. Tong, *Proc. R. Soc. London Ser. B* **265**, 1957 (1998)].
15. N. C. Stenseth, W. Falck, O. N. Bjørnstad, C. J. Krebs, *Proc. Natl. Acad. Sci. U.S.A.* **94**, 5147 (1997).
16. N. C. Stenseth and K.-S. Chan, *Nature* **395**, 620 (1998).
17. D. R. Cox and D. V. Hinkley, *Theoretical Statistics* (Chapman & Hall, London, 1974); H. Akaike, in *Second International Symposium on Information Theory*, B. N. Petrov and F. Csaki, Eds. (Akademiai Kiado, Budapest, 1973); H. Akaike, *IEEE Trans. Autom. Cont.* **AC-19**, 716 (1974).
18. Using the tests described in (*22*) and adopting a 5% significance level, we cannot reject the hypothesis of common slopes and the hypothesis of the common ratio of intercepts for the pairs (L1 and L2) and (L4 and L5). For L6 and L7, we cannot reject the hypothesis of common slopes in the upper regimes and that the lag-2 coefficients may be the same for the two series. However, L6 and L7 appear to have a different lag-1 coefficient in the lower regime. Thus, the evidence of common dynamics for L6 and L7 is somewhat weak. For details, see (*6*).
19. B. Wiken, D. Gauthier, I. Marshall, K. Lawton, H. Hirvonen, *A Perspective on Canada's Ecosystems*, CCEA Occasional Paper No. 14, Ottawa (1996).
20. J. S. Rowe, *Forest Regions of Canada* (publ. 1300, Environment Canada, Canadian Forestry Service, Ottawa, 1972).
21. We have investigated the constraint structure further (Table 1): assuming only a common structure in the lag-1 and lag-2 in the upper regime provides worse fit. This suggests that the additional constraints [found by testing (*6*)] are significant. Relocating L3 to the Pacific-maritime group yields an AIC slightly larger than −1819.38. This implies some fuzziness in the boundary of the three zones. One can classify L3 into the Pacific-maritime zone or the Continental zone. The case of Yukon seems rather clear cut: including it in the Pacific zone substantially increases the AIC and hence provides a poorer description.
22. Supplementary information is available at www.sciencemag.org/feature/data/1037845.shl.
23. J. M. Wallace and D. S. Gutzler, *Mon. Weather Rev.* **109**, 784 (1981).
24. Ecosystem Stratification Working Group, *A National Ecological Framework for Canada* (Agriculture and Agri-Food Canada/Environment Canada, Ottawa, 1995).

25. This work was initiated while several of the authors were at the Centre for Advanced Study at the Norwegian Academy for Sciences and Letters, Oslo. N.C.S. acknowledges generous support from the Norwegian Science Council and the University of Oslo. K-S.C. acknowledges generous support from the University of Iowa. H.T. acknowledges the Engineering and Physical Sciences Research Council (UK) and the University of Hong Kong for support. N.C.S. and H.T. acknowledge support from UK Biotechnology and Biological Sciences Research Council/Engineering and Physical Sciences Research Council (UK) grant 96/MMI09785. R.B., S.B., and C.J.K. acknowledge support from the Natural Sciences and Engineering Research Council of Canada. E.P. acknowledges the National Science Foundation (grant DBI-98-04178) for support. M.C.F. acknowledges the Danish National Research Council for support. We thank A. Kenney for preparing the figure and three anonymous reviewers for providing valuable comments, helping us to clarify our argument.

17 December 1998; accepted 21 June 1999

The Importance of TAR-Modelling for Understanding the Structure of Ecological Dynamics: The Hare-Lynx Population Cycles as an Example

NILS CHR. STENSETH

Centre for Ecological and Evolutionary Synthesis (CEES)
Department of Biology, University of Oslo
P.O. Box 1066 Blindern, N-0316 Oslo, Norway
E-mail: n.c.stenseth@bio.uio.no

In this essay I summarize work I have been involved in together with Howell Tong and his colleagues in the field of statistics. Although briefly mentioning our joint work on the mink-muskrat ecological system in Canada, I do focus on the hare-lynx system of the same sub-continent. I show how statistical modelling has played a key role in the advancing of our ecological insight on that system. I close the essay with some reflections on how statisticians and ecologists (and other scientists) ought to interact in order to fully take advantage of each other's expertise.

Introduction

The regular population cycles in the abundance of Canadian hare and lynx are featured in nearly all biology textbooks (Stenseth 1995a). The abundance is used by ecologists as a prime example of how interactions between predators (the lynx) and prey (the hare) might, depending upon the nature and strength of the interaction (see, e.g., May 1973), lead to regular population cycles (Krebs 2001, Begon et al. 2006). How external factors (such as climate) might affect the dynamics of populations (-and as part of that how external factors might affect the nature and strength of the interactions within ecological systems) has been much debated over the years: Elton (1924), one of the pioneers within the study of population fluctuations, focused on intrinsic factors (i.e., the nature and strength of how individuals of the same species and of different species interact) as the key determinants of the observed population dynamics of natural populations, while Andrewartha and Birch (1954) stressed the importance of extrinsic factors such as climate for understanding why some populations exhibited extensive fluctuations while others did not. Although the pioneers within this debate favoured an either-or categorical view, it is now generally agreed that both intrinsic density-dependent factors and extrinsic density-independent factors determine the observed population dynamics (e.g., Turchin 1995). (Note that the biologists use the term density to mean the population size.) The dynamics of the snowshoe hare (*Lepus americanus*) and the Canadian lynx (*Lynx canadensis*) is a prime example for assessing the relative importance of the two, with climate as the main extrinsic density-independent factor.

In this essay I will summarize studies on the dynamics of the hare-lynx cycle, mostly "seen through the eyes" of the lynx, these studies being in close collaboration with Howell and his statistical colleagues. I will show how statistical modelling can help ecologists better understand the dynamics of systems such as the hare-lynx fluctuations. I also hope to expose the many statistical challenges hidden in such biological systems, the solution to which would help ecologists in their endeavours.

Statisticians know very well about the Canadian lynx data, but mostly the single time series from the MacKenzie River in the northernmost part of Canada. It is less known among statisticians that there are many more time series on the lynx from Canada – see Fig. 1. Below I will demonstrate, by referring to joint work with Howell and others, how the joint analysis of all these data may shed important light on the dynamics of the hare-lynx system of the Canadian sub-continent.

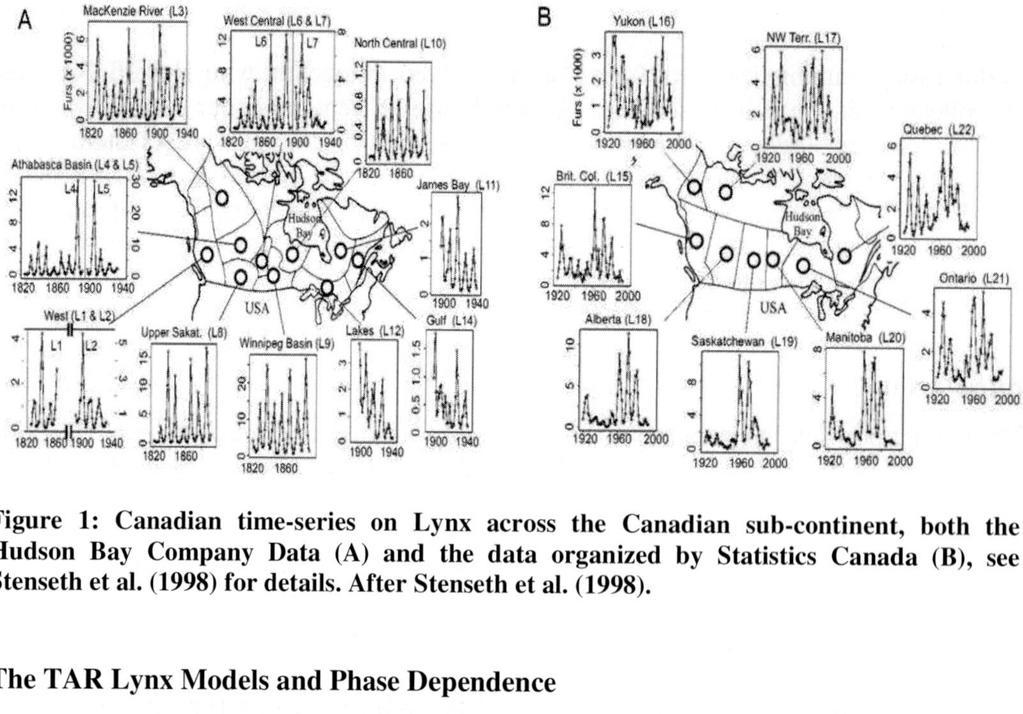

Figure 1: Canadian time-series on Lynx across the Canadian sub-continent, both the Hudson Bay Company Data (A) and the data organized by Statistics Canada (B), see Stenseth et al. (1998) for details. After Stenseth et al. (1998).

The TAR Lynx Models and Phase Dependence

Consider a predator–prey model for the lynx [P_t; $y_t = \ln(P_t)$] and the snowshoe hare [H_t; $x_t = \ln(H_t)$]. The following model may be seen as a Taylor-approximation to various, more specific model formulations (Stenseth et al. 1997):

$$H_{t+1} = H_t \, exp(a_{i,0} - a_{i,1}x_t - a_{i,2}y_t)$$

$$P_{t+1} = P_t \, exp(b_{i,0} - b_{i,1}y_t - b_{i,2}x_t) \qquad (1)$$

where $a_{i,1}$ and $b_{i,1}$ determine the degree of intraspecific regulation in the hare and the lynx (within phase or regime i), respectively; $a_{i,2}$ and $b_{i,2}$ determine the strength of the trophic interaction between the two species; and $a_{i,0}$ and $b_{i,0}$ represent the "intrinsic growth rate" (corresponding to conditions without any significant intraspecific interactions and in the absence of other species).

Unfortunately, we do not have many data on the hare dynamics, the only exception being those hare and lynx data south of the Hudson Bay (see Stenseth et al.

1997). For other parts of Canada we will have to see the hare-lynx dynamics "through the eyes of the lynx" (i.e., by using the lynx data only). For this purpose we need to reduce the ecological two-dimensional model given by Eq. (1) to a delay model in the lynx variable only. Taking logs on both sides of Eq. (1), we may, under reasonable biological assumptions, write

$$y_t = (a_{i,0}\, b_{i,2} + a_{i,1}\, b_{i,0}) + (2 - a_{i,1} - b_{i,1})\, y_{t-1} + (a_{i,1} - b_{i,1} - a_{i,1}b_{i,1} - a_{i,2}b_{i,2} - 1)\, y_{t-2} \tag{2}$$

which is equivalent to an autoregressive model, but written with the biological parameters determining the statistical autoregressive parameters: The lag-one autoregressive parameter, $\beta_{i,1} = 2 - a_{i,1} - b_{i,1}$, is a function of the ecological density dependence in both the hare and the lynx ($a_{i,1}$ and $b_{i,1}$). The lag-two autoregressive parameter, $\beta_{i,2} = a_{i,1} + b_{i,1} - a_{i,1}b_{i,1} - a_{i,2}b_{i,2} - 1$, incorporates the strength of the trophic interaction (i.e., $a_{i,2}b_{i,2}$). Note that the coefficient $\beta_{i,2}$ also may be represented as $\beta_{i,2} = 1 - (\beta_{i,1} + \text{“}intra\text{”}_i) - \text{“}inter\text{”}_i$, where $intra_i$ and $inter_i$ signify the strength of the multiplicative intra- and interspecific (trophic) interactions in the system. Thus, the lag-two autoregressive parameter is influenced by trophic interactions as well as the ecological density dependencies within the interacting species. Assuming that the trophic model is appropriate, this suggests that the ecological mechanism underlying the observed nonlinearity is likely to be related to the interaction between the hare and the lynx.

An interesting ecological question is where the non-linearity is located? Is it primarily located in the $intra_i$ or the $inter_i$ components – or both? Together with Howell and others, I have been involved in the statistical analysis of the Canadian lynx data in order to answer this ecological question. Building upon previous analysis of ours (Stenseth et al. 1997) we (Stenseth et al. 1998) chose for this purpose a TAR modeling approach, i.e.:

$$y_t = \begin{cases} \beta_{1,0} + \beta_{1,1}y_{t-1} + \beta_{1,2}y_{t-2} + \varepsilon_{1,t} & y_{t-d} \le \theta \\ \beta_{2,0} + \beta_{2,1}y_{t-1} + \beta_{2,2}y_{t-2} + \varepsilon_{2,t} & y_{t-d} > \theta \end{cases} \tag{3}$$

where θ is the threshold applicable to the density d years earlier, and $\varepsilon_{i,t}$ are noise-terms with independent and normal distributions, $N(0, \sigma_i^2)$; this model is a SETAR($2;2,\ 2$) model (Tong 1990). The optimal thresholding lag, d_{opt}, was found to be $d_{opt} = 2$. The non-linearity was furthermore found to be located in $\beta_{i,2}$ (see Fig. 2).

The ecological machinery behind the threshold structure of the lynx time series may be sought in the hare–lynx interaction. Fig. 3 depicts the hare–lynx cycle at Rochester, Alberta (Fig. 3 *A*) and at Kluane Lake, Yukon (Fig. 3 *B*); a more stylized hare–lynx cycle together with the threshold level separating the two regimes also is shown (Fig. 3 *C*). The lower regime applies during the lynx increase, and the upper regime applies during the lynx (and hare) decrease. As a result, we call the lower regime the "increase phase" and the upper regime the "decrease phase" (i.e., the non-linearity is due to what may appropriately be termed a phase-dependency).

By combining statistical modelling of 21 time series spanning 175 years (as shown in Fig. 1) with mathematical modelling and experimental data (as shown in Fig. 3D), we have succeeded in drawing the following conclusions: (*i*) The dynamic

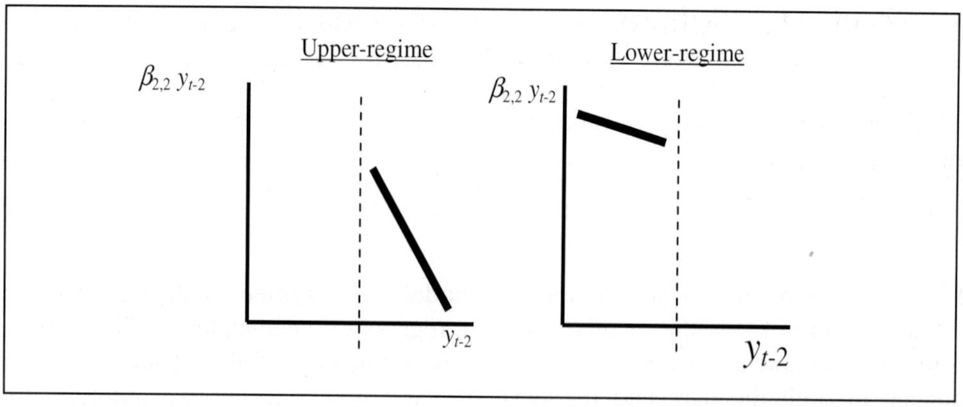

Figure 2: The non-linear structure of the Canadian lynx data. After Stenseth et al. (1998).

patterns (or structure) of the Canadian lynx is both phase- and density-dependent. The density dependence involves both direct and delayed effects. The structure of the dynamics is broadly similar over time (from the early part of the 19th century until the present) and space (the entire Canadian boreal forest region). (*ii*) The density-dependent structure of the lynx time series is consistent with the proposition that the dynamics in the Canadian lynx is governed by processes involved in the trophic interactions between the snowshoe hare. (*iii*) The phase-dependent structure of the lynx time series is primarily a result of the strength of how the hare and lynx interact during the different phases of the population cycle. Finally, (*iv*) density dependence induces the regulatory delays whereas phase dependence induces the nonlinearity. The two act in concert to produce the observed lynx cycle. These ecological conclusions were a direct result of statistical time-series modelling. I would add that it would not have been possible to reach our ecological conclusions without the TAR approach.

In subsequent work, we also showed that the phase-dependency of how the lynx and the hare interact is influenced by the snow-condition (see Stenseth et al. 2004b), another conclusion that was made possible only through a genuine combination of statistical and ecological expertise.

The Geographic Structuring of the Canadian Lynx: The Importance of Statistical Modelling

An ecologically interesting question pertinent to the dynamic structure of the hare-lynx model is this: Is there any geographic structure? And, if there is, what might it be caused by? I have been fortunate to work with Howell and his statistical colleagues on this topic as well. In 1999 we published a paper on the topic in *Science*. We wanted to test whether the lynx data were structured according to the nature of the ecosystem in different parts of Canada – i.e., as indicated by Fig. 4B. I was, before the project, convinced that this

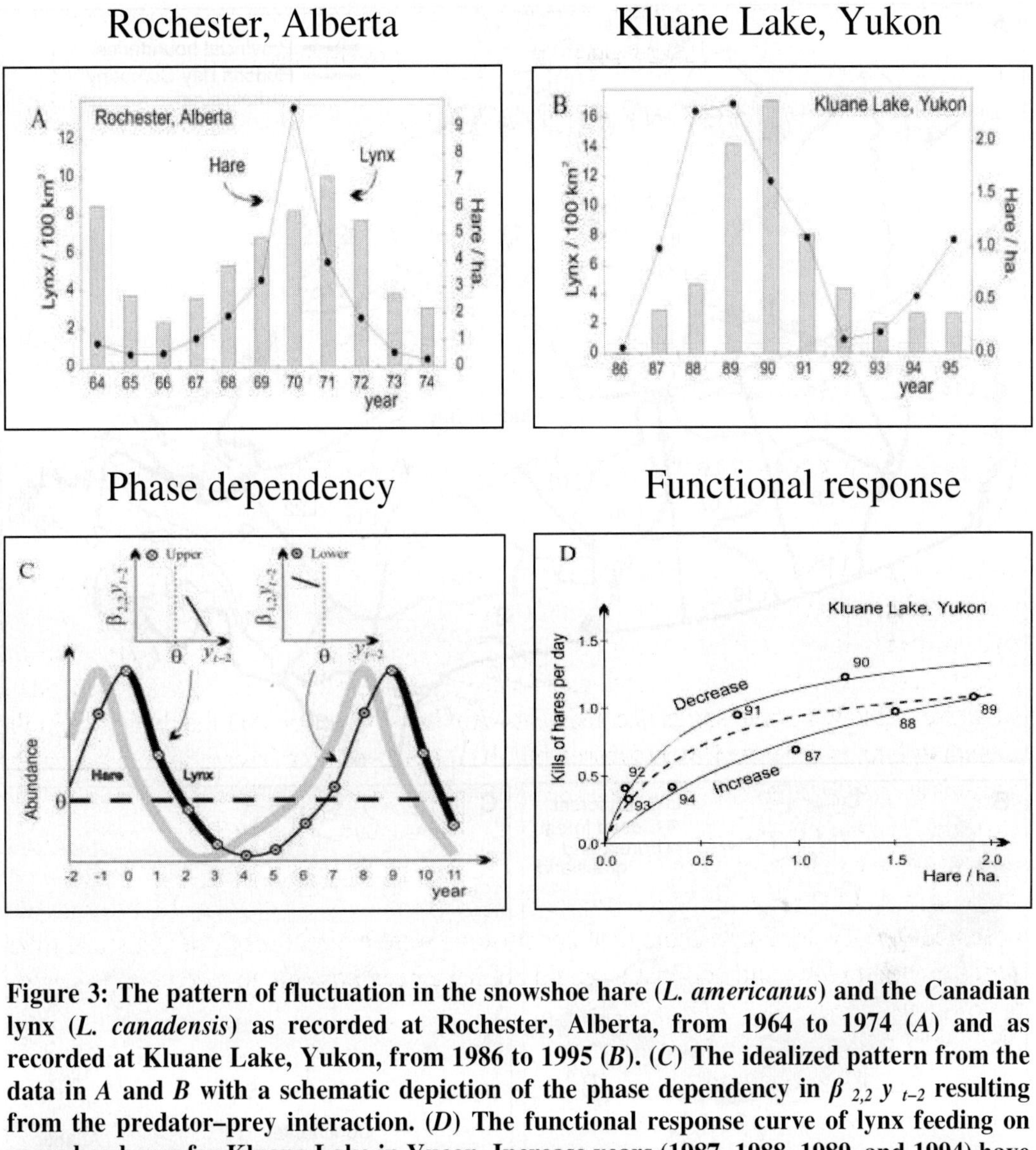

Figure 3: The pattern of fluctuation in the snowshoe hare (*L. americanus*) and the Canadian lynx (*L. canadensis*) as recorded at Rochester, Alberta, from 1964 to 1974 (*A*) and as recorded at Kluane Lake, Yukon, from 1986 to 1995 (*B*). (*C*) The idealized pattern from the data in *A* and *B* with a schematic depiction of the phase dependency in $\beta_{2,2}\, y_{t-2}$ resulting from the predator–prey interaction. (*D*) The functional response curve of lynx feeding on snowshoe hares for Kluane Lake in Yucon. Increase years (1987, 1988, 1989, and 1994) have a different functional response than decrease years (1990, 1991, 1992, and 1993), thus explaining the phase dependency in this system. See Stenseth et al. (1998) for details. After Stenseth et al. (1998).

was going to be the case, but the results of the statistical analysis (see Stenseth et al. 1999) showed otherwise: the structuring was as indicated by Figure 4C – i.e., an Atlantic, a Continental and a Pacific region. We argued that this feature is most likely due to the climate. Again this was a conclusion resulting directly from a collaborative effort involving statisticians, ecologists and climatologists. I would add this: our conclusions would be impossible without the statisticians' effort. It is certainly not strange that the

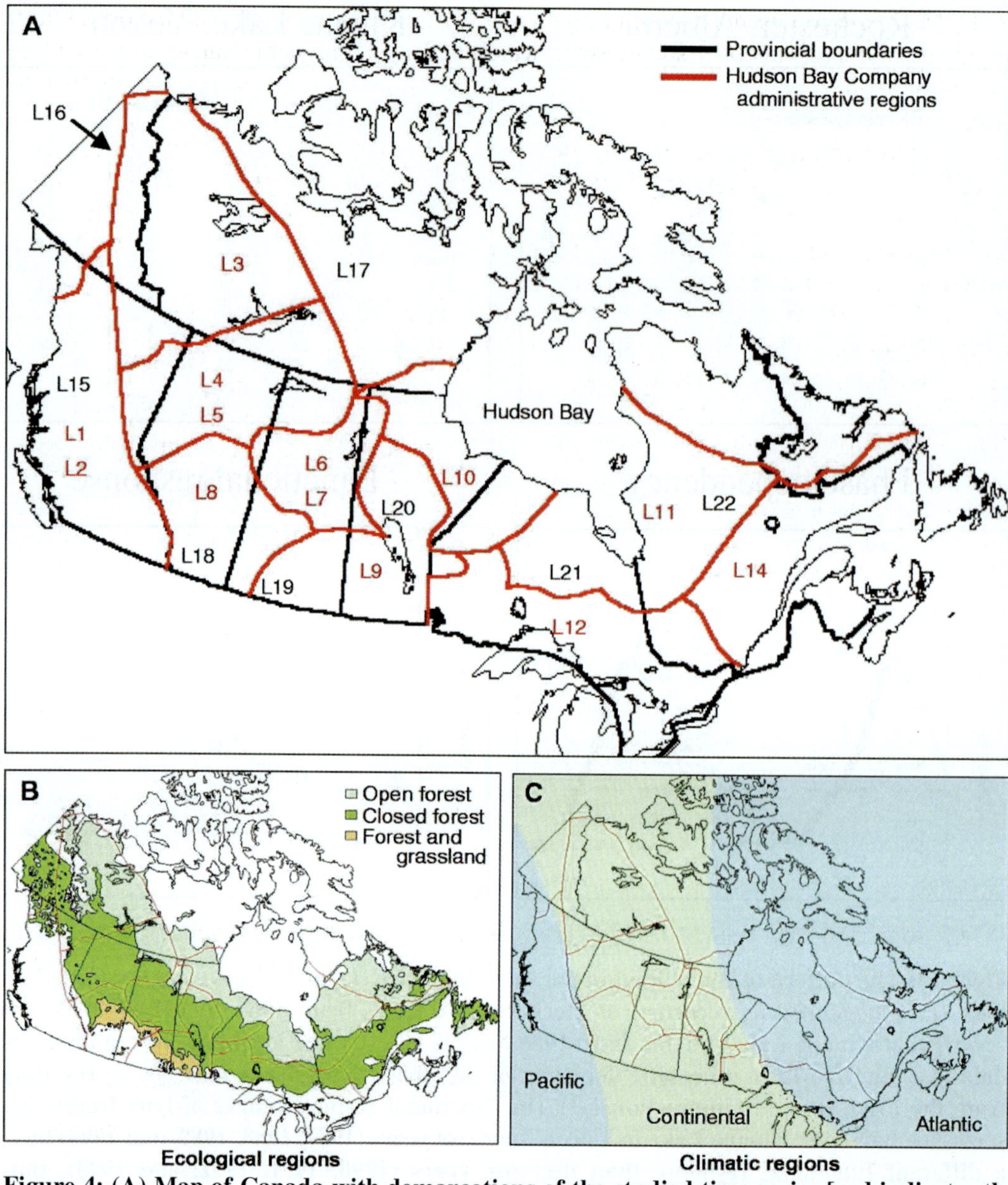

Figure 4: (A) Map of Canada with demarcations of the studied time series [red indicates the Hudson Bay Company time series (see Fig. 1A) and black indicates the recent series (see Fig. 1B)]. (B) Ecological regions of Canada (Ecosystem Stratification Working Group 1995). (C) Climatic regions of Canada (J. W. Hurrell & H. Van Loon 1997). The NAO (essentially a "package of weather"; cf. Stenseth et al. 2003, Stenseth and Mysterud 2005) refers to a meridional oscillation in surface pressures with centers of action near Iceland and over the subtropical Atlantic. When surface pressures are lower than normal near Iceland and higher than normal over the subtropical Atlantic (the positive phase of the NAO), enhanced northerly flow over eastern Canada cools surface temperatures and enhanced southerly flow from the Gulf of Mexico into much of central Canada produces warm surface anomalies. Over the Pacific-maritime region, there is no significant NAO signature. After Stenseth et al. (1999).

Rocky Mountain represents a structuring factor (certainly linked to climate factors), but it is quite surprising to uncover such a structuring barrier – an invisible barrier as we called it (see Rueness et al. 2003; Stenseth et al. 2004a), a feature we found thanks to our interactions with Howell and his statistical colleagues.

In our study (Stenseth et al. 1999) we used the model given by Eq. (3) and looked for common parameters in the models for the various time series. We found strong evidence in support of climate-based properties contributing to the structuring of the lynx dynamics (i.e., Atlantic, Continental and Pacific) – such a structuring was particularly strong for the decrease phase.

We have later followed up this work and shown that there also is a genetic structuring of the lynx populations into an Atlantic, a Continental and a Pacific region (see Rueness et al. 2003). Subsequent analysis suggests that there might be an evolutionary basis for this structuring (see Stenseth et al. 2004a), a structuring caused by climate operating through the snow (see Stenseth et al. 2004b).

Starting with the joint statistical work published in 1999 (Stenseth et al. 1999), we have now been able to reach a fairly comprehensive understanding of the panel of 21 time series of the Canadian lynx. Specifically, the lynx cycle is a direct result of hare-lynx interactions varying structurally in three different regions of Canada, a grouping that is associated with the large-scale climatic effects known to be associated with the North Atlantic Oscillation (see, e.g., Steneth et al. 2003) – a regional climatic effect which seems to be mediated through the nature of the snow.

Conclusions

Above I have provided a synoptic overview of joint work with Howell and his statistical colleagues; the work has provided new insights into the ecology of the hare-lynx cycle. Specifically, we have learnt the following:

- Although we found that both the hare and the lynx are self-regulated, the dynamics of the Canadian lynx is primarily "driven" by interactions between the hare and the lynx (Stenseth et al. 1997);
- Non-linearity in the lynx dynamics is primarily due to the phase dependency in the way lynx and hare interact (i.e., phase dependency; Stenseth et al. 1998);
- A geographic structuring across the Canadian sub-continent is generated by large-scale climate variation across Canada (Stenseth et al. Science, 1999);
- Snow hardness (crustiness) influences the way hare and lynx interact (Stenseth et al. 2004b,a); and
- Genetic geographic structuring is due to climate-influenced ecological differential dynamics (Rueness et al. 2003).

It has been fascinating for me to be an ecologist in this statistical-ecological team; I trust that the statisticians have enjoyed the collaboration as much as I have. Indeed, I have been told by some of my statistician colleagues that the ecological problems (and data) I bring into the collaboration have challenged them with new methodological problems in a way that they otherwise would have not so easily come across.

What Next?

The question "what next?" may be answered at least at two levels – one regarding interaction between statisticians and another regarding the biological issues. I will in closing briefly reflect on both:

- *The interaction between statisticians and ecologists (or any other subject-field scientists):* Building upon my experience in multidisciplinary work (such as joint work involving statisticians and ecologists) I think it is very important that we spend time to get to learn each others' way of thinking. Often this might initially appear to be a waste of time, but my own experience always tells me that such initial investment is always wroth it. It is further important, in my experience, that the two (or more) partners treat each other as equal: e.g., the statistician should not be the assistant of the ecologists performing some statistical testing or other analysis, nor should the ecologist be only helping the statistician in using the correct phrases. And never treat us, biologists, as your clients (which has happened to me elsewhere) – it will not work as a scientific enterprise. They should get involved in truly collaborative work as I have described in this essay; then, there is a basis for further methodological work (to be published in the appropriate statistical journals) as well as further ecological work (to be published in the appropriate ecological journals). In fact, I have always as a working rule in such collaborative work that two types of papers should result from the collaborative work – at least one paper in a statistical journal and at least one paper in a biological journal. This has been the rule in my work with Howell and his statistical colleagues. I do believe this strategy has contributed to the success of our collaboration.
- *The study of the hare-lynx and other similar ecological systems:* Further detailed studies should be carried out on the hare-lynx system, which for sure should include work aiming at a better understanding of how snow affects the interaction between the hare and the lynx. The statistical analysis described in this essay has pointed towards snow as a key environmental factor, but much remains to be done; some work is already under way, stimulated by the work we have done together with Howell and his colleagues. In Canada there is also another system similar to the hare-lynx system – the mink-muskrat – for which much data exist. Again, I have been fortunate in collaborating with Howell and his colleagues on the analysis of that system as well. Among our findings is that which suggests the same invisible (presumably, climatically induced) barrier exists south of the Hudson bay (see, Yao et al. 2000).

This essay is published in a volume primarily read by statisticians and hence written accordingly. It is my hope that you, statisticians, realize that you can be of tremendous help in moving fields such as ecology forward by entering into close collaboration with us, ecologists. In a way, this essay is an invitation to further collaborative work with statisticians – statisticians primarily having the ambition of doing groundbreaking work within your own field of statistics. It is my hope that some of you will respond to this (general) invitation – there are many exciting topics which would benefit from such collaborative work.

Acknowledgements

It has been a great pleasure to work with Howell over the years; altogether we have done six papers together – many of which having been published in the very best journals (and more papers are to come). I thank Howell for many very stimulating discussions and for introducing me to several of his colleagues, some of which have been good collaborative partners. In this essay I have highlighted only the joint work we have done on the hare-lynx ecological system of the Canadian sub-continent. I could have shown other rewording joint work of ours – but that will have to await another essay.

References

1. Andrewartha, H.G. & Birch, L.C. 1954. Distribution and Abundance of Animals. University of Chicago Press, Chicago
2. Begon, M, Townsend, C.R. & Harper, J.L. 2005. Ecology, Fourth Edition. Blackwell Publishing, Oxford
3. Ecosystem Stratification Working Group 1995. A National Ecological Framework for Canada (Agriculture and Agri-Food Canada/Environment Canada, Ottawa)
4. Elton, C.S. 1924. Periodic fluctuations in numbers of animals: their causes and effects. British Journal of Experimental Biology 2: 119–163
5. Hurrell, J.W. & Van Loon, H. 1997. Decadal variations in climate associated with the North Atlantic Oscillation. Clim. Change 36, 301-326
6. Krebs, C.J. 2001. Ecology: the experimental analysis of distribution and abundance. Benjamin Cummings. New York
7. May, R.M. 1973. Complexity and Stability of Model Ecosystems. Princeton Univ. Press, Princeton, New Jersey
8. Rueness, E.K., Stenseth, N.C, O'Donoghue, M., Boutin, S., Ellegren, H. & Jakobsen, K.S. 2003. Ecological and genetic spatial structuring in the Canadian lynx. Nature 425, 69-72
9. Stenseth, N.C. 1995a Snowshoe hare populations: squeezed from below and above. Science 269, 1061-1062
10. Stenseth, N.C. & Mysterud, A. 2005. Weather packages: finding the right scale and composition of climate in ecology. Journal of Animal Ecology 74, 1195-1198
11. Stenseth, N.C., Falck, W., Bjørnstad, O.N. & Krebs, C.J. 1997. Population regulation in snowshoe hare and Canadian lynx: asymmetric food web configurations between hare and lynx. Proceedings of National Academy of Science, Washington 94, 5147-5152.
12. Stenseth, N.C., Falck, W., Chan, K.-S., Bjørnstad, O.N., O'Donoghue, M., Tong, H., Boonstra, R., Boutin, S., Krebs, C.J. & Yoccoz, N.G. 1998. From ecological patterns to ecological processes: phase- and density-dependencies in the Canadian lynx cycle. Proceedings of National Academy of Science, Washington 95, 15430-15435
13. Stenseth, N.C. Chan, K.-S., Tong, H., Boonstra, R., Boutin, R., Krebs, C.J., Post, E., O'Donoghue, M., Yoccoz, N.G., Forchhammer, M.C. & Hurrell, J.W. 1999. Common dynamic structure of Canada lynx populations within three climatic regions. Science 285, 1071-1073

14. Stenseth, N.C., Ottersen, G., Hurrell, J.W., Mysterud, A., Lima, M., Chan, K.-S., Yoccoz, N.G. & Ådlandsvik, B. 2003. Studying climate effects on ecology through the use of climate indices: the North Atlantic Oscillation, El Niño Southern Oscillation and beyond. Proceedings of the Royal Society of London, B 270, 2087-2096

15. Stenseth, N.C., Ehrich, D., Rueness, E.K., Lingjærde, O.C., Chan, K.-S., Boutin, S., O'Donoghue, M., Robinson, D.A., Viljugrein, H. & Jakobsen, K.S. 2004a. The effect of climatic forcing on population synchrony and genetic structuring of the Canadian lynx. Proceedings of National Academy of Science, Washington 101, 6056-6061

16. Stenseth, N.C., Shabbar, A., Chan, K.-S-, Boutin, S., Rueness, E.K., Ehrich, D., Hurrell, J.W., Lingjærde, O.C. & Jakobsen, K.S. 2004b. Snow conditions may create an invisible barrier for lynx. Proceedings of National Academy of Science, Washington 101, 10632-10634

17. Tong, H. 1990. Non-linear time series. Clarendon Press, Oxford.

18. Turchin P. 1995. Population regulation: old arguments and a new synthesis. Pp 19-40 in: Cappuccino, N. & Price, P.W. (eds). Population Dynamics: New Approaches and Synthesis. Academic Press, New York

19. Yao, Q., Tong, H., Finkenstäd, B. & Stenseth, N.C. 2000. Common structure in panels of short ecological time series. *Proceedings of Royal Society of London, B.* **267**, 2459-2467

On Howell Tong's Contributions to Reliability

M. MASOOM ALI*

Department of Mathematical Sciences, Ball State University
Muncie, IN 47306, USA
E-mail: mali@bsu.edu

Professor Howell Tong has distinguished himself for his work in non-linear time series analysis and especially for his seminal work on threshold autoregression, limit cycles and cyclical data. Among his extensive publications over the past four decades, there are two papers on reliability which he had published in the seventies. In this paper, a brief review of his two papers on the reliability function $P(Y < X)$ is made.

1. Introduction

Let X and Y be two independent continuous random variables having distribution functions $F_X(x)$ and $G_Y(y)$, respectively. Then the reliability function P is defined by $P = P(Y \leq X)$.

$$\begin{aligned} P &= P(Y \leq X) \\ &= \int_{-\infty}^{\infty} \int_{-\infty}^{x} dG_Y(y) dF_X(x) \\ &= \int_{-\infty}^{\infty} G_Y(x) dF_X(x) \\ &= E(G_Y(X)). \end{aligned}$$

The problem of estimating the probability that a random variable Y is less than or equal to an independent random variable X, arises in reliability studies. When the random variable Y represents a stress that a device is subjected to in service and the random variable X represents the strength that varies from item to item in the population of devices, then the reliability P, i.e., the probability that a randomly selected device functions successfully, is equal to $P(Y \leq X)$. The same problem also arises in the context of statistical tolerance where Y represents the diameter of a shaft and X denotes the diameter of a bearing that is to be mounted on the shaft. The probability that the bearing fits without interference is then $P(Y \leq X)$. In biometry, Y represents a patient's remaining years of life if treated with drug A and X represents a patient's remaining years when treated with drug B. If the choice of drug is left to the patient, then the person's deliberation will center on whether $P(Y \leq X)$ is less than or greater than 1/2.

Estimation of $P = P(Y \leq X)$ has been considered over the years for many probability distributions (e.g., exponential, normal, gamma, Pareto etc.) with unknown parameters. Maximum likelihood, minimum variance unbiased estimator (MVUE), and Bayes estimators

Dr. M. Masoom Ali is George and Frances Ball Distinguished Professor Emeritus of Statistics and Professor Emeritus of Mathematical Sciences.

of P were found by various authors. Recently Ali, *et al* and other authors in the past considered this problem of estimating the reliability function P. See for example Ali and Woo (2005a), (2005b), Ali, Pal and Woo (2005), Ali, Woo and Pal (2004), Basu (1964), Beg (1980), Church and Harris (1970), Downtown (1973), Ivshin (1996), Iwase (1987), Kelley, Kelley, and Suchany (1976), McCool (1991), Pal, Ali, and Woo (2005), Nadarajah (2003a), (2003b), and Sathe and Verde (1969). Also, many authors considered the problem of finding the distribution of X/Y and $X/(X+Y)$ in order to study the reliability function P. See for example, Ali, Nadarajah, and Woo (2005). In this paper we review the two papers which Professor Howell Tong had published, namely Tong (1974) and Tong (1977).

2. Tong's Publications on Reliability Estimation

Tong published two papers and a letter to the editor in *Technometrics*, namely, Tong (1974), Tong (1977), and Tong (1975), respectively. In his first paper, Tong (1974) considered the problem of estimating $P = P(Y < X)$ where the independent random variables X and Y have exponential distribution with parameters λ and μ, respectively. He obtained the MVUE of P. However, Johnson (1975) in a letter to the editor pointed out an error due to an incorrect upper limit of equation (5). Johnson pointed out that equation (5) holds for $m\bar{y} \le n\bar{x}$ and he added another expression in his letter to the editor for $m\bar{y} > n\bar{x}$. Johnson also pointed out that for the case when μ is known, the expression for the MVUE of P in equation (6) of Tong's 1974 paper needed another term to be added and Johnson in his letter to the editor provided the extra term. Johnson (1975), with a touch of humor, mentioned that he had made a similar type of error in one of his earlier papers (see Johnson (1968) and Sathe and Verde (1969)) and that he hoped that this was not the case with the corrections he had suggested this time!

Let X and Y be two independent negative exponential random variables with parameters λ and μ, respectively. Furthermore, consider two independent random samples $X_1, X_2, \ldots, X_n$ and $Y_1, Y_2, \ldots, Y_m$ from the X- and Y-distributions, respectively, and let

$$\bar{X} = n^{-1}\sum_{i=1}^{n} X_i, \text{ and } \bar{Y} = m^{-1}\sum_{i=1}^{m} Y_i.$$

The minimum variance unbiased estimator $\hat{P}$, of $\Pr\{Y < X\}$ in the final corrected form is given by

$$\hat{P} = \begin{cases} \sum_{s=0}^{n-1}(-1)^s \frac{(m-1)!(n-1)!}{(m-1+s)!(n-1-s)!}\left(\frac{m\bar{y}}{n\bar{x}}\right)^s, & m\bar{y} \le n\bar{x}, \\ \sum_{s=0}^{n-1}(-1)^s \frac{(m-1)!(n-1)!}{(m-1+s)!(n-1-s)!}\left(\frac{m\bar{y}}{n\bar{x}}\right)^s & \\ +(-1)^n \frac{(m-1)!(n-1)!}{(m+n-2)!}\left(1-\frac{n\bar{x}}{m\bar{y}}\right)^{n-1}\left(\frac{m\bar{y}}{n\bar{x}}-1\right)^{m-1}, & m\bar{y} > n\bar{x}. \end{cases}$$

In the case when μ is known, the minimum variance unbiased estimator of P is given by

$$\hat{P} = \frac{\sum_{s=0}^{n-1}(-1)^s(n-1)!(n\mu\bar{x})^{-s}}{(n-s-1)!} + (-1)^n(n-1)!e^{-n\mu x}(n\mu\bar{x})^{-(n-1)}.$$

In his letter to the editor Tong (1975) acknowledged the error in his previous paper [Tong (1974)] and extended his results on exponential distribution to the gamma distributions, namely, where $X \sim \Gamma(N, \lambda)$ and $Y \sim \Gamma(M, \mu)$, where N and M are pre-fixed and λ and μ are unknown, using essentially the same arguments of his 1974 paper [Tong (1974)] to obtain the MVUE of P. If $(X_1, X_2, \ldots, X_n)$ and $(Y_1, Y_2, \ldots, Y_m)$ are independent random samples on X and Y, respectively, then the minimum variance unbiased estimator, $\hat{P}$, of $\Pr\{Y < X\}$ is given by

$$\hat{P} = \begin{cases} K \sum_{i=0}^{N-1} \sum_{j=0}^{Nn-N+i} \left[(-1)^{i+1} \binom{N-1}{i} \binom{Nn-N+i}{j} \right. \\ \left. \cdot \frac{B(M+j, Mm-M)(Y_{.}/X_{.})^i}{(Nn-N+i)} \right], \text{ if } Y_{.} \leq X_{.} \\ 1 - K \sum_{i=0}^{M-1} \sum_{j=0}^{Mm-M+i} \left[(-1)^{i+j} \binom{M-1}{i} \binom{Mm-M+i}{j} \right. \\ \left. \cdot \frac{B(N+j, Nn-N)(X_{.}/Y_{.})^i}{(Mm-M+i)} \right], \text{ if } Y_{.} \geq X_{.}, \end{cases}$$

where

$$X_{.} = \sum_{i=1}^{n} X_i, \quad Y_{.} = \sum_{i=1}^{m} Y_i,$$

$$K = \{B(N, Nn-N)B(M, Mm-M)\}^{-1},$$

with B denoting the beta function.

Professor Tong's second paper, Tong (1977), deals with the estimation of the reliability function P for the Koopman-Pitman-Darmois exponential family Σ. He shows how one can obtain the unique MVUE of P where X and Y are restricted to the s-independent random variables each with a cdf belonging to the exponential family Σ, thus unifying some previous work in the area; e.g., Washio *et al* (1955), Basu (1964), Church and Harris (1970), and Downtown (1973).

Below is the theorem which gives the main result Tong had derived in his paper Tong (1977).

Let $X_1, X_2, \ldots, X_n$ and $Y_1, Y_2, \ldots, Y_m$ be two s-independent random samples on X and Y, respectively. Write the pdf of X in the form

$$f_X(x; \phi) = \exp[-x\phi + c(\phi) + d_1(x)]$$

where ϕ is the only unknown parameter, $c(\phi)$ denote function of ϕ only and $d_1(x)$ is a function of x only. Similarly, the pdf of Y is written in the form

$$f_Y(y; \phi') = \exp[-y\phi' + c'(\phi') + d'_1(y)].$$

Let $u = x_1 + x_2 + \ldots + x_n$ and $\nu = y_1 + y_2 + \ldots + y_m$. Then the conditional distribution of X_1 given u is

$$f(x_1|u) = K_1 \exp[x_1((h_{n-1}(\phi) - \phi) + d_{n-1}(u - x_1) + d_1(x_1)],$$

where K_1 is the normalizing constant. Similarly the conditional distribution of Y_1 given ν is given by

$$g(y_1|\nu) = K_2 \exp[y_1((h'_{m-1}(\phi') - \phi') + d'_{m-1}(\nu - y_1) + d'_1(y_1)],$$

where K_2 is the normalizing constant.

Theorem: Given s-independent random samples $X_1, X_2, \ldots, X_n (n \geq 2)$ and $Y_1, Y_2, \ldots, Y_m$ $(m \geq 2)$ on s-independent r.v.'s X and Y respectively, where the cdf of X and Y belong to Σ [Koopman-Pitman-Darmois exponential family], each with a single unknown parameter, the MVUE of P is

$$\hat{P}(u, \nu) = \int \int_R \theta(x_1, y_1) f(x_1|u) g(y_1|\nu) dx_1 dy_1,$$

where R is the region in which both $f(x_1|u)$ and $g(y_1|\nu)$ are non-zero, and

$$\theta(x_1, y_1) = \begin{cases} 0, & \text{for } x_1 < y_1, \\ 1, & \text{otherwise.} \end{cases}$$

See Tong (1977), pages 54-55 for explanation of the notations and symbols and proof of the theorem.

Tong (1977) then applied this theorem to the estimation of P where X and Y are two s-independent normal random variables each with variances known but means unknown. As a second example, he applied the theorem to the case where X has an exponential distribution with parameter λ and Y has a geometric distribution with parameter μ. This result is applicable to a situation where the time of failure of one component is recorded continuously while that of the other discretely. He also used the main result of the paper to find the MVUE of the cdf's belonging to the family Σ. In particular, he derived the MVUE of the cdf of the exponential distribution.

Between the two papers Tong published on reliability, the contribution of his second paper, namely Tong (1977) is by far the most important contribution.

Acknowledgements

I am very grateful to Professor Kung-Sik Chan of the Department of Statistics at the University of Iowa for inviting me to write this article on the contributions of Professor Howell Tong in the area of reliability on the occasion of Professor Tong's 65th Birthday Anniversary. I am also very thankful to Professor Kung-Sik Chan and a referee for their several constructive suggestions which improved the paper considerably.

References

1. Ali, M. Masoom, Nadarajah, S., and Woo, J. (2005). On the ratio $X/(X+Y)$ for Weibull and Levy distributions. *J. Korean Statist. Soc.* 34(1), 11-20.
2. Ali, M. Masoom and Woo, J. (2005a). Inference on reliability $P(Y < X)$ in a p-dimensional Rayleigh distribution. *Math. Comput. Modelling* 42, 367-373.
3. Ali, M. Masoom and Woo, J. (2005b). Inference on $P(Y < X)$ in a Pareto distribution. *J. Mod. Appl. Statist. Meth.* 4(2), 583-586.
4. Ali, M. Masoom, Pal, M., and Woo. J. (2005). Estimation and testing of $P(Y < X)$ in a two-parameter exponential distribution. *Statistics* 39(5), 415-428.
5. Ali, M. Masoom, Woo, J., and Pal, M. (2004). Inference on reliability $P(Y < X)$ in two-parameter exponential distribution. *Internat. J. Statist. Scis*, 3, 119-125.
6. Basu, A.P. (1964). Estimates of reliability for some distributions useful in reliability. *Technometrics* 6, 215-219.

7. Beg, M. A. (1980). On the estimation of $P(Y < X)$ for two-parameter exponential distribution. *Metrika* 27, 29-34.
8. Church, J.D. and Harris, V. (1970). The estimation of reliability from stress-strength relationships. *Technometrics* 12, 49-54.
9. Downtown, F. (1973). The estimation of $\Pr\{Y < X\}$ in the normal case. *Technometrics* 15, 551-558.
10. Ivshin, V. V. (1996). Unbiased estimators of $P(X < Y)$ and their variances in the case of uniform and two-parameter exponential distributions. *J. Math. Scis.* 81(4), 2790-2793.
11. Iwase, K. (1987). On UMVU estimators of $\Pr(Y < X)$ in the 2-parameter exponential case. *Mem. Fac. Eng. Hiroshima Univ.* 9(3) (Ser 29), 21-24.
12. Johnson, N. L. (1975). Letter to the Editor. *Technometrics*, 10, 429.
13. Johnson, N. L. (1968). Letter to the Editor. *Technometrics*, 17(3), 393.
14. Kelley, G. D., Kelley, J. A., and Suchany, W. R. (1976). Efficient estimation of $P(Y < X)$ in the exponential case. *Technometrics* 18, 359-360.
15. McCool, J.I. (1991). Inference on $P(X < Y)$ in the Weibull case. *Commun. Statist.-Simula.*, 20(1), 129-148.
16. Pal, M., Ali, M. Masoom, and Woo, J. (2005). Inference on $P(Y < X)$ in a generalized uniform distribution. *Calcutta Statist. Assoc. Bull.*, 57(225-226), 35 - 48.
17. Nadarajah, S. (2003a). Reliability for extreme value distributions. *Math. Comput. Modelling* 37, 915-922.
18. Nadarajah, S. (2003b). Reliability for lifetime distributions. *Math. Comput. Modelling* 37, 683-688.
19. Sathe, Y. S. and Verde, S. D. (1969). Minimum variance unbiased estimation of reliability for the truncated exponential distribution. *Technometrics*, 11, 609-612.
20. Tong, H. (1977). On the Estimation of $\Pr\{Y < X\}$ for Exponential Families. *IEEE Trans. Reliab.*, R-26(1), 54-56.
21. Tong, H. (1975). Letter to the Editor. *Technometrics*, 17(3), 393.
22. Tong, H. (1974). A Note on the Estimation of $\Pr\{Y < X\}$ in the Exponential Case. *Technometrics*, 16(4), 625.
23. Washio, Y., Morimoto, H., and Ikeda N. (1955). Unbiased estimation based on sufficient statistics. *Bull. Math. Statist.*, 6, 69-93.

草短露真經
分而治則明
銀光門檻上
晚照更歡迎

A poem (in the Tang style) that Howell wrote on the occasion of his receiving the Guy medal. Below is a translation by Mr. Kwan Yee-Kwong:

My newly-mowed lawn showed me the Way

Non-linear vision finally carried the day.

On my threshold shone a silvery light,

Late-coming, yet a joyous sight.

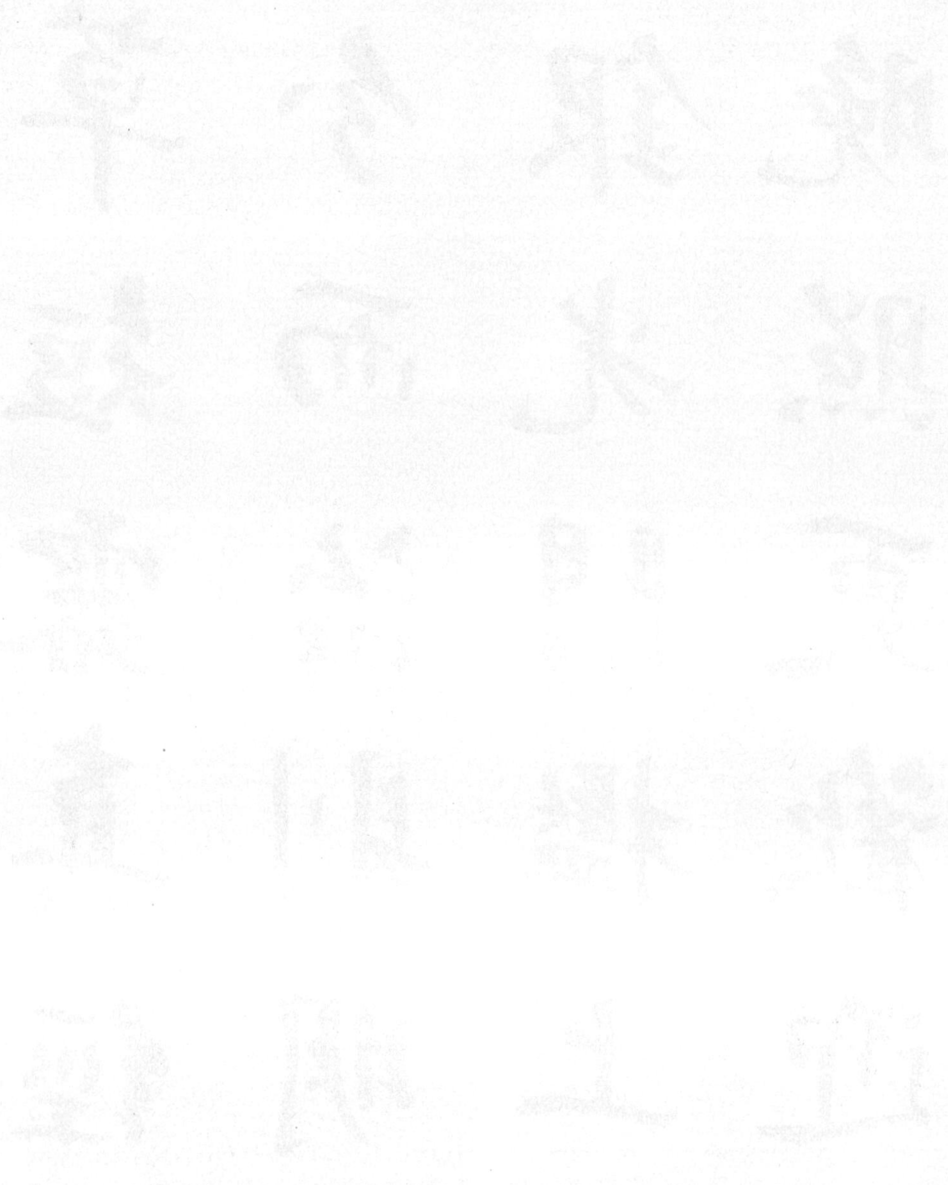

